W9-AFT-585

CONCEPTS
OF STATISTICAL
INFERENCE

CONCEPTS
OF STATISTICAL
INFERENCE

SECOND EDITION

William C. Guenther

Professor of Statistics
University of Wyoming

BRIAR CLIFF COLLEGE
LIBRARY
SIOUX CITY, IOWA

McGRAW-HILL BOOK COMPANY
New York / St. Louis / San Francisco
Düsseldorf / Johannesburg / Kuala Lumpur
London / Mexico / Montreal / New Delhi / Panama
Rio de Janeiro / Singapore / Sydney / Toronto

CONCEPTS OF STATISTICAL INFERENCE

Copyright © 1965, 1973 by McGraw-Hill, Inc.
All rights reserved.   Printed in the United States of America.
No part of this publication may be reproduced, stored in
a retrieval system, or transmitted, in any form or by any
means, electronic, mechanical, photocopying, recording, or
otherwise, without the prior written permission of the
publisher.

1 2 3 4 5 6 7 8 9 0 K P K P 7 9 8 7 6 5 4 3

This book was set in Times New Roman.   The editors were
Brete C. Harrison and Barry Benjamin; the designers
were Emily Harste and Ben Kann; and the production
supervisor was John A. Sabella.   The drawings were
done by John Cordes, J & R Technical Services, Inc.
The printer and binder was Kingsport Press, Inc.

Library of Congress Cataloging in Publication Data

Guenther, William C.
     Concepts of statistical inference.

     1. Mathematical statistics.   I. Title.
QA276.G79   1973   519.5   73–1208
ISBN 0–07–025098–7

QA
276
.979
1973

CONTENTS

68934

Chapter 3
STATISTICS, SAMPLING DISTRIBUTIONS, AND POINT ESTIMATION OF PARAMETERS

Chapter 4
TESTING STATISTICAL HYPOTHESES

Chapter 5
HYPOTHESIS TESTING AND INTERVAL ESTIMATION FOR MEANS AND VARIANCES

Chapter 6
HYPOTHESIS TESTING AND INTERVAL ESTIMATION FOR PARAMETERS OF DISCRETE DISTRIBUTIONS; APPROXIMATE CHI-SQUARE TESTS

Chapter 7
ANALYSIS OF VARIANCE   342

Chapter 8
REGRESSION AND CORRELATION   371

Chapter 9
DISTRIBUTION-FREE INFERENCES   406

Appendix A
SOME ADDITIONAL RESULTS   439

Appendix B

PREFACE

This book was written for a general one-semester (or one quarter) introductory course in statistics. The primary objective has been to present some of the important concepts of statistical inference which are common to all areas of application. Examples and exercises have been drawn from business, psychology, sociology, engineering, agriculture, education, quality control, the biological and biomedical sciences, and games of chance. A minimum amount of emphasis has been placed upon algebraic details, computing, and complicated statistical methods. A few "mathematical" arguments are used (when they are elementary and straightforward) but in many cases results and procedures are justified on an intuitive basis.

No previous statistical experience has been assumed. A working knowledge of high school algebra should provide the student with a sufficient mathematical background. It is, however, a good idea to precede a course taught from this book with a semester of college mathematics, mainly for the experience and maturity to be gained from it. Those who have avoided mathematical and other scientific courses are apt to have difficulty with statistics regardless of the choice of textbook or the ability of the instructor.

Some of the main features of the book are:

1. Emphasis is placed upon selecting an appropriate model and investigating the reasonableness of the conditions under which the statistical procedures are derived. (When performing an actual experiment, it is essential for an investigator to sample so as to make these conditions as realistic as possible.)
2. A number of tables and graphs are presented and used. Many of these do not often appear in an elementary textbook. With the general availability of high speed computers, numerous tables have been prepared in recent years. Frequently such tables enable an investigator to

take a more sophisticated approach to standard statistical problems. In particular, it is now relatively simple to compute the power of a number of tests and to solve sample size problems.

3. A large number of worked examples are included. These are written out in detail to show the student how to attack similar problems in the Exercises and in his practical applications of statistical inference.

4. The Exercise sections contain problems phrased in the language of many different areas of application. Answers to the odd-numbered problems in Chapters 1 to 4 and selected sets of problems in Chapters 5 to 9 are given in the rear of the book, in greater detail than is usual when the problem seems to require this detail. Those exercises for which answers are given are marked by a diamond symbol ◈ in the Exercise sections.

5. To assist the student with the organization and assimilation of the material, a summary has been provided at the end of each of the nine chapters.

In this edition the number of "realistic sounding" exercises has been considerably increased. There is no doubt that real life statistical problems are very good for motivational purposes but unfortunately most such problems are too difficult for a beginning course. Consequently, many of the examples and exercises we have used are simplified versions of real problems. These can appropriately be called "realistic sounding" exercises implying that they sound like the kind of problem that someone might be interested in solving. (Exercises which begin with the phrase "Consider the data" are not very useful for motivational purposes.) Since this type of exercise is difficult to construct, it is no wonder that most textbooks have a shortage of such problems. In addition to nearly 200 Examples, this edition contains more than 500 Exercises which should be a sufficient number to provide the instructor with some choice.

In comparing this edition to the first the following remarks seem to be appropriate:

1. All chapters have been completely rewritten. An attempt was made to say things correctly without making discussions too complicated. Care has been taken to distinguish between statistics (and random variables) and their observed values.

2. The new Chapter 3 is quite different from the old one. Point estimation of parameters is used as the starting point of statistical inference. "Good" estimates of parameters encountered in Chapter 2 are justified on an intuitive basis.

3. Sections 6-5, 6-6, 6-7, and Chapter 9 on distribution-free inferences are new.

4. The usually neglected sample size problems are discussed.

5. As mentioned in the preceding paragraph, many new realistic sounding exercises have been added.

Various approaches can be used successfully for a beginning statistics course. This text follows the "systematic approach" (terminology of the CUPM Panel of Statistics for a course in which basic tools are developed slowly and in some detail before complicated real problems are attacked) and is based upon classical statistics. A major advantage of this path is that needed materials are readily available and well organized. Another possibility is the "data-oriented" approach—teaching principles and concepts by conducting actual experiments and analyzing real problems. Major factors in the success of any approach are the enthusiasm of the instructor and his organization and use of appropriate materials.

After teaching statistics at various levels for a number of years, one is bound to develop some opinions and prejudices. For what it is worth, here is one of mine. A beginning (one semester) course should introduce concepts involved in probability and probability models, estimation and testing hypotheses. Since these principles are the same in all areas of application, the first course should not be area oriented but should demonstrate the universality of statistical inference. After all, there is no guarantee that the student will encounter statistical thinking only in his speciality area and he may even be exposed to it in newspapers and magazines. If it is felt that students need a course specialized for their area, give it to them as a second course.

Undoubtedly this book contains more material than can be covered in three semester hours (or five quarter hours). It is recommended that any course include the first four chapters, Sections 5-1 to 5-5, and Sections 6-1 to 6-3. If more time is available, then the first choices for additions should probably be Sections 5-9, 5-10, 6-8, 6-10 (and possibly 5-7). These latter sections can be followed by other sections selected from Chapters 5 to 9. Since distribution-free techniques seem to be used more than in the past, it is probably advisable to try to cover part of Chapter 9.

Logically a course from this book can be followed by a number of special topics or methods courses. Some of these include analysis of variance, regression and correlation, sampling, time series, distribution-free statistics, quality control, and a second course specifically designed for an individual area of application.

At the University of Wyoming this course has been taught with large lecture sections (preferably handled by an experienced staff member) meeting three times a week and small laboratory or problem sections (taught by graduate students) meeting twice a week. Although this system may not be ideal, it does provide a satisfactory method of handling a great many students, provided the laboratory sections are not too large.

William C. Guenther

CONCEPTS
OF STATISTICAL
INFERENCE

INTRODUCTION

In beginning the study of a new subject such as statistics, one may hope to be presented with either a definition or a description which essentially summarizes the scope of the field. Such statements usually depend upon terms and concepts which are not yet well defined and, as a consequence, may not be too helpful to the student. We could say, at least with partial accuracy, that statistics is a field of endeavor in which data are collected for the purpose of drawing conclusions and making decisions. Some of the difficulties with this definition arise from the inability of the student (because of inexperience) to attach the appropriate interpretation to terms like *data*. (According to the dictionary, data are facts or figures used as a basis of a discussion or decision.) Perhaps more serious inadequacies are due to the lack of indication as to how the data are collected and the omission of clarifying comments about the types of inferences and conclusions which might be made.

Fundamental to statistical analysis is the concept of a random experiment. The term *experiment* could be defined as an operation carried out under controlled conditions for the purpose of discovering or verifying facts. Frequently the terminology *physical experiment* is used to characterize that type of experiment which, when repeated under the same conditions, yields essentially the same result. To illustrate, suppose that the experiment consists of dropping a ball. On the basis of our experience we expect the same result—the ball falls to the ground—every time the performance is repeated. Some experiments may produce different results in spite of all efforts to keep the conditions of performance constant. Classic examples are coin and dice throwing, but the results produced by subjecting human beings, animals, or other objects to the same environmental conditions also fall into this category. Experiments of this latter type are called *random experiments*.

We next mention some specific problems which a statistician might be expected to consider. Medical researchers have discovered a new cure for tuberculosis. They would like to know if the new cure is superior to the old cure already in use. A government testing agency, one of whose missions is public protection, desires information on the results of cigarette smoking. A manufacturing company wants to know if a new machine which produces their product is superior to an old one already in use. A university registrar would like to predict a student's future performance on the basis of high school records, entrance examinations, or other information at his disposal. A television network wants to predict the results of an upcoming election. An educational researcher is interested in determining which of several teaching methods is superior. A seed corn company can market several different hybrid varieties and desires information about yield. A game and fish expert needs to predict the antelope population for the coming hunting season. A manufacturer of television tubes must have information concerning the length of life of his product in order to make a guarantee. The producer of a certain brand of photographic equipment is interested in knowing the fraction of flashbulbs which fail to operate.

In the types of investigations which might be performed in connection with the above examples, a statistician would visualize the totality of measurements (called the *population*) about which information is desired. In the election example this would consist of all votes to be cast for the various candidates. (To obtain a numerical measure in a two-man race, we could count a vote as a " 1 " if it is for a candidate, as a " 0 " if it is for his opponent). In the seed corn example the population would probably be all conceivable yields per acre which might be obtained under a given set of conditions. For the flashbulb problem the population could be the performances of all bulbs (with " 1 " corresponding to a success, " 0 " to a failure) currently available in the producer's warehouse or these plus all those which might be turned out in the future by the manufacturing process operating in its present form. The game and fish expert would be interested in the size of the population (the term is used in the conventional sense of actual numbers). The government agency which is studying cigarette smoking may be interested in two populations, perhaps life spans of smokers and life spans of nonsmokers. One important characteristic of a population is that it must be capable of definition; that is, we should be able to identify those measurements, either real or conceptual, which belong to it.

As has been implied by the above discussion, the major objective of a statistical investigation is to find out something about the population. If the population is small, then it may be feasible to obtain every single measurement. For example, heights of 10 members of a basketball team can be measured very quickly and with little or no cost or inconvenience. Sometimes, however, not even a small population can be collected in its entirety since to do so may be prohibitively expensive. This is particularly true if an item must be destroyed or expended (as with a camera flashbulb) to obtain the measurement in question. Not infrequently the population is so large that a complete census is out of the question. Measurements on items not yet produced or on subjects not yet in existence are not available and, consequently, cannot be used for any purpose.

In those problems for which a complete census is impractical or impossible statistical methods are valuable. Not being able to examine every item for measurement, we draw inferences based upon a part of the population (called a *sample*). It does not take much imagination to see that if a sample is to be useful for the purpose intended, the measurements must be collected in a sensible way. Thus, if we wished to estimate the average age of residents of Los Angeles and took as our sample all first-grade children in six conveniently located elementary schools, the estimate would undoubtedly be very poor. Most statistical analysis is based upon "random" samples which we shall discuss in more detail later. In the current average-age example, it is sufficient to say that a random sample is a sample selected in such a way that every sample of the same size has an equal chance of being selected.

Another distinguishing feature of statistical analyses is that inferences which are drawn on the basis of a sample are usually accompanied by statements indicating the amount of confidence that a statistician has in the results. The basis of such statements depends upon the study of probability (Chap. 1) and the development of probability "models" (Chap. 2). We will, of course, devote considerable space to these topics. For the time being we can regard a model as an idealized representation that we hope characterizes the behavior of an experiment and is sufficiently close to reality to be useful in predicting future behavior. For our purposes it is probably more accurate to say that a model is a set of assumptions that hold true for a class of random experiments. As we shall see, frequently these assumptions lead to a formula (also sometimes referred to as a model) which is used in the process of drawing inferences.

There are times when it is easy to draw inferences on the basis of results from a series of random experiments. If a six-sided die is rolled 30 times and each time the side showing 6 dots appears on top, a fairly obvious conclusion is that the die is loaded or the thrower is an expert at controlling the outcome. The correct conclusion is apt to be reached by everyone with or without statistical training. Unfortunately, most experimental results are somewhat more difficult to analyze so that justifiable conclusions may be drawn. In fact many real-life problems present difficulties even for an acknowledged expert in the field.

The previously discussed example concerning the Los Angeles first graders can be used to illustrate an important principle regarding experimentation and statistical analysis. Generally, before any experimental results are observed some advanced planning is advisable. If the objective of the investigator is to estimate the average age of all residents, then he should plan to use a sampling procedure which can be expected to give him a reasonably good result. Not infrequently an experimenter will collect a mass of data and present it to a statistician for analysis. More often than not on such occasions the statistician will be unable to draw any meaningful conclusions and the time and expense of the experimenter is apt to be wasted. Sometimes such "preliminary" results can be useful in planning a well-designed experiment (requiring further sampling), but as a general rule the time to consult the statistician is before any measurements are made.

One may get the impression that the subject matter of statistics varies for each area of application. It may appear that the general theory and principles are different for engineering, psychology, business, sociology, biology, medicine, education, etc. Although there are some special methods which rarely find application in more than one of these fields, most methods and principles (particularly those which we will discuss) are applicable to any area in which investigations are concerned with random experiments.

As in every field, there are various skills, aptitudes, and tools which a statistician can use to his advantage. We can list the manipulation of mathematical formulas, the ability to organize and analyze data, skill with high-speed computers, and experience with graphs and tables. Perhaps the most important possession for an applied statistician is that nebulous quality which is called "good common sense." Without it all the fancy formulas and computers are apt to be of limited usage. With it and a little patience and perserverance, even those with modest abilities can enjoy some success as a statistician.

This is intended to be an introductory course in statistics. It is not expected that the student will qualify as a professional statistician even if he receives a superior grade. As a parallel it is universally recognized that one does not become an expert physicist, mathematician, economist, or anything else after so brief an exposure. Some reasonable objectives are

1 To introduce students to the language and philosophy of statisticians.
2 To acquaint students with the type of problems that lend themselves to statistical solution.
3 To present enough basic statistical technique so that the student can work some standard-type problems.
4 To enable the student to read and understand the summarized results of statistical experiments performed by others. He may never have to perform an experiment on his own, but he might have to read about the work of others in journals.
5 To interest some students in the further study of statistics.

1

PROBABILITY

1-1
INTRODUCTION

The use of the word "probability" has become so commonplace that almost everyone has a vague notion about its meaning. One frequently hears statements such as the probability of rain is .3, the probability that a sprinter will win a race is 2/3, or the probability that two dice will produce a sum of 7 is 1/6. Two important considerations arise immediately. The first concerns the meaning of such a statement and the second involves the problem of determining the numerical value of a probability. Both of these aspects will be considered in detail.

In discussing probability we shall find it convenient to use the term *event*. Although a mathematical definition of this term will be given later in the chapter, for the present we can consider an event in its everyday context. Usually we think of an event as something that has occurred, is about to occur, or might conceivably occur. To be more specific we could say that an event is one of the possible outcomes associated with a random experiment. Examples of events are rain falling during a given day, a specified sprinter winning a race, and two dice falling so that the sum is 7.

One might consider probability as a measure of likelihood ranging from 0 to 1. If an event is impossible, the weight or measure (or probability) 0 is assigned to it. On the other hand an event that is certain to happen is assigned weight 1. In between these extremes (the interesting and more difficult cases), we endeavor to assign weights so that the more likely an event, the higher its probability. As we shall see, the weights that are assigned may depend upon prior knowledge of the situation, experimental evidence, or just plain intuition. In some situations a formula (or tables derived from the formula) yields numbers which are realistic probabilities for events of interest. If a truckload of pennies were dumped on the pavement (or if a single coin were tossed thousands of times), it would be reasonable to expect that heads would show about 1/2 of the time. It would seem sensible, therefore, to associate the weight 1/2 with the chances of obtaining a head when discussing standard coin-tossing problems. Our choice here is based upon symmetry considerations and intuitive appeal; that is, with balanced coins we have no reason to believe that a head is more or less likely to show than a tail. As we shall soon see, selection of the correct weight is generally much more difficult.

We can use the coin example to emphasize a further point. As we have already implied, the figure 1/2 is associated with long-run behavior. Thus when we say that the probability of a head is 1/2, we do not mean that one out of every two tosses results in a head; we mean, rather, that in the long run we expect heads to show half the time. If we form the

$$\text{Frequency ratio} = \frac{\text{number of heads}}{\text{number of tosses}}$$

we might expect that this fraction would tend to get closer to 1/2 as the number of tosses is increased. That this is indeed what happens has been borne out by actual experience of many investigators. It is, of course, possible to obtain a head every time or nearly every time so that the frequency ratio would be far from 1/2 after a large number of tosses. However, the latter type of situation has practically no chance of occurring and would be classified as a miracle. In other words the above ratio will nearly always stabilize in the vicinity of 1/2 when the coin-tossing experiment is repeated a large number of times. This property of long-run stability has been found to characterize the results of many experiments.

Thus far we have indicated that the probability of an event is a number between 0 and 1 which reflects the chance that the event will happen when some random experiment is performed. We now turn our attention to the task of defining probability. Although we shall usually think in terms of the mathematical definition (given last), we shall present two other definitions with the three being given in the chronological order of development. Not only do all three definitions contribute to the understanding of probability, but also it is usually beneficial to review the thought processes that lead to the final result (in this case, the third definition) we use.

1-2
DEFINITIONS OF PROBABILITY

The first definition of probability arose several hundred years ago from consideration of well-known games of chance with dice, coins, cards, roulette, etc. Let us consider a simple example. Suppose that we have a well-balanced (or symmetric) die with six sides numbered 1 to 6. We may feel that any one of the sides is as likely to show as any other when the die is rolled. If our assumption is correct and we wish to form an estimate of our chances of rolling a 4, then it seems fairly logical to use the ratio 1/6. The fraction is obtained by using for the numerator the number of ways a 4 can be obtained and for the denominator the number of different results that the die can produce. The classical definition of probability was developed from this kind of thought process. That definition is as follows:

Classical Definition *If an experiment can produce n different* (1-1) *mutually exclusive results all of which are equally likely, and if f of these results are considered favorable (or result in event A), then the probability of a favorable result (or the probability of event A) is f/n.*

Events are mutually exclusive if the occurrence of one prevents the occurrence of any of the others. Thus a die, when rolled, can produce six mutually exclusive results, 1, 2, 3, 4, 5, 6. Only one side can be up when the die comes to rest. The probability that event A happens will be denoted by $\Pr(A)$.

EXAMPLE 1-1

A single six-sided symmetric die is rolled. What is the probability of rolling either a 4 or a 5?

Solution

We have agreed that the die can produce six equally likely results. Of these, two are favorable (either a 4 or a 5). Hence $n = 6, f = 2$, and the probability of rolling a 4 or a 5 is 2/6. If we let A be the event that either a 4 or a 5 is rolled, then $\Pr(A) = 2/6$.

EXAMPLE 1-2

From an ordinary deck of cards that has been well shuffled one is drawn. What is the probability that the card is a heart?

Solution

An ordinary deck contains 52 cards. When a drawing is made from a well-shuffled deck, then it is not unreasonable to assume that every card has an equal chance of being drawn or that all 52 possible results are equally likely. Of these, 13 are hearts. Thus $n = 52, f = 13$, and the probability of drawing a heart is 13/52. If we let A be the event of drawing a heart, then we can write $\Pr(A) = 13/52$.

EXAMPLE 1-3

Two coins are tossed. What is the probability that one of the coins shows a head, the other a tail?

Solution

Let H stand for head, T for tail. The possible outcomes can be denoted by HH, HT, TH, and TT where the first letter indicates the result on the first coin, the second letter the result of the second coin. Thus there are four possible outcomes and $n = 4$. Of these, two are favorable to obtaining a head and a tail. Hence, if we can agree that the four results are equally likely, then the desired probability is 2/4. With balanced coins and a fair tossing procedure we have no reason to think that the four results are not equally likely. Letting A be the event one coin shows a head, the other a tail, we may write $\Pr(A) = 2/4$.

A beginning student may list the outcomes by the number of heads, 0, 1, 2, and regard these results as equally likely. He may then conclude that the probability of obtaining one head and one tail is 1/3. One way to check the appropriateness of this choice is to perform the experiment, tossing two coins, a large number of times and examine the frequency ratio for 1 head. He should find that it is close to 1/2, not 1/3.

One objection to the classical definition is that it contains the phrase "equally likely," an undefined term. Attempts to define it usually lead to the use of phrases such as "equally probable," and critics have pointed out that this involves circular reasoning. The objection is not serious, however, since almost everyone has an intuitive feeling of what is meant. We might be hard pressed to define "time," yet we use the word with little or no confusion. It is probable that, owing to inexperience, we might label events as equally likely when actually they are not.

A practical objection to the classical definition is that it depends upon being able to classify outcomes as equally likely. When calculating probabilities with a symmetric die, no difficulty arises. Suppose one has a crude homemade wooden die cut in some odd shape with sides numbered 1 to 6. Due to lack of symmetry, one would have little faith in regarding the outcomes 1 to 6 as equally likely. What is the probability that the side numbered 4 will show if the die is thrown? The classical definition offers no assistance. This type of situation has led to the following definition:

Relative Frequency Definition *Suppose that an experiment is* (1-2) *performed n times with f successes. Assume that the relative frequency f/n approaches a limit as n increases (so that as n increases, so does f). Then the probability of a success is the limit of f/n as n approaches infinity.*

The definition (1-2) implies that we need to keep on performing the experiment forever to determine the probability. This is, of course, impossible. The best we can do is to make an estimate based upon a large n, a fact that might leave us somewhat dissatisfied. Frequently we have nothing better at our disposal.

EXAMPLE 1-4

The unbalanced homemade die was rolled 100 times. The side labeled 1 occurred 12 times, 2 occurred 18 times, 3 occurred 20 times, 4 occurred 27 times, 5 occurred 13 times, and 6 occurred 10 times. What probabilities should be associated with each of the sides?

Solution

The best we can do is to use the relative frequencies obtained from the 100 throws. Thus

Probability of a 1 = 12/100   Probability of a 4 = 27/100
Probability of a 2 = 18/100   Probability of a 5 = 13/100
Probability of a 3 = 20/100   Probability of a 6 = 10/100

If probabilities such as those obtained in Example 1-4 are to be meaningful, the conditions of the experiment should be kept as constant as possible. If the die deteriorates as it is thrown, then it is likely that the probabilities will change and the relative frequencies obtained will not be useful in predicting the future behavior of the die.

A practical use of the relative frequency definition arises in determining insurance rates. In figuring such rates a company may need to know, for example, the probability that a person age 20 lives to be age 70. Several tables giving just such relative frequencies are in existence. These are based upon actual records and are constantly being updated.

For our next definition of probability, we need the notion of a sample space. To illustrate this concept, consider tossing a coin twice. All possible outcomes can be listed in a number of ways. First, they can be denoted by HH, HT, TH, and TT where the first letter indicates the result of the first toss, the second letter the result of the second toss. A second way to characterize the outcomes is by listing the number of heads, 0, 1, 2. Third, one of many other listings of all possible outcomes is obtained by recording the result of the first throw, the result of the second throw, and the distance from which the coin is dropped. In any such listing which includes all possible outcomes, each outcome is called a *sample point,* and the set or collection S of all sample points for a given listing is called the *sample space.*

We shall frequently use pictorial representation as a visual aid to the discussion. Figure 1-1 shows the sample space when the results of both throws are recorded. Figure 1-2 serves the same purpose when outcomes

FIGURE 1-1

A sample space representation for all possible outcomes of two tosses of a coin.

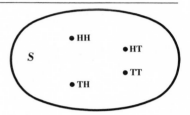

are given in terms of the number of heads obtained. Thus, if we prefer we can think of each outcome as a point in space. The oval-shaped curve around the points serves no purpose other than to indicate that the sample spaces are contained therein.

Having observed that different sample spaces can be used for the same experiment, it is reasonable to ask which one is more suitable. The choice will depend upon the features of the experiment one needs to know in order to work the problem at hand. About the best advice that can be given is to label the points of the sample space so as to include

FIGURE 1-2

A second sample space representation for all possible outcomes of two tosses of a coin using the number of heads obtained.

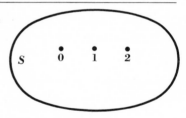

all the necessary information, but no useless information. If we knew nothing about probabilities associated with the coin example, very likely we would first be interested in the representation of Fig. 1-1. Our intuition might tell us that these are mutually symmetric results and should be assigned probabilities of 1/4. Having somehow determined the probabilities that should be associated with the points of the sample space given in Fig. 1-2, perhaps by using the first sample space, we would probably find that the second representation would prove more useful in applied problems.

Any subset *A* of the sample space (part of the original listing) is called an *event*. Figure 1-3 shows an event associated with the sample space of Fig. 1-1. The event *A* represents the outcome resulting in a head on

FIGURE 1-3

A sample space and an event. The event *A*, a subset of the sample space, represents the outcome "head on the second toss."

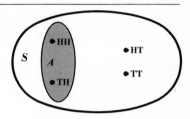

the second toss. To generalize the discussion, suppose that for a particular experiment we have agreed upon a sample space with *n* outcomes. Denote these outcomes by a_1, a_2, ..., a_n as in Fig. 1-4. A subset containing only one sample point is called a *simple* (or *elementary*) event. Hence a_1, a_2, ..., a_n represent *n* elementary events. We observe that the terms *sample point*, *simple event*, and *elementary event* are all synonymous. Any event *A* will consist of one or more simple events. For example, the event *A* of Fig. 1-4 includes the simple events a_2, a_3, and a_5.

FIGURE 1-4

A sample space with *n* simple events.

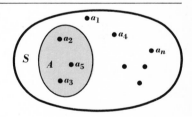

Now assign to each point a_i of *S* a weight w_i subject to

(*a*) $w_i \geq 0$

(*b*) $w_1 + w_2 + \cdots + w_n = \sum_{i=1}^{n} w_i = 1$ (1-3)

All (1-3) implies is that the weights are positive numbers between 0 and 1 inclusive and that the sum of the weights is 1. The capital sigma is standard mathematical notation for sum. (For more on summation notation see Appendix A1.) We now make the following definition:

Mathematical Definition *Let weights satisfying* (1-3) *be assigned* (1-4)
to points of the sample space. Then the probability Pr(*A*) *of any
event A is the sum of the weights of all sample points in A.*

According to the definition each weight is itself a probability since it is
possible for an event A to contain only one sample point.

Although definition (1-4) is well adapted to proving theorems, it is
of no assistance in determining the weights. In choosing the weights,
we have to depend upon experience, intuition, and experimentation.
Frequently we recognize from the symmetry of the simple events of the
sample space that equal weights are reasonable. When this situation
arises, each of the n sample points is assigned weight $1/n$, and definition
(1-4) can be modified to read

Modified Mathematical Definition *The probability* Pr(*A*) *that the* (1-5)
*event A will occur is the ratio of the number of sample points in A
to the total number sample points in S. That is,* Pr(*A*) = *n*(*A*)/*n
where n(A) is the number of sample points in A.*

EXAMPLE 1-5

Four slips of paper numbered 1, 2, 3, 4 are placed in a hat and one slip
is drawn out. What is an appropriate sample space? What weights would
you assign to the sample points?

Solution

In order to simplify the selection of weights, we would probably choose
the sample space consisting of four sample points that could be designated 1, 2, 3, 4. If the slips are the same size so that each has an equal
opportunity of being selected, the weights 1/4, 1/4, 1/4, 1/4 are reasonable.

EXAMPLE 1-6

Suppose that in Example 1-5 the number is recorded, the slip replaced,
and a second slip drawn. What sample space would probably be selected,
and what weights would you assign to the sample points?

Solution

When two slips are drawn as described, the sample space we would probably select contains 16 points that could be designated

(1, 1)	(2, 1)	(3, 1)	(4, 1)
(1, 2)	(2, 2)	(3, 2)	(4, 2)
(1, 3)	(2, 3)	(3, 3)	(4, 3)
(1, 4)	(2, 4)	(3, 4)	(4, 4)

where the first number is the result of the first draw and the second is the result of the second draw. If the slips are well mixed after each draw, it is reasonable to assign a weight of 1/16 to each point.

EXAMPLE 1-7

In Example 1-6 let A be the event that includes all the sample points such that the sum of the numbers from the two draws is 6. What is $\Pr(A)$?

Solution

We have already selected the sample space containing 16 points, each with weight 1/16. Of these (2, 4), (3, 3), and (4, 2) belong to A. Thus, according to definition (1-4) or (1-5), we get $\Pr(A) = 1/16 + 1/16 + 1/16 = 3/16$.

EXAMPLE 1-8

A roulette wheel is divided into 38 equal parts numbered 1 to 36, 0, and 00. Eighteen of the numbers are colored red, eighteen are colored black, and 0 and 00 are colored white. A perfectly built wheel is supposed to be symmetric so that all numbers are equally likely. Suppose that we are interested in a sample space containing the simple events red, black, and white. What weights should be assigned and how does one get them?

Solution

It is probably obvious that one would select weights 18/38, 18/38, and 2/38, respectively. These numbers are obtained by considering the sample space consisting of 38 sample points that could be designated $1, 2, \ldots, 36, 0, 00$. Because of the symmetry of the simple events associated with the latter sample space, it is logical to assign weights of 1/38 to each point. The event "red" contains 18 sample points, the event "black" the same number, and the event "white" contains two sample points. Hence the weights 18/38, 18/38, 2/38. This example is a further demonstration of the comment made earlier in this section. That is, we may first work with a sample space that is not particularly interesting but consists of simple events that are mutually symmetric in order to determine the probabilities associated with the simple events in a second sample space that is of interest.

Thus far we have considered only experiments with a *finite sample space*, that is, a sample space consisting of a finite number of points. Some interesting problems require a countably infinite number of points.

By this we mean that the points of the sample space can be put into one-to-one correspondence with the positive integers. In other words the points can be counted 1, 2, 3, etc., but the counting process never ends. As a simple illustration suppose that a coin is tossed until a head appears. The outcomes could be denoted by H, TH, TTH, TTTH, etc., which can be counted by identifying H with 1, TH with 2, TTH with 3, etc. A sample space which includes a finite number of points or an infinite number that is countable is called a *discrete sample space*. For the moment we will defer the discussion of probability assignment for our example, but the alert reader might guess that weights should be 1/2, 1/4, 1/8, 1/16, etc.

Let us reexamine condition (*b*) of definition (1-3) when the sample space contains a countably infinite number of points $a_1, a_2, a_3, \ldots$. Now an infinite number of weights are required satisfying condition (*a*) of (1-3), and condition (*b*) should be changed to

$$(b') \quad w_1 + w_2 + w_3 + \cdots = \sum_{i=1}^{\infty} w_i = 1 \tag{1-6}$$

From a practical point of view no difficulty is encountered since, in most applications of interest, the sum of the weights from a certain point on in the above series is practically zero. Thus the probability of the event containing sample points $a_1, a_2, \ldots, a_n$ for some n is almost 1, and the probability of the event containing sample points $a_{n+1}, a_{n+2}, \cdots$ is nearly zero. Theoretically we are faced with interpreting the meaning of a sum containing an infinite number of terms. Obviously we cannot write down all terms and form their sum as in the finite case. What is meant by (1-6) is that the sum can be made as close to 1 as we like by taking a sufficiently large number of terms. To illustrate the situation a little more graphically, let us consider the following example. A frog 1 inch away from a pond jumps toward it in such a way that with each jump he reduces the remaining distance by one-half. How many jumps are required to reach the pond? Theoretically, since there is always some distance left to be jumped, the frog never arrives. However, most people would be willing to concede that the frog is at the pond after 50 (or perhaps fewer) jumps since the remaining distance is hardly worth any consideration. When nonnegative weights satisfy (1-6), definition (1-4) is still correct without modification. Definition (1-5) is, of course, useless since it is applicable only when the sample space contains n (a finite number) points.

EXERCISES

◈ 1-1 A symmetric six-sided die is rolled once. Construct a sample space for this experiment and assign weights to the sample points. What is the probability that the number which appears is greater than 4?

1-2   A card is drawn from an ordinary deck that has been well shuffled. What is the probability that the card is an ace, king, or queen?

◈ 1-3   A student is taking an examination which consists of multiple-choice questions with five possible answers (designated by A, B, C, D, and E). If he attempts to guess the answer knowing nothing about a particular question, what is the probability that he gets the correct choice? Suppose that the student knows enough about the subject to eliminate three answers as being incorrect. If he now guesses, what is the probability that he answers correctly?

1-4   From an urn containing four white and two red balls, all the same size, one is to be drawn. Construct a sample space for which equal weights are reasonable. What is the probability of drawing a red ball?

◈ 1-5   A committee consists of five men and two women one of whom must serve as secretary. To determine the unlucky individual a drawing is held. Each person writes his name on a slip of paper, the slips are put in a hat and well mixed, and one is drawn. If this procedure is followed, what is the probability that the secretary is a woman? Suppose that each man, instead of writing his own name, wrote the name of a woman on his slip, while each woman wrote her own name as instructed. Now what is the probability of having a woman secretary?

1-6   An ordinary coin is tossed three times. Construct a sample space with equal weights, and compute the probability that two out of three times the result is heads.

◈ 1-7   Three sprinters A, B, and C race against each other frequently. They have won 60, 30, and 10 percent of the races respectively. What sample space and what weights does this information suggest? What is the probability that A loses the next race?

1-8   A symmetric six-sided die is rolled twice. For this experiment construct a sample space with equal weights. What is the probability of obtaining a total of 7 on the two rolls?

◈ 1-9   Based upon very extensive study, the records of an insurance company reveal that the population of the United States can be classified according to ages as follows: under 20, 35 percent; 20 and over but under 35, 25 percent; 35 and over but under 50, 20 percent; 50 and over but under 65, 15 percent; 65 and over, 5 percent. Suppose that you could select an individual in such a way that everyone in the United States has an equal chance of being chosen. Using the above information, construct a sample space for the individual's age and assign weights to the sample points. What is the probability that an individual is 35 or over?

1-10   A shipment of paint contains 2,000 1-gallon cans of which 800 are white, 500 are yellow, 300 are red, 300 are green, 100 are blue. During transit the cans are accidentally submerged in water and all labels are lost. Upon arrival the cans are placed upon a platform and one is selected and opened. If color is of interest, what sample space and what weights are suggested by the above information? What is the probability that the selected can contains red, white, or blue paint?

◇ 1-11 Suppose the weather bureau classifies each day according to wind condition as windy or calm, according to rainfall as moist or dry, and according to temperature as above normal, normal, or below normal. Records are available for the past 100 years. Construct a sample space suggested by the above information which can be used to characterize the weather on next July 4. How might you obtain weights for your sample points?

1-12 The balls for various games of pool are numbered consecutively from 1 to 15. In addition there is a white cue ball which is unnumbered. These balls are placed in a bag and one is drawn out. Construct an equal-weight sample space for this experiment. Then compute the probability that the selected ball has an even number printed on it.

◇ 1-13 A four-sided die is so constructed that in the long run the side labeled 4 lands on the bottom about twice as frequently as each of the sides labeled 1, 2, 3. If a 4-point sample space is used for the outcome of a single roll, what weights should be used?

1-14 To determine who pays for coffee, three people each toss a coin and the odd person pays. If the coins all show heads or all show tails, they are tossed again. With the discussion following Example 1-8 in mind, construct a sample space that characterizes the manner in which a decision is reached.

◇ 1-15 Extensive records on 3-child families indicate that 13.3 percent have 3 boys, 38.2 percent have 2 boys, 36.7 percent have one boy, and 11.8 percent have 3 girls. If a family plans to have 3 children, what is the probability that 2 or more will be boys?

1-3
THE ADDITION AND MULTIPLICATION THEOREMS

In many probability problems it is convenient to regard an event in terms of two or more simpler events. For example, if a six-sided die is rolled twice, one possible event is 5 on the first roll, 6 on the second roll. Alternatively, the same result can be regarded as a combination of two events where the first event is 5 on the first roll, the second event is 6 on the second roll. As we shall see, it is often easier to evaluate the probability of an event by first finding probabilities of simpler events and then using theorems to compute the probability of the more complicated situation.

Before discussing some of the devices we shall use to simplify the calculation of probabilities, we need to introduce some new terminology and notation. Consider again the sample space associated with tossing a coin twice. Let A be the event "at least one head shows." Another event related to A, say $\bar{A}$, is the event "A does not occur" or "no heads show." The events A and $\bar{A}$ are called complementary events (see Fig. 1-5). If A occurs, $\bar{A}$ does not, and vice versa. Thus, the *complement* of

FIGURE 1-5

Complementary events.

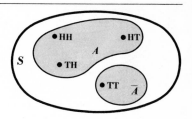

an event A is the event $\bar{A}$ that includes all sample points not in A. Since, according to conditions (1-3), the sum of the weights must be 1, it is obvious that $\Pr(A) + \Pr(\bar{A}) = 1$ or

$$\Pr(A) = 1 - \Pr(\bar{A}) \tag{1-7}$$

Since it is frequently easier to calculate $\Pr(\bar{A})$ than $\Pr(A)$, formula (1-7) is itself a useful device. Having already agreed that the sample points in Fig. 1-5 should each be assigned weights $1/4$, we get for this example $\Pr(A) = 3/4$, $\Pr(\bar{A}) = 1/4$, and the sum of the two probabilities is 1.

Suppose that we throw an ordinary coin and a symmetric six-sided die simultaneously. Let A_1 represent the occurrence of a head on the coin and A_2 represent the occurrence of a 1 or a 2 on the die. The 12-point sample space and events A_1 and A_2 are pictured in Fig. 1-6. The oval includes all the points in A_1, the rectangle all the points in A_2. From the events A_1 and A_2 arise two new events. One of these is "either A_1 or A_2 or both occur" and is called the union of events A_1 and A_2. The *union* of two events A_1 and A_2 is the event that contains all the sample points in A_1 or A_2 or both A_1 and A_2. The union of A_1 and A_2 will be denoted by $A_1 \cup A_2$. In Fig. 1-6 we observe with no difficulty that $A_1 \cup A_2$ contains the eight points H1, H2, H3, H4, H5, H6, T1, T2. The second new event is defined by the condition "both A_1 and A_2 occur" and is called the intersection of the two events A_1 and A_2. Thus, the *intersection* of two events A_1 and A_2 is the event that contains all the sample points in both A_1 and A_2. The intersection of A_1 and A_2 will be denoted by $A_1 A_2$. In Fig. 1-6 $A_1 A_2$ contains the points H1 and H2.

FIGURE 1-6

Sample space for a coin and die thrown simultaneously.

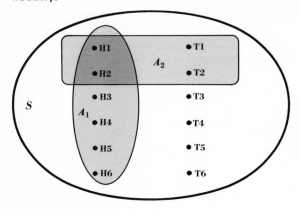

The definitions of union and intersection easily extend to any number k of events. Thus by $A_1 \cup A_2 \cup \cdots \cup A_k$ we mean that at least one of the events $A_1, A_2, \ldots, A_k$ occurs, and $A_1 A_2 \cdots A_k$ means that all k of the events occur.

Another important term we shall need is mutually exclusive (previously defined in dictionary terms). We say that events $A_1, A_2, \ldots, A_k$ are *mutually exclusive* if the events have no sample points in common. As an illustration let us again use the coin-die example and the three events pictured in Fig. 1-7. Thus A_1 contains the points H5 and H6; A_2 contains T1, T2, and T3; and A_3 contains the point H1; the three events have no points in common and are, according to the definition, mutually exclusive.

FIGURE 1-7

Three mutually exclusive events resulting from the simultaneous tossing of a coin and a die.

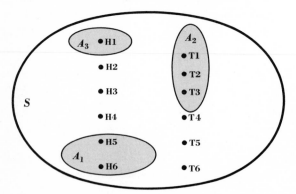

In everyday language, if several events are mutually exclusive, the occurrence of one of the events excludes or prevents the occurrence of any of the other events. In other words, two such events cannot occur simultaneously. We observe that if A_1 and A_2 are mutually exclusive, then A_1A_2 contains no points, and in particular A and $\bar{A}$ are such events.

Consider again the sample space for the coin-die example and the events A_1 and A_2 as defined for Fig. 1-6. It seems reasonable to assign each of the 12 points a weight $1/12$. With this assignment we see by using definition (1-4) that

$$\Pr(A_1A_2) = 1/12 + 1/12 = 2/12$$

$$\Pr(A_1 \cup A_2) = 8(1/12) = 8/12$$

We can also write

$$\Pr(A_1 \cup A_2) = 6(1/12) + 4(1/12) - 2(1/12)$$
$$= \Pr(A_1) + \Pr(A_2) - \Pr(A_1A_2) \qquad (1\text{-}8)$$

The reasoning behind (1-8) is simple. We add the weights of the points for events A_1 and A_2 and subtract the sum of the weights for A_1A_2 because the latter sum has been added twice. Although we have used a specific example, the argument holds for any two events A_1 and A_2. Formula (1-8) is called the *addition theorem*. An important special case arises when A_1 and A_2 are mutually exclusive so that the event A_1A_2 contains no points. Since there are no weights to add up, $\Pr(A_1A_2) = 0$, and (1-8) reduces to

$$\Pr(A_1 \cup A_2) = \Pr(A_1) + \Pr(A_2) \qquad (1\text{-}9)$$

The formulas of the previous paragraph can be extended to handle more than two events. If the events $A_1, A_2, \ldots, A_n$ are mutually exclusive, then with no difficulty we find the generalization of formula (1-9) to be

$$\Pr(A_1 \cup A_2 \cup \cdots \cup A_n) = \Pr(A_1) + \Pr(A_2) + \cdots + \Pr(A_n) \qquad (1\text{-}10)$$

Figure 1-7 can be used to illustrate (1-10) for three events. Having assigned equal weights of $1/12$ to each point, we get

$$\Pr(A_1 \cup A_2 \cup A_3) = 2/12 + 3/12 + 1/12 = 6/12$$

If n events are mutually exclusive and exhaust all possibilities—that is, $A_1 \cup A_2 \cup \cdots \cup A_n = S$—then formula (1-10) yields the useful result

$$\Pr(A_1) + \Pr(A_2) + \cdots + \Pr(A_n) = 1 \qquad (1\text{-}11)$$

For the situation pictured in Fig. 1-7, let A_4 contain all sample points not in $A_1, A_2,$ or A_3. Then

$$\Pr(A_1) + \Pr(A_2) + \Pr(A_3) + \Pr(A_4) = 2/12 + 3/12 + 1/12 + 6/12 = 1$$

It is easy to see that we might save considerable work by using formula (1-11). If we desire $\Pr(A_2) + \Pr(A_3) + \cdots + \Pr(A_n)$, it may be much easier to compute $\Pr(A_1)$ and evaluate the sum as $1 - \Pr(A_1)$.

When $A_1, A_2, \ldots, A_n$ are not mutually exclusive, the generalization of the addition theorem is more complicated. Since we will not have use for such a formula, we will omit this case.

EXAMPLE 1-9

Suppose that we select one card from an ordinary deck that has been well shuffled. Use formula (1-8) to find the probability that the card is a heart or a face card.

Solution

Let A_1 and A_2 be the events "a heart is drawn" and "a face card is drawn," respectively. Having selected a 52-point sample space and assigned weights of $1/52$ to each point, we get

$$\Pr(A_1) = 13/52 \qquad \Pr(A_2) = 12/52 \qquad \Pr(A_1 A_2) = 3/52$$

Hence

$$\Pr(A_1 \cup A_2) = 13/52 + 12/52 - 3/52 = 22/52$$

Of course, since there are 13 hearts and 9 face cards not hearts, we have immediately, using definition (1-4), $\Pr(A_1 \cup A_2) = 22(1/52)$.

EXAMPLE 1-10

Suppose that the probability of a cloudy day is .40, the probability of a windy day is .30, and the probability that a day is both cloudy and windy is .18. What is the probability that a day is either cloudy or windy or both?

Solution

Letting A_1 be the event "a day is cloudy," A_2 be the event "a day is windy," and using formula (1-8), we get

$$\Pr(A_1 \cup A_2) = .40 + .30 - .18 = .52$$

In both the coin-die example associated with Fig. 1-6 and in Example 1-9 we computed $\Pr(A_1 A_2)$ by using the definition of probability. Since this is not always the most convenient way to evaluate $\Pr(A_1 A_2)$, we next develop a formula that is useful for this type of calculation. To do this we will first introduce the concept of conditional probability.

Suppose that five sprinters are entered in a race. Sprinters 1 and 2 represent State University, 3 and 4 represent State College, and 5 represents State Teachers College. We shall assume on the basis of past experience that the weights which are associated with each individual's winning are respectively 3/20, 2/20, 4/20, 3/20, 8/20. Let A_2 be the event that State College wins the race. We easily calculate $\Pr(A_2) = 4/20 + 3/20 = 7/20$. The five-point sample space is pictured in Fig. 1-8

FIGURE 1-8

Five-point sample space for sprinter example.

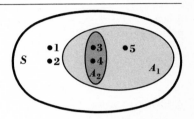

with the number of each point corresponding to that particular sprinter's winning. Now suppose that State University is disqualified and cannot participate because of illegal recruiting practices, thus eliminating sprinters 1 and 2. Letting A_1 be the event "State College or State Teachers College wins the race," we observe that A_1 must occur. We are now interested in calculating the probability that A_2 happens subject to the condition that A_1 is known to have occurred. This probability will be denoted by $\Pr(A_2 | A_1)$ and is called a conditional probability. In words, the symbol $\Pr(A_2 | A_1)$ will be referred to as the "probability of A_2 given A_1." We see that a *conditional probability* is one calculated under a condition that provides further or new information that was not previously known or available. The main problem involves the reassignment of weights in a manner that makes sense and is consistent with the information already available. Since neither 1 nor 2 can win the race, we can use the three-point sample space of Fig. 1-9 with A_1 now playing

FIGURE 1-9

Sample space for sprinter example, given that only 3, 4, and 5 can participate.

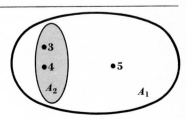

the role of S.   It is reasonable to select new weights that are proportional to the old ones, 4/20, 3/20, 8/20, and according to conditions (1-3), the sum of the weights must be 1.   This will be accomplished if the old weights are divided by $\Pr(A_1) = 4/20 + 3/20 + 8/20 = 15/20$, computed with the original sample space.   Hence the new weights for the sample space of Fig. 1-9 are 4/15, 3/15, 8/15, respectively.   Using these, we find $\Pr(A_2 \mid A_1) = 4/15 + 3/15 = 7/15$, which also can be written

$$\Pr(A_2 \mid A_1) = \frac{4/20}{15/20} + \frac{3/20}{15/20} = \frac{7/20}{15/20} = \frac{\Pr(A_2)}{\Pr(A_1)}$$

Since A_2 and $A_1 A_2$ contain the same sample points, we can also write the previous equation as

$$\Pr(A_2 \mid A_1) = \frac{\Pr(A_1 A_2)}{\Pr(A_1)} \tag{1-12}$$

Another form of formula (1-12) obtained by multiplying both sides of the equation by $\Pr(A_1)$ is

$$\Pr(A_1 A_2) = \Pr(A_1) \, \Pr(A_2 \mid A_1) \tag{1-13}$$

Formula (1-13) is called the *multiplication theorem*.

In the sprinter example A_2 was a subset of A_1.   In most situations where we shall want to use formula (1-13), this will not be the case.   To illustrate consider again the coin-die example of Fig. 1-6.   Suppose we seek the probability that A_2 happens, given that A_1 has already occurred. (That is, the probability of a 1 or 2 on the die given that coin has produced a head.)   Now two points in A_2, T1 and T2, are not in A_1.   Knowing that A_1 has already happened, A_1 in Fig. 1-10 plays the role of S, and the number of sample points in A_2 is reduced from four to two, yielding A_2'.

FIGURE 1-10

Sample space for coin-die example, given the coin shows a head.

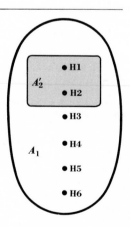

The A_2' of Fig. 1-10 is $A_1 A_2$ in the original sample space. Undoubtedly, we would agree that equal weights of $1/6$ are reasonable for the sample space pictured in Fig. 1-10 so that

$$Pr(A_2 | A_1) = \frac{2}{6} = \frac{2/12}{6/12} = \frac{Pr(A_1 A_2)}{Pr(A_2)}$$

as before.

Although we used specific examples to derive forumlas (1-12) and (1-13), the argument is exactly the same with n sample points $a_1, a_2, \ldots, a_n$ assigned weights $w_1, w_2, \ldots, w_n$. The new weights, given A_1 is certain to occur, are for $i = 1, 2, \ldots, n$

$$w_i' = \frac{w_i}{Pr(A_1)} \qquad \text{if } a_i \text{ is in } A_1$$
$$ = 0 \qquad \text{if } a_i \text{ is not in } A_1 \tag{1-14}$$

Obviously, these new weights satisfy conditions (1-3) since each is non-negative and their sum is 1. Perhaps it should be added that whenever $Pr(A_1)$ appears in the denominator of a fraction, as in formula (1-12) or (1-14), we must require $Pr(A_1) > 0$ in order that the fraction be meaningful.

EXAMPLE 1-11

In Example 1-6 a sample space was constructed for an experiment in which a slip of paper was drawn from a hat containing four numbers, the slip replaced, and a second slip drawn. Use that sample space and let A_1 be the event "the sum is 6" and A_2 be the event "one number is a 2." Find $Pr(A_2 | A_1)$.

Solution

Since A_1 contains three points, $Pr(A_1) = 3/16$. Similarly $A_1 A_2$ contains two points, and $Pr(A_1 A_2) = 2/16$. Thus according to formula (1-12)

$$Pr(A_2 | A_1) = \frac{2/16}{3/16} = 2/3$$

EXAMPLE 1-12

We already know by using definition (1-4) that $Pr(A_1 A_2) = 2/12$ for the coin-die example whose sample space is pictured in Fig. 1-6. Obtain this result by using formula (1-13).

Solution

In computing $Pr(A_1)$ we need consider only a two-point sample space, H and T with weights each $1/2$, since A_1 is the event "a head shows on the coin." Hence $\Pr(A_1) = 1/2$. To find $\Pr(A_2|A_1)$ only a six-point sample space, 1, 2, 3, 4, 5, 6 with weights each 1/6, is necessary since we want the probability that a die shows a 1 or a 2, given that the coin shows a head. (The result on the coin, of course, has no influence upon the result produced by the die.) Thus $\Pr(A_2|A_1) = 2/6$ and

$$\Pr(A_1 A_2) = (1/2)(2/6) = 2/12$$

In the coin-die problem of Example 1-12 we found that $\Pr(A_2|A_1) = 2/6$. From Fig. 1-6 it is easy to see that $\Pr(A_2) = 4/12 = 2/6$. Thus, in this example

$$\Pr(A_2|A_1) = \Pr(A_2) \tag{1-15}$$

In other words, the fact that the coin produces a head does not change the probability of a 1 or a 2 on the die. The latter figure remains 2/6 irrespective of what happens with the coin. We would probably be surprised if this were not the case, since the throwing of a coin and the rolling of a die are unrelated acts and it is difficult to visualize how the result on the coin could possibly affect the result on the die. Whenever formula (1-15) holds, the multiplication theorem (1-13) takes on a simplified appearance, becoming

$$\Pr(A_1 A_2) = \Pr(A_1)\,\Pr(A_2) \tag{1-16}$$

In this case the events A_1 and A_2 are said to be *independent events*. Theoretically, the only way to verify that A_1 and A_2 are independent is to compute $\Pr(A_1)$, $\Pr(A_2)$, and $\Pr(A_1 A_2)$ using the sample space and weights selected as suitable for the given experiment, and then substitute in formula (1-16) and check for equality. However, as is the case with formula (1-13), formula (1-16) is used primarily to find $\Pr(A_1 A_2)$ from the probabilities that appear on the right-hand side of the equation. In order to use this form of the multiplication theorem in this way, we must be able to judge on the basis of our knowledge of the experiment that there is no connection or relationship between the events A_1 and A_2. That is, we must believe that the occurrence or nonoccurrence of A_1 in no way affects the occurrence or nonoccurrence of A_2 and vice versa. Sometimes it will be perfectly clear that events are independent while in other cases it may be difficult or impossible to make the judgment. Of course, the advantage of formula (1-16) over formula (1-13) is that no thought or consideration must be given to the other event when calculating either probability.

EXAMPLE 1-13

A symmetric six-sided die is rolled twice. What is the probability of getting a 2 and a 3 in that order? In any order?

Solution

Let A_1 be the event "a 2 is obtained on the first roll" and A_2 be the event "a 3 is obtained on the second roll." We seek $\Pr(A_1A_2)$. Our judgment tells us that A_1 and A_2 are independent and hence we use formula (1-16). To compute $\Pr(A_1)$ we select a sample space with six points, each with associated weight 1/6, and we find $\Pr(A_1) = 1/6$. The same sample space is appropriate for the second roll, and $\Pr(A_2) = 1/6$. Hence

$$\Pr(A_1A_2) = (1/6)(1/6) = 1/36$$

Letting A_3 and A_4 be "a 3 is obtained on the first roll" and "a 2 is obtained on the second roll," respectively, the same argument yields $\Pr(A_3A_4) = 1/36$. Now, if we let $B_1 = A_1A_2$ and $B_2 = A_3A_4$ we recognize that B_1 and B_2 are mutually exclusive since only one order of results can occur on two rolls. Further, these are the only two orders which yield a 2 and a 3. Hence, we seek

$$\Pr(B_1 \cup B_2) = \Pr(B_1) + \Pr(B_2) = 1/36 + 1/36 = 2/36$$

a result obtained by using the addition theorem for mutually exclusive events.

EXAMPLE 1-14

If a symmetric six-sided die is rolled twice, what is the probability of getting at least one 2?

Solution

Let A_1 and A_2 be the events "a 2 appears on the first roll" and "a 2 appears on the second roll." Then, using the same argument as in Example 1-13, we get

$$\Pr(A_1A_2) = \Pr(A_1)\Pr(A_2) = (1/6)(1/6) = 1/36$$
$$\Pr(A_1\bar{A}_2) = \Pr(A_1)\Pr(\bar{A}_2) = (1/6)(5/6) = 5/36$$
$$\Pr(\bar{A}_1A_2) = \Pr(\bar{A}_1)\Pr(A_2) = (5/6)(1/6) = 5/36$$

Now, we observe that the three events $B_1 = A_1A_2$, $B_2 = A_1\bar{A}_2$, $B_3 = \bar{A}_1A_2$ are the only ways to get at least one 2, and since only one order can be obtained on two rolls, B_1, B_2, and B_3 are mutually exclusive. Thus, we want

$$\Pr(B_1 \cup B_2 \cup B_3) = 1/36 + 5/36 + 5/36 = 11/36$$

obtained by using the addition theorem for mutually exclusive events.

In a number of interesting applications a series of experiments is conducted in such a way that events associated with one experiment are independent of events associated with any other experiment in the series. For example, if a die is rolled 10 times, the result on any one roll in no way affects the result on any other roll, and the 10 results constitute a series of 10 independent events. If $\Pr(A_2|A_1) \neq \Pr(A_2)$, the events A_1 and A_2 are said to be dependent. Example 1-15 demonstrates the use of the multiplication theorem in such a case.

EXAMPLE 1-15

Two cards are dealt from an ordinary deck that has been well shuffled. What is the probability that at least one of the cards is a heart?

Solution

Let A_1 and A_2 be the events "a heart is obtained on the first draw" and "a heart is obtained on the second draw." The three mutually exclusive events that yield at least one heart are $B_1 = A_1 A_2$, $B_2 = A_1 \bar{A}_2$, $B_3 = \bar{A}_1 A_2$. That is, we can get a heart both times, the first time but not the second, or the second time but not the first. For the first draw we would probably select a sample space of 52 points each with weight 1/52. Then, from the definition of probability we easily find $\Pr(A_1) = 13/52$, $\Pr(\bar{A}_1) = 39/52$. Having been dealt one card, 51 cards remain and a 51-point sample space could be used for the second draw, each point being given weight 1/51. The number of hearts remaining for the second draw will be 12 or 13, depending on the result of the first draw. To evaluate $\Pr(A_1 A_2) = \Pr(A_1) \Pr(A_2|A_1)$ by using the multiplication theorem we need $\Pr(A_2|A_1)$. But if a heart was drawn the first time, only 12 of the remaining 51 are hearts, and we easily calculate $\Pr(A_2|A_1) = 12/51$. Thus

$$\Pr(A_1 A_2) = \Pr(A_1) \Pr(A_2|A_1) = (13/52)(12/51) = (1/4)(12/51)$$

Similarly,

$$\Pr(A_1 \bar{A}_2) = \Pr(A_1) \Pr(\bar{A}_2|A_1) = (13/52)(39/51) = (1/4)(39/51)$$
$$\Pr(\bar{A}_1 A_2) = \Pr(\bar{A}_1) \Pr(A_2|\bar{A}_1) = (39/52)(13/51) = (1/4)(39/51)$$

Finally, by the addition theorem for mutually exclusive events, we have

$$\Pr(B_1 \cup B_2 \cup B_3) = (1/4)(12/51) + (1/4)(39/51) + (1/4)(39/51) = 15/34$$

Alternatively, and perhaps somewhat easier, we could use formula (1-7). Let B be the event "at least one card is a heart" so that $\bar{B}$ is the event "neither card is a heart." Then $\Pr(B) = 1 - \Pr(\bar{B})$ and

$$\Pr(\bar{B}) = \Pr(\bar{A}_1 \bar{A}_2) = \Pr(\bar{A}_1) \Pr(\bar{A}_2|\bar{A}_1) = (39/52)(38/51) = 19/34$$

Finally, $\Pr(B) = 1 - (19/34) = 15/34$ as before.

So far, we have used the multiplication theorem for only two events. With n events $A_1, A_2, \ldots, A_n$ it is fairly easy to establish that

$$\Pr(A_1 A_2 \cdots A_n)$$
$$= \Pr(A_1) \, \Pr(A_2 \,|\, A_1) \, \Pr(A_3 \,|\, A_1 A_2) \cdots \Pr(A_n \,|\, A_1 A_2 \cdots A_{n-1}) \quad (1\text{-}17)$$

When the events are independent, then formula (1-17) can be replaced by the simpler result

$$\Pr(A_1 A_2 \cdots A_n) = \Pr(A_1) \, \Pr(A_2) \cdots \Pr(A_n) \quad\quad\quad (1\text{-}18)$$

As with two events, the left-hand side of formulas (1-17) and (1-18) is computed by evaluating the n probabilities on the right-hand side and multiplying.

EXAMPLE 1-16

If three cards are dealt from an ordinary well-shuffled deck, what is the probability that the first two are hearts and the third is a spade?

Solution

Let A_1, A_2, A_3 be the events "the first card is a heart," "the second card is a heart," and "the third card is a spade," respectively. We seek

$$\Pr(A_1 A_2 A_3) = \Pr(A_1) \, \Pr(A_2 \,|\, A_1) \, \Pr(A_3 \,|\, A_1 A_2)$$

From Example 1-15 we have $\Pr(A_1) = 13/52, \Pr(A_2 \,|\, A_1) = 12/51$, but we also need $\Pr(A_3 \,|\, A_1 A_2)$. Given that the first two draws have produced hearts, 50 cards containing 13 spades are available for the third draw. Now we visualize a 50-point space, each with weight 1/50, and $\Pr(A_3 \,|\, A_1 A_2) = 13/50$. Finally,

$$\Pr(A_1 A_2 A_3) = (13/52)(12/51)(13/50) = 13/850 = .0153$$

Example 1-16 is a good illustration of the remarks made in the first paragraph of this section concerning the simplification achieved by regarding an event in terms of simpler events. Had we not done this but instead constructed a sample space corresponding to Fig. 1-6 (for the coin-die example), we would have needed $(52)(51)(50) = 132{,}600$ sample points. After assigning equal weights we would have counted $(13)(12)(13) = 2{,}028$ points corresponding to the event "a heart is dealt, followed by another heart, followed by a spade." Then by the definition of probability, once more we get $(13)(12)(13)/(52)(51)(50) = 13/850$.

EXERCISES

Use the addition and multiplication theorems in preference to the definition.

1-16 If the probability that a person has blond hair is .23, blue eyes is .17, both blue eyes and blond hair is .11, what is the probability that a person has either blond hair or blue eyes or both?

◊ 1-17 Two runners are competing in a 5-mile race. Suppose that the probability that the first runner wins is 3/10 and the probability that the second runner wins is 3/20. What is the probability that one of these two wins the race?

1-18 Suppose five coins are thrown simultaneously and it is known that the probability of obtaining three heads is 10/32, the probability of obtaining four heads is 5/32, and the probability of obtaining five heads is 1/32. With this information find the probability of obtaining three or more heads.

◊ 1-19 A student applies for a fellowship at two different universities. The probability is 1/3 that he will be awarded a fellowship by the first school, 1/4 that he will be awarded a fellowship by the second, and 1/6 that he will be awarded a fellowship by both. What is the probability that the student is awarded at least one fellowship?

1-20 In the World Series the two teams continue to play until one team wins exactly 4 games. Suppose that it has been determined that the probabilities of a series ending in 4, 5, 6, or 7 games are respectively .20, .26, .24, .30. What is the probability that the series ends in 5 or fewer games? 6 or more games?

◊ 1-21 A coin is tossed three times. What is the probability that two out of three times the result is heads?

1-22 Two symmetric six-sided dice are rolled. What is the probability that the total on the two dice is 6?

◊ 1-23 Two cards are dealt from an ordinary well-shuffled deck. What is the probability that one is an ace and one is a king?

1-24 An urn contains six red and four black balls. Two balls are drawn out in succession without replacement. What is the probability that the first is red, the second black? What is the probability of getting a red and a black ball in either order? What is the probability of drawing a white ball? If five balls are drawn out, what is the probability that at least one is red?

◊ 1-25 An ordinary symmetric six-sided die is rolled until a 6 appears. What is the probability that exactly four rolls are required?

1-26 Three cards are dealt from an ordinary well-shuffled deck. What is the probability that two are red and one is black?

◊ 1-27 Suppose it is known that three hunters, say A, B, and C, can kill a pheasant on 1/2, 2/3, and 3/4 of their shots, respectively. What is the probability that they can kill a bird if all three shoot at it simultaneously?

1-28 A hat contains 100 slips of paper numbered 1 to 100. Three are drawn out successively without replacement. What is the probability that two of the numbers are larger than 90?

◇ 1-29 Two coins are tossed and you are told that one of them shows a head. What is the probability that the other is also a head?

1-30 Using the sample space and weights suggested in Exercise 1-10, find the probability that the selected can contains red paint, given that it does not contain white paint.

◇ 1-31 Following Example 1-8 we considered an experiment that consisted of tossing a coin until a head appears. We observed that a sample space of interest might be designated by H, TH, TTH, TTTH, ... where the ith point can be identified with the event "the first head occurs on the ith toss." We further commented that appropriate weights might be 1/2, 1/4, 1/8, 1/16, etc. Justify that choice.

1-32 In Exercise 1-14 we made no attempt to assign weights to the sample points. What should these weights be? *Hint:* This problem is similar to Exercise 1-31.

◇ 1-33 Two cards are dealt from an ordinary well-shuffled deck. What is the probability of drawing a heart or a spade? *Hint:* It is easier to work with the complementary event.

1-4
PERMUTATIONS AND COMBINATIONS

We have seen how it is sometimes simpler to compute probabilities by using the addition and multiplication theorems rather than the definition. In this section we shall discuss some counting formulas that can be used to simplify the calculation of probabilities, especially in cases where enumeration of the sample points is excessive. The counting formulas can be used either with or without the theorems.

Suppose that there are three highway routes between towns A and B and two routes between towns B and C. Then there is a total of six possible routes from A to C traveling through B. With each of the three routes for the first lap, there are two choices for the second lap, giving the total of six. This illustrates a fundamental principle of counting. If a first thing can be done in n_1 ways after which a second thing can be done in n_2 ways, then the two things can be done in $n_1 n_2$ ways. The truth of the principle is readily demonstrated by making a list of the ways the two things can be done. For example, using the numbering of Fig. 1-11, the routes are 1–4, 2–4, 3–4, 1–5, 2–5, 3–5. The principle is readily extended to r things. If a first thing can be done in n_1 ways,

FIGURE 1-11

Highway routes and fundamental counting principle.

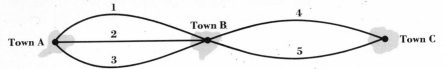

after which a second thing can be done in n_2 ways, ..., after which an rth thing can be done in n_r ways, then all r things can be done in $n_1 n_2 \cdots n_r$ ways.

Consider three things which we shall call A, B, and C. They can be arranged in six ways if we use all three. The arrangements, called *permutations*, are ABC, ACB, BAC, BCA, CAB, CBA. We would like to have a formula for counting the number of permutations of n things using all n of them at a time. The fundamental counting principle provides a means of obtaining the answer. The first thing can be chosen from any of the total of n things available, after which the second can be selected from any of the remaining $n - 1$. Consequently, the first two places in the permutation can be chosen in $n(n - 1)$ ways. With each choice of the first two things, the third can be chosen in $n - 2$ ways. By continuing this line of reasoning, we find that the number of permutations of n things taken n at a time is

$$_nP_n = n(n - 1) \cdots (3)(2)(1) = n! \tag{1-19}$$

where the symbol $n!$ is called n factorial. When $n = 3$, $3! = 3 \cdot 2 \cdot 1 = 6$. agreeing with the result obtained by enumeration. Sometimes we wish to count the number of permutations of n things taken r at a time where r is less than n. With the three letters A, B, C the permutations taken two at a time are AB, AC, BA, BC, CA, CB. We proceed as before, but since there are only r positions to fill in the permutation, the product replacing formula (1-19) contains only r numbers. The number of permutations of n things taken r at a time is

$$_nP_r = n(n - 1) \cdots [n - (r - 1)]$$
$$= n(n - 1) \cdots (n - r + 1) \tag{1-20}$$

Multiplying and dividing the right-hand side of formula (1-20) by $(n - r)!$ results in the more convenient form

$$_nP_r = \frac{n!}{(n - r)!} \tag{1-21}$$

If $r = n$, formula (1-21) reduces to formula (1-19), since $0!$ is defined to be 1.

EXAMPLE 1-17

Use formula (1-21) to obtain the number of permutations of ten things taken three at a time.

Solution

We identify $n = 10$, $r = 3$, $n - r = 7$. Hence

$$_{10}P_3 = \frac{10!}{7!} = \frac{10 \cdot 9 \cdot 8 \cdot 7 \cdot 6 \cdot 5 \cdot 4 \cdot 3 \cdot 2 \cdot 1}{7 \cdot 6 \cdot 5 \cdot 4 \cdot 3 \cdot 2 \cdot 1} = 10 \cdot 9 \cdot 8 = 720$$

Here, it is probably just as easy to argue directly as we did when deriving formula (1-20). We immediately get $10 \cdot 9 \cdot 8 = 720$ choices.

EXAMPLE 1-18

A baseball manager has selected his starting lineup of eight regulars (disregarding the pitcher). As yet he is undecided on his batting order except that the pitcher will bat ninth. To determine his final order for the season, he plans to try out each possible order in one game and select the one which produces the best results. If this were his plan, how many games would be required to give each batting order a trial?

Solution

We need the number of permutations of 8 things taken 8 at a time. This is

$$_8P_8 = 8! = 8 \cdot 7 \cdot 6 \cdot 5 \cdot 4 \cdot 3 \cdot 2 \cdot 1 = 40,320$$

The plan is obviously impractical.

Next, suppose that we want to count the number of permutations of n things taken n at a time when not all n things are distinguishable. To take a simple example, consider the permutations of the letters of the word *book*. Four different letters yield $4! = 24$ different permutations. However, this is too large a number, since some of the permutations look exactly alike because the o's are indistinguishable. To clarify the point, assume that the o's have subscripts (denote them o_1 and o_2) so that we can tell them apart. Then bo_1o_2k and bo_2o_1k are different permutations but if the subscripts are dropped, both look exactly alike. Let P be the number of distinguishable permutations. If, for each of the P permutations, subscripts are added to the o's and o's are permuted in all possible ways, then the total number of permutations of n distinct things taken n at a time is obtained. In our example we get by the fundamental counting principle that this total is $2! P$. But we already know that the total is $4!$, so consequently $2! P = 4!$ or $P = (4!)/(2!)$.

This argument, based upon counting the total number of permutations of n different things taken n at a time in two ways, can be readily generalized. Let n_1 things be alike, n_2 alike, ..., n_k alike such that $n_1 + n_2 + \cdots + n_k = n$. The number of distinct permutations P gives rise to $Pn_1! n_2! \cdots n_k!$ permutations if subscripts are added to the like things and if each of the (formerly) like things is permuted in all possible ways. Thus, for example, the distinguishable permutation "street" yields $st_1re_1e_2t_2$, $st_1re_2e_1t_2$, $st_2re_1e_2t_1$, $st_2re_2e_1t_1$ or a total of $(2!)(2!) = 4$ permutations. Likewise "stteer," "sreett," "reetts," and any other distinct permutation of these six letters yields $(2!)(2!) = 4$ permutations when subscripts are added. Since the total number of permutations of n things taken n at a time is $n!$ we must necessarily have

$$Pn_1! n_2! \cdots n_k! = n!$$

or

$$P = \frac{n!}{n_1! n_2! \cdots n_k!} \tag{1-22}$$

EXAMPLE 1-19

How many distinct permutations can be constructed from the letters of the word "Mississippi"?

Solution

The $n = 11$ letters are composed of $n_1 = 1$ m, $n_2 = 4$ i's, $n_3 = 4$ s's, $n_4 = 2$ p's. Thus, according to the formula (1-22), we get

$$P = \frac{11!}{1!4!4!2!} = \frac{11 \cdot 10 \cdot 9 \cdot 8 \cdot 7 \cdot 6 \cdot 5 \cdot 4 \cdot 3 \cdot 2 \cdot 1}{1 \cdot 4 \cdot 3 \cdot 2 \cdot 1 \cdot 4 \cdot 3 \cdot 2 \cdot 1 \cdot 2 \cdot 1}$$

$$= 11 \cdot 10 \cdot 9 \cdot 7 \cdot 5 = 34{,}650$$

In a number of counting problems order is not important as it is when we count permutations. For example, we may want to know the number of five-card hands that can be constructed from an ordinary deck. A hand dealt AKQJ10 of hearts is the same as one that comes in the order QJ10AK of hearts. Consequently, many permutations lead to the same hand or to the same combination of cards. By *combination*, we mean an unordered grouping of things. We seek a formula for counting the number of combinations of n things taken r at a time. To find that result we again count the total number of permutations in two ways. We already know from formula (1-21) that the number of permutations of n things taken r at a time is $n!/(n - r)!$.   Let $\binom{n}{r}$ denote the number of combinations of n things taken r at a time, the number which we seek.   Another procedure for constructing and counting all permutations of n things taken n at a time would be to write down all $\binom{n}{r}$ combinations and then permute each of the r letters in each combination in all possible ways.   This yields $\binom{n}{r}r!$ permutations which must equal our previous count $n!/(n - r)!$.   Consequently,

$$\binom{n}{r} = \frac{n!}{r!(n - r)!} \tag{1-23}$$

Let us illustrate the construction process with three letters A, B, C.   It is easy to recognize that there are three combinations when the letters are taken two at a time.   These are AB, AC, and BC.   Now permuting all letters of each, we see that combination AB produces permutations AB, BA; combination AC produces permutations AC, CA; combination BC produces permutations BC, CB, so that we get $3 \cdot 2! = 6$ permutations, the same as we obtained earlier.   Similarly, the general result is $\binom{n}{r}r!$ as previously stated.

EXAMPLE 1-20

How many five-card hands can be constructed from a deck of 52 cards?

Solution

The answer is the number of combinations of 52 things taken 5 at a time or

$$\binom{52}{5} = \frac{52!}{5!\,47!} = \frac{52 \cdot 51 \cdot 50 \cdot 49 \cdot 48}{5 \cdot 4 \cdot 3 \cdot 2 \cdot 1} = 2{,}598{,}960$$

EXAMPLE 1-21

What is the probability that a five-card poker hand contains three aces?

Solution

The total number of sample points (one for each hand) is $\binom{52}{5}$ from Example 1-20.   If a hand is dealt from a well-shuffled deck, then equal weights of $1 \left/ \binom{52}{5} \right.$ for each point are reasonable.   If we let A be the event "the hand contains three aces," then to evaluate $\Pr(A)$ we need to count the number of sample points in A. Now three aces can be chosen from four aces in $\binom{4}{3}$ ways. With each choice, the hand can be filled out from the remaining 48 non-aces in $\binom{48}{2}$ ways. Then, by the fundamental counting principle, the total number of hands containing three aces is $\binom{4}{3}\binom{48}{2}$ and the definition of probability yields

$$\Pr(A) = \frac{\binom{4}{3}\binom{48}{2}}{\binom{52}{5}} = \frac{94}{54,145} = .0017$$

EXAMPLE 1-22

A committee of college faculty members consists of three full professors, five associate professors, and two assistant professors. A subcommittee of six is selected by drawing names out of a hat. What is the probability that the subcommittee is composed of two full professors, three associate professors, and one assistant professor?

Solution

There are $\binom{10}{6}$ possible committees. Hence a reasonable sample space for the problem contains $\binom{10}{6}$ points, one for each subcommittee. The method used to make the selection suggests weights of $1 \Big/ \binom{10}{6}$. If we let A be the event "the subcommittee is composed of two full professors, three associate professors, and one assistant professor," then to evaluate $\Pr(A)$ we need to count the number of sample points which fall in A. Since two professors can be chosen in $\binom{3}{2}$ ways, three associate professors in $\binom{5}{3}$ ways, and one assistant professor in $\binom{2}{1}$ ways, there are $\binom{3}{2}\binom{5}{3}\binom{2}{1}$ possible subcommittees and

$$\Pr(A) = \frac{\binom{3}{2}\binom{5}{3}\binom{2}{1}}{\binom{10}{6}} = \frac{2}{7}$$

EXERCISES

1-34 In how many ways can five people be seated in a row? Suppose that three are men and two are women and that their positions are determined by drawing numbers out of a hat. What is the probability that men occupy the end positions?

◈ 1-35 How many distinct permutations can be formed from the letters *AAABBCCCC*? What would be the answer if the letters were all different?

1-36 Eight people are available to form a basketball team. How many ways can the team be formed if each person can play any of the five positions? If five players are chosen from eight, how many ways can this be done?

◈ 1-37 Suppose that of the eight people available in Exercise 1-36 three are over 6 feet tall. If the five players selected to form the team are chosen by drawing names out of a hat, what is the probability that all three six-footers are on the starting five?

1-38 In how many ways can a five-card hand be selected from an ordinary deck so that it contains two aces and two kings?

◈ 1-39 How many different sets of answers are possible for a 10-question true-false examination?

1-40 How many distinct permutations can be formed from the word "statistics"?

◈ 1-41 A hat contains 25 slips of paper numbered 1 to 25. If three are drawn without replacement, what is the probability that all three numbers are less than 10?

1-42 Five cards are dealt from a well-shuffled deck. What is the probability that the cards form a full house (three of one denomination, two of another denomination, such as three aces and two kings)?

◈ 1-43 In a bridge game the declarer holds 5 trump and the dummy has 3. The opponents passed throughout the bidding so that no clue to their suit distribution is available. What is the probability that the outstanding trump are split 3 and 2? 4 and 1? 5 and 0?

1-44 A box contains 8 camera flashbulbs, all of which look alike. From these 8 bulbs, 4 are taken to load a camera. If the box contains 3 defective bulbs, what is the probability that the loaded camera will contain 1 defective bulb?

◈ 1-45 Suppose that you have 10 tickets listing items, amounts, and totals for purchases at a grocery store. It is known that 4 of the tickets contain arithmetic errors, but to find the errors it is necessary to check the tickets. If 5 tickets are selected for checking, what is the probability that exactly 3 of the 4 tickets in error will be checked?

1-46 A company receives a moderately priced piece of equipment in lots of size 25. A sample of 5 pieces is selected and each piece tested.

If the sample contains no defective pieces the lot is accepted. If the sample contains one or more defective pieces, all 25 items are tested before a decision is made. If the lot contains one defective piece, what is the probability that the lot will be accepted without further testing?

◈ 1-47 A senate committee is composed of 3 Democrats, 3 Republicans, and 1 Conservative. A subcommittee of three members is selected by drawing names from a hat. What is the probability that all three parties are represented on the subcommittee?

1-5
RANDOM VARIABLES, PROBABILITY DISTRIBUTIONS, AND EXPECTED VALUE

In the first four sections of this chapter we have encountered a number of random experiments. Our examples have included dice throwing, coin tossing, drawing numbers from a hat, dealing cards from a deck, inspection of items from a shipment of goods, selection of subcommittees from a committee, etc. In applications of probability we are often interested in a number associated with the outcome of a random experiment. Such a quantity whose value is determined by the outcome of a random experiment is called a *random variable*.

Let us consider some examples of random variables. If a single six-sided symmetric die is rolled, the random variable that is usually of interest is the number of spots showing. Of less interest, but also random variables, are the number of spots showing plus 2, 6 minus the number of spots, the number of spots squared, and any other function of the number of spots. Other examples of random variables are the numbers of heads obtained when a coin is tossed twice, the number of aces in a five-card hand dealt from a deck, the number of women on a subcommittee selected by drawing from a hat, the number of boys in a four-child family, the number of defective flashbulbs in a sample selected from a package of 25 such bulbs, and the sum of spots when two dice are rolled.

We will use capital letters as symbols for random variables. For example, with a single six-sided die we could let X be the number of spots showing if the die is rolled. When we speak of the random variable X we will have in mind *all* possible values which an experiment can produce. Small letters, x for example, will be used to denote *one* possible, though unspecified, value which can be achieved. In the die situation x could be one, but only one, of the numbers 1, 2, 3, 4, 5, 6. The symbol $Pr(X \leq 4)$ will be interpreted as the probability that the random experiment produces a numerical value less than or equal to 4. Similarly,

Pr($X \leq r$) means the probability that the experiment yields a value less than or equal to r, and Pr($X = x$) means the probability that the experiment yields *the* value x. Almost always r or x will be one of the numerical values which the experiment can generate. If we define Y to be the number of heads produced when a coin is tossed twice, then Pr($Y = 1$) is interpreted as the probability that one head is obtained and Pr($Y \leq y$) is the probability that the experiment produces y or fewer heads. Here interesting values of y would be one of the numbers 0, 1, or 2. The phrase "the random variable X can take on (or assume) the value x" means "x is one of the values which the random experiment can produce."

A table listing all possible values that a random variable can take on together with the associated probabilities is called a *probability distribution*. The probability distribution of X, where X is the number of spots showing when a six-sided symmetric die is rolled, is given in Table 1-1. We have used the notation $f(x) = \text{Pr}(X = x)$. We note that

TABLE 1-1

Probability distribution for a single die

x	1	2	3	4	5	6
$f(x)$	1/6	1/6	1/6	1/6	1/6	1/6

the sum of all probabilities in a probability distribution is 1, since one of the values produced by a random experiment must occur if the experiment is performed. The same information can be conveniently expressed by the formula

$$f(x) = 1/6 \qquad x = 1, 2, 3, 4, 5, 6$$

A formula, or function $f(x)$, from which Pr($X = x$) can be obtained, is called a *probability function*. As we shall see, many important probability functions are tabulated (so that usually we do not need to construct a table as in Example 1-24), but quite often in another form. Usually tables contain

$$F(r) = \text{Pr}(X \leq r) = \sum_{x \leq r} f(x) \tag{1-24}$$

which is known as the *cumulative distribution* or *distribution function*. Table 1-2 gives the cumulative distribution for the die random variable. Values of $F(r)$ are easily obtained by using Table 1-1. Thus, for example,

$$F(3) = f(1) + f(2) + f(3) = 1/6 + 1/6 + 1/6 = 3/6$$

TABLE 1-2

Cumulative distribution for die random variable

r	1	2	3	4	5	6
$F(r)$	1/6	2/6	3/6	4/6	5/6	1

If Y is the number of heads produced when a coin is tossed twice, and we let

$$g(y) = \Pr(Y = y)$$

and

$$G(r) = \Pr(Y \leqq r) = \sum_{y \leqq r} g(y)$$

then it is easy to verify that the probability distribution and cumulative distribution of Y are as given in Tables 1-3 and 1-4.

TABLE 1-3

Probability distribution of Y

y	0	1	2
$g(y)$	1/4	2/4	1/4

TABLE 1-4

Cumulative distribution of Y

r	0	1	2
$G(r)$	1/4	3/4	1

EXAMPLE 1-23

Suppose that the probability function of a random variable X is $f(x) = x/10$, $x = 1, 2, 3, 4$. List the values that would be included in the probability distribution and the values that would be listed in the cumulative distribution.

Solution

For the probability distribution we need

$$f(1) = 1/10, f(2) = 2/10, f(3) = 3/10, f(4) = 4/10$$

For the distribution function we need

$F(1) = 1/10$
$F(2) = 1/10 + 2/10 = 3/10$
$F(3) = 1/10 + 2/10 + 3/10 = 6/10$
$F(4) = 1/10 + 2/10 + 3/10 + 4/10 = 1$

EXAMPLE 1-24

Given that the cumulative distribution of X is

r	0	1	2	3	4	5
$F(r)$	.0459	.2415	.5747	.8585	.9794	1

Find $\Pr(X \leq 2)$, $\Pr(X > 2)$, $\Pr(X = 2)$, and $\Pr(1 \leq X \leq 3)$.

Solution

We read from the table $\Pr(X \leq 2) = .5747$. Since the experiment must produce either a value less than or equal to 2 or greater than 2 (complementary events), we must have $\Pr(X \leq 2) + \Pr(X > 2) = 1$. Hence

$\Pr(X > 2) = 1 - \Pr(X \leq 2) = 1 - .5747 = .4253$

Since

$\Pr(X \leq 2) = \Pr(X = 0) + \Pr(X = 1) + \Pr(X = 2)$
$\Pr(X \leq 1) = \Pr(X = 0) + \Pr(X = 1)$

we see that

$\Pr(X = 2) = \Pr(X \leq 2) - \Pr(X \leq 1) = .5747 - .2415 = .3332$

Finally

$$\begin{aligned}
\Pr(1 \leq X \leq 3) &= \Pr(X = 1) + \Pr(X = 2) + \Pr(X = 3)\\
&= [\Pr(X = 0) + \Pr(X = 1) + \Pr(X = 2) + \Pr(X = 3)]\\
&\quad - \Pr(X = 0)\\
&= \Pr(X \leq 3) - \Pr(X \leq 0)\\
&= .8585 - .0459 = .8126
\end{aligned}$$

EXAMPLE 1-25

In Example 1-6 a sample space was constructed for an experiment in which a slip of paper was drawn from a hat containing four numbers, the slip replaced, and a second slip drawn. Use that sample space and the weights 1/16 selected at that time to find the probability distribution of the sum of the numbers on the two slips.

Solution

For convenience the 16-point sample space is reproduced in Table 1-5. The number appearing below the description of the point is the sum of the two numbers. Let X be the sum of the two numbers. To obtain

TABLE 1-5

Sample space for two slips of paper drawn from hat			
(1, 1)	(2, 1)	(3, 1)	(4, 1)
2	3	4	5
(1, 2)	(2, 2)	(3, 2)	(4, 2)
3	4	5	6
(1, 3)	(2, 3)	(3, 3)	(4, 3)
4	5	6	7
(1, 4)	(2, 4)	(3, 4)	(4, 4)
5	6	7	8

$Pr(X = x)$ we add the probabilities of all sample points which yield a sum of x. Since the event $X = 2$ contains one point, $X = 3$ contains two points, etc., we easily get the first two rows of Table 1-6.

TABLE 1-6

Probability distribution for sum on two slips

x	2	3	4	5	6	7	8
$f(x)$	1/16	2/16	3/16	4/16	3/16	2/16	1/16

Another important concept in probability theory that had its origin in gambling is expected value. As a simple illustration, suppose that 100 tickets are sold for $1 each to raffle off a watch worth $80. Since one ticket is worth $80 and the other 99 will pay off nothing, the average value of a ticket is

$$\frac{\$80 + \$0 + \$0 + \cdots + \$0}{100} = \$.80$$

Another way to write this fraction is

$$\$80(1/100) + \$0(1/100) + \cdots + \$0(1/100) = \$80(1/100) + \$0(99/100) \tag{1-25}$$

If we let the random variable X be the amount a ticket is worth, then the probability distribution of X is easily found to be the one appearing in Table 1-7. We observe that the right-hand side of formula (1-25) can

TABLE 1-7

Probability distribution of X, the amount a ticket is worth	x	$0	$80
	$f(x)$	99/100	1/100

be obtained from Table 1-7 by multiplying each value of x by the probability associated with that value and summing the resulting numbers.

If an individual participated in only one raffle and with a single ticket, he would receive either $0 or $80. The average amount received by all ticketholders is, as we have already seen, 80 cents. Now suppose that an individual buys one ticket for many such raffles of the type described. After several thousand raffles it is quite likely that he would win a few times. To be specific, suppose that after 10,000 tries he has won 110 times. On the average he has received

$$\frac{\$80(110) + \$0(9,890)}{10,000} = \$80 \left(\frac{110}{10,000}\right) + \$0 \left(\frac{9,890}{10,000}\right)$$

Since 110/10,000 and 9,890/10,000 are relative frequencies, we would undoubtedly expect, after our discussion of the definitions of probability, that these numbers should be close to the probabilities appearing in Table 1-7. The closer the relative frequencies are to the probabilities, the closer the average amount received is to 80 cents. Hence, if one participated in many such raffles, in the long run he would expect to get about 80 cents back for each dollar invested.

With the raffle example in mind, we define expected value as follows: If a random variable X can assume values $x_1, x_2, \ldots, x_k$ with probabilities $f(x_1), f(x_2), \ldots, f(x_k)$, respectively, then the *expected value* of X (sometimes called the expectation of X or the mean of the probability distribution) is

$$E(X) = x_1 f(x_1) + x_2 f(x_2) + \cdots + x_k f(x_k)$$

$$= \sum_{i=1}^{k} x_i f(x_i) = \sum_{i=1}^{k} x_i \, \Pr(X = x_i) \tag{1-26}$$

If the random variable has an infinite number of outcomes $x_1, x_2, x_3, \ldots,$ which are countable, then

$$E(X) = x_1 f(x_1) + x_2 f(x_2) + x_3 f(x_3) + \cdots$$

$$= \sum_{i=1}^{\infty} x_i f(x_i) = \sum_{i=1}^{\infty} x_i \, \Pr(X = x_i) \tag{1-27}$$

As we have seen with the raffle example, $E(X)$ does not have to be equal to a value the random variable can assume nor is it the value we expect to get on any one performance of the experiment. If the experiment is repeated many times and the observed values of the random variable are averaged (the sum is divided by the number of repetitions of the experiment), then we do expect that this average will be close to $E(X)$. It is the latter interpretation that justifies the suggestive name "expected value" for the sum $E(X)$.

The Greek letter μ (mu) is often used to denote the expected value of a random variable. Thus $E(X)$ and μ_X are notations that stand for the same quantity. When there is no possibility for confusion (that is, when it is clear which random variable is under consideration), the subscript is frequently omitted from μ_X. Hence, we can write

$$E(X) = \mu_X = \mu$$

EXAMPLE 1-26

Find the expected value of the die random variable whose probability distribution is given in Table 1-1.

Solution

By formula (1-26) we have

$$E(X) = 1(1/6) + 2(1/6) + 3(1/6) + 4(1/6) + 5(1/6) + 6(1/6) = 21/6 = 7/2$$

EXAMPLE 1-27

If X has probability function $f(x) = x/10$, $x = 1, 2, 3, 4$ (Example 1-23), find the mean of the probability distribution.

Solution

Now we get

$$\mu = E(X) = 1(1/10) + 2(2/10) + 3(3/10) + 4(4/10) = 3$$

EXAMPLE 1-28

A gambling game that was played at small-town celebrations a few years back was known as "chuckaluck." The player bets an amount, say $1, on a number from 1 to 6. To be specific, let us take 2. Then a cage containing two dice is spun. If two 2s appear, the player gets his dollar back with two additional dollars. If only one 2 appears, the player receives one dollar in addition to having his bet returned. If no 2s appear, the house wins. Find the probability distribution of the number of 2s, the probability distribution of the player's winnings, and the player's expected winnings per play.

Solution

In Example 1-14 we found that the probability of getting one 2 is 10/36 and the probability of getting two 2s is 1/36. By subtraction, using formula (1-11), the probability of no 2s is $1 - 10/36 - 1/36 = 25/36$. Let X be the number of 2s; the probability distribution of X is

x	0	1	2
$f(x)$	25/36	10/36	1/36

Let Y be the player's winnings. Then, since $x = 0$ yields $y = -\$1$, $x = 1$ yields $y = \$1$, and $x = 2$ yields $y = \$2$, the probability distribution of Y is

y	$-\$1$	$\$1$	$\$2$
$g(y)$	25/36	10/36	1/36

Finally, the expected value of Y is

$$E(Y) = (-\$1)(25/36) + (\$1)(10/36) + \$2(1/36) = -\$13/36 = -\$.36$$

In other words, if the game were played a large number of times, the player would expect to lose on the average about 36 cents per play. For most games played in gambling casinos, this figure is about 2 to 4 cents on the dollar.

EXAMPLE 1-29

Several times we have considered the experiment which consists of tossing a coin until a head appears. We chose a sample space with sample points designated by H, TH, TTH, TTTH, etc., and assigned weights 1/2, 1/4, 1/8, 1/16, etc. If we let X be the number of tosses required to produce a head, find the probability distribution of X and the expected value of X.

Solution

The sample point H (which occurs with probability 1/2) corresponds to $x = 1$, TH (which occurs with probability 1/4) corresponds to $x = 2$, TTH (which occurs with probability 1/8) corresponds to $x = 3$, etc., and the probability distribution of X is

x	1	2	3	4	5	$\cdots$
$\Pr(X = x)$	1/2	1/4	1/8	1/16	1/32	$\cdots$

Here X can take on a countably infinite number of values. To find the expected value of X, according to formula (1-27) we write

$$E(X) = 1(1/2) + 2(1/4) + 3(1/8) + 4(1/16) + \cdots$$
$$= (1/2)[1 + 2(1/2) + 3(1/2)^2 + 4(1/2)^3 + \cdots]$$

where the sum contains an infinite number of terms. In this book we will make no attempt to develop experts in evaluating this kind of sum. We merely wish to illustrate that a random variable can take on a countably infinite number of values, and $E(X)$ is interpreted as before. A well-known mathematical result is

$$\frac{1}{(1-a)^2} = 1 + 2a + 3a^2 + 4a^3 + \cdots$$

provided $-1 < a < 1$. (For further information see Appendix A4-2.) Hence

$$1 + 2(1/2) + 3(1/2)^2 + 4(1/2)^3 + \cdots = \frac{1}{(1-1/2)^2} = 4$$

and $E(X) = (1/2)[4] = 2$. Thus on the average 2 tosses are required to produce a head.

The definition of expected value given by formulas (1-26) and (1-27) can be written in a more general form. If X is a random variable, so is X^2, $2X + 1$, $3X^3$, or any function $u(X)$. Consequently, it seems natural to define the expected value of $u(X)$ by

$$E[u(X)] = \sum_i u(x_i)f(x_i) = \sum_i u(x_i) \Pr(X = x_i) \tag{1-28}$$

where $\sum_i$ implies that the summation is taken over all i. It can be shown that formula (1-28) is consistent with our previous definition, and obviously the two are the same if $u(X) = X$.

EXAMPLE 1-30

For the probability distribution of X given in Example 1-27, find $E(X^2)$ and $E[(X - \mu)]$.

Solution

The probability function was $f(x) = x/10$, $x = 1, 2, 3, 4$. Then according to formula (1-28) we get

$$E(X^2) = (1)^2(1/10) + (2)^2(2/10) + (3)^2(3/10) + (4)^2(4/10) = 10$$

and since $\mu = 3$,

$$E[(X - \mu)] = (1 - 3)(1/10) + (2 - 3)(2/10)$$
$$+ (3 - 3)(3/10) + (4 - 3)(4/10) = 0$$

We have seen that one interesting characteristic of a probability distribution is the mean. Sometimes, particularly in some statistical applications which we will consider later, it is of interest to have a quantity that measures the variability of the distribution. To illustrate what we have in mind, consider the three random variables X, Y, and Z whose distributions are given in Table 1-8. It is to verify that all three

TABLE 1-8

Three probability distributions with the same mean

x	1	2	3	4	5
$Pr(X = x)$	.05	.1	.7	.1	.05
y	1	2	3	4	5
$Pr(Y = y)$	.2	.2	.2	.2	.2
z	1	2	3	4	5
$Pr(Z = z)$	.4	.1	0	.1	.4

random variables have expected value equal to 3. For the X distribution, values equal to or near the mean occur with high probability and those further away with low probability. For the Y distribution all values occur with equal probability. Finally, for the Z distribution values furthest from the mean occur with high probability and those near the mean with low probability. By a *measure of variability* we mean a quantity that reflects (or measures) the extent to which a random variable tends to be close to its mean. The X distribution exhibits " low " variability when compared to the Z distribution which exhibits " high " variability.

Suppose we attempt to construct measures of variability using the X distribution. It may occur to us to consider $E(X - \mu)$ which appears to be the " average deviation " from the mean. Here we find

$$E(X - \mu) = (1 - 3)(.05) + (2 - 3)(.1) + (3 - 3)(.7)$$
$$+ (4 - 3)(.1) + (5 - 3)(.05) = 0$$

an answer which is no accident. Unfortunately, $E(X - \mu) = 0$ for every random variable X, a result which is easily proved so that this quantity is worthless as a measure of anything.

Perhaps the next logical choice for a measure of variability is the *mean* or *average deviation*. It is defined to be $E(|X - \mu|)$. Let us illustrate with X distribution. (The vertical bars indicate " the absolute value of." Whether positive or negative, the difference $x - \mu$ is always given a positive sign.) We get

$$E(|X - \mu|) = |1 - 3|(.05) + |2 - 3|(.1) + |3 - 3|(.7) + |4 - 3|(.1)$$
$$+ |5 - 3|(.05)$$
$$= 2(.05) + 1(.1) + 0(.7) + 1(.1) + 2(.05) = .4$$

Similarly, $E(|Y - \mu|) = 1.2$ and $E(|Z - \mu|) = 1.8$. Thus the mean deviation measures variability in the sense we desire. It is, however, rarely used because it does not have the nice mathematical properties possessed by the variance, a quantity we consider next.

The *variance* of the distribution associated with a random variable, say X, is

$$\sigma_X^2 = E[(X - \mu)^2] = \sum_i (x_i - \mu)^2 f(x_i) \tag{1-29}$$

For the X distribution of Table 1-8, formula (1-29) becomes

$$E[(X - \mu)^2] = (1 - 3)^2(.05) + (2 - 3)^2(.1) + (3 - 3)^2(.7) + (4 - 3)^2(.1)$$
$$+ (5 - 3)^2(0.5)$$
$$= 4(.05) + 1(.1) + 0(.7) + 1(.1) + 4(.05) = .6$$

Similarly, $E[(Y - \mu)^2] = 2$, $E[(Z - \mu)^2] = 3.4$, and variance also measures variability in the sense we desire. That is, if large deviations from the mean occur with low probability and small deviations occur with high probability, we get a smaller variance than if the reverse is true.

The square root of the variance is called the *standard deviation*. This, too, can be regarded as a measure of variability since it is large or small depending upon whether the variance is large or small. Since we have used the symbol σ_X^2 for the variance of X, σ_X will denote the standard deviation of X. For the three distributions given in Table 1-8, we have $\sigma_X = \sqrt{.6} = .775$, $\sigma_Y = \sqrt{2} = 1.414$, $\sigma_Z = \sqrt{3.4} = 1.845$.

When it is clear which random variable is under consideration, the subscript is frequently omitted from σ_X^2, σ_X. Thus we write σ^2 or σ, whichever is appropriate. The terms "variance of the distribution" and "variance of the random variable" are used interchangeably.

The variance can be expressed in another form that is more adaptable to computation in cases for which the quantities $(x_i - \mu)$ are not integers. In Example A1-8 of Appendix A1 it is shown that

$$E[(X - \mu)^2] = E(X^2) - \mu^2 \tag{1-30}$$

EXERCISES

1-48　Find the probability distribution for the number of heads appearing when three coins are tossed. What is the expected number of heads? Find the variance of the distribution.

◊ 1-49　Two symmetric six-sided dice are rolled. Let $X =$ the sum of the spots. Find the probability distribution of X, the cumulative distribution, and the expected value of X.

1-50 Suppose that X has the probability distribution obtained in Exercise 1-49. A game is played with the following payoffs:
If $x = 7$ or 11, the player wins \$2
   $x = 2, 4,$ or 8, the player wins \$1
   $x = 3, 6, 9,$ or 12, the player loses \$1
   $x = 5$ or 10, the player loses \$2
For this game find the expected winnings per play.

◇ 1-51 The probability function of a random variable X is $f(x) = x/21$, $x = 1, 2, 3, 4, 5, 6$. Make a table of the distribution function. Find the expected value of X and the variance of X.

1-52 The probability function of a random variable X is $f(x) = (x^2 + 1)/35$, $x = 0, 1, 2, 3, 4$. Make a table of the distribution function. Find the mean of the distribution.

◇ 1-53 Suppose that the cumulative distribution of a random variable X is

r	0	1	2	3	4	5
$F(r)$	.13	.27	.53	.84	.92	1

(a) Find $\Pr(X \leq 3)$, $\Pr(X = 3)$, $\Pr(X = 2$ or $3)$.
(b) Find the probability distribution of X.

1-54 Given that the cumulative distribution of a random variable X is

r	0	1	2	3
$F(r)$	1/35	13/35	31/35	1

(a) Find $\Pr(X \leq 2)$, $\Pr(X = 2)$, $\Pr(X > 2)$.
(b) Find the probability distribution of X.
(c) Find the mean and variance of X.

◇ 1-55 Suppose that the chuckaluck game of Example 1-28 is changed by increasing the winning payoffs by \$1. Now what is the expected value of Y?

1-56 A hat contains 10 slips of paper numbered 1 to 10. One slip of paper is to be drawn. Let X denote the number which appears on the slip. Find the probability distribution of X, the expected value of X, and the variance of X.

◇ 1-57 Suppose that in families with four children the probabilities of having 0, 1, 2, 3, 4 boys are respectively 1/16, 4/16, 6/16, 4/16, 1/16. Find the expected number of boys in a family of four children.

1-58 A four-answer multiple-choice examination with five questions is taken by a student who guesses at each answer. Suppose that the probabilities of getting 0, 1, 2, 3, 4, 5 correct answers are 243/1,024, 405/1,024, 270/1,024, 90/1,024, 15/1,024, 1/1,024, respectively. What is the expected number of correct answers?

◈ 1-59 A blindfolded individual is given four different brands of cigarettes to smoke, told the names of the brands, and asked to identify each. Suppose that if the individual possesses no discriminatory ability (that is, he guesses), the probabilities of making 0, 1, 2, 3, 4 correct identifications are respectively 9/24, 8/24, 6/24, 0, 1/24. What is the expected number of correct identifications for such an individual?

1-60 A box contains eight camera flashbulbs. It is known that three of the bulbs are defective (will not work), and the other five are nondefective (satisfactory). Defective and nondefective bulbs look exactly alike. If two bulbs are taken from the box, find the probability distribution of the number of defective bulbs in the two which have been selected. Find the expected number of defective bulbs in the sample of size 2.

◈ 1-61 Suppose that the chuckaluck game of Example 1-28 is played with three dice. If a player's number shows on all three dice, he gets his dollar back with 3 additional dollars. If his number shows on two dice, he gets his dollar back with 2 additional dollars. If his number appears once, he gets his dollar back plus 1 additional dollar. When none of the dice show his number, he loses his money. Find the expected winnings for one play of the game.

1-62 A box contains two silver screws and three black screws, all the same size. They are drawn from the box one at a time without looking into the box and are not replaced. Let X be the number of drawings required to remove both silver screws. Find the probability distribution of X and the expected value of X.

◈ 1-63 In the Exercise 1-14 we discussed an experiment in which three people match coins to determine who pays for coffee. If we let D indicate that a decision is reached on a set of tosses and N indicate that no decision is reached, then we observed that the points of the sample space can be designated by D, ND, NND, $NNND$, etc. In Exercise 1-32 we should have assigned weights of 3/4, (1/4)(3/4), $(1/4)^2(3/4)$, $(1/4)^3(3/4)$, etc. If we let X be the number of sets of tosses required to reach a decision, find the probability distribution of X and the expected value of X. *Hint:* The problem is similar to Example 1-29.

1-6
CONTINUOUS RANDOM VARIABLES

In Sec. 1-5 we were mainly concerned with random variables that could assume a finite number of values. In Example 1-29 we have an illustration of a random variable which can assume a countably infinite number of values. If a random variable X can take on only a finite number of values or a countably infinite number of values, then we call X a *discrete random variable.*

In many interesting problems the random variable under consideration can take on any value within an interval (or perhaps several intervals). A random variable of this type is called a *continuous random variable.* Let us turn to a simple example. Consider the spinner pictured in Fig. 1-12 with the outside scale labeled uniformly from 0 to 1. Assume that it

FIGURE 1-12

A balanced spinner.

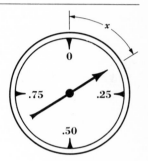

is perfectly balanced, that is, as likely to stop at one place as another if spun. Let X denote the distance on the circular scale from 0 to the arrow end of the spinner. Suppose that A_1 is the event that X is between .50 and .75. Our intuition tells us that .25 would be a good choice for $\Pr(A_1)$, the probability that A_1 happens. We would like to construct a method of assigning weights to an event A which is harmonious with our notions of probability, particularly definition (1-4). In this example it is easy. We define

$\Pr(A) = \Pr(X$ belongs to an interval or several intervals comprising $A)$
  $= $ length of A

We note that $\Pr(A)$ satisfies the conditions

(*a*) $\Pr(A) \geq 0$ for any A
(*b*) $\Pr(S) = 1$ (1-31)

where S is the entire sample space, that is, the whole interval $0 \leq x < 1$. Conditions (1-31) roughly correspond to conditions (1-3). We observe that if A has no length, then $\Pr(A) = 0$. Hence the probability that X takes on any specific value, say $X = .50$, is 0, a fact which is true for any continuous random variable.

We can also regard the $\Pr(A)$ of the spinner example as the area above the horizontal axis and under the curve defined by

$$f(x) = 1 \qquad 0 \leqq x < 1$$
$$ = 0 \qquad \text{elsewhere}$$

between the end points of the interval defining A. Thus if A is the A_1 mentioned above, the interval $.50 < x < .75$, the area under the curve between .50 and .75 is .25. (See Fig. 1-13.) The function $f(x)$ is called a density function and has the properties:

(a) It is nonnegative.

(b) The total area under the curve is 1.

$$(1\text{-}32)$$

FIGURE 1-13

Probability interpreted as area under a curve (uniform density function).

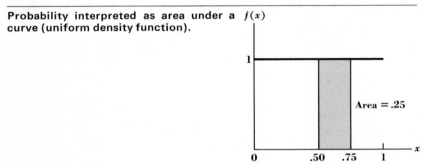

Any function $f(x)$ having properties (1-32) can be used as a density function, and areas under $f(x)$ within intervals can be interpreted as probabilities. Any nonnegative $f(x)$ can be used if the area under the curve is finite. If the area is C instead of 1, construct a new $f(x)$ by dividing the old one by C. Constructing density functions which give realistic probabilities for real-life problems may be quite difficult and is a chore which may profitably utilize a joint effort of a statistician and an experimental scientist. The evaluation of areas under density functions is frequently a difficult numerical problem. Fortunately, tables are available that make evaluation easy in the important cases to be considered in later chapters.

To illustrate the kind of difficulties we may have in selecting the proper density function, let us again consider the spinner example with some slight changes. Suppose it is a requirement that the arrow must be placed at 0 before the spinner is hit or tapped. Also suppose that a game is being played in which it is advantageous to have the spinner stop at .50 or as near to it as possible. Hence, the player would try to tap the spinner just hard enough to get near .50. Now what is the density which yields realistic probabilities? Certainly our earlier choice is inappropriate. One possibility that has appeal (but by no means the only one) is the triangular density shown in Fig. 1-14. Those who are familiar with right triangles and their areas could easily show that

$$\Pr(X \leqq r) = F(r) = 2r^2 \qquad 0 \leqq r \leqq .50$$
$$= 1 - 2(1 - r)^2 \qquad .50 < r < 1 \tag{1-33}$$

With this density we find

$$\Pr(A_1) = \Pr(.50 < X < .75) = \Pr(X \leqq .75) - \Pr(X \leqq .50)$$
$$= 1 - 2(1 - .75)^2 - 2(.50)^2$$
$$= 1 - .125 - .50 = .375$$

as compared with .25 before. Another density would yield yet another value for $\Pr(A_1)$. Before deciding (correctly or incorrectly) that the triangular density is appropriate, it is quite likely that a statistician would like some experimental verification indicating the adequacy of the choice. At any rate, the choice is not clear-cut as it was in the first situation.

FIGURE 1-14

Triangular density function for the spinner problem.

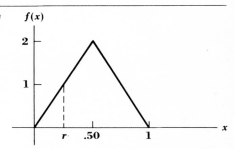

The preceding discussion suggests the following course of action in case a random experiment yields a continuous random variable: By some means (sound mathematical arguments or accumulated experimental evidence) select a density function which associates realistic probabilities with events (intervals) generated by the random experiment. Then to compute the probability that the experiment yields a value in an interval, find the area under the curve within the given interval.

In comparing the continuous situation with the discrete we have observed that (1) intervals are used instead of sample points, (2) weights are associated with intervals instead of with points, (3) probabilities are evaluated by finding areas under curves rather than by summing weights.

The mean and variance of a continuous random variable have definitions quite similar to those we have given for the discrete case. Since these definitions and their evaluations require the use of calculus, they will be omitted. However, the interpretations we have given for mean and variance remain unchanged. For the density pictured in Fig. 1-13 it can be shown that $E(X) = .50$, $\sigma_X^2 = 1/12 = .0833$. For the density of Fig. 1-14 we get $E(X) = .50$, $\sigma_X^2 = 1/24$. It is intuitively reasonable that the mean should be .50 in each case and that the variance of the second distribution should be smaller than the variance of the first.

Several important density functions will be introduced in the next chapter. At that time we will consider the tables that enable us to compute probabilities. Later we will match the density functions with specific problems for which they are appropriate.

EXERCISES

1-64   (a)   Assuming that the density function of Fig. 1-13 is appropriate, compute Pr(.40 < X < .60).

      (b)   Repeat part (a) using the density of Fig. 1-14.

◈ 1-65   (a)   Assuming that the density function of Fig. 1-13 is appropriate, compute Pr(.45 < X < .55).

      (b)   Repeat part (a) using the density of Fig. 1-14.

Summary of Results

I. Definitions of probability (Sec. 1-2)

An *event* can be regarded as something that has occurred, is about to occur, or might conceivably occur. Events are *mutually exclusive* if the occurrence of one event prevents the occurrence of any other event.

Classical Definition *If an experiment can produce n different* (1-1)
mutually exclusive results all of which are equally likely, and if f of these results are considered favorable (or result in A), then the probability of a favorable result (or the probability of event A) is f/n.

Relative Frequency Definition *Suppose that an experiment* (1-2)
is performed n times with f successes. Assume that the relative frequency f/n approaches a limit as n increases (so that as n increases, so does f). Then the probability of a success is the limit of f/n as n approaches infinity. (In actual practice the probability of an event is estimated by the ratio f/n based upon a "sufficiently large" number of repetitions, n.)

In any listing of all possible outcomes for an experiment, each outcome is called a *sample point*, and the set S of all sample points is called the *sample space*. Any part (subset) of the listing is called an *event*.

Mathematical Definition *Assign to each point a_i of the* (1-4)
sample space S a weight w_i subject to the conditions
(a) $w_i \geq 0$ for all i.
(b) $w_1 + w_2 + \cdots + w_n = \sum_{i=1}^{n} w_i = 1$ *with a finite number*
n of sample points.
(b') $w_1 + w_2 + w_3 + \cdots = \sum_{i=1}^{\infty} w_i = 1$ *with a countable infinity of sample points. Then the probability $\Pr(A)$ of an event A is the sum of the weights of all sample points in A.*

Modified Mathematical Definition *Let the sample space S* (1-5)
consist of a finite number of points n each assigned weight $1/n$. Then the probability $\Pr(A)$ of an event A is the ratio of the number of sample points in A to the total number of sample points in S. That is, $\Pr(A) = n(A)/n$ where $n(A)$ is the number of sample points in A.

II. Addition and multiplication theorems (Sec. 1-3)

The event "A does not occur" is called the complement of A and is denoted by $\bar{A}$. A sometimes useful result is

$$\Pr(A) = 1 - \Pr(\bar{A})$$ (1-7)

The event "either A_1 or A_2 occurs" is called the *union* of A_1 and A_2, designated by $A_1 \cup A_2$. The event "both A_1 and A_2 occur" is called the *intersection* of A_1 and A_2, designated by $A_1 A_2$. If A_1 and A_2 have no sample points in common, then the two events are said to be *mutually exclusive*. All three definitions are easily extended to the case of k events.

For any two events A_1 and A_2

$$\Pr(A_1 \cup A_2) = \Pr(A_1) + \Pr(A_2) - \Pr(A_1 A_2) \tag{1-8}$$

If A_1 and A_2 are mutually exclusive, then

$$\Pr(A_1 \cup A_2) = \Pr(A_1) + \Pr(A_2) \tag{1-9}$$

If n events $A_1, A_2, \ldots, A_n$ are mutually exclusive, then formula (1-9) generalizes to

$$\Pr(A_1 \cup A_2 \cup \cdots \cup A_n) = \Pr(A_1) + \Pr(A_2) + \cdots + \Pr(A_n) \tag{1-10}$$

Formulas (1-8) to (1-10) are all versions of the *addition theorem*.

The symbol $\Pr(A_2 \mid A_1)$ denotes the probability that A_2 occurs given that A_1 already has occurred and is called a *conditional probability*. If $\Pr(A_1) \neq 0$, then

$$\Pr(A_2 \mid A_1) = \frac{\Pr(A_1 A_2)}{\Pr(A_1)} \tag{1-12}$$

Formula (1-12) yields

$$\Pr(A_1 A_2) = \Pr(A_1)\Pr(A_2 \mid A_1) \tag{1-13}$$

If A_1 and A_2 are unrelated events so that $\Pr(A_2 \mid A_1) = \Pr(A_2)$, then A_1 and A_2 are said to be *independent*, and formula (1-13) reduces to

$$\Pr(A_1 A_2) = \Pr(A_1)\Pr(A_2) \tag{1-16}$$

With n events $A_1, A_2, \ldots, A_n$ the generalization of formula (1-13) is

$$\Pr(A_1 A_2 \cdots A_n) = \Pr(A_1)\Pr(A_2 \mid A_1)\Pr(A_3 \mid A_1 A_2)$$
$$\cdots \Pr(A_n \mid A_1 A_2 \cdots A_{n-1}) \tag{1-17}$$

and if the events are independent then,

$$\Pr(A_1 A_2 \cdots A_n) = \Pr(A_1)\Pr(A_2) \cdots \Pr(A_n) \tag{1-18}$$

Formulas (1-13) and (1-16) to (1-18) are all versions of the multiplication theorem.

III. Counting formulas (Sec. 1-4)

A *permutation* is an ordered grouping or arrangement of things. The number of permutations of n things taken n at a time is

$$_nP_n = n! \tag{1-19}$$

The number of permutations of n things taken r at a time, $r \leqq n$, is

$$_nP_r = \frac{n!}{(n-r)!} \tag{1-21}$$

The number of permutations of n things taken n at a time if n_1 are alike, n_2 are alike, $\ldots$, n_k are alike is

$$P = \frac{n!}{n_1! \, n_2! \cdots n_k!} \tag{1-22}$$

where $n_1 + n_2 + \cdots + n_k = n$.

A combination is an unordered grouping of things. The number of combinations of n things taken r at a time is

$$\binom{n}{r} = \frac{n!}{r! \, (n-r)!} \tag{1-23}$$

IV. Random variables, probability distributions, and expected value (Sec. 1-5)

Types of experiments which may produce different results when repeated in spite of all efforts to keep the conditions of performance constant are called *random experiments*. A quantity whose value is determined by the outcome of a random experiment is called a *random variable*. If X is a random variable, then

$$F(r) = \Pr(X \leqq r) = \sum_{x \leqq r} f(x) \tag{1-24}$$

is called the *cumulative distribution* or *distribution function* of X. By $\Pr(X \leqq r)$, the probability the X is less than or equal to r, we mean the probability that the associated random experiment yields a value less than or equal to r. A table listing $\Pr(X = x)$ for all possible values of X is called the *probability distribution* of X.

The expected value of X is

$$E(X) = \sum_{i=1}^{k} x_i f(x_i) = \sum_{i=1}^{k} x_i \, \Pr(X = x_i) \tag{1-26}$$

if X has a finite number of outcomes, and

$$E(X) = \sum_{i=1}^{\infty} x_i f(x_i) = \sum_{i=1}^{\infty} x_i \, \Pr(X = x_i) \tag{1-27}$$

if X has a countable infinity of outcomes. Expected value is interpreted as the long-run average of values produced by a random experiment.

The expected value of $u(X)$ is

$$E[u(X)] = \sum_{i} u(x_i) f(x_i) = \sum_{i} u(x_i) \, \Pr(X = x_i) \tag{1-28}$$

The variance of a random variable X is

$$\sigma_X^2 = E[(X - \mu)^2] = \sum_i (x_i - \mu)^2 f(x_i) \tag{1-29}$$

$$= E(X^2) - \mu^2 \tag{1-30}$$

Variance is a measure of variability in the sense that it reflects the extent to which a random variable tends to be close to its mean.

V. Continuous random variables (Sec. 1-6)

If a random variable X can take on only a finite number of values or a countably infinite number of values, it is called a *discrete random variable*. A *continuous random variable* can assume any value within an interval or several intervals.

Associated with a continuous random variable X is a density function $f(x)$ where

(*a*) $f(x)$ is nonnegative

(*b*) The total area under the graph of $f(x)$ and above the x axis is 1

$$\tag{1-32}$$

Areas under density functions are interpreted as probabilities. The choice of the correct density function to yield probabilities for a random variable X is not always easily made and is influenced by experience.

2

SOME IMPORTANT
PROBABILITY MODELS

2-1
INTRODUCTION

When we construct a sample space and assign weights, as we did through-out Chap. 1, we are engaging in a process called *model building*. Those models can appropriately be called probability models and are representations which we hope will be useful in drawing inferences or in predicting results associated with the performance of a random experiment. In dictionary language a model is a set of assumptions. To be more specific in terms of our situation, we assume that certain outcomes (usually numerical results) are of interest and that a certain set of weights give a realistic measure of the chance that these outcomes will occur.

In Chap. 1 we regarded every probability problem separately. Each time we started from " scratch," ignoring the fact that we may have worked essentially the same problem before. Several of the models which we are about to discuss have already been considered for special cases. Now we will generalize the assumptions slightly and obtain results which hold for a whole class of problems. Besides being a more efficient way to proceed, once and for all we can establish properties which hold for every problem in the class. When numerical calculations become prohibitive, it is frequently possible to prepare a set of tables which is of sufficient size to handle most cases one is likely to encounter. Then, the next time we have a problem which we recognize as belonging to the given class, the solution is obtained with a minimum of effort.

Associated with each model will be a random variable, say X. For some of the models we can easily derive the probability function (and hence the probability distribution). For others we give the probability distributions without derivation. The chapter could also be appropriately entitled " Some Important Probability Distributions." The assumption that a given probability function or probability density yields appropriate probabilities is also a way to formulate a model. Thus, the model for the second spinner problem of Sec. 1-6 is the triangular density function pictured in Fig. 1-14. Alternatively, we could say that the model for that problem is the cumulative distribution given by formula (1-33).

2-2
THE BINOMIAL PROBABILITY MODEL

There are several reasons for beginning our study of probability models with the binomial. First, it probably is used in more applications than any of the others to be discussed this chapter. Second, the formula for computing probabilities (the probability function) is easy to derive, yet the derivation is not trivial. Third, it is a straightforward task to investigate the reasonableness of the model assumptions. Finally, computation of probabilities is simple since tables of the binomial distribution are very extensive and easy to read.

Consider a series of experiments which have the following properties:

(a) The result of each experiment can be classified into one of two categories, say, success and failure.

(b) The probability p of a success is the same for each experiment. (2-1)

(c) Each experiment is independent of all the others.

(d) The series consists of a fixed number of experiments, say n.

These conditions appear to be satisfied when a symmetric six-sided die is rolled 10 times, regarding a 1 or 2 as a success and using $1/3$ for the value of p. Now suppose we would like to know the probability of obtaining exactly four successes. One way to achieve the result is SSSSFFFFFF, where S and F denote success and failure. That is, we could be successful the first four times and fail the last six. The probability that this series of events happens is $(1/3)(1/3)(1/3)(1/3)(2/3)(2/3)(2/3)(2/3)(2/3)(2/3) = (1/3)^4(2/3)^6$, a result obtained by recognizing that the 10 events are independent and by using the multiplication theorem for independent events, formula (1-18). Obviously, there are lots of other orders of successes and failures, each order being an event, which will yield exactly four successes in ten rolls of a die (for example, SSSFSFFFFF). The probability of obtaining any one of the orders is $(1/3)^4(2/3)^6$, and the total number of possible orders is $k = 10!/4!6! = \binom{10}{4}$, the number of ways 10 things taken 10 at a time can be permuted if four are alike and six are alike. Now if we let $A_1, A_2, \ldots, A_k$ be the events associated with the k orders, we recognize that these k events are mutually exclusive since the occurrence of one specified order excludes the occurrence of any other order. Hence, the probability of exactly four successes in ten rolls is $\Pr(A_1 \cup A_2 \cup \cdots \cup A_k) = \Pr(A_1) + \Pr(A_2) + \cdots + \Pr(A_k)$ by the addition theorem for mutually exclusive events. Since each of the latter probabilities is $(1/3)^4(2/3)^6$, we have

$$\text{Pr}(4 \text{ successes in } 10 \text{ experiments}) = \binom{10}{4}(1/3)^4(2/3)^6$$

In the die experiment, the same die was rolled each time. Exactly the same experiment was repeated 10 times. Each roll—and in general, each repetition of the experiment—is sometimes referred to as a "trial" of the experiment. The same terminology is used if 10 dice (all symmetric) are thrown simultaneously. That is, each die rolled is called a trial of the experiment even though 10 different (but equivalent) experiments have been performed. Thus each trial is either a separate experiment or a repetition of the same experiment. The kind of trials we are concerned with in this section satisfy conditions (2-1).

We now generalize the discussion about 10 rolls of a die to yield a formula for the probability of exactly x successes in n repetitions or trials with probability of success p. One way to achieve this number of successes is

$$\underbrace{S \cdots S}_{x \text{ times}} \underbrace{F \cdots F}_{n-x \text{ times}}$$

that is, be successful x times in a row, then fail the remaining $n - x$ times. According to the assumptions, each experiment is independent of all the others, and by the multiplication theorem, the probability of achieving the given order is

$$\underbrace{p \cdots p}_{x \text{ times}} \underbrace{q \cdots q}_{n-x \text{ times}} = p^x q^{n-x}$$

where $q = 1 - p$. There are many other orders that yield x successes and $n - x$ failures, the total being $k = n!/x!(n-x)! = \binom{n}{x}$, the number of ways n things taken n at a time can be permuted if x are alike and $n - x$ are alike. The probability associated with each order is $p^x q^{n-x}$. Now let $A_1, A_2, \ldots, A_k$ be the events associated with the k orders; these events are mutually exclusive since only one order can occur at a time. Further by the addition theorem

$$\text{Pr}(A_1 \cup A_2 \cup \cdots \cup A_k) = \text{Pr}(A_1) + \text{Pr}(A_2) + \cdots + \text{Pr}(A_k)$$
$$= p^x q^{n-x} + p^x q^{n-x} + \cdots + p^x q^{n-x}$$
$$= \binom{n}{x} p^x q^{n-x}$$

Hence the probability function for the random variable X, the number of successes in n trials satisfying conditions (2-1), is

$$b(x; n, p) = \binom{n}{x} p^x q^{n-x} \qquad x = 0, 1, 2, \ldots, n \qquad (2\text{-}2)$$

When the probability function of a random variable is given by formula (2-2), X is said to have a binomial distribution. The name *binomial* arises from the fact that $b(x; n, p)$ is a term in the expansion of the binomial $(q + p)^n$. (For further information see Appendix A4-1.)

In most interesting problems involving the binomial model we need a sum of terms rather than individual terms. Appendix B1 gives values of

$$B(r; n, p) = \sum_{x=0}^{r} b(x; n, p) \tag{2-3}$$

$$= b(0; n, p) + b(1; n, p) + \cdots + b(r; n, p)$$

$$= \Pr(r \text{ or fewer successes in } n \text{ trials})$$

$$= \Pr(X \leq r)$$

for some values of p and n. Several extensive tables of the binomial distribution have been published. The largest of these are the Harvard table [3] and the Ordnance Corps table [9]. The Harvard table gives $\Pr(X \geq r)$, instead of $\Pr(X \leq r)$, to five decimal places for

$$n = 1(1)50(2)100(10)200(20)500(50)1000$$

and

$$p = .01(.01).50, \ 1/16, \ 1/12, \ 1/8, \ 1/6, \ 3/16, \ 1/3, \ 3/8, \ 5/12, \ 7/16$$

(Sums of the type $\Pr(X \leq r)$ and $\Pr(X \geq r)$ are sometimes called *left-hand* and *right-hand* sums, respectively.) This notation means that the table entries are all combinations of $n = 1, 2, 3, \ldots, 49, 50, 52, 54, \ldots, 98, 100, 110, \ldots, 190, 200, 220, \ldots, 480, 500, 550, \ldots, 950, 1,000$, and $p = .01, .02, .03, \ldots, .49, .50$, plus the nine fractional values. The Ordnance Corps table also gives $\Pr(X \geq r)$, but to seven decimal places, for $n = 1(1)150$ and $p = .01(.01).50$. Tables for small values of p have been prepared by Robertson [8] for $p = .001(.001).050$ and for roughly the same values of n (plus a few more) contained in the Harvard table and by Weintraub [10] for $p = .00001, .0001(.0001).001(.001).10, \ n = 1(1)100$.

EXAMPLE 2-1

If an ordinary six-sided symmetric die is rolled four times, what is the probability that exactly two 6s occur?

Solution

We first check conditions (2-1).

(*a*) Each roll results in a 6 or not a 6.
(*b*) For each roll $p = 1/6$ remains constant as the probability of a success.
(*c*) Successive rolls are independent.
(*d*) The die is rolled four times.

Hence the conditions of the binomial model seem to be satisfied and X is the number of 6s that occur. We identify $n = 4$, $x = 2$, and evaluate

$$\Pr(X = 2) = b(2; 4, 1/6) = \binom{4}{2}\left(\frac{1}{6}\right)^2\left(\frac{5}{6}\right)^2 = \frac{150}{1,296}$$

EXAMPLE 2-2

If the die in Example 2-1 is rolled four times, what is the probability of obtaining two or fewer 6s?

Solution

By two or fewer 6s we mean zero, one, or two 6s. We can compute

$$\Pr(\text{no 6s}) = \Pr(X = 0) = \binom{4}{0}\left(\frac{1}{6}\right)^0\left(\frac{5}{6}\right)^4 = \frac{625}{1,296}$$

$$\Pr(\text{one 6}) = \Pr(X = 1) = \binom{4}{1}\left(\frac{1}{6}\right)^1\left(\frac{5}{6}\right)^3 = \frac{500}{1,296}$$

$$\Pr(\text{two 6s}) = \Pr(X = 2) = \binom{4}{2}\left(\frac{1}{6}\right)^2\left(\frac{5}{6}\right)^2 = \frac{150}{1,296}$$

Then, the probability of two or fewer 6s is

$$\Pr(X \leq 2) = \Pr(X = 0) + \Pr(X = 1) + \Pr(X = 2)$$

$$= \frac{625}{1,296} + \frac{500}{1,296} + \frac{150}{1,296}$$

$$= \frac{1,275}{1,296}$$

EXAMPLE 2-3

In a 10-question true-false examination, what is the probability of getting 70 percent or better correct by guessing? Exactly 7 out of 10 correct?

Solution

We check conditions (2-1).

(a) Assuming each question is answered, it is right or wrong.

(b) For each question the probability of a correct guess is $p = 1/2$.

(c) If one guesses at each answer, perhaps by throwing a coin, then results for one question are independent of those for any other question. In an actual examination independence can be achieved by constructing questions so that each gives no information about the correctness of any of the others. Undoubtedly a good true-false examination should have this property.

(d) There are 10 questions to be answered.

Hence, the conditions of the binomial model seem to be satisfied and we can let X be the number of correct answers. With $n = 10$, $p = 1/2$, we need

$$Pr(X \geq 7) = Pr(X = 7) + Pr(X = 8) + Pr(X = 9) + Pr(X = 10)$$
$$= 1 - Pr(X \leq 6)$$
$$= 1 - \sum_{x=0}^{6} b(x; 10, 1/2)$$
$$= 1 - .82812 \qquad \text{from Appendix B1}$$
$$= .17188$$

The probability of getting exactly seven correct is

$$Pr(X = 7) = Pr(X \leq 7) - Pr(X \leq 6)$$
$$= .94531 - .82812 \qquad \text{from Appendix B1}$$
$$= .11719$$

EXAMPLE 2-4

In a 20-question, 5-answer multiple-choice examination, what is the probability of getting 6 or more correct by guessing?

Solution

We observe that

(a) Assuming all questions are answered, each is right or wrong.

(b) For each question the probability of a correct guess is $p = 1/5$.

(c) When each answer is a guess (obtained, perhaps, by drawing numbers from a hat), then the answer to any question does not influence the answer to any other question. Thus each question is answered independently of all the others. To achieve this condition in an actual examination, questions should be constructed so that the answer does not depend on any other answers. In many multiple-choice examinations there would be real doubt as to whether the independence condition is satisfied.

(d) There are 20 questions to be answered.

Thus, the random variable $X = $ the number of correct answers appears to have a binomial distribution with $n = 20$, $p = 1/5 = .20$. We want

$$Pr(X \geq 6) = Pr(X = 6) + Pr(X = 7) + \cdots + Pr(X = 20)$$
$$= 1 - Pr(X \leq 5)$$
$$= 1 - \sum_{x=0}^{5} b(x; 20, .20)$$
$$= 1 - .80421 \qquad \text{from Appendix B1}$$
$$= .19579$$

EXAMPLE 2-5

It is estimated that 90 percent of a potato crop is good, the remainder having rotten centers that cannot be detected unless the potatoes are cut open. What is the probability of getting 20 or fewer good ones in a sack of 25 potatoes?

Solution

It is not unreasonable to assume

(*a*)   Each potato can be classified as good or bad.
(*b*)   The probability of getting a good potato remains approximately $p = .90$ from trial to trial, since the crop consists of a very large number of potatoes.
(*c*)   Potatoes are independently good or bad. (Perhaps this is not so for several potatoes coming from the same plant.)
(*d*)   In addition, the number of trials is a fixed number, being $n = 25$.

Thus, the random variable $X = $ the number of good potatoes in the sack appears to have a binomial distribution with $n = 25, p = .90$. We need

$$\Pr(X \leq 20) = \sum_{x=0}^{20} b(x; 25, .90)$$

Since binomial tables give p's only up to .50, the latter sum cannot be read directly from the tables. However, if we let $Y = $ the number of bad potatoes in the sack, then the above assumptions are the same except that now $p = .10$ and Y has a binomial distribution with $n = 25, p = .10$. Since getting 20 or fewer good ones is the same as getting 5 or more bad ones, it is apparent that

$$\Pr(X \leq 20) = \Pr(Y \geq 5) = \sum_{y=5}^{25} b(y; 25, .10)$$

$$= 1 - \Pr(Y \leq 4) = 1 - \sum_{y=0}^{4} b(y; 25, .10)$$

$$= 1 - .90201 \qquad \text{from Appendix B1}$$

$$= .09799$$

Thus, when $p > 1/2$ interchange the roles of p and q, success and failure.

In Examples 2-3 to 2-5 there is some room to doubt that the assumptions (2-1) are satisfied. This is not an uncommon state of affairs in probability problems. One often encounters situations in which the conditions needed to derive the probability distribution are not fulfilled. When this happens, the probability distribution may still give probabilities that are sufficiently close to actual probabilities for practical purposes. Some further experience with the particular random variable is then necessary to verify the adequacy of the model.

In Example 2-5 we encountered a p larger than $1/2$ and found that the roles of both p and q and success and failure had to be interchanged to enter the tables. We used a specific example to convince ourselves that if X has a binomial distribution with given values of n and p, then $Y = n - X$ has a binomial distribution with the same n and probability of success $q = 1 - p$. Then we observed that

$$\Pr(X \leqq r) = \Pr(Y \geqq n - r)$$

$$= 1 - \Pr(Y \leqq n - r - 1)$$

$$= 1 - \sum_{y=0}^{n-r-1} b(y; n, q) \tag{2-4}$$

Expressed another way, this is

$$B(r; n, p) = 1 - B(n - r - 1; n, q) \tag{2-5}$$

Perhaps it is easier to go through the above common-sense procedure each time than it is to remember and use formula (2-5).

Sometimes it is of interest to find the mean and variance of a binomial random variable. To find the expected value of X we would evaluate

$$E(X) = \sum_{x=0}^{n} x \binom{n}{x} p^x q^{n-x}$$

For $n = 2$ this is

$$E(X) = 0 \binom{2}{0} p^0 q^2 + 1 \binom{2}{1} p^1 q^1 + 2 \binom{2}{2} p^2 q^0$$

$$= 0 + 2pq + 2p^2 = 2p(q + p) = 2p$$

and for $n = 3$ we get

$$E(X) = 0 \binom{3}{0} p^0 q^3 + 1 \binom{3}{1} p^1 q^2 + 2 \binom{3}{2} p^2 q^1 + 3 \binom{3}{3} p^3 q^0$$

$$= 0 + 3pq^2 + 6p^2 q + 3p^3$$

$$= 3p(q^2 + 2pq + p^2) = 3p(q + p)^2 = 3p$$

These two special cases suggest that

$$E(X) = np \tag{2-6}$$

a formula which can be verified with a little algebra. The result is intuitively reasonable. If the experiment "a coin is thrown 100 times" is repeated many times, the average number of heads obtained ought to be about $(100)(1/2) = 50$. If a student takes a large number of 25-question multiple-choice examinations, always guessing at the answers, then we might expect that his average number of correct answers ought to be about $25(1/5) = 5$. It can also be shown that the variance of the distribution is

$$\sigma_X^2 = npq \tag{2-7}$$

In the coin experiment we get $\sigma^2 = 100(1/2)(1/2) = 25$ (so $\sigma = 5$), and in the multiple-choice guessing game we find $\sigma^2 = 25(1/5)(4/5) = 4$ (so $\sigma = 2$).

EXERCISES

Use Appendix B1 to evaluate probabilities in the following problems. Where appropriate comment upon the reasonableness of the binomial assumptions (2-1).

◈ 2-1   A well-known baseball player has a lifetime batting average of .300. If he comes to bat five times in the next game, what is the probability that he will get more than two hits? Three hits?

2-2   A four-answer multiple-choice examination has 100 questions. Assuming that a student only guesses and answers every question, what is the probability that he gets 30 or more correct? Suppose that the instructor decides that no grade (number correct) will be passing unless the probability of getting or exceeding that grade by guessing is less than .01. What is the minimum passing grade? Find the mean, variance, and standard deviation of the number of correct answers.

◈ 2-3   A missile manufacturer claims that his missiles are successful 90 percent of the time. The Air Force checks the stock by firing 10 missiles and obtains 5 successes. What is the probability of obtaining 5 or fewer successes if $p = .90$? What conclusion is one apt to draw?

2-4   Suppose that from a very large group of voters 25 are selected independently of one another and asked if they favor a certain proposal. If 40 percent of the voters favor the proposal, what is the probability that a majority of the 25 voters chosen will favor the proposal?

◈ 2-5   Suppose that you believe that you can hit a bull's-eye with a dart one time in ten. If this is the case, compute the probability of scoring 15 or more bull's-eyes in 100 throws. Find the mean, variance, and standard deviation of the number of bull's-eyes in 100 throws.

2-6   It is known that 25 percent of all rabbits inoculated with a serum containing a certain disease germ will contract the disease. If 20 rabbits are inoculated, what is the probability that at least 3 get positive reactions?

◈ 2-7   A standard cure for tuberculosis is successful 30 percent of the time. A new cure is tried on a group of 50 patients and is successful in 29 cases. If the new cure also is successful 30 percent of the time, what is the probability of 29 or more successes in 50 trials? What conclusion is one apt to draw and why?

2-8   Weather bureau records in a certain locality show that 40 percent of the days in April are cloudy. Find the probability that, of the first 20 days of next April, at most 5 will be cloudy.

◈ 2-9   A coin is tossed 3 times and a die is rolled 5 times. Consider a head on the coin a success and a 2 on the die a success. Then, to compute the probability of 4 or more successes in the 8 "experiments," why would the binomial not be appropriate?

2-10   A machine produces bolts. If a bolt is good, the probability that the machine produces a good bolt the next time is .9. However, if the bolt is defective, then the probability that the next bolt is good is .8. To compute the probability that 6 or more of the next 10 bolts are good, why would the binomial not be appropriate?

◈ 2-11   Two tennis players are to play 10 games. Suppose that we would like to compute the probability (by using the binomial model) that one of the players (perhaps our favorite) wins 6 or more games. Recalling that the server has a big advantage in tennis, which condition or combination of conditions is most apt to be violated in the binomial model?

2-12   Ten percent of all jackrabbits have diseased livers (which we will assume is not contagious). During the hunting season a hunter bags 25 jackrabbits. He would like to know the probability that 3 or more have diseased livers. Do the conditions for use of the binomial model seem to be satisfied? Assuming the model is satisfactory, find the required probability.

◈ 2-13   A certain type of automobile battery is guaranteed to last 36 months. If a battery fails to last that long, a price adjustment is made on the purchase of a new battery. The manufacturer claims that 80 percent of the batteries will outlive the guarantee. A dealer receives and sells 50 batteries of this type. Of these 50 what is the probability that at least 38 last longer than 36 months?

2-14   Suppose that 1/4 of all radios taken from the end of a production line will not operate satisfactorily (and are temporarily defective) without further work. If 15 radios are taken from the production line, what is the probability that 6 or more fail to operate in a satisfactory manner?

◈ 2-15   A small town has 100 street lights to maintain. New bulbs are installed in all 100 locations. The supplier claims that 70 percent of his bulbs will last 6 months or longer. Find the probability that 35 or more bulbs fail to last for 6 months.

2-16   In order to select its beer tasters, a brewery gives an applicant a tasting examination.   The applicant is presented with four glasses, one of which contains ale and three of which contain beer, and is asked to identify the one containing ale.   If the procedure is repeated 10 times and the brewery requires 7 or more correct answers for a satisfactory score, what is the probability that an applicant will pass the test if he cannot discriminate and only guesses each time? Suppose that an applicant does possess discriminatory ability and has a certain probability p of making a correct choice on each trial. What is the smallest value of p of those given in Appendix B1 that will guarantee the applicant at least a probability of .85 of passing the test?

2-3
THE NEGATIVE BINOMIAL PROBABILITY MODEL

Our next model is very similar to the one discussed in the previous section. Although there are some important applications of the negative binomial, it does not arise nearly as often as the binomial.

Consider a series of experiments having the following properties:

(*a*)   The result of each experiment can be classified into one of two categories, say, success and failure.
(*b*)   The probability p of a success is the same for each experiment.
(*c*)   Each experiment is independent of all the others.   (2-8)
(*d*)   The series consists of a variable number of experiments with the performance continuing until a fixed number of successes, say c, is achieved.

Only (*d*) differs from assumptions (2-1).   These assumptions appear to be satisfied when we continue to roll a symmetric six-sided die, regarding a 1 or a 2 as a success, until exactly 4 successes have occurred.   Now suppose that we would like to know the probability that exactly 10 rolls are required.   One way to obtain the required probability is by using an argument almost identical to the one used in the binomial case.   The desired result will be obtained with the order SSSFFFFFFS, that is, succeed the first three times, fail the next six, and then succeed.   Obviously, the last experiment must produce a success or four successes would have been achieved sooner.   By the multiplication theorem, the probability of the above sequence is

$$(1/3)(1/3)(1/3)(2/3)(2/3)(2/3)(2/3)(2/3)(2/3)(1/3) = (1/3)^4(2/3)^6$$

Once more, many other sequences, all mutually exclusive events with probability of occurrence $(1/3)^4(2/3)^6$, yield the desired result.   Hence, again we need to count the number of ways the S's and F's can be permuted. Since the last letter must be an S, only the remaining nine letters can be rearranged, the number of rearrangements being $9!/3!6! = \binom{9}{3}$. Consequently,

$$\Pr(10 \text{ trials are required to achieve 4 successes}) = \binom{9}{3}(1/3)^4(2/3)^6$$

The preceding argument is easily generalized. Now we seek a formula for the probability that exactly y experiments are required to produce c successes. One way to achieve c successes in exactly y trials is

$$\underbrace{SS \cdots S}_{c-1 \text{ times}} \underbrace{F \cdots F}_{y-c \text{ times}} S$$

That is, be successful $c - 1$ times, fail the next $n - c$ times, and succeed the last time. By the multiplication theorem, the probability that this happens is

$$\underbrace{pp \cdots p}_{c-1 \text{ times}} \underbrace{qq \cdots q}_{y-c \text{ times}} p = p^c q^{y-c}$$

There are many other orders that yield the cth success on the yth trial. Since the last letter in every sequence must be S, this leaves $c - 1$ of the S's and $y - c$ of the F's for forming permutations. The number that can be constructed is

$$k = \frac{(y-1)!}{(c-1)!(y-c)!} = \binom{y-1}{c-1}$$

the total number of ways to permute $y - 1$ things taken $y - 1$ at a time if $c - 1$ are alike and $y - c$ are alike. The probability associated with each order is $p^c q^{y-c}$. Now let $A_1, A_2, \ldots, A_k$ be the events associated with the k orders; these events are mutually exclusive since only one order can occur at a time. As in the binomial case, we have by the addition theorem

$$\Pr(A_1 \cup A_2 \cup \cdots \cup A_k) = \binom{y-1}{c-1} p^c q^{y-c}$$

Hence the probability function of the random variable Y, the number of trials required to achieve exactly c successes, is

$$b^*(y; c, p) = \binom{y-1}{c-1} p^c q^{y-c} \qquad y = c, c+1, c+2, \ldots \qquad (2\text{-}9)$$

The random variable Y cannot assume a value less than c since at least c repetitions are required to produce c successes. We observe that the random variable can take on a countable infinity of values. When the probability function of a random variable Y is given by formula (2-9), Y is said to have a negative binomial distribution. The name negative binomial arises from the fact that $b^*(y; c, p)$ is a term in the expansion of $p^c(1 - q)^{-c}$. (For further information, see Appendix A4-2.)

Again, most interesting problems involving the model require a sum of terms rather than individual terms. Specifically, it would be useful to have

$$B^*(r; c, p) = \sum_{y=c}^{r} b^*(y; c, p) \tag{2-10}$$

$$= \Pr(r \text{ or fewer trials are required to achieve } c \text{ successes})$$

$$= \Pr(Y \leq r)$$

Fortunately no new tables are needed as such sums are available from binomial tables. We can show that

$$\Pr(Y \leq r) = \Pr(X \geq c) \tag{2-11}$$

where X has a binomial distribution with the same value of p and with n being replaced by r. To be more precise, the result is

$$B^*(r; c, p) = \sum_{y=c}^{r} b^*(y; c, p) = \sum_{x=c}^{r} b(x; r, p) \tag{2-12}$$

$$= 1 - \sum_{x=0}^{c-1} b(x; r, p) \tag{2-13}$$

$$= 1 - B(c - 1; r, p)$$

Thus a left-hand sum of negative binomial probabilities is equal to a right-hand sum of ordinary binomial probabilities.

[The result (2-11), (2-12) is not terribly surprising once it is pointed out. For the sake of argument let us again consider the die example where a 1 or a 2 was called a success. Suppose we set as a goal the obtaining of 4 successes but agree to roll no more than 10 times. Our goal can be achieved in two ways. First, if the fourth success is obtained on either the fourth, fifth, sixth, seventh, eighth, ninth, or tenth roll, the goal is accomplished. The probability that this series of events happens is $\Pr(Y \leq 10)$ where $c = 4$, $p = 1/3$. The second way to achieve the goal in 10 tries is to plan to roll exactly 10 times and count 4, 5, 6, 7, 8, 9, or 10 successes. The probability of the latter series of events is $\Pr(X \geq 4)$ where $n = 10$, $p = 1/3$. In both cases we are considering the event "the goal is achieved in 10 trials" so that the two probabilities must be the same. If we replace 4 by c, 10 by r, and let p be any number between 0 and 1, the argument is readily generalized.]

EXAMPLE 2-6

A student takes a five-answer multiple-choice examination orally. He continues to answer questions until he gets five correct answers. What is the probability that he gets them on or before the twenty-fifth question if he guesses at each answer?

Solution

The reasonableness of assumptions (*a*) to (*c*) of (2-8) was discussed in Example 2-4. Also, (*d*) is satisfied since the number of questions required to achieve five successes is a random variable. Thus we use the negative binomial model with $c = 5$, $r = 25$, $p = 1/5 = .20$, and $Y = $ the number of questions answered. We seek

$$\Pr(Y \leq 25) = \sum_{y=5}^{25} b^*(y; 5, .20)$$

$$= \sum_{x=5}^{25} b(x; 25, .20)$$

$$= 1 - \sum_{x=0}^{4} b(x; 25, .20)$$

$$= 1 - .42067 \qquad \text{from Appendix B1}$$

$$= .57933$$

EXAMPLE 2-7

Consider again the potato crop of Example 2-5. Suppose that a cook needs 20 good potatoes for a meal, and so he selects potatoes at random, cuts them open, and throws away the bad ones. What is the probability that he must cut open more than 25 potatoes?

Solution

The assumptions (*a*) to (*c*) of (2-8) have already been discussed in Example 2-5. Since the number of successes is fixed and the number of trials is a random variable, condition (*d*) is satisfied and the negative binomial with $c = 20$, $p = .9$ is appropriate. Letting $Y = $ the number of potatoes cut open, we need

$$\Pr(Y \geq 26) = 1 - \Pr(Y \leq 25)$$

$$= 1 - \sum_{y=20}^{25} b^*(y; 20, .9)$$

$$= 1 - \sum_{x=20}^{25} b(x; 25, .9)$$

$$= \sum_{x=0}^{19} b(x; 25, .9)$$

Since $p = .9$ in the latter sum, we interchange the roles of success and failure, p and q, and the latter sum is

$$= \sum_{x=6}^{25} b(x; 25, .1)$$

$$= 1 - \sum_{x=0}^{5} b(x; 25, .1)$$

$$= 1 - .96660 \qquad \text{from Appendix B1}$$

$$= .03340$$

EXAMPLE 2-8

To determine who pays for coffee, three people each toss a coin and the odd person pays. If the coins all show heads or all show tails, they are tossed again. What is the probability that a decision is reached in five repetitions or sooner?

Solution

A decision is reached on any trial if the result is one head and two tails or two heads and one tail. To compute the probability of these events, the binomial is appropriate since

(*a*) Each coin will show either a head or a tail.
(*b*) The probability that each coin shows a head is 1/2.
(*c*) The three coins are tossed independently of one another.
(*d*) Exactly three coins are tossed in each repetition.

With $n = 3$, $p = 1/2$ we compute

$$\text{Pr(1 head, 2 tails)} = b(1; 3, 1/2) = \binom{3}{1}\left(\frac{1}{2}\right)\left(\frac{1}{2}\right)^2 = \frac{3}{8}$$

$$\text{Pr(2 heads, 1 tail)} = b(2; 3, 1/2) = \binom{3}{2}\left(\frac{1}{2}\right)^2\left(\frac{1}{2}\right) = \frac{3}{8}$$

and the probability of a decision is $3/8 + 3/8 = .75$.

Next each set of tosses is characterized by these facts:

(*a*) A decision is reached or not reached with each set.
(*b*) The probability of a decision is .75 for all sets.
(*c*) The result of each set is independent of the result for any other set.
(*d*) A variable number of sets is required to produce 1 decision.

Thus the negative binomial with $c = 1$, $p = .75$ is appropriate to determine the probability needed to answer the question. Now Y is the number of sets of tosses required to produce 1 decision and we need

$$\Pr(Y \leqq 5) = \sum_{y=1}^{5} b^*(y; 1, .75)$$

$$= \sum_{x=1}^{5} b(x; 5, .75) \qquad \text{by formula (2-12)}$$

$$= \sum_{x=0}^{4} b(x; 5, .25) \qquad \text{by interchanging success and failure,}$$
$$\qquad\qquad\qquad\qquad\qquad p \text{ and } q$$

$$= .99902 \qquad \text{by Appendix B1}$$

To find the expected value of a negative binomial random variable Y, we would evaluate

$$E(Y) = \sum_{y=c}^{\infty} y b^*(y; c, p)$$

It can be shown that this yields

$$E(Y) = \frac{c}{p} \qquad\qquad (2\text{-}14)$$

Further calculations also lead to

$$\sigma_Y^2 = \frac{cq}{p^2} \qquad\qquad (2\text{-}15)$$

It is easily verified that for the situation in Example 2-8 the average number of sets of tosses required to produce a decision is 4/3 and for the situation of Example 2-6 the student must answer 25 questions on the average to obtain 5 correct answers.

EXERCISES

Use Appendix B1 to evaluate probabilities in the following problems. Where appropriate comment upon the reasonableness of the negative binomial assumptions (2-8).

◈ 2-17 A well-known baseball player has a lifetime batting average of .300. He needs 32 more hits to up his lifetime total to 3,000. What is the probability that 100 or fewer times at bat are required to achieve his goal? Find the expected number of times at bat required to get 32 hits. What is the variance of Y, the number of turns required to get 32 hits?

2-18   A student takes a four-answer multiple-choice examination orally. He continues to answer questions until he gets 10 correct answers. What is the probability that more than 25 questions are required if he guesses at each answer? Find the mean and variance of the number of trials required.

◇ 2-19   A missile manufacturer claims that his missiles are successful 90 percent of the time. The Air Force checks the stock by firing until 4 successes are obtained, and 11 trials are required. What is the probability that 11 or more trials are required if $p = .90$? What conclusion is one apt to draw?

2-20   Suppose it is estimated that 40 percent of a certain large group are strong supporters of a project and are willing to volunteer their services if asked. The remaining 60 percent will decline to volunteer. In order to get five people for a committee, members of the group are contacted one at a time until the commitee is complete. Twenty-one contacts are required. If the 40 percent figure is correct, what is the probability that 21 or more contacts are required? Are you inclined to believe the 40 percent figure? If the 40 percent figure is correct, find the expected number of contacts required.

◇ 2-21   Suppose that you believe that you can hit a bull's-eye with a dart 1 time in 10. If this is the case, compute the probability that more than 100 throws are required to obtain exactly 15 bull's-eyes.

2-22   A scientist needs three diseased rabbits for an experiment. He has 20 rabbits available and inoculates them one at a time with a serum containing the disease germ, quitting if and when he gets 3 positive reactions. If the probability is .25 that a rabbit can contract the disease from the serum, what is the probability that the scientist is able to get 3 diseased rabbits from 20? What is the expected number of inoculations to get 3 positive reactions?

◇ 2-23   To determine who buys coffee, four people toss one coin each. This is repeated until someone has a result different from the other three (in which case he buys). What is the probability of reaching a decision on any one set of tosses? What is the probability that three or fewer sets of tosses are required to reach a decision? (*Hint:* Evaluate without the table.)

2-24   Suppose that 1/4 of all radios taken from the end of a production line will not operate satisfactorily without further work. If radios taken from the line are checked until 6 unsatisfactory radios are found, what is the probability that 15 or fewer radios are examined?

◇ 2-25   Camera flashbulbs are produced independently of one another by a production process. Sampling is to be used to determine whether or not the process is yielding a sufficiently small fraction of defective bulbs. Based upon some requirements specified by the company management, the firm's statistician decides that bulbs should be inspected until 4 or more defectives are found with action being taken to reduce the fraction defective if, and only if, sampling is terminated with 20 items or less. If actually 25 percent of the bulbs are defective, what is the probability that action will be taken? If the fraction defective is 30 percent, what is the answer?

2-4
THE MULTINOMIAL PROBABILITY MODEL

A fairly obvious generalization of the binomial conditions yields the multinomial model. Consider a series of experiments which have the following properties:

(a) The result of each experiment can be classified into one of k categories $C_1, C_2, \ldots, C_k$.
(b) The probabilities of falling into these categories are $p_1, p_2, \ldots, p_k$ for each experiment.　　(2-16)
(c) Each experiment is independent of all the others.
(d) The series consists of a fixed number of experiments, say n.

Thus, if $k = 2$, conditions (2-16) are the same as (2-1). We note that $p_1 + p_2 + \cdots + p_k = 1$, since the result of each experiment must fall into one of the k categories. (In the binomial case we had p, q where $p + q = 1$.) To illustrate a series of experiments satisfying the above conditions, suppose that a symmetric six-sided die is rolled 10 times. Let C_1 be the outcome 1 or 2, C_2 be the outcome 3, 4, or 5, and C_3 be the outcome 6. We identify $p_1 = 2/6$, $p_2 = 3/6$, $p_3 = 1/6$. Now suppose that we would like to know the probability that 3 of the rolls yield C_1, 5 of the rolls yield C_2, and 2 of the rolls yield C_3. One order of outcomes which will yield the result under consideration is

$$C_1 C_1 C_1 \, C_2 C_2 \, C_2 C_2 \, C_2 \, C_3 C_3$$

By the multiplication theorem for independent events, the probability of this sequence is

$$(2/6)(2/6)(2/6)(3/6)(3/6)(3/6)(3/6)(3/6)(1/6)(1/6) = (2/6)^3 (3/6)^5 (1/6)^2$$

As in the binomial case there are many other orders of results which yield the event under consideration. The probability associated with each order is $(2/6)^3 (3/6)^5 (1/6)^2$. The total number of orders which yield the desired result is $10!/3!5!2! = 2,520$, the number of ways to permute 10 things taken 10 at a time if 3 are alike, 5 are alike, and 2 are alike. Since only one order can occur at a time, the 2,520 orders represent

2,520 mutually exclusive events, each with probability of occurrence $(2/6)^3(3/6)^5(1/6)^2$. By the addition theorem, the probability that one of these orders occurs is

$$\frac{10!}{3!\,5!\,2!}\left(\frac{2}{6}\right)^3\left(\frac{3}{6}\right)^5\left(\frac{1}{6}\right)^2 = .0810$$

We now generalize the above discussion to yield a formula for the probability of obtaining x_1 outcomes in category C_1, x_2 outcomes in category C_2, ..., x_k outcomes in category C_k, where

$$x_1 + x_2 + \cdots + x_k = n \tag{2-17}$$

since a total of n experiments is under consideration. One order which produces the result is

$$\underbrace{C_1 \cdots C_1}_{x_1 \text{ times}} \underbrace{C_2 \cdots C_2}_{x_2 \text{ times}} \cdots \underbrace{C_k \cdots C_k}_{x_k \text{ times}}$$

By the multiplication theorem, the probability that this happens is

$$\underbrace{p_1 \cdots p_1}_{x_1 \text{ times}} \underbrace{p_2 \cdots p_2}_{x_2 \text{ times}} \cdots \underbrace{p_k \cdots p_k}_{x_k \text{ times}} = (p_1)^{x_1}(p_2)^{x_2} \cdots (p_k)^{x_k}$$

There are many other orders that yield x_1 results in C_1, x_2 results in C_2, ..., x_k results in C_k, the total being $n!/x_1!\,x_2! \cdots x_k!$, the number of ways to permute n things taken n at a time if x_1 are alike, x_2 are alike, ..., x_k are alike. Each order corresponds to an event, and only one order can occur at a time. The probability associated with each order is $(p_1)^{x_1}(p_2)^{x_2} \cdots (p_k)^{x_k}$. By the addition theorem for mutually exclusive events the probability that one of these orders occurs is

$$f(x_1, x_2, \ldots, x_k) = \frac{n!}{x_1!\,x_2! \cdots x_k!}(p_1)^{x_1} \cdots (p_k)^{x_k} \tag{2-18}$$

The x's can take on any of the values 0, 1, 2, ..., n subject to the restriction (2-17). When the probability function of random variables X_1, X_2, ..., X_k is given by formula (2-18), then X_1, X_2, ..., X_k are said to have a multinomial distribution. The name *multinomial* arises from the fact that the right-hand side of formula (2-18) is a term in the expansion of $(p_1 + p_2 + \cdots + p_k)^n$.

The computation of formula (2-18) is a tedious chore when performed by hand calculations. This type of calculation is easily handled by high-speed computing machines. Tabulation would be cumbersome because of the variety of choices possible for the p's, the x's, k, and n. Fortunately, in the statistical problems for which we will use the multinomial model, satisfactory approximations are available.

EXAMPLE 2-9

If a symmetric six-sided die is rolled five times, what is the probability that the results are a 1, a 2, and three other numbers?

Solution

First we check conditions (2-16).

(*a*) Each throw results in a 1, or a 2, or not a 1 or 2.
(*b*) The probabilities $p_1 = 1/6$, $p_2 = 1/6$, $p_3 = 4/6$ remain constant for each roll.
(*c*) Successive rolls are made independently of one another.
(*d*) The die is rolled 5 times, a fixed number.

Hence the multinomial model seems to be appropriate. Here $x_1 = 1$, $x_2 = 1$, $x_3 = 3$, and from formula (2-18) we have that the probability of obtaining a 1, a 2, and three other numbers is

$$\frac{5!}{1!\,1!\,3!} \left(\frac{1}{6}\right)\left(\frac{1}{6}\right)\left(\frac{4}{6}\right)^3 = \frac{40}{243} = .1646$$

In Sec. 1-5 we defined a random variable as a quantity whose value is determined by the outcome of a random experiment. In the binomial and negative binomial cases our interest was in only one such quantity. With the multinomial model we are concerned with several random variables at one time. The series of experiments generates several numbers which we have denoted by $x_1, x_2, \ldots, x_k$. The random variables of interest are $X_1 =$ the number of outcomes in category C_1, $X_2 =$ the number of outcomes in category $C_2, \ldots, X_k =$ the number of outcomes in category C_k. Formula (2-18) represents the probability that the series of experiments produces a value x_1 for X_1, a value x_2 for $X_2, \ldots$, a value x_k for X_k which could be denoted by $\Pr(X_1 = x_1, X_2 = x_2, \ldots, X_k = x_k)$. The function $f(x_1, x_2, \ldots, x_k)$ which yields probabilities for all $x_1, x_2, \ldots, x_k$ is called a *joint probability function*. The random variables $X_1, X_2, \ldots, X_k$ are said to have a joint probability distribution.

When several random variables are governed by a joint probability distribution, it is frequently of interest to know the probability distribution (or probability function) of each random variable separately. Thus, for example, we may seek the probability function of X_1 all by itself, temporarily paying no attention to the other random variables. Such information would be useful for the die problem of Example 2-9 if we desire $\Pr(X_1 = 1)$ or $\Pr(X_1 \leqq 1)$. These single variable distributions are called *marginal distributions*. In the multinomial case it is easy to see that each random variable by itself has a binomial distribution. For example,

X_1 is the number of occurrences of category C_1 for a series of experiments with the following properties:

(a) The result of each experiment is either C_1 (success) or not C_1 (failure).
(b) The probability of a success is p_1 for each experiment.
(c) Each experiment is independent of all the others.
(d) The number of experiments in the series is n, a fixed number.

Hence X_1 is the type of random variable whose probability function is (2-2). Here $p = p_1$ and the probability function of X_1 is

$$b(x_1; n, p_1) = \binom{n}{x_1} p_1^{x_1}(1 - p_1)^{n - x_1} \qquad x_1 = 0, 1, 2, \ldots, n \qquad (2\text{-}19)$$

By the same argument the probability function of X_i, $i = 1, 2, \ldots, k$ is

$$b(x_i; n, p_i) = \binom{n}{x_i} p_i^{x_i}(1 - p_i)^{n - x_i} \qquad x_i = 0, 1, 2, \ldots, n \qquad (2\text{-}20)$$

From our discussion of the binomial case, we know that

$$E(X_i) = np_i \qquad (2\text{-}21)$$

$$\sigma_{X_i}^2 = np_i(1 - p_i) \qquad (2\text{-}22)$$

EXAMPLE 2-10

For the die that was rolled five times in Example 2-9, find $\Pr(X_1 = 1)$.

Solution

As we have indicated by the above discussion X_1 has a binomial distribution with $p = 1/6$, $n = 5$. Hence

$$\Pr(X_1 = 1) = b(1; 5, 1/6) = \binom{5}{1}\left(\frac{1}{6}\right)^1\left(\frac{5}{6}\right)^4 = \left(\frac{5}{6}\right)^5 = .4019$$

Without recognizing that X_1 has a binomial distribution we would have to evaluate

$\Pr(X_1 = 1, X_2 = 0, X_3 = 4) + \Pr(X_1 = 1, X_2 = 1, X_3 = 3)$

$+ \Pr(X_1 = 1, X_2 = 2, X_3 = 2) + \Pr(X_1 = 1, X_2 = 3, X_3 = 1)$

$+ \Pr(X_1 = 1, X_2 = 4, X_3 = 0)$

That is, compute probabilities for every mutually exclusive event that allows X_1 to be 1, and then sum the results.

Because of formula (2-17) it is possible to rewrite the probability function in terms of only $k - 1$ of the x's. For example, with $k = 2$ we could write (since $x_2 = n - x_1$)

$$f(x_1) = \frac{n!}{x_1!(n - x_1)!} p_1^{x_1} p_2^{n - x_1} \qquad x_1 = 0, 1, \ldots, n$$

(the binomial), and with $k = 3$ we could write (since $x_3 = n - x_1 - x_2$)

$$f(x_1, x_2) = \frac{n!}{x_1! x_2!(n - x_1 - x_2)!} (p_1)^{x_1}(p_2)^{x_2}(p_3)^{n - x_1 - x_2} \qquad x_1 + x_2 \leqq n$$

$$x_1, x_2 = 0, 1, 2, \ldots, n$$

EXERCISES

Comment on the reasonableness of the multinomial assumptions when it is appropriate to do so.

2-26 A card is drawn from an ordinary deck of 52 cards. The result is recorded, the card is replaced in the deck, and the deck is shuffled. This is repeated 10 times. Let X_1 be the number of spades drawn and let X_2 be the number of hearts drawn. Find $\Pr(X_1 = 2, X_2 = 3)$, $\Pr(X_1 \leqq 4)$, and the expected number of spades drawn.

◇ 2-27 In a large university it has been determined that 20 percent of the students live in fraternities and sororities, 30 percent live in dormitories, and 50 percent live in private homes. If a committee of 5 is selected, each person being chosen independently of the others, what is the probability that the committee has one person from a dormitory, one from a private home, and three persons from fraternities and sororities? What is the probability that it contains three or more from fraternities and sororities?

2-28 Suppose that national records reveal that twice as many automobile accidents occur on Saturday and Sunday as on other days of the week. That is, the probability that an accident occurs on Saturday is 2/9, the probability that an accident occurs on Sunday is 2/9, and the probability that an accident occurs on each of the other days of the week is 1/9. From the record file, 50 accident reports are selected independently of one another. The observed distribution of accidents according to days of the week is

Sun.	Mon.	Tues.	Wed.	Thurs.	Fri.	Sat.
10	8	2	7	6	3	14

Write down, but do not attempt to evaluate, an expression for computing the probability that the random variables assume this particular set of values.

◈ 2-29 A production line produces mechanical parts. Of these 50 percent are not defective, 40 percent have one defect, and 10 percent have two defects. What is the probability that of the first 4 parts selected 2 are not defective, 1 has one defect, and 1 has two defects? Find the probability that 2 or more of the 4 parts are not defective.

2-30 An urn contains 10 balls of equal size. Of these 6 are red, 3 are white, and 1 is blue. A ball is drawn, the color written down, and the ball is returned to the urn. After the balls are well mixed the procedure is repeated a second, third, and fourth time. What is the probability that we have recorded 3 red, 1 white, and 0 blue?

◈ 2-31 A psychologist is conducting an experiment with rats. Each rat is presented with a choice of 5 different entrances into a cage containing food. The entrances are all the same size but are painted different colors—red, green, blue, black, and white. If each rat has an equal chance of selecting any entrance and five rats (at different times and independently of one another) are used in the experiment, what is the probability that each entrance is used once? What is the probability that three or more rats use the red entrance?

2-32 Suppose that a new cure for the common cold helps 70 percent of those who take it, makes 20 percent of those who take it feel worse, and produces no effect in the remaining 10 percent. If 5 cold victims are given the cure, what is the probability that 4 are helped and 1 feels worse?

◈ 2-33 According to a theory concerning elementary statistics textbooks, the beginning word of a paragraph is short (1 to 4 letters) 60 percent of the time, intermediate (5 to 8 letters) 20 percent of the time, and long (9 letters or over) 20 percent of the time. If the theory is true, compute the probability that the first word will be short 4 times and long 1 time when 5 paragraphs are selected by thumbing through a book of this type.

2-5
THE UNIFORM PROBABILITY MODEL

Suppose that an experiment can terminate in one of k mutually exclusive ways, all equally likely. A single roll of a symmetric six-sided die is an example of this kind of experiment. Any one of the sides is as likely to show as any other. A more practical situation is created by writing k numbers, one each on k slips of paper of equal size, placing the numbers in a hat or box, mixing the slips well, and then drawing one.

For a six-sided symmetric die, the probability associated with each outcome is $1/6$. If X is the number of spots showing, then the probability function of X is

$$f(x; 6) = 1/6 \qquad x = 1, 2, 3, 4, 5, 6$$

For any experiment where the outcomes are 1, 2, ..., k occurring with equal probabilities, the probability function is

$$f(x; k) = \frac{1}{k} \qquad x = 1, 2, \ldots, k \qquad (2\text{-}23)$$

We shall generalize formula (2-23) slightly by letting the random variable X assume any one of k values with equal probability. This yields

$$f(x; k) = \frac{1}{k} \qquad x = x_1, x_2, \ldots, x_k \qquad (2\text{-}24)$$

A random variable X having the probability function (2-24) is said to have a uniform distribution. (The model assumption is that all values occur with equal probability.)

Extensive tables have been prepared that make it possible to generate observed (sample) values of random variables having the uniform distribution. One such table has been published by the Rand Corporation [7]. A small part of that table appears in Appendix B2. We could construct such a table by using the hat procedure, mentioned at the beginning of the section, with numbers 0, 1, 2, ..., 8, 9. (The table was, of course, prepared in a more efficient manner, using high-speed computers.) After each draw the number would be recorded and replaced before the next draw. Thus, the one-digit numbers appearing in the table are observations of a random variable with probability function

$$f(x; 10) = 1/10 \qquad x = 0, 1, 2, \ldots, 9$$

The table can also be regarded as being composed of a series of two, three, or more digit numbers. Suppose we put our finger over a two-digit number and ask "What is the probability that the number is 27?" Since the probability that the first digit is 2 is 1/10 and the probability that the second digit is 7 is 1/10, and the events "the first digit is 2" and "the second digit is 7" are independent, the answer is 1/100. Similarly the probability of selecting any given two-digit number is 1/100. The table can be regarded as being composed of two-digit numbers which are observations of a random variable with probability function

$$f(x; 100) = 1/100 \qquad x = 0, 1, 2, \ldots, 99$$

Similarly, we can look upon the table as being composed of three-digit numbers which are observations of a random variable with probability function

$$f(x; 1{,}000) = 1/1{,}000 \qquad x = 0, 1, 2, \ldots, 999$$

and so on for four or more digits.

The random number table can be used to generate observed values of a random variable with a uniform distribution when k is not a multiple of

10. Suppose we put our finger over a two-digit number in the table. If it is more than 27 we disregard it. If it is less than or equal to 27, we could ask "What is the probability that the number is 15?" To get the answer let A_1 be the event "the number is less than or equal to 27" and A_2 be the event "the number is 15." We observe that $\Pr(A_1A_2) = \Pr(A_2) = 1/100$, $\Pr(A_1) = 28/100$, and by our conditional probability formula (1-12) we have $\Pr(A_2 \mid A_1) = (1/100)/(28/100) = 1/28$, a result that is intuitively obvious. The answer is the same if 15 is replaced by any number 27 or less. Hence, if we disregard all the numbers larger than 27, the remaining numbers are observations of a random variable with probability function

$$f(x; 28) = 1/28 \qquad 0, 1, 2, \ldots, 27$$

If we disregard zero and all digits larger than k, the remaining numbers are observations of a random variable with probability function (2-23).

The Rand publication includes suggestions for the use of a random number table. A starting position in the table is determined by selecting a seven-digit number in some arbitrary manner. We might, for example, take the last seven digits in the first row of the table. This yields 9274945. We use the first five digits to locate a row (rows are numbered 00000 to 19999). Since we have 92749, which is out of the range, we change the first digit to a 1 (when the number is odd, to a 0 when the number is even) and use row 12749. The last two numbers, 45 in our example, are used to locate a column (there are 50 columns in each line). When the number exceeds 50, we subtract 50 from it to determine the column number. Hence if the last two digits were 77, we would begin in column 27. Since Appendix B2 contains only 200 rows, a five-digit number will suffice for locating the starting point. After using a five-digit number from the table for this purpose (seven-digit with the original table), it is advisable to draw a line through it so that the same starting point is not used in the future. From the starting point we can proceed to read off any size numbers we desire by going down a column, up a column, left to right as we would read a book, right to left as if we were reading backwards, across diagonals, and many other ways. To be consistent we shall read down a column until the bottom of the page is reached. Then we shall proceed to the top of the page and into the next column on the right made up of the same size numbers.

EXAMPLE 2-11

Use the table of random numbers to obtain 15 observed values of a random variable having the probability function (2-23) with $k = 6$. Use 12749 to determine the starting point. We note that the experiment (selecting a number from a table) is equivalent to rolling a die.

Solution

We will use only the integers 1, 2, 3, 4, 5, 6 and disregard 0, 7, 8, 9. We start in row 127 and column 49. It is easy to verify that the first number is 3. Below 3 we find 1, 9 (disregarded), 0 (disregarded), 2, 9 (disregarded), 6, 6, 2, 8 (disregarded), 7 (disregarded), 3, 1. Since the bottom of the page has been reached and only 8 numbers have been obtained, we proceed to the top of column 50. Now we find 4, 7 (disregarded), 5, 3, 1, 1, 5, 3. According to the procedure we have agreed to use, the numbers are 3, 1, 2, 6, 6, 2, 3, 1, 4, 5, 3, 1, 1, 5, 3.

The numbers we have obtained from the table in Example 2-11 constitute an *observed random sample*. If we let X_1 be the first number selected from the table, X_2 the second number selected from the table, ..., X_{15} be the fifteenth number selected from the table, then X_1, X_2, ..., X_{15} are fifteen random variables which take on a different set of values every time the random number table is used differently. We observed one set of 15 numbers starting in row 127, column 49. A different set would be observed starting in row 80, column 18. Still another starting point would yield another set. The set of random variables X_1, X_2, ..., X_{15} is called a random sample from the distribution having the probability function $f(x; 6) = 1/6$, $x = 1$, 2, 3, 4, 5, 6 or, alternatively, a random sample from a uniform distribution. In general X_1, X_2, ..., X_n is called a *random sample* of size n if

(*a*)   each of the X's has the same distribution

and   (2-25)

(*b*)   each experiment which yields a value of X is performed independently of all the others.

Conditions (*a*) and (*b*) are satisfied for the situation of Example 2-11 and, equivalently, for 15 rolls of a symmetric six-sided die. For the binomial model conditions (*b*) and (*c*) of (2-1) are exactly the same as (*a*) and (*b*), respectively, of (2-25) if we regard the result of each experiment as a random variable X (assuming value 1 for a success, 0 for a failure) with probability function

$$f(x) = b(x; 1, p) = p^x(1 - p)^{1 - x} \qquad x = 0, 1$$

(Thus, what we proved in Sec. 2-2 was that the sum of a random sample of such variables has a binomial distribution with constants n and p.) The notation x_1, x_2, ..., x_n is used for an unspecified set of observed values, that is, an observed random sample. Sometimes x_1, x_2, ..., x_n is called a random sample from a (infinite) population, but the implication is that X_1, X_2, ..., X_n satisfy (2-25). Also each x_i is called an observation.

A table of random numbers can be used to generate an observation or an observed random sample from many other distributions besides the uniform. Suppose, for example, we desire an observation x from a binomial distribution with $p = .2$, $n = 10$. We can let 1 and 2 correspond to success, 0, 3, 4, 5, 6, 7, 8, 9 correspond to failure, read 10 numbers from the table, and count the number of successes. It is easy to verify that the binomial conditions are satisfied for the series of experiments. If the 10 numbers from the table are 1, 0, 0, 9, 7, 3, 2, 5, 3, 3, then we have observed $x = 2$. With 3, 7, 5, 4, 2, 0, 4, 8, 0, 5 we have observed $x = 1$.

Ordinarily, in statistical experiments we will not be interested in observed random samples generated by the random number table. Rather we will be interested in observed random samples generated by experimental situations (i.e., the number of defective flashbulbs in a sample of size 10 from a production line). When sampling is performed in real-life problems and the experimenter intends to use procedures based upon random sampling, he should do his best to make conditions (2-25) seem realistic. In practical terms we can rephrase (2-25) as

(a)   each of the n experiments is performed under the same conditions or as nearly so as possible

and   (2-26)

(b)   each experiment yields a result which in no way depends upon or influences the result of any other experiment.

Perhaps one of the main uses of a random number table arises in sampling when it is required to "draw a random sample from a finite population." By this is meant that k objects are under consideration (the population) from which we wish to select a subset of n of the objects (the sample), where $n < k$, in such a way that every sample of size n has the same probability of being selected. A sample so chosen is called a *simple random sample*. To see how this is done suppose a finite population consists of seven individuals or objects (with which we may or may not associate numbers x_1, x_2, ..., x_7). Let the individuals or objects be designated by 1, 2, ..., 7 from which we want to select a simple random sample of size $n = 3$. Start someplace in the table and read numbers until three numbers between 1 and 7 inclusive are obtained. (We will use the same procedure as in Example 2-11 except that repeats will be disregarded.) Suppose we get 3, 5, and 6. Then our simple random sample consists of the objects or individuals designated by 3, 5, 6. We have yet to demonstrate that the probability of drawing any sample is the same as the probability of drawing any other sample if we use this procedure. In selecting the first number for the sample, the possible choices are governed by the uniform distribution with probability function $f(x; 7) = 1/7$, $x = 1, 2, ..., 7$. The probability of getting 3, 5, or 6 on the first selection is 3/7. Suppose we drew 5 first. Then the next selection is to be made from the uniform distribution with probability function $f(x; 6) = 1/6$, $x = 1, 2,$

3, 4, 6, 7. The probability of getting a 3 or a 6 is 2/6. Suppose we draw a 6 second. Then the last selection is made from numbers having the uniform distribution with probability function $f(x; 5) = 1/5$, $x = 1, 2, 3, 4, 7$, and the probability of getting a 3 is 1/5. Consequently, by the multiplication theorem the probability of drawing the sample consisting of objects numbered 3, 5, 6 is $(3/7)(2/6)(1/5) = 3!\,4!/7! = 1 \Big/ \binom{7}{3}$. The argument is the same as for each of the $\binom{7}{3}$ possible samples, so that the probability of drawing each is $1 \Big/ \binom{7}{3}$. Thus the procedure has produced a simple random sample.

The above discussion is easily generalized to a population of k items and a sample of size n. On each selection the distribution of the numbers assigned to the items is uniform. The probability that one of a specified set of n items is drawn the first time is n/k; the probability that one of the remaining $n - 1$ items is selected the second time is $(n - 1)/(k - 1)$; $\ldots$; the probability that the one remaining at the time of the last draw is then selected is $1/[k - (n - 1)]$. Consequently, the probability of drawing any specified set of n items is

$$\frac{n}{k}\frac{n-1}{k-1} \cdots \frac{1}{k-n+1} = \frac{n!\,(k-n)!}{k!} = \frac{1}{\binom{k}{n}}$$

No matter which set of n is selected, the probability of drawing the sample is $1 \Big/ \binom{k}{n}$ for any one of the $\binom{k}{n}$ samples. We observe that the distribution of samples is uniform with probability function

$$f(x; m) = \frac{1}{\binom{k}{n}} \qquad x = S_1, S_2, \ldots, S_m$$

where $m = \binom{k}{n}$ and $S_1, S_2, \ldots, S_m$ are the m possible samples. (Here the random variable X is n dimensional since each sample contains n items.)

EXAMPLE 2-12

Draw a simple random sample of size 5 from items designated by $1, 2, \ldots, 353$ by using Appendix B2 and the five-digit number in the upper left-hand corner of the second page to determine the starting point.

Solution

The five-digit number in the upper left-hand corner of the second page of Appendix B2 is 32179. Thus we start in row 121, column 29. The first three-digit number is 216.   Below 216 we find 777, 761, 769 (all too large and are disregarded).   Next we read 200, 655, 676, 099, 594, 127, 199.

Our sample consists of the items numbered 216, 200, 99, 127, 199, the first five numbers between 1 and 353 inclusive, disregarding any repeats (here there were none).

EXAMPLE 2-13

Suppose a box contains 106 oranges.   We want to select 6 at random, cut them open to determine if they are good or spoiled, and, on the basis of the 6, decide in some way whether or not to purchase the box.   Use Appendix B2 to draw the sample and the five-digit number 56058 to determine the starting point.

Solution

The five-digit number 56058 determines our starting point.   Hence we start in row 160, column 8, with 584.   To number the oranges we could lay them out in a row so that we know which orange goes with which number.   Reading down the column from the starting point we get 338, 292, 004.   Hence orange number 4 is in the sample.   Continuing down the column, 091 is the only other number less than or equal to 106 obtained before the bottom of the page is reached.   Then going to the top of the page, column 11, the first three-digit number is 824.   Proceeding downward looking for numbers less than or equal to 106, we find 047 and 073 before reaching the bottom of the page.   Then we go to the top of the page, column 14, and read 742.   Proceeding downward again we find 018 and 014.   Hence we take the oranges numbered 4, 91, 47, 73, 18, 14 for our simple random sample of size $n = 6$.   In actual practice one would probably use the hat procedure with the oranges.   That is, select oranges one at a time from various places in the box and hope that a simple random sample is obtained.

With a little algebra it can be shown that if X has probability function (2-23), then

$$E(X) = \frac{k + 1}{2} \tag{2-27}$$

and

$$\sigma_X^2 = \frac{k^2 - 1}{12} \tag{2-28}$$

We have verified formula (2-27) in Example 1-26 with a die random variable.

EXERCISES

2-34   Suppose that your class contains 80 students numbered 1 to 80. Select a simple random sample of size 10 following the procedure used in the examples and using the first five-digit number found in row 00030 to determine the starting point in the tables.

◈ 2-35   A committee contains 10 people. It is desired to select 3 of these at random to form a subcommittee to prepare a report. To do this the committee members are each assigned a number from 1 to 10 inclusive and the table of random numbers is used. The starting point is determined by the first five digits in row 00199 and the procedure outlined in the examples is used. Which people are on the subcommittee?

2-36   A game is played with a balanced spinner similar to the one described in Sec. 1-6. The spinner is divided into 10 equal sections numbered 1 to 10 inclusive. The player advances from 1 to 10 steps depending upon where the spinner stops. If the spinner is broken, how can the game be continued with a table of random numbers? Using 07001 to determine the starting point in the table and following the procedure used in the examples, how many steps are awarded on each of the next five turns? (Count 0 as 10.)

◈ 2-37   A game is played with a balanced spinner similar to the one described in Sec. 1-6. The spinner is divided into 10 equal sections with three of the sections labeled " 1," two sections labeled " 2," two sections labeled " 3," one section labeled " 4," one section labeled " 5," and one section labeled " 6." The player advances from 1 to 6 steps depending upon where the spinner stops. If the spinner is broken, how can the game be continued with a table of random numbers? Using 07001 to determine the starting point in the table and following the procedure used in the examples, how many steps are awarded on each of the next five turns?

2-38   The game of monopoly is played with two symmetric six-sided dice. The player advances from 2 to 12 steps according to the sum appearing on the dice when he rolls them at his turn. Suppose that the monopoly game is ready to start but that a pair of dice cannot be found. Describe a simple way to use the random number table and one-digit numbers as a substitute for the dice. Using 14001 to determine the starting point in the table and following the procedure used in the examples, how many steps are awarded on each of the next five turns?

◈ 2-39   The military draft lottery is to be conducted using a table of random numbers. The days of the year are numbered 1 to 366 consecutively (counting February 29 as day 60). The potential draftees are identified by their birthdates and will be eligible for consideration in the order in which their birthdays are selected.

Using 14001 to determine the starting point in the table and following the procedure used in the examples, identify the first three birthdays selected in the lottery.

2-40 A professor is constructing a five-answer multiple-choice examination. The five choices are designated 1, 2, 3, 4, 5 (or A, B, C, D, E). The correct answer is to be assigned to one of these five numbers for each question, each assignment being made according to the uniform distribution. The random number table is to be used to make the assignments. If 6, 7, 8, 9, 0 occur, these are converted to 1, 2, 3, 4, 5, respectively. Using 14001 to determine the starting point in the table and following the procedure used in the examples, determine the answers to the first 10 questions on the professor's examination.

◈ 2-41 A seating arrangement is to be constructed for 73 students in an elementary statistics class. Seats numbered 1 to 73 inclusive are to be used. A random seating order is to be determined from the table of random numbers. The registrar's alphabetically arranged official class list is used to identify the students with numbers. Using 64016 to determine the starting point in the table and following the procedure used in the examples, determine which numbered students will sit in seats 1 to 10 inclusive.

2-42 If X has the probability function given by formula (2-23) with $k = 10$, find the mean and variance of X.

2-6
THE HYPERGEOMETRIC PROBABILITY MODEL

One of the concepts we discussed in the last section was concerned with drawing a sample from a finite number of objects in such a way that every sample of the same size has the same probability of selection. A sample of this type was called a simple random sample and can be obtained by drawing items one at a time without replacement if for every drawing each of the remaining items has an equal chance of being selected. Now we wish to consider an important probability model based upon this kind of sampling.

Let us suppose that a sack of fruit contains six apples and four oranges. We may be interested in the probability of getting two apples and three oranges if we were to draw five pieces of fruit from the sack one at a time without looking. (We assume that for each draw every remaining piece of fruit has an equal chance of being selected.) As we have already demonstrated in Sec. 1-4, knowing how to count combinations makes the calculation of such a probability a fairly easy problem. The total number of ways the drawing can be accomplished is $\binom{10}{5}$, the number of combinations of 10 things taken 5 at a time. If the sample is a simple random sample, then each possible sample has probability $1 \Big/ \binom{10}{5}$ of being selected. In other words, the sample space for our experiment consists of $\binom{10}{5}$ points with each being assigned probability $1 \Big/ \binom{10}{5}$. Now all we have to do is count the number of sample points which correspond to the event of interest. Since 2 apples can be selected from 6 in $\binom{6}{2}$ ways and, with each of these selections, 3 oranges can be picked from 4 in $\binom{4}{3}$ ways, by the fundamental counting principle the total number of ways to draw 2 apples and 3 oranges is $\binom{6}{2}\binom{4}{3}$. Since we have agreed that all $\binom{10}{5}$ sample points should be assigned equal weights, the probability of drawing two apples and three oranges is

$$\frac{\binom{6}{2}\binom{4}{3}}{\binom{10}{5}} = \frac{(6!/2!\,4!)(4!/3!\,1!)}{10!/5!\,5!} = \frac{5}{21}$$

Before attempting to generalize the preceding discussion, we note that the following conditions characterize the situation:

(a) The result of each draw can be classified into one of two categories, say, success (apples) and failure (oranges).
(b) The probability of a success changes on each draw.   (2-29)
(c) Successive draws are dependent.
(d) The drawing is repeated a fixed number of times.

The fact that drawings are made *without replacement* changes (b) and (c) from the binomial situation. Obviously, if an apple is drawn the first time, the probability of drawing an apple the second time is reduced from an original 6/10 to 5/9. Since the outcomes of any draw after the first is affected by what has happened on preceding draws, the outcomes of drawings are dependent events. If the apple or orange is replaced after every draw and the fruit is mixed before the next draw, then the probability of a success remains constant, the results of drawings are independent, and the binomial model is the appropriate one to use.

The conditions satisfied in the fruit example can be generalized as follows:

(a) The drawing is made from N items.

(b) A simple random sample of size n is selected (which implies sampling without replacement).

(c) k of the N items have some characteristic (for example, they may be apples or they may be defectives). By subtraction $N - k$ of the items do not have the characteristic.

(2-30)

With these model assumptions we seek the probability of obtaining x "defectives" in n draws. The total number of ways to draw n items is $\binom{N}{n}$. Since simple random sampling is used, the sample space we use consists of $\binom{N}{n}$ points, one for each sample, each assigned probability $1 \big/ \binom{N}{n}$. To obtain exactly x defectives implies that $n - x$ nondefective items must be drawn from a total of $N - k$ available. The total number of ways to get exactly x defectives and $n - x$ nondefectives is $\binom{k}{x}\binom{N-k}{n-x}$ by the fundamental counting principle. Consequently, because of the equal-weight assignment, the modified mathematical definition of probability yields

$$p(N, n, k, x) = \frac{\binom{k}{x}\binom{N-k}{n-x}}{\binom{N}{n}} \qquad (2\text{-}31)$$

for the probability that x of the n items are defective. The three numbers N, n, k are constants, and $X =$ the number of defective items in the sample is a random variable. A random variable X having probability function (2-31) is said to have a hypergeometric distribution. In the apple-orange illustration we can make the identification $N = 10, n = 5, k = 6, x = 2$. In this case the possible values which X can assume are 1, 2, 3, 4, 5 or $n - (N - k) \leq x \leq n$ (we have to draw at least one apple). The alternative identification $N = 10, n = 5, k = 4, x = 3$ would work just as well with an orange being regarded as a defective. Now we see that the values that X can assume are $0 \leq x \leq k$. Thus x can be no smaller than the larger of the two numbers 0 and $n - (N - k)$, and it can be no larger than the smaller of k and n. Hence, X can take on all integer values such that

$$\text{maximum } [0, n - (N - k)] \leq x \leq \text{minimum } [k, n] \qquad (2\text{-}32)$$

[The limits (2-32) can always be determined without formulas as in the apple-orange example.]

As is the case with the binomial and negative binomial models, interesting problems usually require a sum of probabilities like (2-31). Appendix B3 gives both $p(N, n, k, x)$ and

$$P(N, n, k, r) = \sum_{x = \max[0, n - (N-k)]}^{r} p(N, n, k, x) = \Pr(X \leq r) \qquad (2\text{-}33)$$

for $N = 10$. Because of the importance of this model, particularly in industrial problems, an extensive table has been published by Lieberman and Owen [4], where both $p(N, n, k, x)$ and $P(N, n, k, r)$ are given to six decimal places for $N = 1(1)50(10)100$. (It would appear that the authors are not card players since $N = 52$ is missing from the table.) In order to use Appendix B3 (and the Lieberman-Owen table) it may be necessary to use the relationships

$$p(N, n, k, x) = p(N, k, n, x) \text{ and } P(N, n, k, r) = P(N, k, n, r) \qquad (2\text{-}34)$$

That is, n and k can be interchanged, a result easily verified by writing formula (2-31) in factorial form. (When using the Lieberman-Owen table with $N > 25$, additional "symmetries" found on page 4 of that publication may also be required.)

EXAMPLE 2-14

What is the probability of drawing two or fewer apples from a sack containing six apples and four oranges if five pieces of fruit are selected by simple random sampling?

Solution

The preceding discussion has indicated that the hypergeometric model is appropriate. We have already identified $N = 10$, $n = 5$, $k = 6$. Now, letting X be the number of apples in the sample, we want

$$\Pr(X \leq 2) = \Pr(1 \leq X \leq 2) = P(10, 5, 6, 2)$$
$$= P(10, 6, 5, 2)$$
$$= .261905 \qquad \text{from Appendix B3}$$

Without the table we compute

$$\frac{\binom{6}{1}\binom{4}{4}}{\binom{10}{5}} + \frac{\binom{6}{2}\binom{4}{3}}{\binom{10}{5}} = \frac{1}{42} + \frac{5}{21} = \frac{11}{42} = .261905$$

EXAMPLE 2-15

Ten vegetable cans, all the same size, have lost their labels. It is known that five contain tomatoes and five contain corn. If five are selected at random (simple random sample), what is the probability that all contain tomatoes? What is the probability that three or more contain tomatoes?

Solution

We check conditions (2-30).

(a) There are $N = 10$ items from which to draw.

(b) A simple random sample of size $n = 5$ is selected.

(c) $k = 5$ of the 10 cans are tomatoes.

Hence the hypergeometric model is appropriate and the probability that all 5 are tomatoes is

$$\Pr(X = 5) = p(10, 5, 5, 5) = .003968 \qquad \text{from Appendix B3}$$

where X is the number of cans of tomatoes in the sample. The probability of obtaining three or more cans of tomatoes is 1 minus the probability of drawing two or fewer. Thus we want

$$
\begin{aligned}
\Pr(X \geq 3) &= 1 - \Pr(X \leq 2) \\
&= 1 - P(10, 5, 5, 2) \\
&= 1 - .500000 \qquad \text{from Appendix B3} \\
&= .500000
\end{aligned}
$$

We observed that the binomial model is not appropriate when conditions (2-30) are satisfied because the probability of a success is not constant and results associated with successive draws are dependent. If N is large and n/N is small, then the probability of a success changes very little from draw to draw, and events associated with successive draws are practically independent. Consequently, one would expect that the binomial would give good approximations to hypergeometric probabilities in this situation. For example, if 3,000 out of 10,000 people in a community favor a proposal, then given any specific order of results for the first 10 people selected at random, the probability that the eleventh person selected favors the proposal is near .3 (actually between $2,990/9,990 = .2997$ if all 10 favored and $3,000/9,900 = .3003$ if all 10 opposed). In the potato problem of Example 2-5 we could argue that actually the hypergeometric model is appropriate. However, since the crop consists of millions of potatoes, for all practical purposes the binomial is a satisfactory model.

It has been verified by actual comparisons that the approximation

$$P(N, n, k, r) \cong \sum_{x=0}^{r} b\left(x; n, \frac{k}{N}\right) \tag{2-35}$$

is reasonably good if $n/N \leq .1$, $k \geq n$, $N > 50$ (for $N \leq 50$ we could use the Lieberman-Owen table). Hence the terms in the hypergeometric sum $\Pr(X \leq r)$ are replaced by the corresponding terms in the binomial sum $\Pr(X \leq r)$. If $k < n$, then interchange the roles of n and k in formula (2-35), getting

$$P(N, n, k, r) \cong \sum_{x=0}^{r} b\left(x; k, \frac{n}{N}\right) \tag{2-36}$$

and require $k/N \leq .1$.

EXAMPLE 2-16

Suppose an organization contains 90 people of whom 45 are Republicans. A simple random sample of 5 is selected to form a committee. What is the probability that 3 or fewer members of the committee are Republicans?

Solution

We check conditions (2-30).

(a) There are $N = 90$ items from which to draw.
(b) A simple random sample of size $n = 5$ is selected.
(c) $k = 45$ of the 90 people are Republicans.

Hence the hypergeometric model is appropriate and we want $Pr(X \leq 3) = P(90, 5, 45, 3)$ where X is the number of Republicans on the committee. This probability is not available in Appendix B3. Since $n/N = 5/90 < .1$, we can use approximation (2-35). We get

$$P(90, 5, 45, 3) \cong \sum_{x=0}^{3} b(x; 5, .50) = .81250 \qquad \text{from Appendix B1}$$

The Lieberman-Owen table (available in most university libraries) yields .819643 for the actual value. Hence for this problem the correct value differs from the approximated value of about .007. In many applications this would be sufficient accuracy.

EXAMPLE 2-17

A company receives a moderately priced piece of equipment in lots of size 100. A simple random sample of 10 pieces is selected and each piece tested. If the sample contains 0 or 1 defective pieces the lot is accepted. If the sample contains 2 or more defective pieces, all 100 items are tested before a decision is made. Assuming that the lot contains 5 defective pieces, what is the probability that all 100 will be inspected?

Solution

We check conditions (2-30).

(a) There are $N = 100$ items to draw from.
(b) A simple random sample of size $n = 10$ is selected.
(c) $k = 5$ of the 100 items are defective.

Hence the hypergeometric model is appropriate and we want

$$Pr(X \geq 2) = 1 - Pr(X \leq 1) = 1 - P(100, 10, 5, 1)$$

where X is the number of defectives in the sample. Again this probability is not available in our tables (but could be easily computed). Since $k < n$ and $k/N = 5/100 < .1$, we can use approximation (2-36). We get

$$P(100, 10, 5, 1) \cong \sum_{x=0}^{1} b(x; 5, .10) = .91854 \qquad \text{from Appendix B1}$$

Hence $\Pr(X \geq 2) \cong 1 - .91854 = .08146$. This time the Lieberman-Owen table yields .076857 and the approximation is off by less than .005.

If the exact probability is immediately available in a table, it is senseless to use an approximation. However, when the range of entries for existing tables is exceeded (or tables are not available) and the conditions for the use of an approximation are satisfied, then the approximate probabilities would be obtained and would be satisfactory for many applications.

With a little algebra it can be shown that if X has a hypergeometric probability function, then

$$E(X) = n \frac{k}{N} = np \tag{2-37}$$

and

$$\sigma_X^2 = \frac{nk(N - k)(N - n)}{N^2(N - 1)} = npq \frac{N - n}{N - 1} \tag{2-38}$$

We note the similarity to (2-6) and (2-7), the corresponding results in the binomial case. For large N the fraction $(N - n)/(N - 1)$ will be near 1 and (2-38) will be about the same as (2-7), a result not unexpected after our discussion concerning the binomial approximation to the hypergeometric.

EXERCISES

In each case discuss the appropriateness of the hypergeometric model. Use Appendix B3 when possible.

◈ 2-43   The names of five men and five women are written on slips of paper and placed in a hat.   After thorough mixing, four names are drawn. What is the probability that two names belong to men and two to women?   What is the probability of obtaining two or fewer men's names?

2-44   A 5-card hand is dealt from a well-shuffled deck of 52 cards.   What is the probability that it contains 3 aces?   What is the probability that 3 or more cards in the hand are aces?   What is the expected number of aces in a 5-card hand?

◈ 2-45   From a crate of 100 oranges, some of which are frozen, 10 are selected at random.   Suppose that it is necessary to cut open an orange to determine whether or not it is frozen.   In order to find the probability that 7 or more oranges out of the 10 are frozen if the box contains 20 frozen oranges, we could use hypergeometric tables. What tabled value would we seek? (The Lieberman-Owen table gives .000392.) Approximate the probability by using the binomial distribution. What conclusion is one apt to draw if the sample of 10 contains 7 frozen oranges?

2-46   In Exercise 2-45 find the mean and variance of X, the number of frozen oranges in the sample, if the box contains 20 frozen oranges.

◇ 2-47   A box of 10 screws contains 3 silver-colored screws and 7 black screws. If a simple random sample of size 4 is selected, what is the probability that all 3 silver screws are in the sample? What is the probability that the number of silver screws in the sample is 3 or fewer? 2 or fewer?

2-48   The game of sheepshead is played with an ordinary deck of cards after all cards with denominations 2, 3, 4, 5, and 6 have been removed. In other words, the sheepshead deck contains 32 cards. Of the kings in a deck of cards, 3 have two eyes, and 1 has one eye. An 8-card hand is dealt from a well-shuffled sheepshead deck. To find the probability that the hand contains 0 or 1 two-eyed kings by using tables, we would seek which tabled value? (The actual probability is .853226.) Approximate the probability by using the binomial distribution. The needed value is not in Appendix B1 but the calculation is easy.

◇ 2-49   An urn contains six red balls and four black balls from which a simple random sample of five balls is drawn. Let X be the number of red balls in the sample. Exhibit the probability distribution of X in table form.

2-50   A box of 10 flashbulbs contains 3 defectives which cannot be identified by observation. If a sample of 6 bulbs is removed from the box, what is the probability that the sample contains 2 or fewer defectives?

◇ 2-51   An auditor is investigating the accounts receivable for a particular firm. From a total of 500 accounts he selects a simple random sample of size 25 to be checked in detail. If 50 of the accounts contain errors, what is the probability that the auditor's sample will contain 5 or more of these with incorrect balances?

2-52   A psychologist has available 50 volunteers for a learning experiment. Of these 20 have IQs which are 120 or more. If he selects a simple random sample of size 10 for his experiment, what is the probability that 5 or more of his sample have IQs no less than 120?

◇ 2-53   A rabbit grower is interested in purchasing some breeding stock from another grower. He is offered a lot of 10 rabbits at an acceptable price but he does not want to buy if too many have a certain liver disease which cannot be detected without killing the animal. The two growers agree upon the following mutually acceptable plan: Two rabbits will be killed and inspected and the transaction will be completed if and only if no diseased livers are found. If one of the 10 rabbits has the ailment, what is the probability that the transaction will be made? If 5 of the 10 have the ailment, then what is the probability that the transaction will be made?

2-7
THE POISSON PROBABILITY MODEL

A model often used for computing probabilities associated with the number of " successes " occurring within a time interval of given length or within a region of space of given size is the Poisson probability model. As an illustration experience has shown that the Poisson is an excellent model to use for computing probabilities associated with the number of calls coming into a telephone switchboard during a period of time of fixed length. This information could be used to determine whether or not present facilities are adequate. We may know that the board can handle up to 20 calls per minute without being overtaxed. If the probability of 21 or more calls within a minute is sufficiently small, the board will provide satisfactory service. If the probability is not small, then perhaps more lines or operators are needed. Some other random variables for which the Poisson has been used to evaluate probabilities are

1   The number of automobile deaths per month in a large city
2   The number of bacteria in a given culture
3   The number of red blood cells in a specimen of blood
4   The number of meteorites located on an acre of desert land
5   The number of typing errors per page
6   The number of defects in a manufactured article
7   The number of atoms disintegrating per second from radioactive material
8   The number of buzz-bomb hits on a square mile of London in 1944
9   The number of calls an individual receives per day
10 The number of deaths from horse kicks per year for each army corps in the Prussian Army over a period of 20 years

   In each of the above situations the following assumptions seem reasonable:

(a)   Events (successes) that occur in one time interval (or region of space) are independent of those occurring in any other non-overlapping time interval (or region of space).

(b)   For a small time interval (or region of space) the probability that a success occurs is proportional to the length of the time interval (or size of the region of space).   (2-39)

(c)   The probability that two or more successes occur in a very small time interval (or region of space) is so small that it can be neglected.

By using the assumptions (2-39), it may be shown by a mathematical argument that the probability of exactly x successes occurring in a time interval of given length (or region of space of a given size) is

$$p(x; \mu) = \frac{e^{-\mu}\mu^x}{x!} \qquad x = 0, 1, 2, \ldots \qquad (2\text{-}40)$$

where $e = 2.71828 \cdots$ and μ is the average number of successes occurring in the time interval of given length (or in the region of given size). A random variable X which has the probability function (2-40) is said to have a Poisson distribution. Appendix B4 contains cumulative sums

$$P(r; \mu) = \sum_{x=0}^{r} p(x; \mu) = \Pr(X \leq r) \qquad (2\text{-}41)$$

for some values of μ. A fairly extensive table has been published by Molina [5]. His table gives $\Pr(X \geq r)$ to at least six decimal places for $\mu = .001(.001).01(.01).3(.1)15(1)100$. Another good table was published by the General Electric Company [1]. For $\mu < 2$ this table contains more entries than the Molina table.

It is not as easy as with some of the preceding models to convince oneself that the model assumptions (2-39) are satisfied. Sometimes it is obvious that the independence condition is not satisfied. For example, we might be tempted to use the Poisson to compute the probability distribution of the number of corn borers found in a hill of corn. A little reflection reveals that events (successes) are not independent, since insects are usually hatched in batches. As we have previously mentioned, however, models sometimes give fairly accurate probabilities even though some of the assumptions are not satisfied. To pass final judgment on the appropriateness of the model, one has to rely on accumulated experimental evidence.

EXAMPLE 2-18

If a person receives five calls on the average during a day, what is the probability that he will receive fewer than five calls tomorrow? Exactly five calls?

Solution

According to previous discussion, experience has shown that the Poisson model is appropriate for this situation. The average number of calls per day, $\mu = 5$, is given. Letting X be the number of calls the person will receive tomorrow, we need

$$\Pr(X \leq 4) = P(4; 5) = \sum_{x=0}^{4} p(x; 5) = .44049 \qquad \text{by Appendix B4}$$

to answer the first question. The probability of receiving exactly five calls is

$$\Pr(X \leq 5) - \Pr(X \leq 4) = P(5; 5) - P(4; 5)$$
$$= .61596 - .44049 \qquad \text{by Appendix B4}$$
$$= .17547$$

That is, the probability of receiving exactly five calls is equal to the probability of receiving five or fewer minus the probability of receiving four or fewer.

EXAMPLE 2-19

A secretary claims that she averages one error per page. A sample page is selected at random from some of her work, and five errors are counted. What is the probability of her making five or more errors on a page if her claim is correct?

Solution

Perhaps the errors made are not independent of one another. If one or two errors occurring relatively near each other disturb the secretary, she may be inclined to make more errors relatively soon. However, letting X be the number of errors on a page and assuming that the Poisson is appropriate, the probability is

$$\Pr(X \geq 5) = \sum_{x=5}^{\infty} p(x; 1) = 1 - \sum_{x=0}^{4} p(x; 1)$$
$$= 1 - P(4; 1)$$
$$= 1 - .99634 \qquad \text{by Appendix B4}$$
$$= .00366$$

In view of the small probability we might be inclined to conclude one of the following:

1. The Poisson model is correct as used and a near miracle has occurred.
2. The model is correct but the wrong average value μ has been used.
3. The model is incorrect.

Probably 2 is the most plausible.

Besides being a probability model in its own right, the Poisson is sometimes used to approximate binomial probabilities when n is large and p is close to 0 or 1. Of course, it would make no sense to approximate a binomial sum that can be read directly from the tables. The individual terms of the binomial are replaced with the corresponding terms from the Poisson with $\mu = np$. Thus, for example, we can write

$$\sum_{x=0}^{r} b(x; n, p) \cong \sum_{x=0}^{r} p(x; np) \qquad (2\text{-}42)$$

EXAMPLE 2-20

A life insurance company has found that the probability is .00001 that a person in the 40 to 50 age bracket dies during a year period from a certain rare noncontagious disease. If the company has 100,000 policyholders in this group, what is the probability that it must pay off more than four claims during a year because of death from this cause?

Solution

The binomial model is not unreasonable since

(a) A person either dies from the disease during the year or he does not.
(b) The records give $p = .00001$, which we shall assume is constant for each person in this group.
(c) Presumably, whether a person dies from the disease in no way affects what happens to another person in the group.
(d) The number of trials is $n = 100,000$.

Letting X be the number of claims, the probability of more than four claims is

$$\Pr(X \geqq 5) = \sum_{x=5}^{100,000} b(x; 100,000, .00001)$$

$$= 1 - \sum_{x=0}^{4} b(x; 100,000, .00001)$$

Since $\mu = np = 100,000(.00001) = 1$, we use the approximation

$$\sum_{x=0}^{4} b(x; 100,000, .00001) \cong \sum_{x=0}^{4} p(x; 1)$$

$$= .99634 \qquad \text{by Appendix B4}$$

Hence $\Pr(X \geqq 5) \cong 1 - .99634 = .00366$.

For a Poisson random variable X it can be shown that

$$E(X) = \mu \qquad (2\text{-}43)$$

so that we are justified in using the symbol μ for the constant appearing in the probability function. The variance is given by

$$\sigma_X^2 = \mu \qquad (2\text{-}44)$$

Thus the Poisson distribution enjoys the unusual property of having the same mean and variance.

EXERCISES

When it seems necessary to do so, comment upon the reasonableness of the Poisson model.

2-54　If a typist makes two errors per page on the average, what is the probability of her typing a page with no errors? With one error?

◇ 2-55　A city has, on the average, five traffic deaths per month. What is the probability that this average is exceeded in any given month?

2-56　A taxicab company has on the average 10 flat tires per week. During the past week they had 20. Assuming that the Poisson model is appropriate, what is the probability of having 20 or more flats during a week? Would you suspect foul play?

◇ 2-57　An intercontinental ballistic missile has 10,000 parts. The probability that each part does not fail during a flight is .99995, and parts work independently of one another. If any one part does not work the flight is a failure. What is the probability of a successful flight?

2-58　An automobile insurance company has found that the probability of paying off on a policy during a year is .001. What is the probability that a company has to pay 15 or more claims next year if it holds 10,000 policies?

◇ 2-59　A certain kind of cookie contains on the average three raisins. If a cookie of this type is selected to be eaten, what is the probability that it contains five or more raisins?

2-60　In setting type for galley proofs a printing company makes on the average two errors per page. If a page of its work is examined, what is the probability that it will contain five or more errors?

◇ 2-61　Suppose that a certain brand of automobile has on the average six major defects (those which require corrective work) when it is delivered to a dealer. What is the probability that a dealer receiving one automobile will get one with two or fewer major defects? Ten or more major defects?

2-8
THE STANDARD NORMAL PROBABILITY MODEL

All the random variables considered in Secs. 2-2 to 2-7 were discrete random variables and the models of those sections could be called *discrete probability models*. In Sec. 1-6 we introduced continuous random variables. At that time we observed that in order to calculate probabilities of events associated with a continuous random variable, one way to preceed was to (1) select an appropriate density function and (2) interpret areas under the curve generated by the density function as probabilities. The density function so selected can be called a *continuous probability model*. The model assumptions are merely that the chosen density function yields realistic probabilities for the random experiment under consideration.

In Sec. 1-6 we considered only the uniform density and the triangular density. Both were interesting in connection with the spinner problem but, unfortunately, have very limited usage in statistical applications. In this and in the next two sections we will discuss some continuous distributions which arise frequently in subsequent chapters. Whenever these models are used we will list the assumptions that must be satisfied if reasonable probabilities are to be obtained. (Frequently we can say more than "the chosen density seems to yield realistic probabilities.") In this chapter our main objectives are to become acquainted with these models and to learn how to evaluate probabilities from tables in Appendix B.

A random variable Z whose density function is

$$\phi(z) = \frac{1}{\sqrt{2\pi}} e^{-z^2/2} \qquad -\infty < z < \infty \tag{2-45}$$

is said to have a standard normal distribution. (Here e and π are well-known constants which to five decimal places are $e = 2.71828, \pi = 3.14159$.) It is instructive to note some of the geometric properties of the density function (2-45). The graph of $\phi(z)$ appears in Fig. 2-1. As z gets further away from $z = 0$, the curve approaches, but never touches, the horizontal axis. The curve is symmetrical about the vertical axis through $z = 0$. In other words, if the graph were folded along the vertical axis, the part of the curve to the right of the axis would fall on top of the curve to the left of the axis. As Fig. 2-1 indicates, the random variable Z can take on any

FIGURE 2-1

Curve for the standard normal density function showing tabled value p and table entry z_p. The total area under the curve is 1.

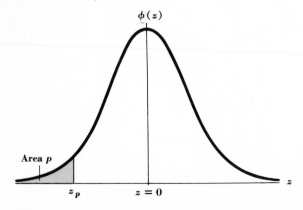

value between minus infinity and plus infinity. However, the tables which we shall discuss in the next paragraph reveal that practically all the area under the curve is between -3 and $+3$, so that the probability that Z is less than -3 or greater than $+3$ is very small. Since the total area under the curve is 1, the area on each side of the vertical axis is 1/2. Because of the symmetry it is necessary to give tabulations on only one side of the vertical axis. However, to make the use of Appendix B5 a little easier, entries are given for both positive and negative values of z_p.

Appendix B5 contains values of p such that

$$Pr(Z < z_p) = p \qquad (2\text{-}46)$$

This means that the probability that the random variable Z is smaller than some specified z_p is equal to p. The geometrical relationship between z_p and p is illustrated in Fig. 2-1. As an example, suppose we want to know the probability that Z is less than -1.12, that is, $Pr(Z < -1.12)$. Proceed down the left-hand column of the first page of Appendix B5 to $z_p = -1.11$, then across the row to the column under .02. We find $p = .1314$, so that $Pr(Z < -1.12) = .1314$.

EXAMPLE 2-21

Find the probability that Z is less than 1.96.

Solution

We look on the second page of Appendix B5 opposite the row 1.9 and under column .06. The table yields $Pr(Z < 1.96) = .97500$. We note that $Pr(Z < 1.96) + Pr(Z > 1.96) = 1$, since $Z > 1.96$ and $Z < 1.96$ are mutually exclusive events, one of which must happen $[Pr(Z = 1.96) = 0]$. Consequently, $Pr(Z > 1.96) = .02500$ by subtraction. These area relationships are indicated in Fig. 2-2.

FIGURE 2-2

Standard normal curve showing $Pr(Z < 1.96)$ and $Pr(Z > 1.96)$ as areas (Example 2-21).

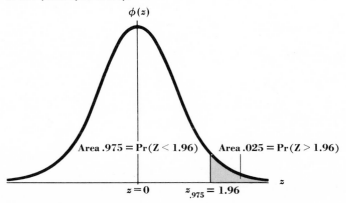

EXAMPLE 2-22

Find the probability that Z is between -1.12 and 1.96.

Solution

We seek the shaded area pictured in Fig. 2-3.   Since the events $Z < -1.12$ and $-1.12 < Z < 1.96$ are mutually exclusive,

$$\Pr(Z < -1.12) + \Pr(-1.12 < Z < 1.96) = \Pr(Z < 1.96)$$

In other words,

$$\Pr(-1.12 < Z < 1.96) = \Pr(Z < 1.96) - \Pr(Z < -1.12)$$
$$= .9750 - .1314$$
$$= .8436$$

It is geometrically obvious that the shaded area is obtained by subtraction.

FIGURE 2-3

Standard normal curve showing
$\Pr(-1.12 < Z < 1.96)$ **as an area (Example 2-22).**

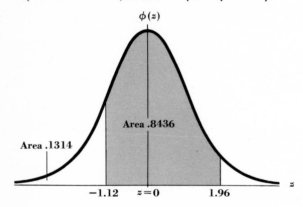

EXAMPLE 2-23

Find $z_{.95}$.

Solution

We want to find the value of z below which lies $.95$ of the area under the curve as indicated in Fig. 2-4.   From Appendix B5 we find

$$\Pr(Z < 1.64) = .9495 \qquad \Pr(Z < 1.65) = .9505$$

Since $.95$ is halfway between $.9495$ and $.9505$, we use $z_{.95} = 1.645$.

FIGURE 2-4

Standard normal curve showing $z_{.95}$ (Example 2-23).

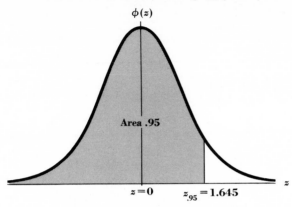

EXAMPLE 2-24

Find a value of Z, say z, such that $\Pr(-z < Z < z) = .95$.

Solution

We seek the shaded area in Fig. 2-5. Since the curve is symmetrical, the

FIGURE 2-5

Standard normal curve showing z such that $\Pr(-z < Z < z) = .95$ (**Example 2-24**).

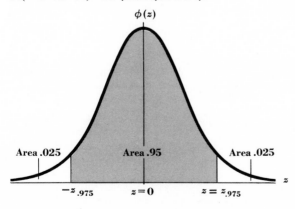

area is .025 under each tail of the curve. Hence the area below z is .975, and $z = z_{.975} = 1.96$ from Example 2-21.

One practical use of the standard normal is to approximate a sum of binomial probabilities.   If a random variable X has a binomial distribution with given values of n and p, then

$$Z' = \frac{X - np}{\sqrt{npq}} \qquad (2\text{-}47)$$

has approximately a standard normal distribution.   (This is a consequence of the Central Limit theorem, discussed in the next chapter.)   Specifically, this means that

$$\Pr(X \leq r) = \Pr\left(\frac{X - np}{\sqrt{npq}} \leq \frac{r - np}{\sqrt{npq}}\right)$$

$$= \Pr\left(Z' \leq \frac{r - np}{\sqrt{npq}}\right)$$

$$\cong \Pr\left(Z < \frac{r - np}{\sqrt{npq}}\right) \qquad (2\text{-}48)$$

Let us consider an illustration.

EXAMPLE 2-25

A symmetric six-sided die is thrown 1,000 times.   What is the probability that 170 or more 6s occur?

Solution

We have already discussed the adequacy of the binomial model in Example 2-1 for throwing a die a fixed number of times.   Here $n = 1,000$, $p = 1/6$, $q = 5/6$, and the exact probability is

$$\Pr(170 \leq X \leq 1,000) = \sum_{x=170}^{1,000} b(x; 1,000, 1/6)$$

where X is the number of 6s.   The exact probability can be read directly from the Harvard table [3] and is .40158.   To use the approximation, we write

$$\Pr(170 \leq X \leq 1,000) = \Pr\left[\frac{170 - 1,000(1/6)}{\sqrt{1,000(1/6)(5/6)}} \leq \frac{X - 1,000(1/6)}{\sqrt{1,000(1/6)(5/6)}}\right.$$

$$\left. \leq \frac{1,000 - 1,000(1/6)}{\sqrt{1,000(1/6)(5/6)}}\right]$$

$$= \Pr(.283 \leq Z' \leq 70.7)$$

$$\cong \Pr(.283 < Z < 70.7)$$

$$= \Pr(Z < 70.7) - \Pr(Z < .283)$$

$$= 1 - \Pr(Z < .283)$$

From Appendix B5 we find

$Pr(Z < .28) = .6103$ $Pr(Z < .29) = .6141$

Thus, to two decimal places $Pr(Z < .283) = .61$ and $Pr(170 \leqq X \leqq 1,000) \cong .39$.

If one has access to the good binomial tables mentioned in Sec. 2-2 (Harvard [3], Ordnance Corps [9]), the above approximation would rarely be used. It is, of course, ridiculous to go through the calculations of Example 2-25 to obtain an approximate result (perhaps not even correct in the second decimal place) when the exact result (to 5 or 7 decimal places) can be read from a table in a fraction of the time. Perhaps our best justification for working the example is to illustrate a practical use of the standard normal. (We will have much better illustrations later.) The above approximation would most likely be used when (1) n is larger than 1,000, so that the range of the Harvard table is exceeded, and (2) p is between .05 and .95. For smaller or larger p, the Poisson approximation demonstrated in Example 2-20 is apt to be better. A slight improvement on the approximation is usually obtained by decreasing the lower limit and raising the upper limit by $1/2$ in the binomial sum before doing the remainder of the arithmetic. (In Example 2-25 replace 170 by 169.5 and 1,000 by 1,000.5). However, for $n > 1,000$ the improvement is very slight and rarely worthwhile in a practical problem. In Example 2-25 the improved approximation yields a probability of .405.

In Sec. 1-6 we have already observed that the interpretations which we give to mean and variance are applicable to both continuous and discrete random variables. In the case of the standard normal it may be shown that $E(Z) = 0, \sigma_Z^2 = 1$.

EXERCISES

2-62   Use Appendix B5 to find (a) $Pr(Z < 1.37)$, (b) $Pr(-.67 < Z < 1.37)$, (c) $Pr(Z > 1.00)$, and (d) the probability that Z differs from 0 by at least 1.

◈ 2-63   Use Appendix B5 to find (a) $z_{.99}$, (b) $z_{.90}$, (c) $z_{.025}$, and (d) $z_{.05}$.

2-64   Use Appendix B5 to find z such that $Pr(-z < Z < z) = (a)$ .90, (b) .98, (c) .99.

◈ 2-65   Find (a) $Pr(.32 < Z < 1.65)$, (b) $Pr(-1.57 < Z < .68)$, and (c) $Pr(-2.32 < Z < -1.64)$.

2-66   A symmetric six-sided die is thrown 2,880 times. What is the probability that the number of 6s is between 450 and 500, inclusive?

◈ 2-67   A coin is to be thrown 1,000 times. What is the probability of getting 490 or more heads? Approximate the answer and compare it to the exact result of .74667.

2-68 Consider again Example 2-7. Suppose that 900 good potatoes are needed. Approximate the probability that more than 1,000 potatoes need to be cut open.

◈ 2-69 The random variable X has a binomial distribution with $n = 100$, $p = .10$. Approximate $\Pr(X \leq 12)$ both with and without adding 1/2 to the limits of the binomial sum, and compare the results with the exact probability given in Appendix B1.

2-9
THE NORMAL PROBABILITY MODEL

In Sec. 2-8 we discussed a random variable Z which has a standard normal distribution and observed that Z has mean $\mu = 0$ and variance $\sigma^2 = 1$. The density function (2-45) is a special case of

$$f(x; \mu, \sigma^2) = \frac{1}{\sigma\sqrt{2\pi}}\, e^{-(x-\mu)^2/2\sigma^2} \qquad -\infty < x < \infty \qquad (2-49)$$

A random variable X with density function (2-49) is said to have a normal distribution. It can be shown that the mean and variance of X are respectively μ and σ^2 so that the usage of these symbols is consistent with the notation we introduced in Sec. 1-5. If $\mu = 0$, $\sigma^2 = 1$, then X has a standard normal distribution since formulas (2-49) and (2-45) are the same (except that x is used in one case, z in the other).

The graph of the density function (2-49) is shown in Fig. 2-6. It is very similar to the one for the standard normal. The curve is symmetrical about the vertical axis through $x = \mu$. Hence μ determines the location of

FIGURE 2-6

Normal distribution with mean μ and standard deviation σ.

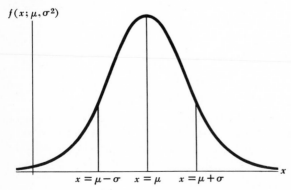

$f(x; \mu, \sigma^2)$

$x = \mu - \sigma$ $x = \mu$ $x = \mu + \sigma$

FIGURE 2-7

FIGURE 2-7

Effect of σ on shape of curve.

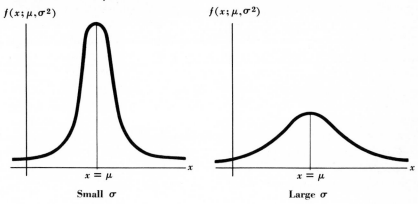

the center of the distribution.   Practically all the area under the curve is
between $\mu - 3\sigma$ and $\mu + 3\sigma$; thus, σ controls the shape of the curve.   If σ
is small, most of the area under the curve is near μ.   The larger σ, the more
the area is spread out (see Fig. 2-7).

It is often assumed that a random variable is normally, or approximately
normally, distributed.   Such an assumption may be supported by accumulated experimental evidence.   (Another justification appearing in
the next chapter is based upon the Central Limit theorem.)   Experience
has shown that the normal with $\mu = 0$ is a good model to use to describe
the behavior of aiming errors arising when darts are thrown at a vertical
line through the origin (see Fig. 2-8).   The random variable X is the miss

FIGURE 2-8

**The random variable X is the distance from the
target line to the dart.**

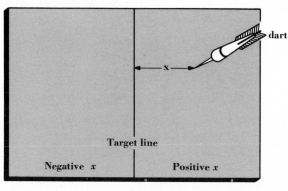

distance, that is, the distance from the target line to the dart. Other more practical examples of random variables that can be adequately characterized by a normal distribution include heights of individuals belonging to a certain definable population, length of machined parts turned out on a lathe, yields per acre of a variety of wheat, IQs of individuals, scores on college entrance examinations, net weights of certain kinds of packaged or canned food, blood pressures of healthy individuals belonging to a well-defined group (i.e., American males in the 30 to 40 year old bracket), scores produced by various psychological tests, and certain performance characteristics of manufactured items (i.e., the amount of electricity consumed in 1 hour by a 100-watt bulb).

Fortunately, no additional tables are needed to evaluate probabilities associated with a normal distribution. If X has a normal distribution, then it can be shown that

$$Z = \frac{X - \mu}{\sigma} \tag{2-50}$$

has a standard normal distribution. Thus, for example, we can write

$$\Pr(X < r) = \Pr\left(\frac{X - \mu}{\sigma} < \frac{r - \mu}{\sigma}\right)$$

$$= \Pr\left(Z < \frac{r - \mu}{\sigma}\right) \tag{2-51}$$

FIGURE 2-9

Geometric illustration of equation (2-51).

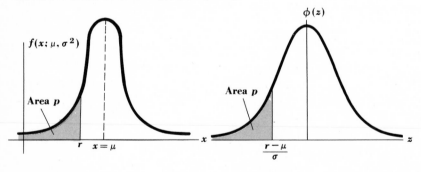

Geometrically, this means that a partial area under any normal curve is equivalent to a partial area under the standard normal (see Fig. 2-9), a quantity which can be read from Appendix B5.

EXAMPLE 2-26

Suppose that it is known that IQs for adult Americans are normally distributed with mean $\mu = 100$ and standard deviation $\sigma = 10$. Find the probability that a randomly selected individual has an IQ between 90 and 120. If an individual with an IQ of 130 or over is classified as a genius, what is the probability that a random selection yields a genius?

Solution

Let X be the individual's IQ. Then we want

$$\Pr(90 < X < 120) = \Pr\left(\frac{90 - 100}{10} < \frac{X - 100}{10} < \frac{120 - 100}{10}\right)$$

$$= \Pr(-1 < Z < 2)$$

$$= \Pr(Z < 2) - \Pr(Z < -1)$$

$$= .9773 - .1587 \qquad \text{by Appendix B5}$$

$$= .8186$$

To find the probability that we have selected a genius we need

$$\Pr(X > 130) = \Pr\left(\frac{X - 100}{10} > \frac{130 - 100}{10}\right)$$

$$= \Pr(Z > 3)$$

$$= 1 - \Pr(Z < 3)$$

$$= 1 - .99865 \qquad \text{by Appendix B5}$$

$$= .00135$$

The first figure obtained above can be interpreted as meaning that .8186 (or 81.86 percent) of adult Americans have IQs between 90 and 120. Similarly, we conclude that .00135 (or .135 percent) of adult Americans fall in the genius category.

EXAMPLE 2-27

The amount of electricity used by a certain brand of 100-watt light bulbs during a 10-minute period is normally distributed with mean 50 units and standard deviation 4 units. If a bulb uses 59 or more units of electricity in the 10-minute period, it is classified as a defective. What is the probability that a bulb selected for testing will be classified as a defective?

Solution

Let X be the number of units of electricity used by a 100-watt light bulb in a 10-minute period. Here $\mu = 50$ and $\sigma = 4$. We seek

$$\Pr(X > 59) = \Pr\left(\frac{X - 50}{4} > \frac{59 - 50}{4}\right)$$

$$= \Pr(Z > 2.25)$$

$$= 1 - \Pr(Z < 2.25)$$

$$= 1 - .98778 \qquad \text{by Appendix B5}$$

$$= .01222$$

Alternatively, we can interpret .01222 (or 1.222 percent) as the fraction of defective 100-watt bulbs produced under the brand name.

In comparing the normal model to the " real-life " situations of Examples 2-26 and 2-27, we find a discrepancy worthy of comment. In the model the random variable X can assume values from minus infinity to infinity, while in the problems that is obviously impossible. For example, the IQ scale of measuring usually begins with zero as a smallest possible value. If so, it is meaningless to consider events like $X < -10$ or $X < 0$. However, as is frequently the case (and as we have observed before), the model will not perfectly characterize the situation but is only a reasonable approximation to it. The event $X < 0$ is possible in the model but $\Pr(X < 0)$ is practically zero. In the real-life situation $\Pr(X < 0)$ is exactly zero. To be strictly correct, we must admit that the normal model only approximately describes the behavior of all the random variables mentioned in this section. The approximation is, however, satisfactory for most practical purposes.

EXERCISES

2-70 With the information given in Example 2-26, find the probability that a randomly selected individual has an IQ greater than 115.

◈ 2-71 Suppose that a machine turns out bolts whose diameters are normally distributed with mean $\mu = 1/2$ inch and standard deviation $\sigma = .01$ inch. What is the probability that a randomly selected bolt has a diameter between .48 and .52 of an inch? If bolts are useless (and called defectives) when the diameter is less than .47 of an inch or greater than .53 of an inch, what fraction of the bolts produced by the machine are defective?

2-72 When a well-known variety of wheat is used in a certain area, wheat yields are normally distributed with a mean of $\mu = 35$ bushels per acre and standard deviation $\sigma = 8$ bushels per acre. What fraction of yields will be in excess of 50 bushels?

◇ 2-73 A company that sells frozen shrimp prints "Contents 12 Ounces" on the package. Actually, contents of packages are normally distributed with a mean of 12 ounces and a standard deviation of .5 of an ounce. If a package of shrimp is purchased by a customer, what is the probability that he receives 11.5 ounces or less? 12.98 ounces or more?

2-74 An expert bridge team has found that when participating in national tournaments their average score per session is 60.0 percent and their scores have a standard deviation of 4.00 percent. If their scores are normally distributed, what is the probability that in a given session they score 65 percent or more?

◇ 2-75 In the second spinner example of Sec. 1-6 we used a triangular distribution as a model. For that distribution it was given that $\mu = .50$, $\sigma^2 = 1/24$. Using the triangular model we found $Pr(.50 < X < .75) = .375$. Now, suppose that we use the normal model with the same mean and variance as the triangular model. What is $Pr(.50 < X < .75)$?

2-76 A large university gives entrance examinations to all prospective students. The scores have a normal distribution with an average of 500 and a standard deviation of 100. What fraction of the scores exceed 580? If the university can only accommodate 40 percent of the students and the 40 percent with the highest scores are chosen, what is the score that separates the successful applicants from the unsuccessful?

◇ 2-77 Suppose that the amount of vaccine required to immunize human beings against smallpox is normally distributed with mean .250 of an ounce and standard deviation .040 of an ounce. Increasing the dosage increases the chances of a successful vaccination. What is minimum dosage required to produce success in 99 percent of the cases?

2-78 A psychologist raises rats to be used in his research. He has found that the length of time required for a rat to reach maturity is normally distributed with a mean of 240 days and a standard deviation of 30 days. What fraction of the rats require 270 or more days to reach maturity? At what age are 95 percent of the rats mature?

2-10
THE t, CHI-SQUARE, AND F
PROBABILITY MODELS

In this section we will consider three other important continuous probability models. Since we shall have no use for the density function formulas, they will not be given. Applications are topics of later chapters, and at that time we will discuss model assumptions. For the present we will be concerned with graphs of the density functions and the use of tables.

The density function for the t distribution contains a constant v, called the degrees of freedom. In statistical applications v will be a known integer. For every different value of v, we have a different t distribution. A typical graph for a t density function is given in Fig. 2-10. It looks very much like a standard normal curve, being symmetric about the vertical axis through 0. In comparing the two, the area under the t curve is larger in tails, smaller in the middle. The larger the value of v, the closer the t curve comes to matching the standard normal curve. In fact, with $v = \infty$ the two curves coincide.

FIGURE 2-10

A typical t curve showing tabulated value $t_{v;\,p}$.

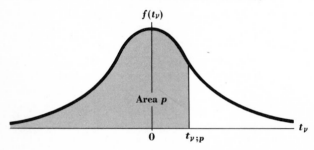

For the random variable we will use the symbol T_v, the subscript indicating the degrees of freedom. The table in Appendix B6 contains values associated with 38 different t distributions. Specifically the table gives $t_{v;\,p}$ such that

$$\Pr(T_v < t_{v;\,p}) = p \qquad (2\text{-}52)$$

for some $p > .50$. The geometrical relationship between p and $t_{v;\,p}$ is indicated in Fig. 2-10. Since the distribution is symmetrical about 0, we observe that

$$t_{v;\,p} = -t_{v;\,1-p} \qquad (2\text{-}53)$$

a relationship which we can use for $p < .50$ (as demonstrated in Example 2-29). Finally, we observe that the bottom row of the table ($v = \infty$) gives values z_p, defined by formula (2-46), for the standard normal. The latter fact means that it is often easier to find z_p for a given p from the t table than from the standard normal table. For example $z_{.90} = t_{\infty;.90} = 1.282$, which is read directly from the table and requires no interpolation.

Another good table of $t_{v;\,p}$ has been published by Owen [6]. He gives this quantity to four decimal places for $p = .75, .90, .95, .975, .99, .995$ and $v = 1(1)100(2)150, 200(100)1000, \infty$.

EXAMPLE 2-28

Find $t_{10;.975}$ and $t_{\infty;.975}$.

Solution

In Appendix B6 under $p = .975$ and opposite $v = 10$ we find $t_{10;.975} = 2.228$. Thus $\Pr(T_{10} < t_{10;.975}) = .975$. Similarly we get $t_{\infty;.975} = 1.960$. We observe that $1.960 = z_{.975}$ as found in Example 2-21.

EXAMPLE 2-29

Find $t_{10;.025}$.

Solution

Because of the symmetry of the distribution which leads to formula (2-53), we have

$$t_{10;.025} = -t_{10;.975} = -2.228$$

This relationship is illustrated in Fig. 2-11.

FIGURE 2-11

Symmetry property of t curve. The area under the curve to the left of $t_{10;.025}$ is equal to the area to the right of $t_{10;.975}$.

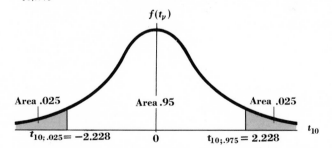

EXAMPLE 2-30

Find $t_{10;p}$ such that $\Pr(-t_{10;p} < T_{10} < t_{10;p}) = .90$.

Solution

The geometric relationships are shown in Fig. 2-12. Both tails under the density must contain an area of .05. Hence, since 95 percent of the area is below $t_{10;p}$, $p = .95$, $t_{10;.95} = 1.812$, $-t_{10;.95} = t_{10;.05} = -1.812$.

FIGURE 2-12

t curve showing $t_{10;.95}$ and illustrating that
$\Pr(-t_{10;.05} < T_{10} < t_{10;.95}) = .90.$

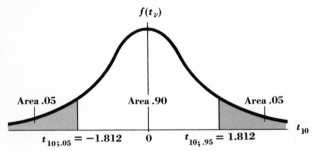

The density function for the chi-square distribution, like the t distribution, depends upon an integer-valued constant v, again called the degrees of freedom and again a known number in applications.   A typical curve for a chi-square density (except for $v = 1, 2$) appears in Fig. 2-13.   We will use Y_v as the symbol for a chi-square random variable.   The graph indicates that Y_v can assume only nonnegative values; that is, all the area under the curve is to the right of 0.   Further, the curve approaches but never touches the horizontal axis, indicating that Y_v can assume infinitely large values.   Unlike the t and normal curves, there is no symmetry about a vertical axis.

Appendix B7 contains values associated with 37 different chi-square distributions.   The table gives $\chi^2_{v;\,p}$ such that

$$\Pr(Y_v < \chi^2_{v;\,p}) = p \tag{2-54}$$

for some values of p.   The best table of $\chi^2_{v;\,p}$ has been prepared by Harter [2].   He gives this quantity to six decimal places for 23 values of p and $v = 1(1)150(2)330.$

FIGURE 2-13

A typical chi-square curve showing tabulated value
$\chi^2_{v;p}.$

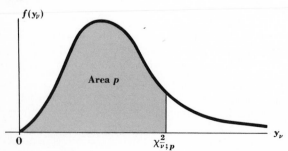

EXAMPLE 2-31

Find $\chi^2_{10;.05}$ and $\chi^2_{10;.95}$.

Solution

From the row of the table in Appendix B7 labeled $v = 10$ we read $\chi^2_{10;.05} = 3.94$ and $\chi^2_{10;.95} = 18.31$. Thus

$$\Pr(Y_{10} < 3.94) = .05 \qquad \Pr(Y_{10} < 18.31) = .95$$

We also have

$$\Pr(3.94 < Y_{10} < 18.31) = .90$$

as illustrated in Fig. 2-14.

FIGURE 2-14

Chi-square curve showing area relationships of Example 2-31.

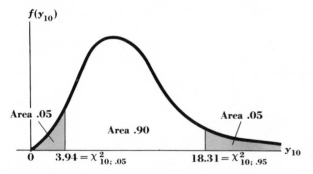

The table in Appendix B8, entitled "Chi-square Divided by Degrees of Freedom," yields essentially the same kind of information available in Appendix B7. Specifically, the latter table gives $\chi^2_{v;p}/v$ such that

$$\Pr\left(\frac{Y_v}{v} < \frac{\chi^2_{v;p}}{v}\right) = p \tag{2-55}$$

Obviously, if $\chi^2_{v;p}$ is available then $\chi^2_{v;p}/v$ can be easily found and vice versa. If there is any advantage of one form of tabulation over the other, it would be due to the fact that $\chi^2_{v;p}/v$ approaches 1 as v increases, making interpolation easy for large degrees of freedom.

EXAMPLE 2-32

Find $\chi^2_{10;.05}/10$ and $\chi^2_{10;.95}/10$ using Appendix B8.

Solution

With the table it is easy to verify that $\chi^2_{10;.05}/10 = .394$, $\chi^2_{10;.95}/10 = 1.83$. We observe that these are the same numbers one gets by dividing $\chi^2_{10;.05} = 3.94$ and $\chi^2_{10;.95} = 18.31$ by 10.

The F probability model has a density function depending upon two integer-valued constants, say ν_1 and ν_2, both called degrees of freedom and both known in statistical applications. A typical curve for an F density (except for $\nu_1 = 1, 2$) looks very much like a chi-square density (see Fig. 2-15). We will use the symbol F_{ν_1,ν_2} to denote a random variable with an

FIGURE 2-15

A typical F curve showing tabulated value $f_{\nu_1, \nu_2; p}$.

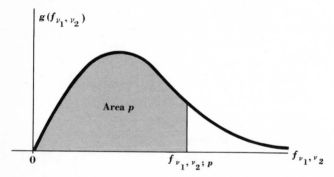

F density (or F distribution). As with chi-square, the random variable can assume values between zero and infinity. Appendix B9 contains $f_{\nu_1, \nu_2; p}$ where

$$\Pr(F_{\nu_1, \nu_2} < f_{\nu_1, \nu_2; p}) = p \tag{2-56}$$

for 17 choices of p and a number of combinations of ν_1 and ν_2. Our table requires a total of 14 pages and contains values associated with 480 different F distributions.

EXAMPLE 2-33

Find $f_{10,5;.05}$ and $f_{10,5;.95}$.

Solution

Here $v_1 = 10$, $v_2 = 5$ and values for $f_{10,5;p}$ appear on the third page of Appendix B9. With no difficulty we should read $f_{10,5;.05} = .301$, $f_{10,5;.95} = 4.74$. Obviously, we also have $\Pr(.301 < F_{10,5} < 4.74) = .90$.

In conclusion we remark that all our tables given in Appendix B1 and B3 to B9 give information about the cumulative distribution of a random variable. If we let X be the random variable associated with the model, then for the binomial, hypergeometric, Poisson, and normal cases, the tables give $\Pr(X \leq r)$ for various values r which are given or chosen. For the t, chi-square, and F distributions the tables give values of r such that $\Pr(X \leq r) = p$ for some selected values of p. One main advantage of giving all tabulations in terms of the cumulative distribution is uniformity. Having learned how to read one table of the cumulative form, we can then read every table given in the cumulative form (including others not yet discussed).

EXERCISES

◈ 2-79　Find (a) $t_{20;.99}$, (b) $t_{15;.05}$, (c) $t_{30;.50}$.

2-80　Find $t_{20;p}$ so that it satisfies $\Pr(-t_{20;p} < T_{20} < t_{20;p}) = .95$.

◈ 2-81　Find (a) $\chi^2_{50;.01}$, (b) $\chi^2_{50;.99}$.

2-82　Find (a) $\chi^2_{50;.01}/50$, (b) $\chi^2_{50;.99}/50$ from Appendix B8. Divide the answers for Exercise 2-81 by 50 and verify that the same results are obtained.

◈ 2-83　Find (a) $\chi^2_{24;.40}$, (b) $\chi^2_{24;.70}$. (*Hint:* Use Appendix B8.)

2-84　Find (a) $\chi^2_{300;.10}$, (b) $\chi^2_{500;.95}$.

◈ 2-85　Find (a) $f_{5,10;.01}$, (b) $f_{5,10;.99}$, (c) $f_{40,40;.50}$.

2-86　Use Appendix B6 to find (a) $z_{.99}$, (b) $z_{.05}$, and (c) z_p which makes $\Pr(-z_p < Z < z_p) = .99$.

REFERENCES

1　General Electric Company, Defense Systems Department: "Tables of the Individual and Cumulative Terms of Poisson Distribution," D. Van Nostrand Company, Inc., Princeton, N.J., 1962.

2　Harter, H. Leon: "New Tables of the Incomplete Gamma-function Ratio and of Percentage Points of the Chi-square and Beta Distributions," Superintendent of Documents, U.S. Government Printing Office, Washington, D.C., 1964.

3　Harvard University Computation Laboratory: "Tables of the Cumulative Binomial Probability Distribution," Harvard University Press, Cambridge, Mass., 1955.

4　Lieberman, G. J., and D. B. Owen: "Tables of the Hypergeometric Probability Distribution," Stanford University Press, Stanford, Calif., 1961.

5   Molina, E. C.: "Poisson's Exponential Binomial Limit," D. Van Nostrand Company, Inc., Princeton, N.J., 1949.

6   Owen, D. B.: "Handbook of Statistical Tables," Addison-Wesley Publishing Company, Inc., Reading, Mass., 1962.

7   Rand Corporation: "A Million Random Digits with 100,000 Normal Deviates," The Free Press of Glencoe, New York, 1955.

8   Robertson, William H.: "Tables of the Binomial Distribution Function for Small Values of p," Sandia Corporation, Albuquerque, N.M., 1960. (Available from the Office of Technical Services, Department of Commerce, Washington, D.C.)

9   U.S. Army Ordnance Corps: "Tables of Cumulative Binomial Probabilities," Ordnance Corps Pamphlet ORDP20-1, September 1952. (A new edition was published in January 1971 by U.S. Army Materiel Command, same title, pamphlet AMCP706-109.)

10   Weintraub, Sol: "Tables of the Cumulative Binomial Probability Distribution for small values of p," The Free Press of Glencoe, New York, 1963.

Summary of Results

I. The binomial probability model (Sec. 2-2)

The probability function of X, the number of successes in n trials, is

$$b(x; n, p) = \binom{n}{x} p^x q^{n-x} \qquad x = 0, 1, 2, \ldots, n \tag{2-2}$$

The binomial conditions (2-1) must be satisfied. Appendix B1 gives

$$B(r; n, p) = \Pr(X \leqq r) = \sum_{x=0}^{r} b(x; n, p) \tag{2-3}$$

To enter the table with $p > 1/2$, interchange roles of p and q, success and failure, or use the formula

$$B(r; n, p) = 1 - B(n - r - 1; n, q) \tag{2-5}$$

For a binomial random variable

$$E(X) = np \tag{2-6}$$

$$\sigma_X^2 = npq \tag{2-7}$$

II. The negative binomial probability model (Sec. 2-3)

The probability function of Y, the number of trials required to achieve c successes, is

$$b^*(y; c, p) = \binom{y-1}{c-1} p^c q^{y-c} \qquad y = c, c+1, \ldots \tag{2-9}$$

The negative binomial conditions (2-8) must be satisfied. Negative binomial sums can be evaluated from binomial tables since

$$B^*(r; c, p) = \Pr(Y \leqq r) = \sum_{y=c}^{r} b^*(y; c, p) \tag{2-10}$$

$$= \Pr(X \geqq c) = \sum_{x=c}^{r} b(x; r, p) \tag{2-11}$$

$$= 1 - \sum_{x=0}^{c-1} b(x; r, p) \tag{2-13}$$

For a negative binomial random variable

$$E(Y) = \frac{c}{p} \tag{2-14}$$

$$\sigma_Y^2 = \frac{cq}{p^2} \tag{2-15}$$

III. The multinomial probability model (Sec. 2-4)

The probability function of $X_1, X_2, \ldots, X_k$, the number of outcomes in categories $C_1, C_2, \ldots, C_k$ for a series of experiments satisfying the multinomial conditions (2-16), is

$$f(x_1, x_2, \ldots, x_k) = \frac{n!}{x_1! \, x_2! \cdots x_k!} p_1^{x_1} p_2^{x_2} \cdots p_k^{x_k} \qquad (2\text{-}18)$$

where $x_1 + x_2 + \cdots + x_k = n$. Each of $X_1, X_2, \ldots, X_k$ by itself has a binomial distribution. For each X_i

$$E(X_i) = np_i \qquad (2\text{-}21)$$

$$\sigma_{X_i}^2 = np_i(1 - p_i) \qquad (2\text{-}22)$$

IV. The uniform probability model (Sec. 2-5)

The probability function of X, the uniform random variable, is

$$f(x; k) = \frac{1}{k} \qquad x = 1, 2, \ldots, k \qquad (2\text{-}23)$$

or, slightly more generally,

$$f(x; k) = \frac{1}{k} \qquad x = x_1, x_2, \ldots, x_k \qquad (2\text{-}24)$$

We defined $X_1, X_2, \ldots, X_k$ to be a random sample from a distribution if

(a) each of the X's has the same distribution and
(b) each experiment which yields a value of X is performed   (2-25)
    independently of all the others.

In more practical terms we could replace (2-25) by

(a) each of the n experiments is performed under the same
    conditions or as nearly so as possible

and   (2-26)

(b) each experiment yields a result which in no way depends
    upon or influences the result of any other experiment.

If n objects are drawn (without replacement) from k objects in such a way that every sample of size n has an equal chance of being selected, then the sample so chosen is called a simple random sample. The table of random numbers (Appendix B2) can be used to draw random samples, in particular simple random samples. For X having probability function (2-23)

$$E(X) = \frac{k + 1}{2} \qquad (2\text{-}27)$$

$$\sigma_X^2 = \frac{k^2 - 1}{12} \qquad (2\text{-}28)$$

V. The hypergeometric probability model (Sec. 2-6)

The probability function of X, the hypergeometric random variable, is

$$p(N, n, k, x) = \frac{\binom{k}{x}\binom{N-k}{n-x}}{\binom{N}{n}} \qquad (2\text{-}31)$$

Conditions (2-30) must be satisfied, and x must be such that the combination symbols make sense. Appendix B3 gives

$$P(N, n, k, r) = \Pr(X \leqq r) \qquad (2\text{-}33)$$

To approximate sums of hypergeometric probabilities we may use

$$P(N, n, k, r) \cong \sum_{x=0}^{r} b\left(x; n, \frac{k}{N}\right) \qquad (2\text{-}35)$$

if $k \geqq n, \dfrac{n}{N} \leqq .1, N > 50$ or

$$P(N, n, k, r) \cong \sum_{x=0}^{r} b\left(x; k, \frac{n}{N}\right) \qquad (2\text{-}36)$$

if $k < n, \dfrac{k}{N} \leqq .1, N > 50.$

For the hypergeometric random variable

$$E(X) = np \qquad (2\text{-}37)$$

$$\sigma_X^2 = npq \frac{N-n}{N-1} \qquad (2\text{-}38)$$

where $p = k/N$.

VI. The Poisson probability model (Sec. 2-7)

The probability function of X, the Poisson random variable, is

$$p(x; \mu) = \frac{e^{-\mu}\mu^x}{x!} \qquad x = 0, 1, 2, \ldots \qquad (2\text{-}40)$$

The Poisson conditions (2-39) should seem reasonable, but we are apt to rely on experience to justify use of the model. Appendix B4 gives

$$P(r; \mu) = \sum_{x=0}^{r} p(x; \mu) = \Pr(X \leqq r) \qquad (2\text{-}41)$$

To approximate binomial sums for small values of p we can use

$$\sum_{x=0}^{r} b(x; n, p) \cong \sum_{x=0}^{r} p(x; np) \qquad (2\text{-}42)$$

which is particularly good if $p \leq .05$ and fairly good for $.05 < p \leq .10$. For the Poisson random variable

$$E(X) = \mu \tag{2-43}$$

$$\sigma_X^2 = \mu \tag{2-44}$$

VII. The standard normal probability model (Sec. 2-8)

The density function of Z, the standard normal random variable, is given by formula (2-45). This graphs into a bell-shaped curve symmetric about the origin. Appendix B5 gives values of p such that

$$\Pr(Z < z_p) = p \tag{2-46}$$

for various z_p. To approximate binomial sums for $.10 < p < .90$ we can use

$$\Pr(X \leq r) = \sum_{x=0}^{r} b(x; n, p)$$

$$\cong \Pr\left(Z < \frac{r - np}{\sqrt{npq}}\right) \tag{2-48}$$

VIII. The normal probability model (Sec. 2-9)

The density function of X, the normal random variable, is given by formula (2-49) and depends upon two constants μ and σ. The graph is bell-shaped, it is symmetric about the vertical line through $x = \mu$, and its shape is determined by σ. Frequently use of the model is justified on the basis of accumulated experience. Probabilities are obtained from Appendix B5 since

$$\Pr(X < r) = \Pr\left(Z < \frac{r - \mu}{\sigma}\right) \tag{2-51}$$

and the latter probability is obtained as described in VII. For the normal random variable

$$E(X) = \mu$$

$$\sigma_X^2 = \sigma^2$$

IX. The t, chi-square, and F probability models (Sec. 2-10)

The density function of T_v, the t random variable, has a graph that is similar to the standard normal curve, being symmetric about the origin (but having more area under the "tails"). Appendix B6 gives $t_{v; p}$ such that

$$\Pr(T_v < t_{v; p}) = p \tag{2-52}$$

As v increases $t_{v; p}$ gets closer to z_p, as can be verified from the tables. Because of the symmetry property we have

$$t_{v; p} = -t_{v; 1-p} \tag{2-53}$$

The density of Y_v, the chi-square random variable, plots into a curve which starts at 0, rises to a maximum, and then approaches 0. Appendix B7 gives $\chi^2_{v;\,p}$ such that

$$\Pr(Y_v < \chi^2_{v;\,p}) = p \tag{2-54}$$

and Appendix B8 gives $\chi^2_{v;\,p}/v$ such that

$$\Pr\left(\frac{Y_v}{v} < \frac{\chi^2_{v;\,p}}{v}\right) = p \tag{2-55}$$

The density of $F_{v_1,\,v_2}$, the F random variable, plots into a curve similar to the one obtained for chi-square. Appendix B9 gives $f_{v_1,\,v_2;\,p}$ such that

$$\Pr(F_{v_1,\,v_2} < f_{v_1,\,v_2;\,p}) = p \tag{2-56}$$

3

STATISTICS, SAMPLING DISTRIBUTIONS, AND POINT ESTIMATION OF PARAMETERS

3-1
INTRODUCTION

In the preceding chapter we introduced a number of very useful probability models. The examples and exercises usually required the evaluation of a probability. Before making any computations, we not only attempted to justify the use of the model but also identified certain constants which appear in the probability function (or density function). These constants, called parameters, were essential in order to do calculations or to enter tables. For any given problem they are fixed numbers, but they may (and usually do) take on different values when another problem is considered. The parameters encountered in Chap. 2 were n and p with the binomial; c and p with the negative binomial; $n, p_1, p_2, \ldots, p_k$ with the multinomial; k with the uniform; N, n, and k with the hypergeometric; μ with the Poisson; μ and σ^2 with the normal; v with the t and chi-square; and v_1 and v_2 with F. When we use the term *parameter*, usually we will have in mind a constant which appears in the probability function or density function associated with the model. In statistical literature "parameter" is frequently used in a slightly broader context to mean any constant associated with a probability distribution whether or not it appears in a formula. In this sense there exists an unending list of parameters for every distribution of which the most interesting are usually those that qualify under the first definition and μ and σ^2. The context of the discussion will generally indicate which of the two meanings is appropriate.

For the probability problems of Chap. 2 all parameters were known. In many situations of practical interest this is not the case. For the potato crop mentioned in Example 2-5 it is quite likely that no one will know the fraction of potatoes which are good. We would probably have general agreement that the binomial is a reasonable model to use for calculating probabilities associated with the number of good potatoes in a sample of size 25. Choosing an appropriate value of p, however, may be the subject of some debate and worthy of investigation. The information about the new cure for tuberculosis given in Exercise 2-7 suggests that the fraction p of patients cured is not the same as for the old standard cure. Perhaps investigators would like to draw some inferences about p. In Example 2-17 we assumed that a lot of 100 items contained 5 defective pieces and used the hypergeometric to calculate the probability that a simple random sample of size 10 items selected from the lot contains 2 or more defectives. In a real-life problem the actual number of defectives in the lot would be unknown. It is unlikely that a secretary such as the one mentioned in Example 2-19 would know μ, the average number of errors per page. Even if she did, a prospective employer may expect some evidence to support or disprove any claim concerning μ she may make. In the dart example associated with Fig. 2-8, we may believe that the miss distance X is normally distributed. We may even accept the statement that the mean of the distribution is zero. However, before we could make any practical use of these assumptions we would have to know σ^2 for a given situation. In general this parameter would be unknown. We could probably believe that bolts turned out by the machine of Exercise 2-71 have diameters which are normally distributed. However, in order to make reasonable statements about the fraction of defective bolts produced, we would have to face the fact that μ and σ^2 would usually be unknown.

The discussion of the preceding paragraphs indicates one major difference between probability and statistical inference. In probability problems parameters are known constants, while in statistical problems they are unknown constants about which inferences are drawn.

To be more specific concerning the types of inferences to be considered, let us again use the potato problem for an illustration. Sometimes we may be interested in obtaining an estimate of the parameter. If we are going to purchase a large shipment of potatoes, our estimate of p may help determine the price we are willing to pay. Two kinds of estimates are in common usage. The first, called a *point estimate*, yields one value which is our best guess for the parameter based upon available information. As a point estimate for p, the fraction of good potatoes, we may get, for example, $\hat{p} = .81$. The second kind of estimate is called an *interval estimate* and yields a range such as .76 to .85 which we think includes or captures the true but unknown p. This chapter will include some point estimates, but interval estimation will be discussed in later chapters. Another major type of statistical inference is based upon *hypothesis testing*. By this we will mean that a decision is made between two different statements regarding parameters. For example, suppose that a potato wholesaler claims that .90 of this potatoes are good. We might believe him, but it is possible that we would require some evidence to support the claim before we accepted it as the truth. Stated another way, we would try to decide if the "hypothesis" that $p = .90$ is reasonable, or alternatively, if the "hypothesis" $p < .90$ is reasonable. Beginning in the next chapter hypothesis testing will be discussed in considerable detail.

Inferences about parameters are based upon one or more observed values of a random variable which has a probability distribution depending upon the unknown parameters. If we are interested in the binomial parameter p, as in the potato problem, we can perform a series of n experiments in such a way that the binomial conditions (2-1) seem to be reasonably well satisfied. Then, inferences can be based upon x, the observed value of the random variable X, the number of successes in n experiments, whose distribution depends upon p. Generally our procedures will require that a random sample X_1, X_2, ..., X_n be observed. [It would be helpful to reread (2-25), (2-26), and the accompanying discussion.] Then, inferences will be made using some function of these variables (usually an intuitively appealing choice) such as the sum $Y = X_1 + X_2 + \cdots + X_n$ or the sum divided by n, say $\overline{X} = (X_1 + X_2 + \cdots + X_n)/n$. These functions are themselves random variables, called statistics. In other words, a statistic is a function of one or more random variables. In terms which are perhaps more meaningful, we could say that a *statistic* is a random variable which is observed for the purpose of drawing inferences about unknown parameters. The observed value of a statistic is determined from observed values of a random sample.

One of the standard problems of probability theory is to derive the distributions of statistics. These derivations are based upon the assumptions that the sampling is random and that a specific probability model is appropriate. Although we will omit the mathematical details, we will stress the underlying assumptions and do our best to see that they are satisfied when we intend to use a given statistic. Failure to do so may result in drawing inferences which are worthless and even ridiculous.

The probability distribution of a statistic is sometimes called a *sampling distribution*. The binomial can be regarded as the sampling distribution of the number of successes in n trials of a series of experiments satisfying the binomial conditions (2-1). The negative binomial can be regarded as the sampling distribution of the number of trials required to achieve c successes when the probability of a success is constant from trial to trial and trials are independent. The hypergeometric can be regarded as the sampling distribution of the number of good items in a simple random sample of size n drawn from a population of N items containing k good ones. The Poisson can be regarded as the sampling distribution of the number of times an event happens in a given time interval (or volume of space) given that the average number of times the event happens is μ and assuming that the Poisson is a reasonable model. The standard normal, the t, the chi-square, and the F are sampling distributions for statistics to be encountered in this and some of the following chapters.

In the Introduction we made an attempt to define the field of statistics and observed that the definition was unsatisfactory. Now we are in position to improve upon our former statement. We will define *statistical inference* as a field of endeavor concerned with drawing conclusions about distributions by using observed values of random variables which are governed by these distributions. Most of the time the inferences will be drawn about one or more parameters. Although this is a much better definition than the previous one, it too will be more meaningful after some further experience is obtained in this and the following chapters.

3-2
POINT ESTIMATION OF
PARAMETERS FOR SOME DISCRETE
DISTRIBUTIONS

Suppose that a random sample of 25 potatoes selected from the previously mentioned crop yields 20 good ones. A reasonable and intuitively obvious choice for an estimate of p, the fraction of good potatoes, is the fraction observed in the sample which here is $20/25 = .8$. This suggests that to estimate p for a series of experiments satisfying the binomial conditions (2-1), we should perform n experiments, count the observed number of successes x, and use the fraction x/n to estimate p. Since x is an observed value of a random variable X, x/n is an observed value of a random variable X/n. According to our definition, both X and X/n are statistics. Further, the distribution of X which is binomial depends upon the parameter p.

In order to distinguish between the random variable used in estimation and its observed value, it is customary to refer to the former as an estimator and to the latter as an estimate. Thus, in the previous paragraph X/n is an estimator, x/n is an estimate.

Having proposed a reasonable estimate, it is natural to seek information about the behavior of the associated random variable. Does the estimator have good properties? If so, what are they? Can we find a "better" estimate? The investigation of such questions has received considerable attention in the literature of mathematical statistics. Here we will be content to quote some of those results.

For a binomial random variable X we know from formula (2-6) that $E(X) = np$. If the average value of X is np, our intuition tells us that the average value of X/n ought to be $np/n = p$. That is,

$$E\left(\frac{X}{n}\right) = p$$

[For a formal justification of $E(cX) = cE(X)$, where c is a constant, see Appendix A1 and, in particular, Example A1-9. The result follows immediately from the properties of sums and summation notation.]

In other words, the average value of X/n is p, the parameter we are seeking to estimate. This seems like a very desirable property for an estimator to have. A random variable is said to be an *unbiased estimator* of a parameter if the average or expected value of the random variable is equal to the parameter. The corresponding observed value is called an *unbiased estimate*.

In addition to being unbiased, a good estimate should tend to be close to the parameter that is being estimated. That is, the probability that the estimator is near the parameter should be large. The measure of closeness most commonly used is the variance. In general, we usually prefer to use unbiased estimators with variance as small as possible. These two concepts form the basis of the definition we shall use for a "best" estimator.

Now let us formalize the previous discussion. Suppose that we know that a random variable X has a given probability function (or density) depending upon an unknown parameter θ. Thus, we may know that X has a binomial distribution, but we do not know the correct value of p. Let T be a statistic that is a function of one or more random variables having the given probability function. If $E(T) = \theta$, then T is an *unbiased estimator* of θ. If in addition to being unbiased, $\sigma_T^2 = E[(T - \theta)^2]$ is less than or equal to the variance of any other unbiased estimator (in general there is more than one), then T is called the "best" estimator of θ. The quotation marks, which will be dropped henceforth, are used to indicate the arbitrariness of the definition. Others may choose to define best in a different way. We could, for example, say that an estimator is best if it has smaller variance than any other statistic used as an estimator for the unknown parameter, thus not requiring the estimator to be unbiased. (Also, an estimate which is best according to our definition may not exist. In such cases other intuitively appealing estimates can frequently be found. Exercise 3-14 illustrates a problem of this type.)

Returning to the binomial situation, it can be shown that

$$T = \frac{X}{n} \tag{3-1}$$

is the best estimator of p. In this case our intuition was very good.

Since the multinomial is a generalization of the binomial, we would expect very similar results. It may be shown that for this model $E(X_i/n) = p_i$ for $i = 1, 2, \ldots, k$, and hence

$$T_i = \frac{X_i}{n} \qquad i = 1, 2, \ldots, k \tag{3-2}$$

are unbiased estimates of $p_1, p_2, \ldots, p_k$.

Next consider the negative binomial model and the random variable Y which has probability function $b^*(y; c, p)$. Our intuition, which worked so well in the binomial case, would again tell us to use the number of successes divided by the number of trials, or c/Y. However, this estimator is not unbiased. It can be shown that the best estimator of p is

$$T = \frac{c-1}{Y-1} \tag{3-3}$$

Admittedly c/y will usually be very close to $(c-1)/(y-1)$, the best estimate.

In the case of the hypergeometric model with probability function $p(N, n, k, x)$, it is frequently of interest to estimate k (perhaps the number of defective items in a lot). Since we have from formula (2-37) that $E(X) = nk/N$, it follows immediately that $E(NX/n) = k$, so that

$$T = N\frac{X}{n} \tag{3-4}$$

is an unbiased (and best) estimator of the parameter k. Since k is necessarily an integer and $N(x/n)$ may not be, one may prefer to round the estimate to the nearest integer. We observe that formula (3-4) is just the number of items multiplied by the fraction of defectives found in the sample, an intuitively obvious choice. Also it is apparent, but perhaps less useful, to note that

$$T = \frac{X}{n} \tag{3-5}$$

is the best estimator of the fraction of defective items.

After performing an experiment (or series of experiments) and obtaining the observed values of the random variables, it is obviously a very easy problem to calculate estimates associated with estimators (3-1) to (3-5). (The more difficult problem is to perform the experiments in such a way that the model conditions seem reasonable.) Even though some of our estimators were labeled as "best," it might occur to us that we should make an investigation to determine just how good the estimate happens to be. One approach would be to evaluate the probability that the estimator differs from the parameter by no more than a specified amount, say d. In other words, find

$$\Pr(-d \leqq T - \theta \leqq d) = \Pr(\theta - d \leqq T \leqq \theta + d) \tag{3-6}$$

In general the probability (3-6) depends upon the parameters of the distribution and specifically upon the sample size. Let us illustrate with an example.

EXAMPLE 3-1

A coin is tossed 100 times and 44 heads are recorded. Find the best estimate of p, the probability of obtaining a head on single throw. Find the probability that the best estimator will differ from p by no more than .08 if $p = .40$, if $p = .50$, and if $p = .60$.

Solution

We recognize that the binomial conditions are satisfied and identify $n = 100$ with observed number of successes $x = 44$. With no difficulty the estimate is $t = x/n = 44/100 = .44$ which, according to formula (3-1) and the accompanying discussion, is a best estimate.

The probability (3-6) is

$$\Pr\left(-.08 \leqq \frac{X}{100} - p \leqq .08\right) = \Pr(100p - 8 \leqq X \leqq 100p + 8)$$

where $d = .08$. This probability obviously depends upon p. With $p = .40, .50, .60$ we get, respectively,

$$\Pr(32 \leqq X \leqq 48) = \sum_{x=32}^{48} b(x; 100, .40) = .95770 - .03985 = .91785$$

$$\Pr(42 \leqq X \leqq 58) = \sum_{x=42}^{58} b(x; 100, .50) = .95569 - .04431 = .91138$$

$$\Pr(52 \leqq X \leqq 68) = \sum_{x=52}^{68} b(x; 100, .60) = \sum_{x=32}^{48} b(x; 100, .40) = .91785$$

where the sums are evaluated from Appendix B1.

In the binomial case we can replace formula (3-6) by

$$\Pr\left(p - d \leqq \frac{X}{n} \leqq p + d\right) = \Pr(np - nd \leqq X \leqq np + nd) \tag{3-7}$$

where X has a binomial distribution with parameters n and p. Now suppose that for a given d we seek the minimum value of n such that the probability (3-7) is at least a high fraction γ. That is, we would like to have the inequality

$$\Pr[n(p - d) \leqq X \leqq n(p + d)] \geqq \gamma \tag{3-8}$$

be satisfied with the smallest possible value of n. The solution will depend on the value of p which is selected for the calculations. Since p is unknown, it would appear that a solution cannot be obtained. However, there are two reasonable alternatives at our disposal. First, on the basis of prior experience we may have a rough idea as to what p ought to be. In coin-tossing experiments perhaps p should be near .50 even for procedures that favor heads or favor tails. We might believe that $.40 \leqq p \leqq .60$ would cover all cases one is likely to encounter. In the potato example past experience may indicate that $p \geqq .80$. The results of

Example 3-1 suggest that the probability (3-7) assumes a minimum when $p = .50$ and increases as p gets further away from .50.   It can be shown that this is indeed the case.   Consequently, we can take for p the value closest to .50 in the range of reasonable values.   If n works for the worst possible case, it will work for any p.   This means take $p = .50$ in the coin problem and $p = .80$ in the potato example.   For the second alternative, we could always use $p = .50$ in inequality (3-8).   If n meets the requirement for .50, then the requirement will be satisfied for any p.   The disadvantage of the second alternative is that if the actual p is far from .50 (say .10 or .90), an unnecessarily large n will be obtained.

Having selected p, inequality (3-8) can be solved for minimum n by trial with a good binomial table.   We would increase n until the inequality is satisfied.   An approximate solution can be obtained by using the normal approximation to the binomial, formula (2-48).   To use that result we rewrite formula (3-7) as

$$\Pr\left(-\frac{nd}{\sqrt{npq}} \leq \frac{X-np}{\sqrt{npq}} \leq \frac{nd}{\sqrt{npq}}\right) \cong \Pr\left(-\frac{nd}{\sqrt{npq}} < Z < \frac{nd}{\sqrt{npq}}\right) \qquad (3\text{-}9)$$

and to make the latter probability $\geq \gamma$ we need $nd/\sqrt{npq} \geq z_{(1+\gamma)/2}$. (Figure 2-5 is appropriate with .95 being replaced by γ.)   The last inequality yields

$$n \geq \frac{pqz_{(1+\gamma)/2}^2}{d^2} \qquad (3\text{-}10)$$

EXAMPLE 3-2

For the coin problem of Example 3-1 find the minimum n such that the probability is at least .90 that the estimator differs from p by no more than .08.

Solution

Since $p = .50$ will yield a larger n than any other p and would have to be included in any reasonable range for p, we use $p = .50$.   We identify $d = .08$, $\gamma = .90$, $(1 + \gamma)/2 = .95$.   To obtain an approximate solution we use formula (3-10), getting

$$n \geq \frac{(.50)(.50)z_{.95}^2}{(.08)^2} = \frac{(.25)(1.645)^2}{.0064} = 105.7 \qquad \text{or} \quad n = 106$$

(To obtain an exact solution for inequality (3-8) would require a good binomial table (i.e., the Ordnance Corps table [3]).   By trial we would find the smallest n such that

$$\Pr(.42n \leq X \leq .58n) = \sum_{x=.42n}^{.58n} b(x; n, .50) \geq .90$$

From Example 3-1 we already know that for $n = 100$ we have $\Pr(42 \leq X \leq 58) = .91138$ so that $n = 100$ is large enough. With the Ordnance Corps table one finds with $n = 99$

$$\Pr(41.58 \leq X \leq 57.42) = \Pr(42 \leq X \leq 57) = .89264$$

It can be verified that no smaller n will satisfy the inequality. Hence the exact answer is $n = 100$. In most practical problems it would be immaterial as to whether we use 100 or 106.)

EXAMPLE 3-3

For the potato problem find the minimum n such that the probability is at least .90 that the estimator differs from p by no more than .08.

Solution

Suppose we agree that $p \geq .80$. In this range the value nearest to .50 is .80, our choice for p in either (3-8) or (3-10). Again $d = .08$, $\gamma = .90$, $(1 + \gamma)/2 = .95$. For formula (3-10) we get

$$n \geq \frac{.80(.20)(1.645)^2}{(.08)^2} = 67.65 \qquad \text{or } n = 68$$

[For the exact solution we need

$$\Pr(.72n \leq X \leq .88n) = \sum_{x=.72n}^{.88n} b(x; n, .80)$$

$$= \sum_{x=.12n}^{.28n} b(x; n, .20) \geq .90$$

With the Ordnance Corps table we would find for $n = 63$

$$\Pr(7.56 \leq X \leq 17.64) = \Pr(8 \leq X \leq 17) = .88697$$

and for $n = 64$

$$\Pr(7.68 \leq X \leq 17.92) = \Pr(8 \leq X \leq 17) = .91591$$

so that actually $n = 64$ meets the requirement. Again, in a practical problem the approximate solution is sufficiently accurate. Only if observations were extremely expensive would it pay to obtain the exact solution.]

Sample size problems for the other estimators of this section are very similar to the binomial case, and we shall omit those topics. With the above discussion one of our main objectives has been to illustrate that sample size considerations may be worthwhile. Even without that discussion it should be intuitively obvious that the estimate is improved (in the sense that it tends to be closer to the parameter) by increasing the sample size.

EXAMPLE 3-4

Suppose that we wish to estimate p, the fraction of defective items associated with a manufacturing process. In practice we would observe a random sample of n items (selected so that the binomial conditions seem realistic), count the number of defectives, and compute x/n. Since we do not have a process available for demonstration purposes, let us substitute the table of random numbers. In place of items we will use the first 40 numbers in the first row of Appendix B2 (row 00000). If we regard 0 or 1 as a defective item, what is the best estimate of p?

Solution

The manner in which the table of random numbers was constructed allows the binomial conditions to be satisfied. We have

(*a*) A one-digit number is either a 0 or 1 or it is not.
(*b*) The probability that any digit is either a 0 or 1 is constant with $p = .2$. (With a real machine we would not know p.)
(*c*) Any one-digit number has been determined independently of all others.
(*d*) Our series of experiments consists of $n = 40$ observations of one-digit numbers.

From the table of Appendix B2, row 00000, we read 10097 ... 09117 as the first 40 one-digit numbers. The number of 0s and 1s is 9, our value of x. The best estimate of p based upon our sample is $x/n = 9/40 = .225$.

EXERCISES

◊ 3-1 A four-sided unsymmetric die is rolled 100 times to estimate the probability p that the side labeled 1 falls on the bottom. If a 1 is counted 28 times, what is the best estimate of the unknown probability p? Find the probability that the best estimator will differ from p by no more than .10 if $p = .20$. If $p = .40$. If $p = .50$. Which value of p minimizes this probability, and what is that minimum value?

3-2 In Exercise 3-1 rework the latter part of the problem concerning the probabilities if the die is rolled 50 times instead of 100.

◈ 3-3   Consider again the four-sided unsymmetric die of Exercise 3-1. Suppose that it is reasonable to assume that p, the probability of a 1, is no less than .10 and no greater than .30. Use the approximation (3-10) to find the minimum n such that the probability is at least .95 that the estimator differs from p by no more than .05.

3-4   In Exercise 2-7 a standard cure for tuberculosis was successful 30 percent of the time. The probability p of a success for a new cure is apparently somewhat higher. It is desired to estimate p with the minimum n such that the probability is at least .90 that the estimator differs from p by no more than .10. If all we can say about the value of p is $p \geq .30$, what is the minimum n? Use the approximation (3-10).

◈ 3-5   In Exercise 2-19 the Air Force fired missiles until 4 successes were obtained, and 11 trials were required. The negative binomial model seemed to be appropriate. Use that information to find the best estimate of p, the probability that a missile is successful.

3-6   A production line process turns out bolts, a fraction of which require rework. To estimate this fraction bolts are selected at random (or in such a way that the model conditions seem realistic) until 10 requiring rework are obtained. If the tenth bolt requiring rework is found on the seventy-third selection, what is the best estimate of the fraction of bolts requiring rework?

◈ 3-7   A box contains 1,000 camera flashbulbs. A simple random sample of size 25 is selected, and each bulb in the sample is tested. If 2 defectives are found in the sample, find best estimates of the fraction of defectives in the box (including the sample) and the number of defectives in the box.

3-8   A simple random sample of 200 voters selected from a list of eligible voters in a community yields 110 Democrats. Find a best estimate of the proportion of Democrats in the community.

◈ 3-9   Consider again Exercise 2-28 concerning automobile accidents and days of the week. Now assume that $p_1, p_2, \ldots, p_7$, the fractions of accidents which occur on Sunday through Saturday, are unknown. Use the information given in that exercise to find unbiased estimates of the p's.

3-10   As in Example 3-4 use the table of random numbers in place of actual sampling to estimate p in a problem for which the binomial is the appropriate model. Start in column 1 of the row numbered 00010, read one-digit numbers horizontally from left to right and regard an 8 or 9 as a success. Using the first 50 numbers (all of row 00010), what is the best estimate of p?

◈ 3-11   Suppose that a simple random sample of size n is drawn from k items numbered consecutively from 1 to k where k is unknown. The size of the number selected on each draw has a uniform distribution with probabilities $1/k$, $1/(k-1), \ldots, 1/(k-n+1)$, respectively. If we let $X_{(n)}$ be the largest number appearing in the simple random sample of size n, it can be shown that the sampling distribution of $X_{(n)}$ has the probability function

$$g(x_{(n)}) = \frac{\binom{x_{(n)} - 1}{n - 1}}{\binom{k}{n}} \qquad x_{(n)} = n, n + 1, \ldots, k$$

and that the best estimate of k is $t = [(n + 1)/n]x_{(n)} - 1$. This estimate is of interest in the so-called locomotive problem. In this problem it is known that a railroad company numbers its locomotives in order from 1 to k. We do not know k, but after observing the numbers on n locomotives, which pass our observation point, we would like to estimate k. Suppose that the numbers of the first five locomotives that we see are 42, 33, 92, 25, 5. Based upon this sample, find the best estimate of the number of locomotives used by the railroad company. In order to make this estimate a realistic choice, what assumption would have to be fulfilled? (We can justify the estimate on an intuitive basis. There are some missing numbers below the smallest observed number, some between the smallest and second smallest, ..., some between the next to largest and the largest, and some above the largest. It seems reasonable to use for an estimate $x_{(n)}$ + the average number of missing numbers in the n intervals formed by the n observed numbers. Below $x_{(n)}$, there are $x_{(n)} - n$ missing numbers. On the average each interval contains $(x_{(n)} - n)/n$. Hence $t = x_{(n)} + (x_{(n)} - n)/n$, the above best estimate.)

3-12 In Exercise 3-11 the numbers 1 to k could be numbers which appear on a fleet of taxicabs owned by a company in a given city. If that is the case and we observe taxis numbered 35, 114, 73, 84 passing our position, what is the best estimate of the number of cabs operated by the company? What assumption must be made to make this estimate a realistic choice?

◇ 3-13 As a promotional scheme, a business establishment is giving away a $1,000 prize to the customer who comes closest to guessing the exact number of slips of paper in a large container. Each customer is told that the slips are numbered consecutively from 1 on up to the total number in the container. Then he is allowed to withdraw and examine five slips for the purpose of improving the accuracy of his guess. Suppose that the five numbers selected by a customer are 761, 127, 4,216, 2,743, and 5,520. What should he use for an estimate? (*Hint:* See Exercise 3-11.)

3-14 A lake contains an unknown number of fish, say N. In order to get an estimate of N, k fish are caught and tagged, and then released. After waiting for the tagged fish to disperse, a second sample of

n fish are caught, and x of them have tags. Under the assumption of simple random sampling, the probability of obtaining x tagged fish is the hypergeometric probability $p(N, n, k, x)$ where N is the only unknown. An intuitively reasonable choice for N is the value for which the probability of getting the observed value x is largest. (This is called a maximum likelihood estimate of N.) With a little algebra it can be shown that we should take for $\hat{N}$ the largest integer not exceeding nk/x. (The estimate is not unbiased.) Actually we can justify this estimate on an intuitive basis. If we assume that the fraction of tagged fish in the sample is the same as the fraction of tagged fish in the total population, we get $x/n = k/N$ which yields $\hat{N} = kn/x$. Suppose that 100 tagged fish were released into the lake. If later a sample of 200 fish was observed to contain 4 with tags, what is the estimate of N? (This procedure could also be used to estimate the number of animals, e.g., antelope, that inhabit a certain region.)

3-3
POINT ESTIMATION OF MEANS

Two of the probability models introduced in Chap. 2 depended upon the parameter $\mu = E(X)$, called the mean of the distribution. These were the Poisson and the normal, the first a discrete model and the second continuous. We have already cited a number of situations for which the Poisson or the normal is a reasonable model and indicated that in many applications μ would be unknown. If we let $X_1, X_2, \ldots, X_n$ be a random sample from a distribution [recall (2-25) and (2-26)], then one logical estimator to use for μ is

$$\bar{X} = \frac{X_1 + X_2 + X_3 + \cdots + X_n}{n} = \frac{1}{n}\sum_{i=1}^{n} X_i \tag{3-11}$$

called the *sample mean*. From observed values $x_1, x_2, \ldots, x_n$ we can calculate a specific number

$$\bar{x} = \frac{x_1 + x_2 + \cdots + x_n}{n} = \frac{1}{n}\sum_{i=1}^{n} x_i \tag{3-12}$$

which we will call the *observed sample mean*. For both the Poisson and the normal cases $\bar{x}$ is a best estimate. As with estimates of the previous section, the calculation of $\bar{x}$ is straightforward. Once more the more difficult problem is to ensure that the conditions of random sampling are satisfied.

Even if neither the Poisson nor the normal is appropriate, we may still be interested in the mean of the probability distribution. The chuckaluck game of Example 1-28 provides an illustration. With symmetric dice we had no difficulty in calculating the expected winnings. However, suppose that we suspect that unsymmetric (or loaded) dice have been substituted into the game. Now, we do not know the exact probability distribution of a player's winnings and cannot evaluate the expected winnings μ. The best we can do is to substitute an estimate based upon the results of a number of games. Again the sample mean is a reasonable estimator, and it is unbiased regardless of the type of distribution which governs the behavior of the random variable.

EXAMPLE 3-5

Suppose that 10 rolls of a six-sided die yield 6, 4, 4, 5, 2, 5, 3, 4, 6, 1. Find the observed sample mean.

Solution

We have

$$\bar{x} = \frac{6 + 4 + 4 + 5 + 2 + 5 + 3 + 4 + 6 + 1}{10} = \frac{40}{10} = 4$$

As long as the condition of the die is unchanged and rolls are independent, then the conditions of random sampling are satisfied. If the die is symmetric, then the mean of the distribution is 3.5 (Example 1-26). If the die is loaded, then μ is unknown. In either case 4 is an unbiased estimate of μ.

EXAMPLE 3-6

An individual records the number of telephone calls he receives per day. From his records he selects 5 days at random and on these days he received 6, 3, 8, 2, 5 calls respectively. Estimate the mean of the distribution.

Solution

In Chap. 2 we indicated that the Poisson is a reasonable model to use for the number of calls received per day. If the individual is at home very little on Saturday and Sunday, undoubtedly a different μ would be applicable for weekends. We could probably believe that the one value of μ holds for Monday through Friday. If so, and the above figures are for weekdays, the conditions of random sampling are satisfied. As our estimate of the weekday mean we get

$$\bar{x} = \frac{6 + 3 + 8 + 2 + 5}{5} = \frac{24}{5} = 4.8$$

EXAMPLE 3-7

In Example 2-27 we considered the amount of electricity used by a certain brand of 100-watt bulbs and decided that X = the number of units of electricity used in a 10-minute period had a normal distribution. If 8 bulbs are tested by burning each of them for 10 minutes and the amounts of electricity consumed are 54, 52, 45, 53, 51, 47, 47, 52, find the best estimate of the unknown mean μ.

Solution

The performance of any bulb is independent of the performance of any others. If the bulbs are selected in some way that seems to be random, we would probably believe that the conditions of random sampling are satisfied. The best estimate of μ is

$$\bar{x} = \frac{54 + 52 + 45 + 53 + 51 + 47 + 47 + 52}{8} = 50.1$$

Another unbiased estimator for the mean of a normal distribution is X_M = middle value in order of magnitude of a random sample X_1, X_2, ..., X_n. If n is odd there is one middle value while if n is even there will be two. In the latter case, we usually take for x_M the average of the two middle values. For example if the observed sample is 7, 2, 5, then $x_M = 5$, while if the observed sample is 7, 2, 5, 9 we get $x_M = (5 + 7)/2 = 6$. Since $\bar{X}$ is the best estimator, X_M will have the larger variance of the two.

When X_1, X_2, ..., X_n is a random sample from a normal distribution, it can be shown that the sampling distribution of $\bar{X}$ is also normal with the same mean but with variance $1/n$th as large as the original distribution. That is, for the $\bar{X}$ distribution we have

$$\mu_{\bar{X}} = E(\bar{X}) = E(X) = \mu \tag{3-13}$$

and

$$\sigma_{\bar{X}}^2 = \frac{\sigma_{\bar{X}}^2}{n} = \frac{\sigma^2}{n} \tag{3-14}$$

In fact, with a little algebra it can be shown that formulas (3-13) and (3-14) hold for a random sample from any distribution (but for a nonnormal distribution the distribution of $\bar{X}$ will not be normal). Although we will omit the derivation of these formulas, in the next section we will demonstrate that they hold for a specific situation. A graphical comparison of the original normal density function and the $\bar{X}$ density function appears in Fig. 3-1. Because the sample size appears in the denominator of formula (3-14), it is apparent that the larger n becomes the higher the probability that $\overline{X}$ is close to μ.

If a simple random sample of size n is drawn (without replacement) from N items, formula (3-13) still holds. However, formula (3-14) must be replaced by

$$\sigma_{\overline{X}}^2 = \frac{\sigma_X^2}{n}\left(\frac{N-n}{N-1}\right) \tag{3-15}$$

Most of our inferences will be based upon random samples from a distribution, which means that formula (3-14) will be the appropriate formula to use.

FIGURE 3-1

Densities of X and $\overline{X}$ for a normal model.

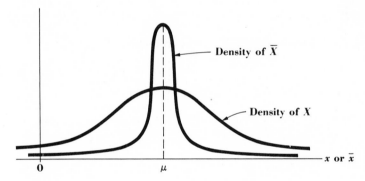

The evaluation of probabilities for $\overline{X}$ based upon random samples from a normal distribution can be accomplished by using formula (2-51). That is,

$$\Pr(\overline{X} < r) = \Pr\left(\frac{\overline{X}-\mu}{\sigma/\sqrt{n}} < \frac{r-\mu}{\sigma/\sqrt{n}}\right)$$

$$= \Pr\left(Z < \frac{r-\mu}{\sigma/\sqrt{n}}\right) \tag{3-16}$$

EXAMPLE 3-8

Consider again the distribution of IQs for adult Americans (Example 2-26). We assumed that the distribution was normal with $\mu = 100$ and $\sigma = 10$. If a random sample of 25 individuals is selected, what is the probability that the sample mean of their IQs is less than 99?

Solution

We have $\mu_{\bar{X}} = \mu_X = 100$, $\sigma_{\bar{X}} = \sigma/\sqrt{n} = 10/\sqrt{25} = 2$, $r = 99$. Hence, according to formula (3-16) we get

$$\Pr(\bar{X} < 99) = \Pr\left(Z < \frac{99 - 100}{2}\right)$$

$$= \Pr(Z < -.5)$$

$$= .3085 \qquad \text{by Appendix B5}$$

The fact that $\bar{X}$ is normally distributed when random samples from a normal distribution are used enables us to solve some sample size problems. For example, suppose we seek the minimum n such that the probability is at least γ that $\bar{X}$ differs from μ by no more than a small amount, say d units. In other words, n must satisfy

$$\Pr(-d < \bar{X} - \mu < d) \geqq \gamma \qquad (3\text{-}17)$$

for a given d and a given γ. We can rewrite inequality (3-17) as

$$\Pr\left(\frac{-d\sqrt{n}}{\sigma} < \frac{\bar{X} - \mu}{\sigma/\sqrt{n}} < \frac{d\sqrt{n}}{\sigma}\right) = \Pr\left(-\frac{d\sqrt{n}}{\sigma} < Z < \frac{d\sqrt{n}}{\sigma}\right) \geqq \gamma \qquad (3\text{-}18)$$

The latter inequality will be satisfied if $d\sqrt{n}/\sigma \geqq z_{(1+\gamma)/2}$. Solving for n we find

$$n \geqq \left[\frac{\sigma z_{(1+\gamma)/2}}{d}\right]^2 \qquad (3\text{-}19)$$

When σ is known (not usually the case in practice), no difficulty is encountered in using inequality (3-19) to find n. If σ is unknown, we could choose d in terms of σ units (instead of the original units of the random variable). For example, if we specify $d = (1/2)\sigma$, then $d/\sigma = 1/2$ (or $\sigma/d = 2$), and again all quantities on the right-hand side of inequality (3-19) are known.

EXAMPLE 3-9

In Example 2-27 the random variable X was the amount of electricity consumed in a 10-minute period by a certain brand of 100-watt bulbs. We assumed that X was normal with mean $\mu = 50$ and standard deviation $\sigma = 4$. Suppose as in Example 3-7 we wish to obtain a point estimate of μ now assumed unknown. Also let us suppose that we are willing to believe that $\sigma = 4$ (perhaps from results associated with an earlier version of this bulb). How large a sample do we need in order to make the probability at least .95 that $\bar{X}$ differs from μ by no more than 1 unit?

Solution

We identify $d = 1$, $\sigma = 4$, $\gamma = .95$, $(1 + \gamma)/2 = .975$. Then $z_{.975} = 1.960$, and we need

$$n \geq \left[\frac{4(1.96)}{1} \right]^2 = 61.47$$

The minimum sample size which meets the requirements is $n = 62$.

In the Poisson case it is almost always preferable to use the sampling distribution of $Y = X_1 + X_2 + \cdots + X_n = \sum_{i=1}^{n} X_i$, rather than distribution of $\overline{X}$ when drawing inferences about μ. While $\overline{X}$ has a nonstandard distribution, the random variable Y has a Poisson distribution with mean $n\mu$. That this should be the case is intuitively reasonable. If the number of telephone calls X that a person receives in a day has a Poisson distribution with an average of 2, then it seems logical to expect that the number of calls he receives in 5 days (also the number of successes in an interval of time, 5 times as long) has a Poisson distribution with an average 5 times as large, or $5(2) = 10$. Sample size problems in the Poisson case are similar to those of the binomial and will be omitted.

EXAMPLE 3-10

Under the assumption that an individual receives on the average 2 telephone calls per day, compute the probability that in 5 days selected at random he receives no more than 8 calls.

Solution

In Example 3-6 we discussed the requirements of random sampling. Assuming these are fulfilled, then $Y = $ sum of the number of calls received in 5 days has a Poisson distribution with mean $5(2) = 10$. We seek

$$\Pr(Y \leq 8) = \sum_{y=0}^{8} p(y; 10)$$

$$= .33282 \quad \text{by Appendix B4}$$

EXERCISES

⬦ 3-15 In Exercise 2-76 we discussed an entrance examination given to all prospective students at a university. The scores were normally distributed with mean 500 and standard deviation 100. Now suppose we consider another distribution, the one for scores of students actually admitted to the university. An observed random sample of 10 student records reveals that their scores were 545, 508, 446, 490, 771, 722, 498, 441, 659, 670. What is the best estimate of the mean score for admitted students?

3-16 During a 12-month period, the number of twin births per month recorded in a hospital are 2, 0, 1, 1, 0, 0, 3, 2, 1, 1, 0, 1. Suppose we believe that this is an observed random sample from a Poisson distribution. What is the best estimate of the average number of twin births per month?

◈ 3-17 The table of random digits yields the following observed random sample (taken from the numbers 1, 2, ..., 100): 10, 37, 8, 99, 12, 66, 31, 85, 63, 73. Find an unbiased estimate of the mean of the distribution from which this sample was taken. Use formula (2-27) to find the actual value of μ.

3-18 Large-size cans of tomato juice have "net weight 46 ounces" printed on the label. An observed random sample of 10 cans of a certain brand yields the following weights in ounces: 45.5, 44.7, 44.1, 44.5, 47.3, 46.8, 44.6, 45.2, 45.3, 45.0. What is the best estimate of the average weight per can?

◈ 3-19 A hybrid seed corn company claims that its product will yield on the average 120 bushels of corn to the acre. A random sample of 25 acres is found to have an observed sample mean of 115 bushels to the acre. If it is reasonable to assume that yields are normally distributed and that the standard deviation is 5 bushels per acre, the same as for other varieties, what is the probability of obtaining a sample average of 115 bushels or less? What conclusion might be drawn?

3-20 In Exercise 2-73 we were informed that the weight of the contents which is found in packages of frozen shrimp is normally distributed with a mean of 12 ounces and a standard deviation of .5 of an ounce. If 16 packages are purchased, what is the probability that the sample mean of these weights exceeds 12.25 ounces?

◈ 3-21 Consider again the bridge team of Exercise 2-74. Recall that their scores were normally distributed with a mean of 60.0 percent and a standard deviation of 4.00 percent. Suppose they participate in an event consisting of four sessions. If we assume that their four scores constitute a random sample of their results, what is the probability their sample mean is 62.00 percent or more?

3-22 Suppose that in Exercise 3-18 we are willing to assume that weights are normally distributed with a standard deviation of .5 of an ounce. What is the minimum sample size required in order to make the probability at least .90 that the sample mean differs from the true average contents by no more than .1 of an ounce?

◈ 3-23 In Exercise 3-15 we were interested in the distribution of entrance examination scores for students actually admitted to the university. Suppose that those scores are normally distributed but with unknown mean μ and unknown standard deviation σ. We wish to estimate μ by having the sample size large enough so that the probability is at least .90 that the estimator differs from μ by no more than .25σ. What is the minimum sample size required?

3-24 Let us assume that for a healthy individual of age 20, the number of years he can expect to live is a random variable with a normal distribution. Suppose we wish to estimate the mean of that distribution by examining past records of an insurance company. (Of course, the mean may change as living conditions change, but we will ignore this complication.) How large a sample of individual files need be examined if it is desired to have a probability of at least .99 that the estimator differs from the mean by no more than .1σ?

◇ 3-25 Recall Exercise 1-59 and the blindfolded individual who was told the names of four brands of cigarettes and asked to identify each. In that exercise we assumed that the individual possessed no discriminatory ability and found that the average number of correct identifications was 1. Now assume that the individual does have some talent for making correct identifications so that the probability distribution of that former exercise is no longer appropriate. To estimate μ the experiment is repeated 50 times. The observed results are 0 identifications 7 times, 1 identification 8 times, 2 identifications 29 times, and 4 identifications 6 times. What is an unbiased estimate of the mean of the unknown probability distribution?

3-26 If a typist makes 1 error per page on the average, what is the probability that a random sample of 10 pages of her work will contain more than 12 errors?

◇ 3-27 It has been found that a certain kind of cloth has on the average one defect for every 10 yards of material. Assuming that the number of defects per yard has a Poisson distribution, what is the probability that a purchase consisting of 50 yards contains three or fewer defects?

3-4
SOME CHARACTERISTICS OF THE SAMPLING DISTRIBUTION OF THE SAMPLE MEAN

In Sec. 3-3 we observed that the sample mean is a best estimator of μ in the normal and Poisson models. In addition, it was stated that $\overline{X}$ is an unbiased estimator for the mean of any distribution and that the variance of the $\overline{X}$ distribution is always $1/n$th the variance of the original distribution governing the behavior of each X [formulas (3-13) and (3-14)].

Now we will demonstrate that this latter statement is true and observe a remarkable property of the sampling distribution of $\overline{X}$.

Consider again the die random variable whose probability distribution is given in Table 1-1. Since the die under consideration was six-sided and symmetric, the appropriate model led to the probability function $f(x) = 1/6$, $x = 1, 2, 3, 4, 5, 6$, a special case of the uniform distribution. It is easy to verify by direct calculation that $E(X) = 3.5$ (Example 1-26) and $\sigma_X^2 = 35/12$, results which can also be obtained from formulas (2-27), (2-28) with $k = 6$. If the die is rolled twice, then the sample space which we would most likely consider (Exercise 1-8) consists of the 36 sample points appearing in Table 3-1, where the first number represents the outcome on the first roll and the second the outcome on the second roll. It

TABLE 3-1

Enumeration of all possible samples of size 2 obtained by rolling a die twice	1, 1	2, 1	3, 1	4, 1	5, 1	6, 1
	1, 2	2, 2	3, 2	4, 2	5, 2	6, 2
	1, 3	2, 3	3, 3	4, 3	5, 3	6, 3
	1, 4	2, 4	3, 4	4, 4	5, 4	6, 4
	1, 5	2, 5	3, 5	4, 5	5, 5	6, 5
	1, 6	2, 6	3, 6	4, 6	5, 6	6, 6

is reasonable to assign weights of 1/36 to each point in the sample space. The number pairs of Table 3-1 represent all possible observed random samples of size 2 that can be obtained by rolling a die. If the two numbers in each pair are added together and each sum divided by 2, we get the 36 sample means of Table 3-2. In other words, these are the 36 possible

TABLE 3-2

Sample means for the samples of Table 3-1	1.0	1.5	2.0	2.5	3.0	3.5
	1.5	2.0	2.5	3.0	3.5	4.0
	2.0	2.5	3.0	3.5	4.0	4.5
	2.5	3.0	3.5	4.0	4.5	5.0
	3.0	3.5	4.0	4.5	5.0	5.5
	3.5	4.0	4.5	5.0	5.5	6.0

values of $\bar{x}$ arising from Table 3-1. Since the weight 1/36 is associated with each sample point of Table 3-1, this weight is also associated with each $\bar{x}$ appearing in Table 3-2. From the latter table we can construct the sampling distribution of $\overline{X}$ which is presented in Table 3-3. The probabilities $h_2(\bar{x})$ are obtained by adding the weights associated with

TABLE 3-3

Sampling distribution of $\bar{X}$ obtained from Table 3-2

$\bar{x}$	1.0	1.5	2.0	2.5	3.0	3.5	4.0	4.5	5.0	5.5	6.0
$h_2(\bar{x})$	1/36	2/36	3/36	4/36	5/36	6/36	5/36	4/36	3/36	2/36	1/36

each value of $\bar{x}$. For example, $\Pr(\bar{X} = 4.0) = 1/36 + 1/36 + 1/36 + 1/36 + 1/36 = 5/36$.

The mean of the sampling distribution of $\bar{X}$ computed from Table 3-3 is

$$\mu_{\bar{X}} = E(\bar{X}) = 1.0(1/36) + 1.5(2/36) + 2.0(3/36) + 2.5(4/36) + 3.0(5/36)$$
$$+ 3.5(6/36) + 4.0(5/36) + 4.5(4/36) + 5.0(3/36) + 5.5(2/36)$$
$$+ 6.0(1/36)$$
$$= 126/36 = 3.5 = \mu_X = E(X)$$

By using the definition of variance on Table 3-3, we get

$$\sigma_{\bar{X}}^2 = (1.0 - 3.5)^2(1/36) + (1.5 - 3.5)^2(2/36) + (2.0 - 3.5)^2(3/36)$$
$$+ (2.5 - 3.5)^2(4/36) + (3.0 - 3.5)^2(5/36) + (3.5 - 3.5)^2(6/36)$$
$$+ (4.0 - 3.5)^2(5/36) + (4.5 - 3.5)^2(4/36) + (5.0 - 3.5)^2(3/36)$$
$$+ (5.5 - 3.5)^2(2/36) + (6.0 - 3.5)^2(1/36)$$
$$= \frac{35}{24} = \frac{35}{12}\left(\frac{1}{2}\right) = \frac{\sigma_X^2}{2}$$

Hence, for one relatively simple situation we have demonstrated a method of constructing the $\bar{X}$ distribution and verified that formulas (3-13) and (3-14) hold as previously claimed.

By following the above enumeration procedure we could construct the $\bar{X}$ distribution for samples of size 3, 4, 5, etc. Of course, the larger the sample size, the more tedious the enumeration becomes. For the case $n = 3$ the counterparts of Tables 3-1 and 3-2 contain 216 pairs and $\bar{x}$'s, respectively. Even so, it is not too difficult to verify that the probability distribution of $\bar{X}$ is the one given in Table 3-4. For the sampling

TABLE 3-4

Sampling distribution of $\bar{X}$ for a die random variable with $n = 3$

$\bar{x}$	1	4/3	5/3	2	7/3	8/3	3	10/3
$h_3(\bar{x})$	1/216	3/216	6/216	10/216	15/216	21/216	25/216	27/216
$\bar{x}$	11/3	4	13/3	14/3	5	16/3	17/3	6
$h_3(\bar{x})$	27/216	25/216	21/216	15/216	10/216	6/216	3/216	1/216

distribution of Table 3-4 it can be verified that $E(\overline{X}) = 3.5$, $\sigma_{\overline{X}}^2 = (35/12)$ $(1/3) = \sigma_X^2/3$.

As we already know, there is no point to constructing the sampling distribution (Tables 3-3 and 3-4 and others) for the purpose of finding $\mu_{\overline{X}}$ and $\sigma_{\overline{X}}^2$. Our formulas give us those results in terms of μ_X and σ_X^2. One practical use of such tables would be to evaluate probabilities for $\overline{X}$ such as $\Pr(\overline{X} \leqq r)$. We have performed enough calculations in this section to realize that for the die random variable and large n, say $n = 2{,}000$, the construction of the $\overline{X}$ distribution is impractical. (High-speed computers could be used, but undoubtedly such expensive machines ought to be engaged in more profitable exercises.) As an alternative, it seems reasonable to seek an approximation. The following discussion is directed toward that goal.

Let us consider the graphs of $f(x)$, $h_2(\overline{x})$, and $h_3(\overline{x})$ for the symmetric die problem. These appear in Figs. 3-2 to 3-4. Obviously, as n increases the number of points on the graph increases. The interesting observation to be made from the third graph is that if the dots were connected by drawing a smooth curve through them, that curve would look very much like a normal density function. If we were to draw curves

FIGURE 3-2

Probability function for a six-sided symmetric die.

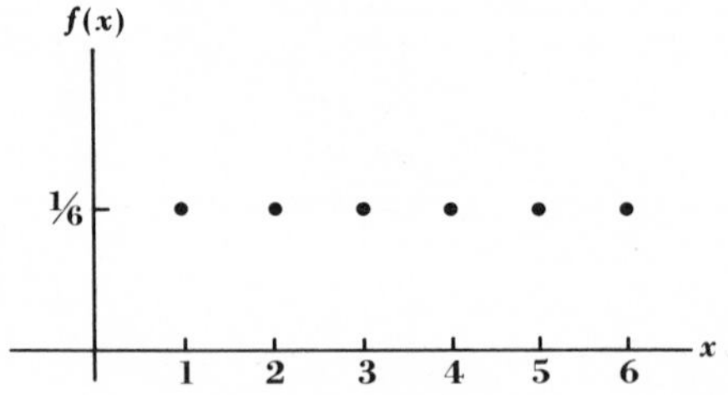

$f(x)$

FIGURE 3-3

Probability function of $\overline{X}$ for samples of size 2.

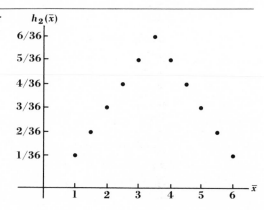

$h_2(\overline{x})$

FIGURE 3-4

Probability function of $\overline{X}$ for samples of size 3.

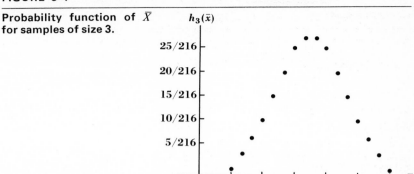

for $n = 4$, $n = 5$, etc., we would get pictures that look even more like a normal density. This might lead us to suspect that probabilities for $\overline{X}$ could be approximated by finding an area under a normal curve, provided n is too small. In other words, it appears that a normal model with the right μ and σ^2 could be used as an approximation to the actual model. This is indeed the case even though $\overline{X}$ is a discrete random variable which has nonzero probabilities only at isolated points while the normal random variable associates probabilities with intervals. Our suspicions are confirmed by one of the most remarkable theorems in mathematical literature called the *Central Limit theorem*. The theorem is as follows:

Let $X_1, X_2, \ldots, X_n$ be a random sample from a distribution with mean μ and variance σ^2. Then the distribution of

$$Z' = \frac{\overline{X} - \mu}{\sigma/\sqrt{n}} = \frac{\overline{X} - \mu_{\overline{X}}}{\sigma_{\overline{X}}} \tag{3-20}$$

approaches the standard normal distribution as n increases.

The approximation is good even for relatively small n in a number of situations and, of course, improves as n increases. It is impossible to give a minimum n which will hold for all distributions. In the die problem Fig. 3-4 suggests that here a good approximation may be obtained with $n = 3$. In other cases we may need an n of 20 or more. We recall from Sec. 3-3 that if the distribution of X is normal, no approximation is involved. In other words, then (3-20) has exactly a standard normal distribution for every n. Thus, perhaps we might expect that the more nearly normal the original distribution, the better the approximation provided by the Central Limit theorem.

One other observation is worth making. Both the statement of the Central Limit theorem and the discussion involving random samples produced by a die imply that the distribution of $\overline{X}$ is approximately normal with mean $\mu = \mu_{\overline{X}}$ and variance $\sigma_{\overline{X}}^2 = \sigma^2/n$.

EXAMPLE 3-11

A six-sided symmetric die· is rolled 105 times.   Approximate the probability that $\overline{X}$ is between 3 and 4 inclusive.

Solution

We have already found $\mu_X = \mu_{\overline{X}} = 3.5$ and $\sigma_X^2 = 35/12$.   For $n = 105$ we get $\sigma_{\overline{X}}^2 = \sigma^2/n = (35/12)(1/105) = 1/36$.   Then $\sigma_{\overline{X}} = 1/6$ and

$$\Pr(3 \leqq \overline{X} \leqq 4) = \Pr\left(\frac{3 - 3.5}{1/6} \leqq \frac{\overline{X} - 3.5}{1/6} \leqq \frac{4 - 3.5}{1/6}\right)$$

$$= \Pr(-3 \leqq Z' \leqq 3)$$

$$\cong \Pr(-3 < Z < 3)$$

$$= \Pr(Z < 3) - \Pr(Z < -3)$$

$$= .998650 - .001350 \qquad \text{from Appendix B5}$$

$$= .997300$$

Thus the desired probability is approximately .997.

EXAMPLE 3-12

In Example 3-10 we assumed that an individual receives on the average 2 telephone calls per day.   Using the Central Limit theorem, approximate the probability that in 50 days selected at random his sample average will exceed 2.25.

Solution

A Poisson random variable has mean μ and variance also equal to μ [formulas (2-43) and (2-44)].   Hence $\mu_{\overline{X}} = 2$, $\sigma_{\overline{X}} = \sigma_X/\sqrt{n} = \sqrt{\mu}/\sqrt{n} = \sqrt{2}/\sqrt{50} = 1/5$.   We want

$$\Pr(\overline{X} > 2.25) = \Pr\left(\frac{\overline{X} - 2}{1/5} > \frac{2.25 - 2}{1/5}\right)$$

$$= \Pr(Z' > 1.25)$$

$$\cong \Pr(Z > 1.25)$$

$$= \Pr(Z < -1.25)$$

$$= .1056 \qquad \text{from Appendix B5}$$

(The exact probability can be found in the General Electric table [1] and is .1072.)

The Central Limit theorem, as we have stated it, implies that each X_i has the same distribution. A slightly more general formulation of the theorem is sometimes used to argue that random variables arising from frequently encountered sampling situations are normally (or approximately normally) distributed. The more general statement is as follows:

Let $X_1, X_2, \ldots, X_n$ be independent random variables having distributions with means $\mu_1, \mu_2, \ldots, \mu_n$ and variances $\sigma_1^2, \sigma_2^2, \ldots, \sigma_n^2$. Let

$$Y = X_1 + X_2 + \cdots + X_n$$

$$\mu^* = \mu_1 + \mu_2 + \cdots + \mu_n$$

$$\sigma^{*2} = \sigma_1^2 + \sigma_2^2 + \cdots + \sigma_n^2$$

Then the distribution of

$$Z' = \frac{Y - \mu^*}{\sigma^*} \tag{3-21}$$

approaches the standard normal distribution as n increases.

The theorem implies that Y is approximately normally distributed with mean μ^* and standard deviation σ^*. It is then contended that some random variables can be considered as a sum Y of certain other random variables and that the more general formulation of the theorem justifies considering Y to be approximately normal. For example, we might argue that an individual's IQ is the result of a sum of variables affecting intelligence. These could include health, home background, emotional stability, climate, educational opportunities, etc. Admittedly, this is a reasonably sound argument, but when it comes to justifying the adequacy of a model, there is probably no substitute for experience.

We note that if $X_1, X_2, \ldots, X_n$ all have the same distribution so that $\mu_1 = \mu_2 = \cdots = \mu_n = \mu$ and $\sigma_1^2 = \sigma_2^2 = \cdots = \sigma_n^2 = \sigma^2$, then formula (3-21) reduces to

$$Z' = \frac{Y - n\mu}{\sigma\sqrt{n}} \tag{3-22}$$

But (3-22) is exactly the same as (3-20), as can be verified by multiplying numerator and denominator of (3-20) by n. It is sometimes more convenient to use (3-22) as we did in formula (2-47) when discussing the normal approximation to the binomial. With discrete random variables which can assume only integer values (i.e., the binomial and the Poisson), the approximation may be improved slightly by subtracting .5 from the lower limit on the sum and adding .5 to the upper limit. For these cases we can write

$$\Pr(r_1 \leqq Y \leqq r_2) = \Pr(r_1 - .5 \leqq Y \leqq r_2 + .5)$$

$$= \Pr\left(\frac{r_1 - .5 - n\mu}{\sigma\sqrt{n}} \leqq \frac{Y - n\mu}{\sigma\sqrt{n}} \leqq \frac{r_2 + .5 - n\mu}{\sigma\sqrt{n}}\right)$$

$$= \Pr\left(\frac{r_1 - .5 - n\mu}{\sigma\sqrt{n}} \leqq Z' \leqq \frac{r_2 + .5 - n\mu}{\sigma\sqrt{n}}\right)$$

$$\cong \Pr\left(\frac{r_1 - .5 - n\mu}{\sigma\sqrt{n}} < Z < \frac{r_2 + .5 - n\mu}{\sigma\sqrt{n}}\right) \qquad (3\text{-}23)$$

EXERCISES

3-28   From a table of random numbers, 33 numbers are selected from those between 1 and 10 inclusive. Hence, the probability function of X, the number appearing on any given selection, is (2-23) with $k = 10$ (the uniform distribution). The mean and variance are given by formulas (2-27) and (2-28). Approximate the probability that $\bar{X}$ for a sample of size 33 is greater than or equal to 6.

◆ 3-29   A game is played with a cup containing 15 six-sided symmetric dice. The cup is turned over so that all the dice fall onto the table. Then the random variable $Y = $ the sum of spots on the 15 dice is observed. Use formula (3-23) to approximate $\Pr(Y \geqq 60)$. (*Hint:* $r_1 = 60, r_2 = 90$.)

3-30   In Exercise 3-27 we considered a kind of cloth that on the average had $\mu = .1$ defects per yard. We assumed that the number of defects per yard had a Poisson distribution. Use formula (3-23) to approximate the probability that 1,000 yards will contain 95 or fewer defects. (Here $r_1 = 0$, $r_2 = 95$.) Suppose that $\mu = 1$, $n = 100$ (instead of $\mu = .1$, $n = 1,000$). Now approximate the probability that 100 yards contain 95 or fewer defects. (Observe that in using the normal approximation to the Poisson that it depends on $n\mu$. Thus if $n\mu = 100$, we get the same answer to the above problem for $n = 1,000$, $\mu = .1$, for $n = 100$, $\mu = 1$, and for $n = 1$, $\mu = 100$. This means that the accuracy of the approximation depends upon the size of $n\mu$, not on n. Incidentally, here $\sum_{x=0}^{95} p(x; 100)$ can be read from the Molina table [2] or the General Electric table [1], and the exact probability is .33119.)

◆ 3-31   Let X be the length of life in hours of a certain kind of television tube. Suppose it is known that on the average this kind of tube will last 1,000 hours, and the standard deviation of X is 100 hours. There is good evidence that the distribution of X is not normal. If 100 new tubes are purchased, what is the approximate probability that the sample mean will be less than or equal to 985 hours?

3-32   Let X be the maximum wind velocity recorded at the local weather station during a 24-hour day. Records reveal that $\mu_X = 15$ miles per hour and $\sigma_X = 15$ miles per hour. We are convinced that X does not have a normal distribution. Suppose that we decide to look up the maximum wind velocities recorded on the last 25 July 4's. Approximate the probability that the average of these 25 numbers will exceed 21 miles per hour.

◈ 3-33   Suppose that the average family income in the United States is $8,000 per year with a standard deviation of $2,000 per year. A random sample of 100 families is selected by a computer. If the average family income of such a sample were to be computed, what could we use for the probability that the sample average exceeds $8,300 if we believe that incomes are not normally distributed?

3-34   For the second spinner problem of Sec. 1-6 we proposed using the triangular density pictured in Fig. 1-14. We were given that $E(X) = .50$ and $\sigma_X^2 = 1/24$. Assuming that this is the correct density, approximate the probability that 24 spins produces a sample mean which is within .05 of the distribution mean.

◈ 3-35   Suppose that we wish to estimate the average age at death of heavy smokers. We believe that we can get a random sample from available hospital records and from records maintained in several other places. We also think that it is highly unlikely that X, the age at death of a heavy smoker, has a normal distribution. Although the standard deviation of X is unknown, experience of earlier investigators indicates that it is no more than 10 years, a figure we decide to use. How large a sample of records should be inspected in order to make the probability at least .95 that $\overline{X}$ differs from the unknown μ by no more than 2 years? [*Hint:* The discussion of the previous section leading to formula (3-19) is still applicable.]

3-36   Three slips of paper numbered 1, 2, and 3 are placed in a hat. A slip is drawn and the number is recorded. Then the slip is returned to the hat, the slips are mixed, and a second number is drawn and recorded. Follow the procedure used in this section to verify that formulas (3-13) and (3-14) hold in this particular case.

◈ 3-37   Repeat Exercise 3-36 if the first slip is not replaced before the second one is drawn and verify that formulas (3-13) and (3-15) hold in this particular case.

3-5
POINT ESTIMATION OF VARIANCE AND STANDARD DEVIATION

Of the probability distributions which we considered in Chap. 2, only the normal contained the parameter $\sigma^2 = E[(X - \mu)^2]$. We have already mentioned a number of situations for which the normal model seems to be appropriate, but in most applications σ^2 would be an unknown parameter. Since the variance is a measure of variability, it is frequently desirable to have an estimate of that constant even if the random variable under consideration does not have a normal distribution.

There are some problems in which variability is a very important consideration. It was used to motivate our definition of "best" estimate of a parameter, low variability of estimators being desirable. Whether an estimator is best or not, it is usually helpful to know its standard deviation or to have an estimate of its standard deviation. Another example arises from the manufacturing of machined parts. It is of little help to know that the average 1/2-inch nut fits the average 1/2-inch bolt. Unless most nuts fit on most bolts, the machining process is turning out a product that is too variable to be of much practical use. World War II–type anti-aircraft shells were timed to explode when the shell intercepted the path of the enemy aircraft. If timing devices operated with large variability, most of the shells would be expended harmlessly. Not always is small variability desirable. If a professor had to choose between two methods of teaching statistics, one which produces final scores with an average of 90 and standard deviation 1 and a second which produces final scores with an average of 50 and standard deviation 15, he would probably prefer the latter. The low variability associated with the first method may indicate that the course is too simple. (In fact if the course were made simple enough, final scores could average 100 with standard deviation 0.)

If we let $X_1, X_2, \ldots, X_n$ be a random sample from a distribution, then

$$S^2 = \frac{(X_1 - \overline{X})^2 + (X_2 - \overline{X})^2 + \cdots + (X_n - \overline{X})^2}{n - 1} = \frac{\sum_{i=1}^{n} (X_i - \overline{X})^2}{n - 1}$$

$$(3\text{-}24)$$

is called the *sample variance*. From observed values $x_1, x_2, \ldots, x_n$ we can calculate a specific number

$$s^2 = \frac{\sum_{i=1}^{n} (x_i - \bar{x})^2}{n - 1} \tag{3-25}$$

called the *observed sample variance*. The symbols S and s, used to denote the square roots of formulas (3-24) and (3-25), are called the *sample standard deviation* and the *observed sample standard deviation* respectively. As we shall demonstrate a little later in this section, the estimator S^2 is unbiased. The result

$$E(S^2) = \sigma^2 \tag{3-26}$$

always holds regardless of the distribution of X. (If a simple random sample of size n is drawn from N items, then $(N - 1)S^2/N$ has average value σ^2.) If in addition the distribution of X is normal, then S^2 is the best estimator of σ^2.

EXAMPLE 3-13

In Example 3-5 ten rolls of a six-sided die produced 6, 4, 4, 5, 2, 5, 3, 4, 6, 1 for which we found $\bar{x} = 4$. Compute the observed sample variance and the observed sample standard deviation.

Solution

Using formula (3-25) we get

$$s^2 = [(6 - 4)^2 + (4 - 4)^2 + (4 - 4)^2 + (5 - 4)^2 + (2 - 4)^2 + (5 - 4)^2$$
$$+ (3 - 4)^2 + (4 - 4)^2 + (6 - 4)^2 + (1 - 4)^2]/9$$
$$= [4 + 0 + 0 + 1 + 4 + 1 + 1 + 0 + 4 + 9]/9 = 24/9 = 2.67$$

Taking the square root we have

$$s = \sqrt{2.67} = 1.63$$

for the observed sample standard deviation.

Formula (3-25) is not well adapted to hand calculations or for use with a desk calculator. (It is satisfactory for use with a high-speed computer.) It produced s^2 in Example 3-13 with very simple arithmetic because each difference $x_i - \bar{x}$ was an integer. To obtain a more convenient computational formula, expand each square in the numerator of s^2. This gives

$$s^2 = \frac{x_1^2 - 2\bar{x}x_1 + \bar{x}^2 + \cdots + x_n^2 - 2\bar{x}x_n + \bar{x}^2}{n - 1}$$

$$= \frac{x_1^2 + x_2^2 + \cdots + x_n^2 - 2\bar{x}x_1 - 2\bar{x}x_2 - \cdots - 2\bar{x}x_n + \bar{x}^2 + \cdots + \bar{x}^2}{n - 1}$$

$$= \frac{\sum_{i=1}^{n} x_i^2 - 2\bar{x}(x_1 + x_2 + \cdots + x_n) + n\bar{x}^2}{n - 1}$$

$$= \frac{\sum_{i=1}^{n} x_i^2 - 2\bar{x}(n\bar{x}) + n\bar{x}^2}{n - 1}$$

$$= \frac{\sum_{i=1}^{n} x_i^2 - n\bar{x}^2}{n - 1} \tag{3-27}$$

$$= \frac{\sum_{i=1}^{n} x_i^2 - n(\sum_{i=1}^{n} x_i/n)^2}{n - 1}$$

$$= \frac{n\sum_{i=1}^{n} x_i^2 - (\sum_{i=1}^{n} x_i)^2}{n(n - 1)} \tag{3-28}$$

Both formulas (3-27) and (3-28), particularly the latter, are easily evaluated on a desk calculator.

EXAMPLE 3-14

Use formula (3-28) to find s^2 for the sample given in Example 3-13.

Solution

Either with a desk calculator or by our own arithmetic we get

$$\sum_{i=1}^{10} x_i = 6 + 4 + 4 + 5 + 2 + 5 + 3 + 4 + 6 + 1 = 40$$

$$\sum_{i=1}^{10} x_i^2 = 6^2 + 4^2 + 4^2 + 5^2 + 2^2 + 5^2 + 3^2 + 4^2 + 6^2 + 1^2 = 184$$

Then, substituting in formula (3-28), we have

$$s^2 = \frac{10(184) - 40^2}{10(9)} = \frac{1{,}840 - 1{,}600}{90} = \frac{240}{90} = 2.67$$

as before.

In general, the statistic S^2 will have a different sampling distribution for every different X distribution used to obtain the random sample. If the X's have a normal distribution with variance σ^2, then it can be shown that

$$Y_{n-1} = \frac{(n-1)S^2}{\sigma^2} \tag{3-29}$$

has a chi-square distribution with $n - 1$ degrees of freedom. To evaluate $\Pr(S^2 < r)$ we could write

$$\Pr(S^2 < r) = \Pr\left[\frac{(n-1)S^2}{\sigma^2} < \frac{(n-1)r}{\sigma^2}\right]$$

$$= \Pr\left[Y_{n-1} < \frac{(n-1)r}{\sigma^2}\right] \tag{3-30}$$

EXAMPLE 3-15

Consider again the distribution of IQs for adult Americans (Examples 2-26, 3-8). We assumed that the distribution was normal with $\mu = 100$ and $\sigma = 10$. If a random sample of size 25 individuals is selected, what is the probability that the sample variance of their IQs is less than 138.33?

Solution

With $n = 25, n - 1 = 24, \sigma^2 = 100$, we get from formula (3-30)

$$\Pr(S^2 < 138.33) = \Pr\left[Y_{24} < \frac{(24)(138.33)}{100}\right]$$

$$= \Pr(Y_{24} < 33.20)$$

$$= .90 \quad \text{from Appendix B7}$$

Our chi-square table is not well adapted to evaluating a probability like (3-30). Interpolation was avoided because 33.20 was a tabulated value. Fortunately, in applications where we will use (3-29), the tables will be quite adequate.

In Sec. 3-3 we learned that $Z = \sqrt{n}(\overline{X} - \mu)/\sigma$ has a standard normal distribution if the random sample comes from a normal distribution with mean μ and variance σ^2. The Central Limit theorem tells us that $\sqrt{n}(\overline{X} - \mu)/\sigma$ has approximately a standard normal distribution even if the X distribution is not normal. Perhaps we would expect a similar result to hold for $Y_{n-1} = (n - 1)S^2/\sigma^2$. Unfortunately, this is not the case. Apparently the distribution of $(n - 1)S^2/\sigma^2$ depends heavily on the form of the X distribution. This means that a probability like (3-30) evaluated from a chi-square table could be considerably in error if the X distribution does not approximate a normal distribution. Such an error is demonstrated in Fig. 3-5. The solid curve represents the density of Y_{n-1} when

FIGURE 3-5

Error in $\Pr(Y_{n-1} < c)$ when sampled distribution is distinctly nonnormal.

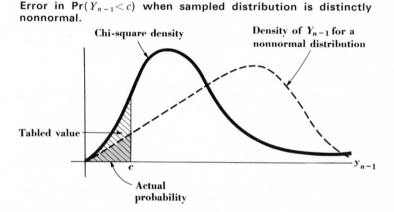

sampling from a normal distribution, while the dotted curve may represent the density if the original sampled distribution is distinctly nonnormal. In comparing $\Pr(Y_{n-1} < c)$, for the two cases, where c is a given number, the graph shows that the difference can be very great. Consequently, any probabilities or inferences based upon the assumption that Y_{n-1} has a chi-square distribution can be ridiculous unless the model governing the behavior of our experimental values is normal or approximately so.

To demonstrate that S^2 is an unbiased estimator of σ^2, consider again the sample space of Table 3-1 arising from two rolls of a die. Let us compute the observed sample variance associated with each of the 36 possible samples. With $n = 2$, $n - 1 = 1$, we calculate $s^2 = (x_1 - \bar{x})^2 + (x_2 - \bar{x})^2$. We get

$$s^2 = (1 - 1)^2 + (1 - 1)^2 = 0 \qquad \text{for 1, 1}$$
$$s^2 = (1 - 1.5)^2 + (2 - 1.5)^2 = .5 \qquad \text{for 1, 2}$$
$$s^2 = (1 - 2)^2 + (3 - 2)^2 = 2 \qquad \text{for 1, 3}$$

and so on. The 36 values of s^2 are given in Table 3-5. Since a probability of 1/36 was assigned to each possible sample of size 2 given in Table 3-1, it is reasonable to associate a weight of 1/36 with each value of s^2 given in Table 3-5. From the latter table we can construct the sampling

TABLE 3-5

Enumeration of values of s^2 for samples given in Table 3-1					
0	.5	2.0	4.5	8.0	12.5
.5	0	.5	2.0	4.5	8.0
2.0	.5	0	.5	2.0	4.5
4.5	2.0	.5	0	.5	2.0
8.0	4.5	2.0	.5	0	.5
12.5	8.0	4.5	2.0	.5	0

distribution of S^2 given in Table 3-6, the probabilities being obtained by adding the weights associated with each value of s^2. The average value of S^2 is

$$E(S^2) = 0(6/36) + .5(10/36) + 2.0(8/36) + 4.5(6/36) + 8.0(4/36)$$
$$+ 12.5(2/36)$$
$$= 105/36 = 35/12 = \sigma_X^2 = \sigma^2$$

as we wished to demonstrate.

TABLE 3-6

Sampling distribution of S^2 obtained from Table 3-5

s^2	0	.5	2.0	4.5	8.0	12.5
$g(s^2)$	6/36	10/36	8/36	6/36	4/36	2/36

Using the sampling distribution given in Table 3-6, it is easy to construct the distribution of S. (Replace s^2 by s and 0, .5, 2.0, 4.5, 8.0, 12.5 by their square roots.) We could then verify that for this example $E(S) \neq \sigma$. Generally S will be a biased estimator of σ. However, unless n is small the amount of bias is not apt to be of practical significance. In other words, s will ordinarily be a fairly good estimate of σ.

EXERCISES

3-38 In Exercise 3-15 we found that the observed sample mean was $\bar{x} = 575$. Use formula (3-25) to find the observed sample variance based upon the 10 student records, and then find s. If you have access to a desk calculator, also use formula (3-28) to find s^2.

◇ 3-39 Subtract 570 from each of the 10 sample values given in Exercise 3-15. Then find s^2 for the new set of 10 numbers, and verify that the answer is the same as the one obtained in Exercise 3-38. (This demonstrates that s^2 does not depend upon the origin one chooses for his numbers. Changing the origin from 0 to 570 results in smaller numbers for arithmetic purposes. Any number near $\bar{x} = 575$ would accomplish the same purpose. With a desk calculator or a computer, it is generally unnecessary to change the origin.)

3-40 For the sample values obtained in Exercise 3-16 compute s^2. Observe that in the Poisson case $E(S^2) = \sigma^2 = \mu$, so that s^2, like $\bar{x}$, is an unbiased estimate of μ. We know, however, that $\overline{X}$ is the best estimator, so that the variance of the $\overline{X}$ distribution is smaller than the variance of the S^2 distribution.

◇ 3-41 In Exercise 3-18 we found that the observed sample mean was $\bar{x} = 45.3$. Use formula (3-25) to find the observed sample variance based upon the 10 weights. If you have access to a calculator, also use formula (3-28) to find s^2.

3-42 Subtract 45.0 from the 10 sample weights given in Exercise 3-18. Then find s^2 for the new set of 10 numbers, and verify that the answer is the same as obtained in Exercise 3-41. (See comment in Exercise 3-39.)

◇ 3-43 Find s^2 and s for the 50 sample values given in Exercise 3-25. Use formula (3-28) to find s^2, and observe that there are a number of repeats in $x_1, x_2, \ldots, x_{50}$. (Thus, the arithmetic is quite simple.)

3-44 A scientist has conducted an experiment and has obtained 20 numbers which he believes constitute an observed random sample from a normal distribution. His secretary calculates $\sum_{i=1}^{20} x_i = 200$, $\sum_{i=1}^{20} x_i^2 = 2,720$. Find $\bar{x}$, s^2, and s.

◇ 3-45 Suppose that the experiment described in Exercise 3-25 is repeated 100 times (instead of 50) and yields $\sum_{i=1}^{100} x_i = 200$, $\sum_{i=1}^{100} x_i^2 = 540$. Find $\bar{x}$, s^2, and s.

3-46 For the situation described in Exercise 3-36 show that $E(S^2) = \sigma^2$ and $E(S) \neq \sigma$.

◇ 3-47 For the situation described in Exercise 3-37 show that $E[(N-1)S^2/N] = \sigma^2$.

3-48 A bowl contains ten slips of paper. The number 1 is written on four slips, 2 on three slips, 3 on two slips, and 4 on one slip. Let X be the number on one slip of paper drawn from the bowl by a blindfolded individual. (*a*) Find the probability distribution of X, $E(X) = \mu$, and σ^2. (*b*) Suppose the number is recorded, the slip is replaced, and a second number is drawn and recorded. Find the probability distribution of $\overline{X}$ for samples of size 2 drawn in this manner, and show that $E(\overline{X}) = E(X)$, $\sigma_{\overline{X}}^2 = \sigma^2/2$. (*c*) Find the probability distribution of S^2 and show that $E(S^2) = \sigma^2$.

◇ 3-49   In Examples 2-27 and 3-9 the random variable X was the amount of electricity consumed in a 10-minute period by a certain brand of 100-watt bulbs. We assumed that the distribution of X was normal with mean 50 and standard deviation 4. If a random sample of 11 such bulbs are to be used and their consumption to be observed, what is the probability of obtaining a sample variance greater than 29.296?

3-50   In Exercise 3-18 the random variable was the contents of a can of tomato juice. Let us assume that their weights are normally distributed with a standard deviation of .5 of an ounce. If a random sample of 10 weights is to be observed, what is the probability that they yield a sample variance less than .47?

REFERENCES

1   General Electric Company, Defense Systems Department: "Tables of the Individual and Cumulative Terms of Poisson Distribution," D. Van Nostrand Company, Inc., Princeton, N.J., 1962.

2   Molina, E. C.: "Poisson's Exponential Binomial Limit," D. Van Nostrand Company, Inc., Princeton, N.J., 1949.

3   U.S. Army Ordnance Corps: "Tables of Cumulative Binomial Probabilities," Ordnance Corps Pamphlet ORDP20-1, September 1952. (See Ref. 9 at the end of Chap. 2 for information about a new edition.)

Summary of Results
STATISTICS, SAMPLING
DISTRIBUTIONS, AND ESTIMATION

I. Some definitions (Sec. 3-1).

By the term *parameter* we usually mean a constant which appears in the probability function or density function associated with the model.

An unknown parameter is frequently estimated by one value, called a *point estimate*. One objective of statistical inference (a primary concern of this chapter) is to obtain good point estimates.

A *statistic* is a random variable which is observed for the purpose of drawing inferences about unknown parameters. The probability distribution of a statistic is called a *sampling distribution*.

II. Point estimation for some discrete distributions (Sec. 3-2)

A random variable is an *unbiased estimator* of a parameter if its expected value is equal to the parameter. An observed value of an estimator is called an *estimate*. If an estimator is unbiased and has variance smaller than any other unbiased estimator of a parameter, it is called a " best " estimator.

If X is a binomial random variable, then a best estimator of p is

$$T = \frac{X}{n} \tag{3-1}$$

If $X_1, X_2, \ldots, X_k$ are multinomial random variables, the unbiased estimators of $p_1, p_2, \ldots, p_k$ are

$$T_i = \frac{X_i}{n} \tag{3-2}$$

If Y has a negative binomial distribution, then the best estimator of p is

$$T = \frac{c - 1}{Y - 1} \tag{3-3}$$

If X has a hypergeometric distribution, then the best estimator of k is

$$T = N \frac{X}{n} \tag{3-4}$$

For the binomial case we can make the probability at least γ that the estimator X/n will be within d units of p by selecting

$$n \geqq \frac{pqz^2_{(1+\gamma)/2}}{d^2} \tag{3-10}$$

In inequality (3-10) we would use for p the value nearest .50 among those p which seem to be reasonable possibilities.

If a simple random sample of size n is drawn from k items numbered 1 to k, then the best estimate of k (Exercise 3-11) is

$$t = \frac{n+1}{n} x_{(n)} - 1$$

where $X_{(n)}$ is the largest number appearing in the sample.

To estimate the size of a biological population, say N, catch k of the species and tag them. After they have been released and allowed to disperse, capture a simple random sample of size n and observe the number which have tags, say x. Then a good estimate to use for N is (Exercise 3-14)

$$t = \text{largest integer not exceeding } \frac{nk}{x}$$

III. Point estimation of means (Sec. 3-3)

If $X_1, X_2, \ldots, X_n$ is a random sample from a distribution, then

$$\overline{X} = \frac{X_1 + X_2 + \cdots + X_n}{n} \tag{3-11}$$

is called the sample mean. If X has either a normal distribution or a Poisson distribution, $\overline{X}$ is a best estimator of $E(X) = \mu$.

If $X_1, X_2, \ldots, X_n$ is a random sample from any distribution, we have

$$\mu_{\overline{X}} = E(\overline{X}) = E(X) = \mu \tag{3-13}$$

(that is, $\overline{X}$ is an unbiased estimator of μ) and

$$\sigma_{\overline{X}}^2 = \frac{\sigma_X^2}{n} = \frac{\sigma^2}{n} \tag{3-14}$$

If the X distribution is normal, then the $\overline{X}$ distribution is normal. Hence

$$\Pr(\overline{X} < r) = \Pr\left(Z < \frac{r - \mu}{\sigma/\sqrt{n}}\right) \tag{3-16}$$

For the normal case we can make the probability at least γ that the estimator $\overline{X}$ will be within d units of μ by selecting

$$n \geq \left[\frac{\sigma z_{(1+\gamma)/2}}{d}\right]^2 \tag{3-19}$$

where d is selected in σ units.

If $X_1, X_2, \ldots, X_n$ is a random sample from a Poisson distribution with mean μ, then $Y = X_1 + X_2 + \cdots + X_n$ has a Poisson distribution with mean $n\mu$. In this case it is frequently more convenient to use Y, rather than $\overline{X}$, when drawing inferences about μ.

IV. Characteristics of the sampling distribution of the sample mean.

In addition to formulas (3-13) and (3-14) mentioned above in III, if n is not too small the distribution of $\overline{X}$ is approximately normal. Hence formula (3-16), also given in III, is approximately true. This is a consequence of the Central Limit theorem.

If $X_1, X_2, \ldots, X_n$ is a random sample from a discrete distribution where only integer values of X are possible, we have because of the Central Limit theorem

$$\Pr(r_1 \leqq Y \leqq r_2) \cong \Pr\left(\frac{r_1 - .5 - n\mu}{\sigma\sqrt{n}} < Z < \frac{r_2 + .5 - n\mu}{\sigma\sqrt{n}}\right) \qquad (3\text{-}23)$$

V. Point estimation of variance and standard deviation (Sec. 3-5)

If $X_1, X_2, \ldots, X_n$ is a random sample from a distribution, then

$$S^2 = \frac{\sum_{i=1}^{n} (X_i - \overline{X})^2}{n - 1} \qquad (3\text{-}24)$$

is called the sample variance. For any distribution S^2 is unbiased, and if the distribution is normal S^2 is a best estimator. A computing formula for the observed value of S^2, say s^2, is

$$s^2 = \frac{n \sum_{i=1}^{n} x_i^2 - \left(\sum_{i=1}^{n} x_i\right)^2}{n(n - 1)} \qquad (3\text{-}28)$$

If the distribution of the X's is normal, then

$$Y_{n-1} = \frac{(n - 1)S^2}{\sigma^2} \qquad (3\text{-}29)$$

has a chi-square distribution with $n - 1$ degrees of freedom. If the distribution is not normal, then probability statements based upon Y_{n-1} may be inaccurate.

4

TESTING STATISTICAL HYPOTHESES

4-1

INTRODUCTION

We now begin the study of a type of problem that is probably more important than any other considered in statistical inference, the testing of statistical hypotheses. We already know something about this new topic. In the discussion of Sec. 3-1 it was mentioned that the basic problem involves making a choice between two different statements regarding one or more parameters. We have also learned that inferences about parameters are based upon one or more observed values of a random variable whose distribution depends upon those parameters. Actually, we have made decisions of this type before. In Exercise 2-7 we were informed that a standard cure for tuberculosis is successful 30 percent of the time. A new cure is successful a fraction p of the time, where p is unknown, and experimentation has produced 29 cures among 50 treated patients. Then, letting X be the number of successes, we were asked to compute $\Pr(X \geq 29)$ under the assumption that $p = .30$ for the new cure. Since the calculation yielded a probability of .00004, we concluded that the new cure must produce a higher fraction of successes. In other words, on the basis of the observation $x = 29$ we decided that $p > .30$ was more reasonable than $p = .30$. A similar type of problem was encountered in Example 2-19. In this case a secretary claimed that on the average she makes one error per page. A sample page of her work contained five errors. Then, letting X be the number of errors per page and assuming $\mu = 1$, we found $\Pr(X \geq 5) = .00366$. We concluded that the observation $x = 5$ supported $\mu > 1$ rather than $\mu = 1$. In both examples a decision was made between two different possibilities for a parameter, possibilities suggested by the particular problem.

In discussing the concepts of hypothesis testing we will find that it is convenient to make a number of definitions. Two terms that deserve immediate attention are *hypothesis* and *test*. A statistical hypothesis

162

is usually defined to be an assumption about the distribution of a random variable. For the kind of problems we are going to consider it would be more informative to say that a *hypothesis* is an assumption that one or more parameters (usually associated with distributions of given or known form) belong to a given set. In the tuberculosis example both the assumption $p = .30$ and the assumption $p > .30$ are hypotheses. Another assumption made in that problem, the assumption that X has a binomial distribution, could also be a hypothesis, but generally statistical inference is concerned with assumptions about the parameters. In the secretary problem both the assumption $\mu = 1$ and the assumption $\mu > 1$ are hypotheses. We also assumed that X has a Poisson distribution but, as in the first example, this is usually not the type of hypothesis that is considered. A *test* of a statistical hypothesis is a rule indicating when to reject the hypothesis. Since testing problems will be concerned with choosing between two hypotheses, say between $p = .30$ and $p > .30$, the rule will enable us to specify which of the two is more reasonable. The test rule might be, for example: reject $p = .30$ in favor of $p > .30$ if $x \geq 21$.

We now turn to some examples of hypothesis testing. In this chapter our main objectives are to introduce and discuss the important concepts and to become familiar with the vocabulary. In later chapters we will make a systematic study of a number of standard hypothesis testing situations.

4-2
AN EXAMPLE USING BINOMIAL PARAMETER p

Let us return to the potato problem first discussed in Example 2-5. For convenience let p be the fraction of bad (or defective) potatoes (rather than the fraction of good ones). If we let X be the number of bad potatoes in a sample of size n, then our previous discussion indicates that the probability function of X is

$$b(x; n, p) = \binom{n}{x} p^x (1 - p)^{n-x} \qquad x = 0, 1, 2, \ldots, n \qquad (4\text{-}1)$$

Now, suppose that no one knows the real value of p but a potato wholesaler claims that $p = .10$ (actually, he would probably claim that $p \leq .10$) and has based his price on that figure. A purchasing agent for a large grocery chain wants to buy several dozen carloads but before agreeing to the price, he would like to believe that the claim is not unreasonable. In statistical language the agent desires to test the hypothesis $p = .10$. If the results of his investigation cast doubt on the wholesaler's claim, the agent will be led to conclude that $p > .10$. This latter conclusion (or assumption) is called the *alternative hypothesis*. We shall designate the hypothesis by H_0 and the alternative hypothesis by H_1. In statistical terminology we would say that the agent is interested in testing

$$H_0 : p = .10 \qquad \text{against} \qquad H_1 : p > .10$$

that is, the hypothesis $p = .10$ against the alternative hypothesis $p > .10$.

Having formulated the problem, we next seek a procedure for reaching a decision. It seems reasonable to observe a sample of n potatoes, count the number of bad ones, x, and reject the hypothesis in favor of the alternative if x is too large. In order to carry out this plan we must first specify the sample size and then determine how many bad ones are regarded as too many to let us believe that $p = .10$. Later we will discuss the determination of sample size, but for the moment suppose we arbitrarily select $n = 25$. Since we have made this choice, the probability function of X is

$$b(x; 25, p) = \binom{25}{x} p^x (1 - p)^{25 - x} \qquad x = 0, 1, 2, \ldots, 25 \qquad (4\text{-}2)$$

and if H_0 happens to be true so that $p = .10$, all parameters are now known. We still must make another arbitrary decision in selecting those x which cast doubt on the claim. Hardly anyone would believe that $p = .10$ if 23, 24, or 25 bad potatoes are in the sample. Not many people would be willing to believe the hypothesis if $x = 15$. Nearly everyone would feel that $p = .10$ is not unreasonable if $x = 2$. Suppose that we decide that we will reject $p = .10$ if x takes on a value of 6 or more. In other words, our test rule is: reject $p = .10$ in favor of $p > .10$ if $x \geqq 6$. Let us examine some of the consequences of our test procedure.

First, if the hypothesis is true so that $p = .10$, then

$$\Pr(X \geqq 6) = \sum_{x=6}^{25} b(x; 25, .10)$$

$$= 1 - \sum_{x=0}^{5} b(x; 25, .10)$$

$$= 1 - .96660 \qquad \text{from Appendix B1}$$

$$= .03340$$

Thus, the probability of rejecting the hypothesis when it is true is .03340. In other words, if $p = .10$ our procedure would lead us to an incorrect decision in about 3 out of every 100 samples. A hypothesis would be rejected when it is true. This kind of mistake is known as a *Type I error*. The probability of committing a Type I error is called the *level of significance* and is denoted by the Greek letter α. Here $\alpha = .03340$. The set of outcomes for the experiment which lead to rejection of the hypothesis

($x = 6, 7, 8, \ldots, 25$) is called the *critical region* (also rejection region, region of rejection). The remaining set of outcomes ($x = 0, 1, 2, 3, 4, 5$) is sometimes called the *acceptance region*.

Next, let us suppose that the true value of p is .20, not .10. Then, again all parameters in formula (4-2), the probability function of X, are known and

$$
\begin{aligned}
\Pr(X \geq 6) &= \sum_{x=6}^{25} b(x; 25, .20) \\
&= 1 - \sum_{x=0}^{5} b(x; 25, .20) \\
&= 1 - .61669 \\
&= .38331
\end{aligned}
$$

Hence, the probability that we reject $p = .10$ is .38331. Of course, since the hypothesis is now false we would much prefer to have the above probability near 1 instead of equal to .38331. The probability that we accept the hypothesis using the test rule is .61669. By accepting a false hypothesis, we commit another kind of mistake called a *Type II error*. Frequently the probability of making a Type II error is designated by the Greek letter β (here $\beta = .61669$ if $p = .20$).

The four possible situations suggested by the above discussion, together with the accompanying consequences, are summarized in Table 4-1.

TABLE 4-1

Decision versus Hypothesis

	Decision	
	Accept	**Reject**
Hypothesis true	Correct decision	Type I error
Hypothesis false	Type II error	Correct decision

Another important concept is power. The *power* of the test is defined to be the probability of rejecting the hypothesis. In our discussion we have already evaluated two values of power. We found power $= .03340$ if $p = .10$ and power $= .38331$ if $p = .20$. The power of our test in the potato example is

$$
\Pr(X \geq 6) = \sum_{x=6}^{25} b(x; 25, p) \tag{4-3}
$$

which holds for all p of interest under both H_0 and H_1. We can evaluate formula (4-3) for any p in the binomial table. Other values obtained from Appendix B1 are

$$\text{Power} = 1 - \sum_{x=0}^{5} b(x; 25, .25) = 1 - .37828 = .62172 \qquad \text{if } p = .25$$

$$= 1 - \sum_{x=0}^{5} b(x; 25, .30) = 1 - .19349 = .80651 \qquad \text{if } p = .30$$

$$= 1 - \sum_{x=0}^{5} b(x; 25, .40) = 1 - .02936 = .97064 \qquad \text{if } p = .40$$

$$= 1 - \sum_{x=0}^{5} b(x; 25, .50) = 1 - .00204 = .99796 \qquad \text{if } p = .50$$

Of course, power could be obtained for many more values of p by consulting more extensive tables. Plotting power versus p on a graph yields Fig. 4-1. An ideal power curve would be of height 1 for all values of the

FIGURE 4-1

Power curve for the test of $H_0: p = .10$ against $H_1: p > .10$ with $n = 25$.

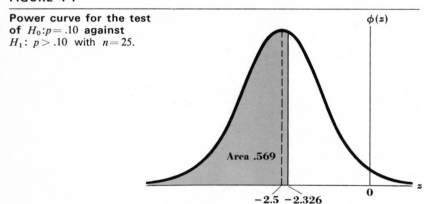

parameter specified by H_1 and of height 0 for values specified by H_0 (see Fig. 4-2). That is, if the hypothesis is true, we would always like to accept; and if the hypothesis is false we would always like to reject. To accomplish this ideal state of affairs the sample would have to include all (or nearly all) of the population, which is generally impossible. As we will see, the power curve can be made more nearly like the ideal one by increasing the sample size. This is not surprising, however, for as everyone knows without studying statistics, the more information available, the better the chance of making a good decision.

Some statisticians prefer to work with the *operating characteristic* (OC) which is 1 minus the power. The operating characteristic curve is an "upside down" power curve. Sometimes the format of existing tables makes it slightly more convenient to work with OC. In this book we will always use power.

FIGURE 4-2

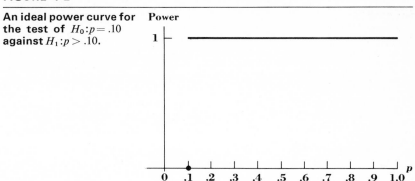

An ideal power curve for the test of $H_0:p=.10$ against $H_1:p>.10$.

In constructing the test for the potato example it appears that we arbitrarily selected the critical region to be $x \geq 6$. Usually, the statistician selects α instead and finds the corresponding critical region. Common choices for α are .05 and .01, although consideration of the consequences of a Type I error may lead to some other choice. Suppose we had selected $\alpha = .05$ for the potato problem. The binomial table gives with

$$p = .10, n = 25$$
$$\Pr(X \geq 6) = .03340$$
$$\Pr(X \geq 5) = .09799$$

illustrating that this was an impractical choice for α. We can either take the critical region $x \geq 6$ with $\alpha = .03340$ or the critical region $x \geq 5$ with $\alpha = .09799$. The difficulty arises from the fact that X has a discrete distribution. As we shall see, if the statistic which is observed for the test has a continuous distribution, then an α of .05 or .01 or any other choice can be achieved.

Returning to the purchasing agent, suppose that he examines a sample of 25 potatoes and observes $x = 5$ bad ones. If we use the test rule given earlier in the section, this means that we accept H_0 since $x = 5$ is in the acceptance region. This does not mean that we have proved that $p = .10$. Probably $x = 5$ would be a reasonable result if $p = .11$ or $p = .12$ or $p = .15$. What we can say is that the sample has not furnished enough evidence to reject $p = .10$, and until further information is available we will probably make our practical decisions (buy the potatoes at the quoted price) as though this is the correct value of p. From a statistical point of view we could make a much more convincing statement if the agent observes $x = 12$. Now since $\Pr(X \geq 12)$ is practically 0 if $p = .10$, it is a virtual certainty that $p > .10$ and almost everyone would be willing to admit that we have "proved" that H_1 is correct. Later in this chapter we will have some summarizing comments along the same lines.

EXERCISES

◈ 4-1 Suppose that we had selected the critical region $x = 7, 8, 9, \ldots, 25$ for the potato example. Verify that α has been reduced to .00948. With this new critical region, what is the power of the test when $p = .20, .25, .30, .40, .50$? Compare with the values obtained with $x \geq 6$ and $\alpha = .03340$. What is the effect on the power curve when α is decreased? What is the effect on the β's associated with $p > .10$?

4-2 Suppose that we had selected a sample size of $n = 100$ for the potato example. Verify that the critical region $x = 16, 17, \ldots, 100$ has associated with it an α of .03989. If $p = .10$ is rejected when $x \geq 16$, find the power of the test when $p = .20, .25, .30, .40, .50$. What does increasing n appear to do the power curve? To the probability of making a Type II error?

4-3
SOME COMMENTS ON SAMPLE SIZE

When statistical experiments are conducted, it is not uncommon to select the sample size haphazardly or by means of an arbitrary choice (as in Sec. 4-2). We next demonstrate that sometimes the conditions of the problem will determine a minimum sample size.

Let us consider once more the purchasing agent and the potato problem. Suppose that if $p = .10$ as claimed, then the agent is satisfied with the wholesaler's price. He would like to be reasonably sure that his test procedure does not lead to rejection if $p = .10$, and he decides that the probability of rejection should be less than or equal to .05. On the other hand, if p is somewhat larger than .10 he may lose money if he pays the quoted price. If p is as poor as .25, he would like to reject $H_0 : p = .10$ with fairly high probability, say at least .90. The agent wishes to determine the minimum sample size and the corresponding test rule that satisfies these conditions.

We observed in the last section that $H_0 : p = .10$ should be rejected when x is large, say $x \geq d$. The probability of rejection (the power) is

$$\Pr(X \geq d) = \sum_{x=d}^{n} b(x; n, p) \tag{4-4}$$

The above conditions imposed by the agent imply that

$$\text{Power} \leq .05 \qquad \text{if } p = .10 \tag{4-5}$$

and

$$\text{Power} \geq .90 \qquad \text{if } p = .25 \tag{4-6}$$

which can be rewritten as

$$\sum_{x=d}^{n} b(x; n, .10) \leq .05 \qquad (4\text{-}7)$$

and

$$\sum_{x=d}^{n} b(x; n, .25) \geq .90 \qquad (4\text{-}8)$$

(See Fig. 4-3 for a graphical representation of the requirements.) We know from Sec. 4-2 that with $n = 25$ and $d = 6$ the first condition (4-5) is satisfied but the second (4-6) is not since power $= .03340$ if $p = .10$ and power $= .62172$ if $p = .25$. Apparently a larger n is required, and so we might next try $n = 50$. The binomial table in Appendix B1 reveals that d has to be at least 10 to satisfy inequality (4-7) and

$$\sum_{x=10}^{50} b(x; 50, .10) = 1 - \sum_{x=0}^{9} b(x; 50, .10)$$
$$= 1 - .97546 = .02454 < .05$$

[We take the smallest d possible since a larger value makes it more difficult to satisfy inequality (4-8).] The power at $p = .25$ is

$$\sum_{x=10}^{50} b(x; 50, .25) = 1 - \sum_{x=0}^{9} b(x; 50, .25)$$
$$= 1 - .16368 = .83632 < .90$$

Since the second condition is still not satisfied we would probably try a larger value of n. At the moment it is not our objective to become experts in the determination of sample size but rather to indicate that such calculations may enter into the planning stage of a statistical experiment. Consequently, we merely state that the purchasing agent should use

FIGURE 4-3

Power requirements imposed by the purchasing agent.

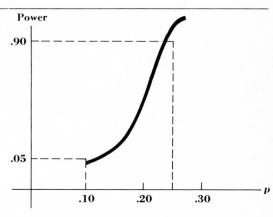

$n = 55$ potatoes and reject $H_0 : p = .10$ if $x \geqq 10$. (The answer is found by inspection with a good binomial table.)

The problem we have described in this section is a special case of one that arises in a field called statistical quality control under the title of sampling inspection by attributes. In such problems p is the fraction defective produced by a manufacturing process and X is the number of defectives in a random sample of n items.

4-4

AN EXAMPLE INVOLVING THE MEAN μ

Suppose that a smoker is hopelessly enslaved by the cigarette habit. Even though he is aware that he may contract lung cancer, it is impossible for him to quit. Suppose further that it has been definitely established by scientists that if cigarettes average 30 milligrams or more of nicotine, it is a virtual certainty that lung cancer will develop in the user. The smoker is willing to take his chances on any brand for which the average μ is less than 30 milligrams. He is particularly interested in brand A, his favorite. If tests on a random sample of 100 brand A cigarettes yield $\bar{x} = 26$ milligrams, what decision should the smoker make?

Certainly he will use brand A if $\bar{x}$ is small enough. Had $\bar{x}$ turned out to be 10, he would have no doubts. An $\bar{x}$ of 31 would not encourage him to use brand A.

One way to proceed would be to follow the scheme used in the tuberculosis and secretary problems. That is, calculate $\Pr(\bar{X} \leq 26)$ under the assumption that $\mu = 30$ and decide in favor of brand A if this is small. To evaluate the probability we could assume that the distribution of nicotine contents is normal (very likely a good assumption), but because of the Central Limit theorem such an assumption is unnecessary. We recall from Chap. 3 that similar calculations depended upon σ. To make things simpler we will assume that $\sigma = 8$ milligrams, a figure suggested by earlier studies of this type. (In the next chapter we will find a practical solution to the problem of unknown σ.) Then $\sigma_{\bar{X}} = \sigma/\sqrt{n} = 8/\sqrt{100} = .8$, and the $\bar{X}$ distribution is normal (or at least approximately so) with mean 30 (our assumption) and standard deviation .8. We evaluate

$$\Pr(\bar{X} \leqq 26) = \Pr(\bar{X} < 26)$$

$$= \Pr\left(\frac{\bar{X} - 30}{.8} < \frac{26 - 30}{.8}\right)$$

$$\cong \Pr(Z < -5)$$

$$= \text{practically } 0$$

More extensive tables of the standard normal reveal that $\Pr(Z < -5) = .0000003$. In view of the low probability the most obvious conclusion is that $\mu < 30$ for brand A. The practical consequences of this decision are that the smoker will use brand A.

From the standpoint of a hypothesis testing problem, a decision is to be made between $\mu = 30$ and $\mu < 30$. The former we will call the hypothesis and the latter the alternative hypothesis. In statistical terminology we are interested in testing

$$H_0 : \mu = 30 \qquad \text{against} \qquad H_1 : \mu < 30$$

A reasonable statistic to consider in drawing inferences about μ is $\overline{X}$, the sample mean. If $\bar{x}$ is sufficiently small we will reject H_0 and believe H_1. Suppose that before observing $\bar{x}$ the smoker decides that he is willing to risk a one in a hundred chance of using an unsafe brand. That is, if $\mu = 30$ is actually the true average, the probability of accepting H_0 is to be .01. The smoker has arbitrarily selected $\alpha = .01$ as his significance level. The critical region will consist of all $\bar{x}$ such that $\bar{x} \leq \bar{x}_0$ with $\bar{x}_0$ being determined so that $\Pr(\overline{X} \leq \bar{x}_0) = .01$ if $\mu = 30$. Actually, it is more convenient to use the statistic

$$Z = \frac{\overline{X} - 30}{.8} \tag{4-9}$$

rather than $\overline{X}$, since under the assumption that $\mu = 30$, Z has a standard normal distribution, and probabilities for Z are readily available in Appendix B5. Obviously, if $\bar{x}$ is small so is z, and we might just as well use the critical region $z \leq z_0$ where $\Pr(Z \leq z_0) = .01$. From the last row of Appendix B6 we see that $z_0 = -2.326$. Hence, the convenient way to express the critical region is

$$z = \frac{\bar{x} - 30}{.8} < -2.326 \tag{4-10}$$

Figure 4-4 shows the density of Z if $H_0 : \mu = 30$ is true, the critical region, and the significance level. Consequently the smoker will observe a random sample of size 100 cigarettes, record nicotine contents $x_1, x_2, \ldots, x_{100}$, calculate $\bar{x}$ and then z, and reject H_0 if $z < -2.326$. With an observed $\bar{x} = 26$ calculations yield $z = -5$, and the conclusion is that H_1 is true (brand A is safe).

The critical region (4-10) is based upon the statistic (4-9). The probability that H_0 is rejected is

$$\text{Power} = \Pr\left(\frac{\overline{X} - 30}{.8} < -2.326\right) \tag{4-11}$$

If $\mu = 30$, then equation (4-11) reduces to $\Pr(Z < -2.326) = .01$, the significance level chosen by the smoker. Now suppose that $\mu = 28$ milligrams for brand A. What is the value of equation (4-11)? In

FIGURE 4-4

Density of Z if $H_0:\mu = 30$ is true, showing significance level .01 and critical region $z < -2.326$.

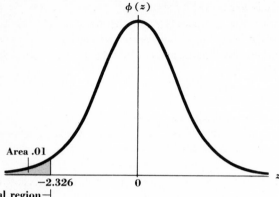

the next chapter we will discuss the use of some graphs which yield the probability immediately. For the present we note that if $\mu = 28$ (and not 30), then the random variable which has a standard normal distribution is $Z = (\overline{X} - 28)/.8$, not statistic (4-9). [Denoting $(\overline{X} - 30)/.8$ by Z, our symbol for a standard normal random variable is not entirely satisfactory. To be strictly consistent Z should be used only if $\mu = 30$ since $Z = (\overline{X} - \mu)/(\sigma/\sqrt{n})$ has a standard normal distribution only if μ is the mean of the X distribution. If the mean is 28, then $Z = (\overline{X} - 30)/.8$ has a normal distribution with mean $(28 - 30)/.8 = -2.5$ and variance $\sigma^2 = 1$. The density of Z and the critical region are shown in Fig. 4-5.] With a

FIGURE 4-5

Density of $Z = (\overline{X} - 30)/.8$ if $\mu = 28$ showing critical region $z < -2.326$ and power of the test.

little manipulation it is easy to overcome the difficulty. We can write formula (4-11) as

$$\text{Power} = \Pr\!\left(\frac{\overline{X} - 28}{.8} + \frac{28 - 30}{.8} < -2.326\right)$$

$$= \Pr(Z - 2.50 < -2.326)$$

$$= \Pr(Z < .174)$$

$$= .569 \qquad \text{from Appendix B5}$$

In general, if the mean is μ, the same algebra yields

$$\text{Power} = \Pr\!\left(\frac{\overline{X} - \mu}{.8} + \frac{\mu - 30}{.8} < -2.326\right)$$

$$= \Pr\!\left(Z < \frac{30 - \mu}{.8} - 2.326\right) \tag{4-12}$$

From formula (4-12) we get

$$\text{Power} = \Pr(Z < -2.326) = .01 \qquad \text{if } \mu = 30$$

$$= \Pr(Z < -1.076) = .141 \qquad \text{if } \mu = 29$$

$$= \Pr(Z < .174) = .569 \qquad \text{if } \mu = 28$$

$$= \Pr(Z < 1.424) = .923 \qquad \text{if } \mu = 27$$

Similarly, any value of μ gives a specific probability. Plotting power against μ yields the curve shown in Fig. 4-6.

Now suppose that the smoker would like to be reasonably sure that if μ is as small as 28, he will conclude $\mu < 30$ and adopt the brand A. To be specific let us assume that he requires power $\geq .90$ if $\mu = 28$. (See

FIGURE 4-6

Power curve for test of $H_0\!:\!\mu = 30$ **against** $H_1\!:\!\mu < 30$ **with** $n = 100$ **and** $\alpha = .01$.

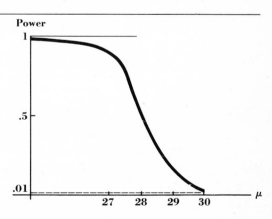

Fig. 4-7). In other words, if $\mu = 30$ the probability is .01 that the brand will be adopted while if $\mu = 28$ the probability of adoption is at least .90. Of course, H_0 should still be rejected if $\bar{x}$ is small. The problem is to find the minimum sample size and the corresponding critical region. (This is essentially the same problem proposed in Sec. 4-3 for binomial p.)

FIGURE 4-7

Power requirements imposed by the smoker.

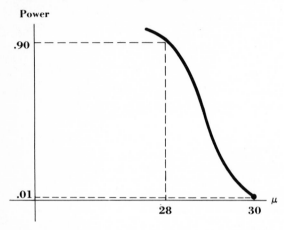

We already know that $n = 100$ is not large enough since our previous calculations gave power $= .569$ with $\mu = 28$. Again we comment that our current interest is to point out that sample size problems exist rather than to become experts in solutions of these problems. Here, it may be shown (see Sec. 5-3) that $n = 208$, and the corresponding critical region is

$$\frac{\bar{x} - 30}{8/\sqrt{208}} < -2.326 \tag{4-13}$$

Let us examine the consequences of Type I and Type II errors. If a Type I error is made, the conclusion is that the mean is less than 30 when in reality it is not. Presumably this would be a very serious kind of mistake to make, since the smoker would adopt a brand of cigarettes that would endanger his life. Although the probability α was set at .01, this may be way too high when a life hangs in the balance. A figure of .001, or even .0001, would be more appealing. A Type II error is made when the observed $\bar{x}$ leads to the conclusion that $\mu = 30$ when actually $\mu < 30$. Brand A cigarettes would not be used even though they satisfy the requirement set up by the smoker. Perhaps the smoker would then test another brand. At any rate, the consequences of a Type II error do not seem to be very serious. Thus, in this example it is likely that a very small probability would be chosen for α and very little attention would be given to power or to values of β for those μ in the alternative hypothesis $H_1 : \mu < 30$.

EXERCISES

◈ 4-3   Suppose that $\alpha = .05$ had been selected in the cigarette problem. Verify that the critical region $(\bar{x} - 30)/.8 < -1.645$ has associated with it a significance level of .05. With this new critical region compute the power when $\mu = 27, 28, 29, 30$ and compare with the values obtained when $\alpha = .01$. What appears to be the effect on power when α is increased?

4-4   Suppose that $n = 25$, not 100 in the cigarette problem. What is the critical region obtained using $\alpha = .01$? Compute the power when $\mu = 27, 28, 29, 30$. What does decreasing n do to the power curve? To the probability of making a Type II error?

4-5
A TWO-SIDED ALTERNATIVE

In the potato example we decided to choose between $H_0 : p = .10$ and $H_1 : p > .10$. Someone might inquire "What, if anything, should be done with those values of p less than .10?" Perhaps a reasonable solution would be to replace the hypothesis by $H_0 : p \leq .10$ so that all possible values of p fall into one of two sets, satisfactory values ($p \leq .10$) and unsatisfactory values ($p > .10$). It would not make sense to put this formerly neglected set under H_1, thus having satisfactory and unsatisfactory p's grouped together in the same hypothesis. In the cigarette problem we chose between $H_0 : \mu = 30$ and $H_1 : \mu < 30$. Since the primary goal was to distinguish between "safe" and "unsafe" averages, it would probably make sense to replace the hypothesis by $H_0 : \mu \geq 30$. Hence, again every possible value of the parameter would fall into one of two sets dictated by the goals of the problem. (When more than one value of the parameter is possible under H_0, our former definition of significance level is meaningless. In the next chapter we will see how that definition should be changed to cover these cases.) Whether or not we change H_0 as indicated above, alternatives of the type $p > .10$, $\mu < 30$ (or more generally $\theta < \theta_0$ or $\theta > \theta_0$, where θ_0 is a known constant) are called *one-sided alternatives*. The accompanying test is called a *one-sided test* or a *one-tailed test*. The name is attributed to the fact that in cases which we consider, H_0 is rejected either for large observed values of a statistic or small observed values but not both.

Some problems require alternatives of the type $\theta \neq \theta_0$ (which includes both $\theta < \theta_0$ and $\theta > \theta_0$). These are called *two-sided alternatives.* In such cases we seek a test of

$$H_0 : \theta = \theta_0 \quad \text{against} \quad H_1 : \theta \neq \theta_0$$

For standard situations which we will consider the hypothesis H_0 is rejected if an observed value of a statistic is either "too small" or "too large." Let us consider an example.

A tomato juice cannery attempts to put 46 ounces (as stated on the label) in each can. The automatic measuring device puts in X ounces, a random variable which is normally distributed. If the average content is below 46 ounces, the company may get into trouble with the government inspectors for false labeling. On the other hand, if the average content is above 46 ounces, the company will make less profit. From the company's viewpoint the only satisfactory value for the average is $\mu = 46$ ounces, and the set $\mu \neq 46$ contains the unsatisfactory averages. In order to determine whether or not the weighing process is operating satisfactorily, the plant statistician would test

$$H_0 : \mu = 46 \quad \text{against} \quad H_1 : \mu \neq 46$$

As we already observed in Sec. 4-4, a reasonable statistic to use in drawing inferences about μ is the sample mean $\overline{X}$. Here both small and large $\bar{x}$ would support H_1 and lead to the rejection of H_0. To simplify the discussion, we will assume that $\sigma = .5$ of an ounce, a figure derived from earlier investigations.

To continue with the above example, suppose that the plant statistician plans to use a random sample of 25 cans to conduct his test. Then $\sigma_{\overline{X}} = .5/\sqrt{25} = .1$. Once more it is more convenient not to use $\overline{X}$ but rather

$$Z = \frac{\overline{X} - \mu}{\sigma_{\overline{X}}} = \frac{\overline{X} - 46}{.1} \tag{4-14}$$

a random variable which has a standard normal distribution when $\mu = 46$. As in the cigarette example, we will express the critical region in terms of z, but first we need a significance level. The plant statistician selects $\alpha = .05$ because he knows that this is the figure that the government inspector uses. We have already concluded that the critical region should consist of small and large $\bar{x}$. In terms of z, this means small and large z. Even with this information various critical regions are possible. One appealing choice is

$$z = \frac{\bar{x} - 46}{.1} < -1.960 \quad \text{and} \quad z = \frac{\bar{x} - 46}{.1} > 1.960 \tag{4-15}$$

a region obtained by associating half of the significance level ($\alpha/2 = .025$) with each tail of the Z distribution (see Fig. 4-8). In other words, if

FIGURE 4-8

Density of Z if $H_0 : \mu = 46$ is true showing significance level and critical region $z < -1.960$ and $z > 1.960$.

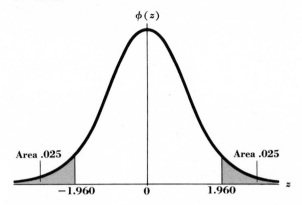

$\mu = 46$ the probability that we reject with small $\overline{X}$ is .025 and the probability that we reject with large $\overline{X}$ is .025. Essentially this implies that we are equally concerned with the consequences associated with having μ either too small or too large. (If we risk heavy penalties from the government by failing to detect $\mu < 46$, it may be preferable to make the left tail area greater than the right. We might, for example, make the two probabilities .04 and .01 yielding critical region $z < -1.751$ and $z > 2.326$. Throughout the book we will use the "equal-tails" choice.)

Having determined the critical region, the plant statistician conducts his experiment and observes $\bar{x} = 46.18$. The observed value of the statistic given by formula (4-14) is

$$z = \frac{46.18 - 46}{.1} = 1.8$$

Since $z = 1.8$ does not fall in the critical region, we accept H_0. We have not proved that $\mu = 46$. A better way to state our conclusion is to say that the observed sample has not provided sufficient evidence to contradict $\mu = 46$. Lacking such evidence, we continue to operate as though $\mu = 46$, a decision which implies that corrective action is unnecessary.

If we use the critical region (4-15), then the power of the test is

$$\text{Power} = \Pr\left(\frac{\overline{X} - 46}{.1} < -1.960\right) + \Pr\left(\frac{\overline{X} - 46}{.1} > 1.960\right) \qquad (4\text{-}16)$$

If $\mu = 46$, then this obviously reduces to

$$\Pr(Z < -1.960) + \Pr(Z > 1.960) = .05,$$

the significance level selected by the plant statistician. For other values of μ, we repeat the algebra which led to formula (4-12) in the smoker example. We get

$$\text{Power} = \Pr\left(\frac{\overline{X} - \mu}{.1} + \frac{\mu - 46}{.1} < -1.960\right) + \Pr\left(\frac{\overline{X} - \mu}{.1} + \frac{\mu - 46}{.1} > 1.960\right)$$

$$= \Pr\left(Z < \frac{46 - \mu}{.1} - 1.960\right) + \Pr\left(Z > \frac{46 - \mu}{.1} + 1.960\right) \quad (4\text{-}17)$$

From formula (4-17) we get that the power is

$\Pr(Z < -1.960) + \Pr(Z > 1.960) = .025 + .025 = .05$ if $\mu = 46$

$\Pr(Z < -.960) + \Pr(Z > 2.960) = .1685 + .0015 = .1700$ if $\mu = 45.9$

$\Pr(Z < .040) + \Pr(Z > 3.960) = .5160 + 0 = .5160$ if $\mu = 45.8$

$\Pr(Z < 1.040) + \Pr(Z > 4.960) = .8508 + 0 = .8508$ if $\mu = 45.7$

   Similarly, using

$\mu = 46.1$ yields   power $= .1700$

$\mu = 46.2$ yields   power $= .5160$

$\mu = 46.3$ yields   power $= .8508$

Plotting power against μ yields Fig. 4-9. We note that the curve is symmetric about $\mu = 46$. This is due to the fact that the critical region was determined by associating $\alpha/2 = .025$ with each tail of the Z distribution.

If we were to specify that the power had to be at least .98 if μ is either 45.8 or 46.2, then we would have another sample size problem on our hands. Our calculations show that $n = 25$ is too small. In the next

FIGURE 4-9

Power curve for test of $H_0:\mu = 46$ against $H_1:\mu \neq 46$ with $n = 25$ and $\alpha = .05$.

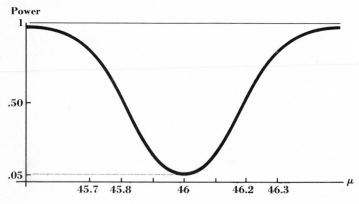

chapter we will find that the minimum sample size required is $n = 100$. The corresponding critical region is

$$z = \frac{\bar{x} - 46}{.5/\sqrt{100}} < -1.960 \qquad \text{and} \qquad z = \frac{\bar{x} - 46}{.5/\sqrt{100}} > 1.960$$

Let us suppose that the plant statistician selected his sample and computed $\bar{x} = 46.18$ before he decided upon his hypothesis and alternative. Since $\bar{x}$ is larger than 46, he may be tempted to test

$$H_0 : \mu = 46 \qquad \text{against} \qquad H_1 : \mu > 46$$

selecting an alternative suggested by the sample. (Of course, with our previous information such a choice makes no sense. Without that information, the statistician may be attempting to formulate a realistic problem.) If he rejects H_0 in favor of the new H_1, he will "prove" that the mean is larger than 46 ounces. Only large $\bar{x}$ and large $z = (\bar{x} - 46)/.1$ support the new alternative hypothesis $H_1 : \mu > 46$. A reasonable critical region to use (with $\alpha = .05$) is $z = (\bar{x} - 46)/.1 > 1.645$, since if $\mu = 46$, $\Pr(Z > 1.645) = .05$. (See Fig. 4-10.) Now the observed value of Z is $z = (46.18 - 46)/.1 = 1.8$, which falls in the critical region and leads to the conclusion that the average weight per can is above 46 ounces. This demonstrates that if an alternative suggested by a sample is chosen and the same sample is used for the test, we are more apt to prove that the alternative is true. In general, it is not an acceptable procedure to use sample results to help formulate the hypothesis testing problem and then to use the same sample to conduct the test. Sometimes an investigator may conduct preliminary experiments in an attempt to construct a reasonable problem, but having done so he should obtain a new sample if he intends to draw inferences.

FIGURE 4-10

Density of Z if $H_0 : \mu = 46$ is true showing significance level of .05 and critical region $z > 1.645$.

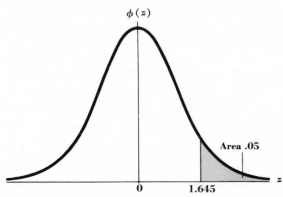

EXERCISES

◈ 4-5 Suppose that $\alpha = .10$ in the tomato juice problem. Verify that $(\bar{x} - 46)/.1 < -1.645$ and $(\bar{x} - 46)/.1 > 1.645$ is a critical region with this α. With this new critical region compute the power when $\mu = 45.7, 45.8, 45.9, 46, 46.1, 46.2, 46.3$ and compare with the values obtained using $\alpha = .05$. What does increasing α do to the power curve? How does it affect the probability of a Type II error?

4-6 Suppose that $n = 100$ (not 25) in the tomato juice problem. Find the appropriate critical region using $\alpha = .05$. Then compute the power when $\mu = 45.7, 45.8, 45.9, 46, 46.1, 46.2, 46.3$ and compare with the values obtained when $n = 25$. What does increasing n do to the power curve? To the probability of making a Type II error?

4-6
SOME GENERAL COMMENTS

As we have demonstrated in this chapter, the first step in a hypothesis testing problem involves the selection of the hypothesis and the alternative. In the examples a parameter associated with a probability model was unknown. Possible values of this parameter were divided into two sets, one called "satisfactory" values and the other "unsatisfactory" (or "safe" and "unsafe"). The assumption that a parameter belonged to a given set was called a hypothesis. Since two sets were involved in each problem, we had two hypotheses, one of which we called the *hypothesis* and the other the *alternative hypothesis*. As these are merely names, it does not matter which name we attach to which set. Sometimes a specific value suggested by the problem is a good choice for the set included under H_0. For example, the potato wholesaler claims $p = .10$, or the secretary claims her average is $\mu = 1$ error per page. Because of the nature of the conclusions which can be drawn from the results of a test, it is often good procedure (when possible and when one can afford to do it) to choose as the hypothesis the conclusion that one hopes to disprove. Thus the alternative (usually a one-sided alternative) states what the experimenter would like to prove. The cigarette problem is a good illustration. Here we selected $H_0 : \mu = 30$ and $H_1 : \mu < 30$, and an observed $\bar{x}$ led to a very satisfactory conclusion. Since $\bar{x}$'s this small are almost impossible to obtain if $\mu = 30$, we have overwhelming confidence in $H_1 : \mu < 30$. Suppose we had made the choice $H_0 : \mu < 30$ against $H_1 : \mu = 30$ (or $\mu \geq 30$). Then only large values of $\bar{x}$ would cause us to reject H_0. Very likely an $\bar{x}$ of 30.4 would be regarded as support for H_0 [since $\Pr(\bar{X} > 30.4) = .31$]. Certainly the smoker would be somewhat reluctant to use brand A if the observed $\bar{x}$ were 30.4. The same principle was involved in the tuberculosis example for which we chose $H_0 : p = .30$ and $H_1 : p > .30$, hopeful of establishing that p for the new cure is more than .30. As we shall demonstrate later, different people can select a different pair H_0, H_1 depending upon the individual's point of view.

The original selection of the hypothesis and the alternative in the cigarette problem placed the burden of proof on the cigarette. This is frequently the position we would like to adopt when considering a new and somewhat expensive machine to replace an old one still in good working condition. Unfortunately, in some experimental situations available procedures permit only a standard selection of H_0 and H_1, whether or not they happen to be interesting. Sometimes the desirable choice may be unreasonable on the basis of economic or realistic considerations. In the cigarette problem the possibility exists that no brand will be acceptable unless we modify the testing procedure. Ruling out all brands may leave the smoker with a choice of either inviting lung cancer or dying of nervous convulsions.

In summary, there are no hard and fast rules for selecting H_0 and H_1. Sometimes it is very easy, other times not so easy. A combination of experience and " good common sense " will usually lead us to reasonable choices.

A second major decision which has to be made in hypothesis testing involves the selection of α and n, or alternatively the selection of α and another condition which will yield a minimum sample size. More often than not α is selected somewhat arbitrarily with .01 and .05 being the most common choices. We have seen with the cigarette example that consequences of a Type I error may affect the choice of the significance level. Sometimes the experimenter may be influenced by the reasonableness of the hypothesis in determining the magnitude of α. If nearly everyone is convinced that male and female high school seniors in the United States have the same average IQ (the hypothesis), only very unusual experimental results would attract attention and a small α should be selected. On the other hand, many people would be willing to believe that high school seniors and Bushmen of the same age from the interior of Australia do not have the same average IQ. Consequently, in testing the hypothesis that such averages are equal, mildly unusual values would be sufficient justification for rejection, and a larger α could be used. In the potato problem we observed that the undesirability of making a Type II error with quality as poor as $p = .25$ led to the selection of a high power (or a low β) for that alternative. To summarize, it is difficult to give clear-cut statements which can be applied to determine α and n (or α and β at a specific alternative). This being the case, the decision comes down to a judgment that someone must make.

In testing the hypotheses of this chapter, the choice of statistic was justified on intuitive grounds. If " best " statistics (as defined in Chap. 3) exist, then good testing procedures are usually based upon such statistics.

Critical regions were also selected on an intuitive basis. It is conceivable that different statistics and different critical regions could be used (given the same α) to test the same hypothesis. Determining which of several tests is best in a given situation is a mathematical problem beyond the level of our discussion. In the following chapters we shall in general use tests which have good properties (often a good power curve), but we shall continue to rely on intuitive justification.

Tests discussed in this chapter and in later chapters are obtained using power considerations. In the potato problem $H_0 : p = .10$ was rejected if $x \geqq d$ where d and perhaps n were determined by arbitrarily selected power conditions. This is the traditional but by no means the only approach. Another possibility is to determine d and n so that costs (or losses) are on the average minimized. In so doing the arbitrary power conditions are replaced by another set of conditions, requiring personal judgment, and a new (and usually more difficult) problem is encountered. In studying statistical inferences, it is probably advantageous to master traditional approaches first and to investigate newer and more complicated methods after some maturity is achieved.

4-7
AN OUTLINE FOR HYPOTHESIS TESTING

When testing a statistical hypothesis, it is good procedure to follow an outline. For the beginning student it is a means of replacing apparent chaos by order. For those with more experience it provides a systematic approach to a type of problem solving and may be comparable to an airline pilot's checklist. For the investigator who wishes to present his results to others, it is a vehicle of intelligible reporting.

There are a number of variations which can be used in constructing an outline. Provided all the essential information is given in a logical fashion, the outline will probably prove to be satisfactory. Of course, before anything else we must have a problem which implies a statistical hypothesis testing situation. In a textbook the problem will be described in an exercise or in an example. In an actual experimental situation, the statistician will have to provide his own description. Given the problem, then the solution outline should include the following steps:

1 State the hypothesis and the alternative hypothesis. If appropriate, add comments supporting this particular choice.
2 Choose α, the significance level, and n, the sample size. Alternatively, specify two power conditions (instead of just α) which will determine a minimum n.

3 Select a statistic whose sampling distribution depends upon the parameter (or parameters) used in the statement of the hypothesis and alternative. If certain assumptions are satisfied, then the sampling distribution of the statistic will be known. These assumptions should be listed so that the experimenter can attempt to make them seem realistic when he observes his sample. We may know the assumptions from our study of Chap. 2. In some cases the necessary assumptions for the use of a statistic will be listed without derivation.

4 Find the critical region and, if necessary, the sample size. Sometimes the choice of critical region will be intuitively obvious. Critical regions will be given for a number of standard situations.

5 Perform the experiment and observe the one or more random variables. Compute the observed value of the statistic selected in step 3.

6 Draw the conclusions. If the observed value of the statistic falls in the critical region, reject the hypothesis; otherwise, accept it. If not obvious, state any practical consequences of the decision.

Since human communication is not always as effective as it should be, do not hesitate to include sufficient detail in the outline. The use of complete sentences is usually helpful. Now let us illustrate the outline by using the three examples discussed earlier in the chapter.

EXAMPLE 4-1

Outline the solution of the potato problem of Sec. 4-2.

Solution

1 The problem suggests that we test $H_0 : p = .10$ against $H_1 : p > .10$. The figure $p = .10$ is based upon the wholesaler's claim which was used to determine price.

2 A sample size $n = 25$ was selected. The significance level α was chosen to be as large as possible but smaller than .05.

3 The statistic used is X, the number of bad potatoes in a sample of size 25. If the hypothesis is true, the sampling distribution of X is a binomial with parameters $n = 25$, $p = .10$ provided that

   (a) Each potato can be classified as bad or good.

   (b) The probability of obtaining a bad potato remains constant from trial to trial.

   (c) Potatoes are independently bad or good.

   (d) The number of trials is fixed.

   In Example 2-5 we decided that the assumptions are reasonable. (If the hypothesis is false, X still has a binomial distribution but with another value of p.)

4 Having decided that large x contradict $H_0 : p = .10$, we find by inspection in Appendix B1 that the critical region is $x = 6, 7, 8, \ldots, 25$,

or equivalently $x \geq 6$. The actual significance level (obtained from the table) is $\alpha = .03340$.

5 At this time the experiment is conducted. That is, 25 potatoes are examined so that the conditions in step 3 seem realistic. We observed $x = 5$. No further calculations are necessary.

6 Since the observed value of the statistic does not fall in the critical region, we accept H_0. The practical consequences of this decision are that the purchasing agent will buy the potatoes at the quoted price.

If the power conditions of Sec. 4-3 are used to determine n, then we would make a few changes in the outline. Specifically, we could write

1 Same as before.

2 It is desired to reject no oftener than .05 of the time if $p = .10$. If p is as poor as .25, the purchasing agent would like to reject at least .90 of the time.

3 The statistic used is X, the number of bad potatoes in a sample of size n, yet to be determined. For every n and p the random variable X has a binomial distribution provided the conditions (a), (b), (c), (d) discussed above are satisfied.

4 By inspection in large binomial table we find that the minimum n is 55 and the accompanying critical region is $x \geq 10$. (The actual significance level read from the table is $\alpha = .04442$.)

5 At this time 55 potatoes are examined. Since no x was given, let us assume that 8 bad potatoes are found.

6 Same as before.

EXAMPLE 4-2

Outline the solution for the cigarette nicotine problem of Sec. 4-4.

Solution

1 The problem suggests that we test $H_0 : \mu = 30$ against $H_1 : \mu < 30$. If we reject H_0 we can be reasonably sure that brand A is safe.

2 A sample size of $n = 100$ was selected. Although a smaller figure may be more appropriate, the significance level $\alpha = .01$ was chosen arbitrarily.

3 The statistic used is $Z = (\bar{X} - 30)/\sigma_{\bar{X}}$. The random variable Z has a standard normal distribution provided that

   (a) The cigarettes are selected at random.

   (b) Nicotine contents have a normal distribution.

   (c) The distribution has mean $\mu = 30$ milligrams and standard deviation $\sigma (= 8$ milligrams).

We try to collect the sample in a manner which makes (a) seem reasonable, probably taking one cigarette from a pack and selecting each pack from a different stock. With n as large as 100 the Central Limit theorem makes assumption (b) unnecessary. The assumption $\sigma = 8$ in (c) may not be realistic, but is made to simplify the problem (the case of unknown σ will be considered in the next chapter).

4 The critical region found from Appendix B6 is $z < -2.326$ (as shown in Fig. 4-4).

5 At this time the nicotine contents would be obtained. Our calculations gave $\bar{x} = 26$. We further compute $\sigma_{\bar{x}} = \sigma/\sqrt{n} = 8/\sqrt{100} = .8$ and

$$z = \frac{26 - 30}{.8} = -5$$

6 Since the observed value of the statistic falls in the critical region, H_0 is rejected. This means that brand A is regarded as safe. Not only is H_0 rejected at the .01 level of significance, but it is also rejected at the .0000003 level of significance. It is generally a good procedure to report the smallest significance level at which H_0 is rejected (subject to the entries of available tables).

Now suppose that the smoker had specified in advance that he wanted the power to be at least .90 if $\mu = 28$. Then we could write

1 Same as before.
2 A significance level $\alpha = .01$ is selected. In addition the smoker would like to reject at least .90 of the time if $\mu = 28$.
3 Same as before.
4 The minimum sample size is calculated to be $n = 208$ (Example 5-5). The critical region is $z = (\bar{x} - 30)/\sigma_{\bar{x}} < -2.326$ as before.
5 Now a random sample of size 208 cigarettes would be used. Suppose we get $\bar{x} = 28.35$. Then $\sigma_{\bar{x}} = 8/\sqrt{208} = 8/14.42 = .5548$ and

$$z = \frac{28.35 - 30}{.5548} = -2.97$$

6 Since z is in the critical region $H_0 : \mu = 30$ is rejected and brand A is adopted. If H_0 is true, then $\Pr(Z < -2.97) = .00205$, a very small probability.

EXAMPLE 4-3

Outline the solution for the tomato juice problem of Sec. 4-5.

Solution

1 The problem suggests that we test $H_0 : \mu = 46$ against $H_1 : \mu \neq 46$ where all unsatisfactory averages are included in the alternative.
2 A sample size of $n = 25$ was selected. The significance level $\alpha = .05$ is chosen because this is the same α used by the government inspector.

3   The statistic used is $Z = (\overline{X} - 46)/\sigma_{\overline{X}}$. The sampling distribution of Z is standard normal provided:
(a)   The cans of tomato juice are selected at random.
(b)   The weights of the contents have a normal distribution.
(c)   The distribution has mean $\mu = 46$ ounces and standard deviation $\sigma(=.5$ of an ounce).
In order to make (a) reasonable, one can might be selected off the production line every 2 minutes. Assumption (b) is not really necessary because of the Central Limit theorem. The assumption $\sigma = .5$ in (c) was made to simplify the problem.

4   The critical region found from Appendix B6 is $z < -1.960$ and $z > 1.960$ as pictured in Fig. 4-8.

5   The random sample of contents is now observed. Our calculations gave $\bar{x} = 46.18$. We also find $\sigma_{\overline{X}} = \sigma/\sqrt{n} = .5/\sqrt{25} = .1$ and

$$z = \frac{46.18 - 46}{.1} = 1.8$$

6   Since the observed value of the statistic does not fall in the critical region, H_0 is accepted. This implies that we operate as though the desired $\mu = 46$ is being maintained. In other words, no corrective action is taken.

EXERCISES

◇ 4-7   Write out the outline for the problem described in Exercise 4-1. Use the largest α possible that is less than .01 and assume that the observed value x is 9.

4-8   Write out the outline for the problem described in Exercise 4-5. That is, use $\alpha = .10$. Then take $\bar{x} = 46.18$ for the observed value of the statistic.

Summary of Results
TESTING STATISTICAL HYPOTHESES

I. Definitions

Hypothesis: An assumption about the distribution of a random variable, an assumption about one or more parameters of a distribution.

Test: A rule indicating when to reject a hypothesis.

Alternative hypothesis: The conclusion one draws if the hypothesis is rejected (also an assumption about parameters).

Type I error: The kind of mistake made when a true hypothesis is rejected.

Level of significance: The probability of committing a Type I error (or, as we see later, the maximum of such probabilities).

Critical region: The set of outcomes for an experiment which lead to rejection of the hypothesis.

Acceptance region: That set of outcomes not included in the critical region and which lead to acceptance of the hypothesis.

Type II error: The kind of mistake made when a false hypothesis is accepted.

Operating characteristic: The probability of accepting the hypothesis.

Power: The probability of rejecting the hypothesis (1 − the operating characteristic).

Power curve: A graph of values of power.

II. General comments

Generally, the hypothesis and the alternative hypothesis are selected before any experimentation is performed.

Sample size determination and sampling procedures should be considered in the planning stage of an experiment (not as an afterthought).

When possible, it is good procedure to have as the alternative state what the experimenter would like to prove.

When a hypothesis is accepted, we usually do not regard this as proof that the hypothesis is true. Until further evidence is available, we may, however, operate under the assumption that the hypothesis is true.

Rejection of a hypothesis is a more satisfying statistical decision since this is usually regarded as " proof " that the alternative hypothesis is true.

Good common sense is a great asset in practical applications of hypothesis testing.

5
HYPOTHESIS TESTING
AND INTERVAL ESTIMATION
FOR MEANS AND VARIANCES

5-1
INTRODUCTION

The material in Chap. 4 can be classified as a "general discussion." Examples were used to illustrate the basic concepts of hypothesis testing and to introduce part of the vocabulary used by statisticians. No attempt was made to develop skills or procedures that might help us work specific problems. The latter is the major goal of this and the following chapters. In the remainder of the text we will make a systematic study of some of the standard hypothesis testing situations.

When one is beginning the study of a field (here, hypothesis testing), the new terminology may at first seem like a major stumbling block. Undoubtedly, it is too much to expect complete mastery of terms like hypothesis, alternative, level of significance, power, critical region, test, acceptance region, sample size, random sample, Type I error, and Type II error after the short discussion of the previous chapter. However, these words are used repeatedly in the remainder of the text and after a few more exposures should become part of our working vocabulary.

We have discussed just enough hypothesis testing to realize that we need more experience if we are to work problems, experience that will help us select hypotheses, alternatives, significance levels, sample sizes, test statistics, and critical regions. The remainder of our study of statistical inference is aimed at providing such experience. Most of the essential ideas have already been presented. Now it is a matter of organizing those ideas to fit specific cases.

The procedures which we will describe are based upon sound mathematical principles. In particular, most of the tests have the property that their power curves are in general better than power curves associated with other tests of the same hypothesis against the same alternative. The mathematical justification of these facts is a topic for more advanced courses in statistics. Our main interest lies in learning how and when to use certain standard techniques.

188

Another possible title for this chapter could be "Some Statistical Inferences for Normal Distributions." The tests to be discussed are derived under the assumption that random samples are drawn from normal distributions. Experience has shown that the assumption is not crucial for the types of inferences which we are going to make about means, provided the sample size is not too small. What this implies is that probability statements involving means will be approximately correct even if the underlying model is not normal. (The Poisson, which also contains μ, will be discussed in Chap. 6.) A similar statement cannot be made concerning inferences about variances. [We have already discussed the latter statement in connection with formula (3-29).] Now the accuracy of probability statements depends upon having normal (or nearly normal) distributions.

5-2
TESTING HYPOTHESES CONCERNING A MEAN WHEN σ^2 IS KNOWN

As we have previously indicated, the variance will not be known in most real-life testing situations. This assumption, which will be eliminated in Sec. 5-5, does permit some simplification which can be used to advantage for introducing new concepts.

For a mean μ the three most used choices for hypothesis and alternative were illustrated in the examples of the preceding chapter. The cigarette problem of Sec. 4-4 is a special case of testing

$$H_0 : \mu = \mu_0 \quad \text{against} \quad H_1 : \mu < \mu_0 \tag{5-1}$$

in which $\mu_0 = 30$ milligrams. Frequently, one may be interested in testing

$$H_0 : \mu = \mu_0 \quad \text{against} \quad H_1 : \mu > \mu_0 \tag{5-2}$$

This situation was demonstrated in Sec. 4-5 with the tomato juice problem and $\mu_0 = 46$ ounces, although it was incorrectly used because the alternative was suggested by the sample. Finally, the correct analysis at the tomato juice cannery led to testing

$$H_0 : \mu = \mu_0 \quad \text{against} \quad H_1 : \mu \neq \mu_0 \tag{5-3}$$

with $\mu_0 = 46$ ounces. As our examples have demonstrated, μ_0 takes on a value suggested by the particular problem.

In Sec. 4-5 we remarked that it may make more sense to replace (5-1) by

$$H_0 : \mu \geq \mu_0 \quad \text{against} \quad H_1 : \mu < \mu_0 \tag{5-4}$$

and (5-2) by

$$H_0 : \mu \leq \mu_0 \quad \text{against} \quad H_1 : \mu > \mu_0 \tag{5-5}$$

thus grouping all "satisfactory" values of μ in one set, "unsatisfactory" values in another set. If these replacements are made, test procedures are unchanged. We must, however, redefine significance level as the maximum value of the power when H_0 is true. This maximum will occur when $\mu = \mu_0$, the boundary value between the two sets. Further comments in terms of an example will be given later in the section.

When the variance is known, the statistic

$$Z = \frac{\bar{X} - \mu_0}{\sigma/\sqrt{n}} \qquad (5\text{-}6)$$

is used for all of the above testing situations. The sampling distribution of Z is standard normal provided (a) $X_1, X_2, \ldots, X_n$ is a random sample from a distribution (the requirement used in the definition of $\bar{X}$), (b) that distribution is normal, and (c) the distribution has mean μ_0 and variance σ^2. As we have observed, because of the Central Limit theorem assumption (b) is unnecessary if n is not too small.

When testing (5-1) or (5-4), only small $\bar{x}$, and hence small z, support the alternative H_1. Consequently, H_0 is rejected when z is small. If the significance level is α, then the critical region is

$$z = \frac{\bar{x} - \mu_0}{\sigma/\sqrt{n}} < z_\alpha \qquad (5\text{-}7)$$

where $\Pr(Z < z_\alpha) = \alpha$ if $\mu = \mu_0$ (see Fig. 5-1). This is the same as Fig. 4-4 except that in that picture $\alpha = .01$. Similarly, only large $\bar{x}$, and large z, support the alternative H_1 of (5-2) and (5-5). A test with significance level α is obtained by rejecting H_0 when

$$z = \frac{\bar{x} - \mu_0}{\sigma/\sqrt{n}} > z_{1-\alpha} \qquad (5\text{-}8)$$

FIGURE 5-1

Density of Z (standard normal) if $\mu = \mu_0$, showing significance level α and critical region $z < z_\alpha$.

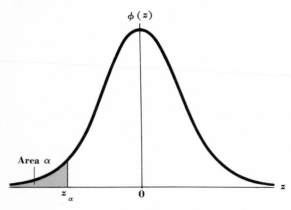

FIGURE 5-2

Density of Z (standard normal) if $\mu = \mu_0$, showing significance level α and critical region $z > z_{1-\alpha}$.

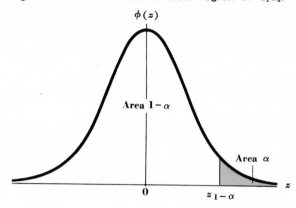

where $\Pr(Z > z_{1-\alpha}) = \alpha$ if $\mu = \mu_0$ (see Fig. 5-2). Figure 4-10 conveys the same information with $\alpha = .05$. Finally, when testing (5-3), both small and large $\bar{x}$, and hence small and large z, provide support for H_1. In other words, it is reasonable to expect that both small and large values of z will fall in the critical region. The intuitively appealing critical region which we shall use is

$$z = \frac{\bar{x} - \mu_0}{\sigma/\sqrt{n}} < z_{\alpha/2} \quad \text{and} \quad z = \frac{\bar{x} - \mu_0}{\sigma/\sqrt{n}} > z_{1-\alpha/2} \tag{5-9}$$

the "equal-tails" choice. If $\mu = \mu_0$, then $\Pr(Z < z_{\alpha/2}) + \Pr(Z > z_{1-\alpha/2}) = \alpha/2 + \alpha/2 = \alpha$ (see Fig. 5-3). Figure 4-8 is the same except that in that picture $\alpha = .05$. Because of the symmetry of the standard normal density curve, we have $z_{\alpha/2} = -z_{1-\alpha/2}$.

EXAMPLE 5-1

A large corporation uses thousands of light bulbs every year. The brand that has been used in the past has an average life of 1,000 hours with a standard deviation of 100 hours. A new brand is offered to the corporation at a price far lower than the one they are paying for the old brand. It is decided that they will switch to the new brand unless it is "proved" with a level of significance of $\alpha = .05$ that the new brand has smaller average life than the old brand. A random sample of 100 new-brand bulbs is tested, yielding an observed sample mean of 985 hours. Assuming that the standard deviation for the new brand is the same as for the old, what conclusion should be drawn and what decision should be made?

FIGURE 5-3

Density of Z (standard normal) if $\mu = \mu_0$, showing significance level α and critical region $z < z_{\alpha/2}$ and $z > z_{1-\alpha/2}$.

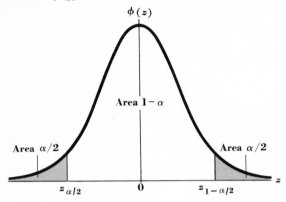

$\phi(z)$

Area $1-\alpha$

Area $\alpha/2$

Area $\alpha/2$

$z_{\alpha/2}$ 0 $z_{1-\alpha/2}$ z

Solution

We shall follow the outline proposed in Sec. 4-7.

1 If we let μ be the unknown average associated with the new brand, then the problem suggests that we test $H_0 : \mu = 1{,}000$ (or $H_0 : \mu \geq 1{,}000$) against $H_1 : \mu < 1{,}000$. If H_0 is rejected, then the observed sample average " proves" that the new-brand mean is lower than the old-brand mean.

2 The significance level $\alpha = .05$ and the sample size $n = 100$ have already been chosen.

3 The statistic we will use is $Z = (\bar{X} - 1{,}000)/(\sigma/\sqrt{n})$, which has a standard normal distribution provided:

(a) The sample of light bulbs tested was randomly selected.

(b) The length of life for new-brand light bulbs is normally distributed.

(c) The distribution has mean $\mu_0 = 1{,}000$ and standard deviation $\sigma = 100$ hours.

We are given that the sample was randomly selected. Condition (b) may not be satisfied, but this is immaterial with a sample this large.

4 The critical region given by (5-7) is

$$z = \frac{\bar{x} - 1{,}000}{\sigma/\sqrt{n}} < z_{.05} = -1.645$$

(Fig. 5-1 applies with $\alpha = .05$.)

5 At this time it would be appropriate to observe the sample results. We have $\sigma = 100$, $n = 100$, $\bar{x} = 985$. Thus

$$z = \frac{\bar{x} - 1{,}000}{\sigma/\sqrt{n}} = \frac{985 - 1{,}000}{100/\sqrt{100}} = \frac{-15}{10} = -1.5$$

6   Since the observed value of the statistic does not lie in the critical region, the hypothesis that the new-brand average is 1,000 (or more) is not rejected. The practical consequences of this conclusion are that the company will switch to the new brand.

The power of test used in Example 5-1 is $\Pr[(\overline{X} - 1,000)/10 < -1.645]$. This, of course, yields .05 if $\mu = 1,000$. Values of power for other μ could be computed as in Sec. 4-4 (or better yet, as in Sec. 5-3). Such calculations would allow us to draw the power curve shown in Fig. 5-4. If we use as $H_0 : \mu \geq 1,000$, the hypothesis is true if the actual value of μ is any number greater than or equal to 1,000. With each such μ there is a probability of rejection. Hence, it does not make sense to talk about *the* probability of rejecting H_0 (our former definition of significance level) since there are many such probabilities. From the graph we see that the largest of these values is .05, associated with $\mu = \mu_0 = 1,000$. If we are going to permit μ (or any parameter θ) to assume more than one value under H_0, then the significance level should be defined as the maximum value of the power when H_0 is true. In all cases which we will consider, this maximum is α and occurs at $\mu = \mu_0$ (or in general at $\theta = \theta_0$), the boundary point between the two sets.

FIGURE 5-4

Power curve associated with the test used in Example 5-1.

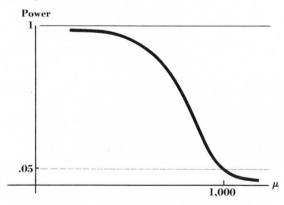

Returning to Example 5-1, we recognize that the problem has been oversimplified. In an actual situation other considerations such as cost per hour or side benefits from extra good service may influence us to make a different decision. Let us consider a slightly different problem concerning new-brand light bulbs which will lead to some major changes in the solution.

EXAMPLE 5-2

The large corporation of Example 5-1 is considering switching to a new brand which costs about the same as its present old brand but supposedly (at least according to the salesman) has a longer average life. The corporation has enjoyed a good relationship with the old-brand company, and is reluctant to change unless there is outstanding evidence to support the new-brand salesman's claim. Now what hypothesis and alternative is likely to be selected? If α is chosen as .01 and 100 new-brand light bulbs yield $\bar{x} = 1,020$, what action should be taken?

Solution

We again follow the outline.

1. If we let μ be the unknown average associated with the new brand, then the problem suggests that we now test $H_0 : \mu = 1,000$ (or $H_0 : \mu \leqq 1,000$) against $H_1 : \mu > 1,000$. The new brand is not adopted unless the sample supports H_1. The burden of proof is thus placed upon the new brand.

2. Again we have $n = 100$, but α is now .01.

3. The comments are the same as in Example 5-1. We still use

$$Z = \frac{\overline{X} - 1,000}{\sigma/\sqrt{n}}$$

4. This time H_0 is rejected if z is large. The critical region given by (5-8) is $z = (\bar{x} - 1,000)/(\sigma/\sqrt{n}) > z_{.99} = 2.326$.

5. Now we perform the sampling and calculate

$$z = \frac{\bar{x} - 1,000}{\sigma/\sqrt{n}} = \frac{1,020 - 1,000}{100/\sqrt{100}} = \frac{20}{10} = 2$$

6. Again the observed value of the statistic does not lie in the critical region, and it is concluded that the sample provides insufficient evidence to conclude that the new brand has an average life longer than 1,000 hours. Consequently the company will continue to use the old brand.

EXERCISES

Use the recommended outline.

◆ 5-1   A company that sells frozen shrimp prints "Contents 12 Ounces" on each package. It is undesirable to have the average contents μ

be either less or greater than 12 ounces. An observed random sample of 25 packages yields an average of 11.83 ounces. Suppose it is known from past experience that the distribution of package weights has a standard deviation of .5 of an ounce. If we take $\alpha = .05$, what conclusion should be drawn concerning the mean of the distribution of package weights?

5-2　A manufacturer of flashlight batteries claims that his product has an average life in excess of 30 hours. A company is willing to buy a very large shipment of batteries but only if it is convinced that the claim is true. An observed random sample of 36 batteries produces an observed sample mean of 40.0 hours. If it is assumed that the distribution of lengths of life for these batteries has a standard deviation of 5 hours, is it likely that the company will make the purchase?

◈ 5-3　The amount of electricity used by a certain brand of 100-watt bulb during a 10-minute period is normally distributed. A merchandising chain will buy a large number of these bulbs if it can be convinced that the mean of the normal distribution is less than 50 units. Based upon prior investigations it is reasonable to assume that $\sigma = 4$ units. In order to make a decision the merchandising chain experiments with 49 bulbs and finds $\bar{x} = 48.9$ units. The appropriate hypothesis is to be tested with a significance level of .05. What action should the chain take?

5-4　Consider again the 100-watt bulbs and the merchandising chain discussed in Exercise 5-3. Suppose that the chain will make the purchase unless sample results indicate that the mean of distribution is in excess of 50 units. Using $\alpha = .05$, what action will be taken if a sample of size 49 yields $\bar{x} = 50.7$ units?

◈ 5-5　A psychologist raises rats to be used in his research. He has found that for a standard variety the length of time required for a rat to reach maturity is normally distributed with a mean of 240 days and a standard deviation of 30 days. A new variety that is supposed to mature faster is being considered as a replacement. Some inconvenience and expense would be involved to replace the old variety by the new. However, the psychologist decides that it will be worthwhile to make the change unless he cannot believe that the new variety average is 220 days. His decision will be based upon observations with 16 rats. If he assumes that $\sigma = 30$ days and uses $\alpha = .10$, what is the appropriate action if the observed sample mean is 229 days?

◈ 5-6　Suppose that the psychologist of Exercise 5-5 decides that he will change varieties if it can be "proved" that the new variety has a mean less than 240 days. Using the same α and the same sample results, test the appropriate hypothesis and recommend the course of action.

◈ 5-7   A corn grower has been planting variety "long ear" for many years. It has been found that this variety has an average yield of 80 bushels per acre with a standard deviation of 10 bushels per acre. A new variety "fat kernel" has been placed upon the market. Due to the higher cost of seed the grower decides he cannot switch to "fat kernel" unless it can be "proved" that the average yield exceeds 90 bushels per acre. Experimentation is conducted with 25 randomly selected acres, and a sample average of 93 bushels is obtained. If the grower decides to use $\alpha = .05$ and assumes $\sigma = 10$ bushels per acre, what action will he take?

5-8   Suppose that the average course grade in statistics is 50 when the subject is taught meeting three times a week in large lecture sections. A statistics department is considering the addition of extra problem sessions meeting twice a week. However, since some expense is entailed, it is felt that the addition is not worthwhile unless there is reasonable evidence to indicate that the course average can be raised above 60. In order to make a decision a randomly selected class of 64 students is taught with five-meeting system and the average of their course grades is 61.7. Assuming $\sigma = 10$, the same standard deviation associated with the three-meeting method, what conclusion is reached with $\alpha = .05$?

◈ 5-9   One part of the entrance examination given at the University of Wyoming to incoming freshmen is on the mathematics. For the years 1940 to 1960 the distribution of math scores for Wyoming high school graduates had a mean of 50 and a standard deviation of 10. Since 1960 the school systems in Wyoming have been gradually converting to the "new math." In 1970 an average score of 48.2 was found for 400 Wyoming high school graduates. If the 1970 test is substantially the same as those given in former years and if it is judged that the old σ is still applicable, what conclusion should be drawn with $\alpha = .01$?

5-10   A company manufactures tires for automobiles. Under a standard set of driving conditions its brand will average 20,000 miles with a standard deviation of 1,000 miles. A new process for making tires is being considered, but the company cannot afford to convert to the new process unless there is outstanding evidence to indicate that a larger average will be achieved. It is decided to use the significance level .01 and a sample of size 100. Assuming that the standard deviation for the new process is the same as for the old, what action should be taken if the sample yields $\bar{x} = 20,200$ miles?

◈ 5-11   A fisherman decides that he needs a line that will test more than 10 pounds if he is to be able to catch the size fish he desires. He tests 16 pieces of brand F line for breaking strength and finds an observed sample mean of 10.4. If it is known that $\sigma = .5$ of a pound and a significance level of .01 is used, what can he conclude about brand F?

5-3
POWER OF TESTS ABOUT MEANS

We have already demonstrated one method of calculating power of tests for the mean based upon the statistic $Z = (\overline{X} - \mu_0)/(\sigma/\sqrt{n})$. For the hypothesis testing situations (5-1) or (5-4) with critical region (5-7), the power is

$$\Pr\left(\frac{\overline{X} - \mu_0}{\sigma/\sqrt{n}} < z_\alpha\right) = \Pr\left(Z < z_\alpha + \frac{\mu_0 - \mu}{\sigma/\sqrt{n}}\right) \tag{5-10}$$

where Z has a standard normal distribution. Actually formula (5-10) is only a slight generalization of formula (4-12). Similarly, for the hypothesis testing situations (5-2) or (5-5) with critical region (5-8), the power can be written

$$\Pr\left(\frac{\overline{X} - \mu_0}{\sigma/\sqrt{n}} > z_{1-\alpha}\right) = \Pr\left(Z > z_{1-\alpha} + \frac{\mu_0 - \mu}{\sigma/\sqrt{n}}\right) \tag{5-11}$$

Formulas (5-10) and (5-11) can be combined in one formula holding for both cases, the result being

$$\text{Power} = \Pr\left(Z < z_\alpha + \frac{|\mu - \mu_0|}{\sigma/\sqrt{n}}\right) \tag{5-12}$$

provided μ is restricted to be a value in the alternative (the interesting cases). If α is replaced by $\alpha/2$, then for all practical purposes formula (5-12) gives very nearly correct probabilities when the alternative is two-sided and the critical region is (5-9).

Although formula (5-12) is easy to evaluate, the values of power can be read from the graphs of Appendix B10 provided $\alpha = .005, .01, .025, .05$. The quantity

$$\delta = \frac{|\mu - \mu_0|\sqrt{n}}{\sigma} \tag{5-13}$$

is used to enter the graph on the horizontal axis and power is read from the vertical axis. The curve labeled $f = \infty$ is the one which applies to our tests based upon the Z statistic. The other curves are for the t test, which will be discussed in Sec. 5-5. As with formula (5-12), μ must be in the range of the alternative. The graphs are constructed to give power for the one-sided tests, but they can be used for two-sided tests with practically no error if $\delta \geq 1$ and with a maximum error of $\alpha/2$ if $\delta < 1$. To use the graphs for two-sided tests, enter Appendix B10 with one-half of the significance level, $\alpha/2$.

EXAMPLE 5-3

In discussing the smoker problem in Sec. 4-4, we found using the standard normal tables and $\alpha = .01$,

Power $= .141$ if $\mu = 29$
Power $= .569$ if $\mu = 28$
Power $= .923$ if $\mu = 27$

Verify these figures with the proper graph of Appendix B10.

Solution

We recall that $n = 100$, $\sigma = 8$, $\mu_0 = 30$. Then we get for formula (5-13)

if $\mu = 29$ $\delta = |29 - 30| \sqrt{100}/8 = 1.25$
if $\mu = 28$ $\delta = |28 - 30| \sqrt{100}/8 = 2.50$
if $\mu = 27$ $\delta = |27 - 30| \sqrt{100}/8 = 3.75$

Turning to the page of Appendix B10 labeled $\alpha = .01$, we see that the vertical line through $\delta = 1.25$ crosses the curve $f = \infty$ opposite power $= .14$. Similarly, with $\delta = 2.50, 3.75$ we find power $= .57, .92$.

EXAMPLE 5-4

In Sec. 4-5 for the tomato juice example, we found using the standard normal tables and $\alpha = .05$,

Power $= .170$ if $\mu = 46.1, 45.9$
Power $= .516$ if $\mu = 46.2, 45.8$
Power $= .851$ if $\mu = 46.3, 45.7$

Verify these figures with the proper graph of Appendix B10.

Solution

We recall that $n = 25$, $\sigma = .5$, $\mu_0 = 46$. Then we get for formula (5-13)

if $\mu = 46.1$ $\delta = |46.1 - 46| \sqrt{25}/.5 = 1$
if $\mu = 46.2$ $\delta = |46.2 - 46| \sqrt{25}/.5 = 2$
if $\mu = 46.3$ $\delta = |46.3 - 46| \sqrt{25}/.5 = 3$

We find the page labeled $\alpha = .025$ ($\alpha/2 = .05/2 = .025$), which gives power for a two-sided test with $\alpha = .05$. From the $f = \infty$ curve we find that $\delta = 1, 2, 3$ yields power $= .17, .52, .85$, respectively. Since 45.9, 45.8, 45.7 again yield $\delta = 1, 2, 3$, the same values of power are obtained.

One of the most important uses of power curves arises from sample size considerations. We recall from Chap. 4 the discussion concerning the minimum sample size required if the test is to satisfy two power conditions. To be more specific, suppose that if $\mu = \mu_0$ we want the power (significance level) to be α while if $\mu = \mu_1$, a value in the alternative, we want the power to be at least $1 - \beta$ (see Fig. 5-5). We seek the mini-

FIGURE 5-5

Power curve satisfying two specified conditions.

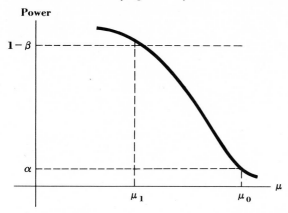

mum n which will meet these requirements. For the tests of Sec. 5-2 the solution is quickly found from the graphs of Appendix B10. In formula (5-13) $\mu = \mu_1$, μ_0, and σ are known from the problem, and δ can be read from the appropriate graph being the value which makes the power $1 - \beta$. (In other words, we use the graph backwards entering with power and reading δ, instead of vice versa.) Then, solving formula (5-13) for n yields

$$n = \left(\frac{\sigma \delta}{\mu - \mu_0} \right)^2 \qquad (5\text{-}14)$$

where we use $\mu = \mu_1$. The desired solution is the next largest integer.

EXAMPLE 5-5

In Example 5-3 we found that $n = 100$ produced a power of .57 when $\mu = 28$. If we retain $\alpha = .01$, how large does n have to be to make the power be at least .90 if $\mu = 28$?

Solution

We have $\sigma = 8$, $\mu_0 = 30$, $\mu_1 = 28$. Turning to the page of Appendix B10 labeled $\alpha = .01$, we find that the horizontal line through Power $= .90$ intersects the $f = \infty$ curve at $\delta = 3.6$. Hence we calculate

$$n = \left[\frac{8(3.6)}{28 - 30} \right]^2 = (14.4)^2 = 207.36$$

The next largest integer is $n = 208$, the required sample size.

EXAMPLE 5-6

Using $\alpha = .05$ and $n = 25$, we found in Example 5-4 that for $\mu = 46.2$ the power of the test was .516 (or .52 rounded to two decimal places). How large should n be to raise this to .98?

Solution

We have $\sigma = .5$, $\mu_0 = 46$, $\mu_1 = 46.2$. Turning to the page of Appendix B10 labeled $\alpha = .025$ (one-half the significance for a two-sided test), we find from the $f = \infty$ curve that $\delta = 4$ is needed to acheive a power of .98. Then, formula (5-14) gives

$$n = \left[\frac{.5(4)}{46.2 - 46} \right]^2 = (10)^2 = 100$$

Hence the minimum sample size is $n = 100$.

Another formula yielding the same result as formula (5-14) is obtained by using formula (5-10). If the power at $\mu = \mu_1$ is to be at least $1 - \beta$, then we must have

$$\Pr\left(Z < z_\alpha + \frac{\mu_0 - \mu_1}{\sigma/\sqrt{n}} \right) \geq 1 - \beta$$

which implies that

$$z_\alpha + \frac{\sqrt{n}(\mu_0 - \mu_1)}{\sigma} \geq z_{1-\beta} = -z_\beta$$

Solving for n yields

$$n \geq \left[\frac{\sigma(z_\alpha + z_\beta)}{\mu_1 - \mu_0} \right]^2 \tag{5-15}$$

Thus, if we use formula (5-15), we calculate $\delta = -(z_\alpha + z_\beta)$ rather than reading this quantity from the graph. Formula (5-15) also holds both for one-sided cases and approximately for the two-sided case if α is replaced by $\alpha/2$. If we rework Example 5-5 using the latter formula, we have $z_\alpha = z_{.01} = -2.326$, $z_\beta = z_{.10} = -1.282$, and we get $n \geq (14.432)^2 = 208.28$, or $n = 209$. The minor difference (from the previously obtained

$n = 208$) is due to our inability to reproduce and read graphs more accurately.

EXERCISES

◈ 5-12   In Exercise 5-1 find the power if $\mu = 11.8$ ounces. The company might want to be fairly certain that a change in average will be detected if the mean weight has shifted .2 of an ounce. Suppose that a power of at least .99 is required if $\mu = 11.8$. If the original significance level $\alpha = .05$ is retained, what is the minimum sample size required to satisfy both power conditions?

5-13   For the problem of Exercise 5-2 find the power if $\mu = 32$ hours using a significance level of .01. What is the minimum sample size needed to raise the power at $\mu = 32$ to at least .90?

◈ 5-14   For the problem of Exercise 5-3, find the probability that the merchandising chain will buy the 100-watt bulbs if the average amount of electricity consumed is 48 units. What is the minimum sample size required to make this probability at least .90 if the significance level is kept at .05?

5-15   In Exercise 5-4 find the probability that the merchandising chain will not buy the 100-watt bulbs if the average amount of electricity consumed is 51 units. Suppose the chain insists that this probability should be at least. 90. With a significance level of .05, what is the minimum sample size needed to meet the given conditions?

◈ 5-16   Consider again the psychologist of Exercise 5-5. What is the probability that he will not switch to the new variety of rats if their average age at maturity is 230 days? In order to raise this probability to at least .90 with $\alpha = .10$, what is the minimum number of rats required for experimentation? [*Hint:* Use formulas (5-12) and (5-15).]

5-17   In Exercise 5-7 find the probability that the corn grower will switch to "fat kernel" if the average yield is 92 bushels per acre. With $\alpha = .05$, what is the minimum number of acres required for experimentation if this probability is to be at least .80?

◈ 5-18   Consider again the mathematics examination discussed in Exercise 5-9, and suppose that the average score has changed by 2 units. What is the probability that such a change will be detected with a sample of 400? With $\alpha = .01$, how large a sample would be required to make the probability at least .80 of detecting a change of 1 unit in the average score?

5-19   In Exercise 5-11 suppose that on the average the line actually has a breaking strength of 10.4 pounds. What is the probability that the fisherman will conclude that brand F meets his requirements? To raise this probability to at least .90 with $\alpha = .01$ requires what minimum sample size?

5-4

CONFIDENCE INTERVALS

In the examples and exercises concerning hypothesis testing, the problem under consideration has suggested a value (or values) for a hypothesis and an alternative. If a problem does not imply such information, perhaps we will be interested in obtaining either a point estimate (discussed in Chap. 3) or an interval estimate. Even if a hypothesis is tested, one or both of these estimates may be useful, particularly if the hypothesis is rejected. Since interval estimates are closely related to tests, we shall generally discuss the two topics in conjunction with one another.

Suppose that the sample of light bulbs in Example 5-1 produces an observed sample mean of 950 hours (instead of 985). The hypothesis that $\mu = 1,000$ (or $\mu \geq 1,000$) would then be rejected. Of course, our best guess for the average life is the point estimate $\bar{x} = 950$ hours. Since estimates may vary considerably from sample to sample, we would probably find it more satisfying to know that we are fairly confident that the unknown average μ is between 930.4 and 969.6. Let us see how the interval and the degree of confidence are obtained.

Regardless of the actual value of μ, the random variable $Z = (\bar{X} - \mu)/(\sigma/\sqrt{n})$ has a standard normal distribution if the X distribution is normal. This statement is even approximately correct for nonnormal distributions provided n is not too small. One of many probability statements which we could write is

$$\Pr\left(-1.960 < \frac{\bar{X} - \mu}{\sigma/\sqrt{n}} < 1.960\right) = .95 \tag{5-16}$$

which is a special case of

$$\Pr\left(-z_{1-\alpha/2} < \frac{\bar{X} - \mu}{\sigma/\sqrt{n}} < z_{1-\alpha/2}\right) = 1 - \alpha \tag{5-17}$$

Consider the two inequalities in formula (5-16). One is $-1.960 < (\bar{X} - \mu)/(\sigma/\sqrt{n})$. Since inequalities, like equations, can be multiplied by a positive number, we can write $-1.960(\sigma/\sqrt{n}) < \bar{X} - \mu$. The same number can be added to or subtracted from each side of an inequality without changing the sense of the inequality. Hence, if $-1.960(\sigma/\sqrt{n}) < \bar{X} - \mu$ is satisfied, so is $-\bar{X} - 1.960(\sigma/\sqrt{n}) < -\mu$. Finally, if both sides of an inequality are multiplied by a negative number (in particular by -1), the sense is reversed. Thus, if $-\bar{X} - 1.960(\sigma/\sqrt{n}) < -\mu$, then $\bar{X} + 1.960(\sigma/\sqrt{n}) > \mu$ or $\mu < \bar{X} + 1.960(\sigma/\sqrt{n})$. (These rules for manipulation of inequalities can be quickly recalled by numerical illustrations.

For example, the inequality $2 < 4$ remains true in sense if both sides are divided by 2, becoming $1 < 2$. However, if both sides are divided by -2, it becomes $-1 > -2$, with the sense reversed.) Similarly, the second inequality $(\overline{X} - \mu)/(\sigma/\sqrt{n}) < 1.960$ can be rewritten $\overline{X} - 1.960(\sigma/\sqrt{n}) < \mu$. Thus

$$-1.960 < \frac{\overline{X} - \mu}{\sigma/\sqrt{n}} < 1.960 \quad \text{and} \quad \overline{X} - 1.960\frac{\sigma}{\sqrt{n}} < \mu < \overline{X} + 1.960\frac{\sigma}{\sqrt{n}}$$

are equivalent inequalities. That is, if the former is true, then so is the latter, and vice versa. Consequently we can rewrite formula (5-16) as

$$\Pr\left(\overline{X} - 1.960\frac{\sigma}{\sqrt{n}} < \mu < \overline{X} + 1.960\frac{\sigma}{\sqrt{n}}\right) = .95$$

In the light bulb example we had $\sigma = 100$, $\sqrt{n} = 10$, so that $1.960(\sigma/\sqrt{n}) = 19.60$. Hence we can write

$$\Pr(\overline{X} - 19.60 < \mu < \overline{X} + 19.60) = .95$$

Both end points of the interval $(\overline{X} - 19.60, \overline{X} + 19.60)$ are random variables, and the interval is a random interval. Now suppose the sample results are obtained, and we get $\bar{x} = 950$. We can then calculate a specific interval $(950 - 19.6, 950 + 19.6)$ or $(930.4, 969.6)$. This specific interval is called a *confidence interval*. The probability .95 is known as the *confidence coefficient*. Similarly, formula (5-17) can be rewritten as

$$\Pr\left(\overline{X} - z_{1-\alpha/2}\frac{\sigma}{\sqrt{n}} < \mu < \overline{X} + z_{1-\alpha/2}\frac{\sigma}{\sqrt{n}}\right) = 1 - \alpha \qquad (5\text{-}18)$$

The interval $(\overline{X} - z_{1-\alpha/2}\,\sigma/\sqrt{n},\ \overline{X} + z_{1-\alpha/2}\,\sigma/\sqrt{n})$ is a random interval from which, after observing $\bar{x}$ from a particular random sample, we can calculate the confidence interval

$$\left(\bar{x} - z_{1-\alpha/2}\frac{\sigma}{\sqrt{n}},\ \bar{x} + z_{1-\alpha/2}\frac{\sigma}{\sqrt{n}}\right) \qquad (5\text{-}19)$$

which has confidence coefficient $1 - \alpha$.

It is easy to evaluate the interval (5-19) as we shall demonstrate in the next example, but first let us discuss the interpretation attached to a confidence interval like (930.4, 969.6). This particular interval may or may not contain μ, the average length of life for the new brand of light bulb. Another experimenter with another observed random sample might get $\bar{x} = 960$ and an interval (940.4, 979.6), also calculated from formula (5-19) with $1 - \alpha = .95$. This second interval also may or may not capture μ. Likewise another experiment will produce still another interval. What we do know is that if many, many intervals are calculated using formula (5-19) with the same $1 - \alpha$ (here .95), then in the long run about $1 - \alpha$ of the intervals will contain μ and about α of the intervals will not. Our interval-forming procedure is somewhat analogous to throwing intervals at a point target in such a way that the target will be covered by the intervals a fraction $1 - \alpha$ of the time.

EXAMPLE 5-7

Using the information given in Sec. 4-4, compute a confidence interval with confidence coefficient .99 for the average nicotine content of brand A cigarettes.

Solution

From Sec. 4-4 we have $\bar{x} = 26$, $\sigma = 8$, $n = 100$. Since the confidence coefficient $1 - \alpha = .99$, we have $\alpha = .01$, $\alpha/2 = .005$, $1 - \alpha/2 = .995$. From the last row of Appendix B6 we find $z_{.995} = 2.576$. Thus, the interval (5-19) becomes $[26 - 2.576(8/\sqrt{100}), 26 + 2.576(8/\sqrt{100})]$, which reduces to $(23.94, 28.06)$.

Since we used the same statistic $Z = (\bar{X} - \mu)/(\sigma/\sqrt{n})$ to obtain a confidence interval for μ and to test hypotheses about μ, we might expect that the interval and one of the tests are equivalent. Suppose that in the light bulb example we had tested $H_0 : \mu = 1{,}000$ against $H_1 : \mu \neq 1{,}000$ with $\alpha = .05$. With an observed sample mean of $\bar{x} = 950$ we find $z = -5$ leading to rejection of H_0. The confidence interval obtained with $\bar{x} = 950$ was $(930.4, 969.6)$, an interval which does not contain $\mu_0 = 1{,}000$. On the other hand with an observed sample mean of $\bar{x} = 985$ we find $z = -1.5$, which does not lie in the critical region $z < -1.960$ and $z > 1.960$, and the corresponding confidence interval is $(985 - 19.6, 985 + 19.6) = (965.4, 1004.6)$ which does contain $\mu_0 = 1{,}000$. The conclusion to be drawn is that provided the same α is used the test of $H_0 : \mu = \mu_0$ against $H_1 : \mu \neq \mu_0$ accepts H_0 if, and only if, the interval (5-19) contains μ_0. Thus, if we so desired, to test H_0 in this case we could compute the interval rather than the observed value of the Z statistic.

The length of the interval given by formula (5-19) is

$$\bar{x} + z_{1-\alpha/2}\frac{\sigma}{\sqrt{n}} - \left(\bar{x} - z_{1-\alpha/2}\frac{\sigma}{\sqrt{n}}\right) = 2z_{1-\alpha/2}\frac{\sigma}{\sqrt{n}} \tag{5-20}$$

We note two rather obvious facts. First, the larger $z_{1-\alpha/2}$, the longer the confidence interval. The higher the probability we associate with the interval, the longer it becomes for fixed n. This is only common sense, since the longer the interval becomes, the more confident we can be that it contains μ. However, it is not as useful to know that μ is between 0 and 5,000 as it is to know μ is between 940 and 960. Second, we observe that the length can be decreased by increasing n. It is reasonable to inquire how large n must be in order to make the length of the interval no greater than L, a specified number. To meet this requirement we must have

$$2z_{1-\alpha/2}\frac{\sigma}{\sqrt{n}} \leqq L$$

which gives

$$n \geqq \left(\frac{2\sigma z_{1-\alpha/2}}{L}\right)^2 \tag{5-21}$$

In other words, we have solved another sample size problem.

EXAMPLE 5-8

In Example 5-7 what is the minimum sample size required to make the length of the interval no greater than 2 milligrams?

Solution

From Example 5-7 we have $\sigma = 8$, $z_{1-\alpha/2} = z_{.995} = 2.576$. We also have $L = 2$ milligrams. Using formula (5-21), we calculate

$$n \geqq \left[\frac{2(8)(2.576)}{2}\right]^2 = 424.8$$

Hence the minimum sample size is $n = 425$.

The confidence interval given by formula (5-19) has some desirable mathematical features. First, for a fixed α and n it is the shortest possible interval which can be based upon the Z statistic. It is easy to verify that longer intervals can be obtained by starting with unequal tail areas in formulas (5-16) and (5-17). Second, the probability that the random interval contains wrong values of μ (any constant other than the correct μ) is smaller than for any other random interval which can be obtained for chosen α and n.

One-sided confidence intervals can be obtained by starting with a probability statement like

$$\Pr\left(\frac{\overline{X} - \mu}{\sigma/\sqrt{n}} < z_{1-\alpha}\right) = 1 - \alpha$$

which can be converted to

$$\Pr\left(\overline{X} - z_{1-\alpha}\frac{\sigma}{\sqrt{n}} < \mu\right) = 1 - \alpha$$

Intervals so obtained can be related to the one-sided testing situations. However, we shall not pursue this topic.

EXERCISES

◈ 5-20   Use the information in Exercise 5-1 to find a confidence interval with confidence coefficient .95 for the mean of the distribution of package weights. Does the interval include 12? What is the relationship of the interval to the hypothesis testing problem of Exercise 5-1? How large would n have to be to reduce the length of the interval to .1 of an ounce or less?

5-21   Use the information of Exercise 5-2 to obtain a confidence interval with confidence coefficient .99 for the mean length of life of flashlight batteries. What is the minimum sample size required to make the length of the interval no greater than 2 hours?

◈ 5-22   Use the information of Exercise 5-3 to obtain a confidence interval with confidence coefficient .90 for the average amount of electricity consumed during a 10-minute period by the brand of 100-watt bulbs. What is the minimum sample size required to make the length of the interval no greater than 1 unit?

5-23   Use the information of Exercise 5-4 to obtain a confidence interval with confidence coefficient .90 for the average amount of electricity consumed during a 10-minute period by the brand of 100-watt bulbs. What is the minimum sample size required to make the length of the interval no greater than 2 units?

◈ 5-24   Use the information in Exercise 5-5 to obtain a confidence interval with confidence coefficient .95 for the mean length of time required for the new variety of rat to reach maturity. What is the minimum sample size required to make the length of the interval no greater than 15 days?

◈ 5-25   Use the information of Exercise 5-7 to obtain a confidence interval with confidence coefficient .90 for the mean yield per acre for "fat kernel." What is the minimum sample size required to make the length of the interval no greater than 5 bushels? Suppose that this new sample size yields $\bar{x} = 93$. Exhibit the new confidence interval.

5-26   Use the information of Exercise 5-8 to obtain a confidence interval with confidence coefficient .95 for the mean course grade when statistics is taught with the extra problem sessions. What is the minimum sample size required to make the length of the interval no greater than 2.5 units? Suppose that this new sample size yields $\bar{x} = 61.7$. Exhibit the new confidence interval.

◈ 5-27   Use the information of Exercise 5-9 to obtain a confidence interval with confidence coefficient .99 for the mean score obtained on the mathematics examination. Does this interval contain 50? What is the relationship of the interval to the hypothesis testing problem of Exercise 5-9? What is the minimum sample size required to make the length of the interval no greater than 2 units? Suppose that the new sample size yields $\bar{x} = 48.2$. Exhibit the new confidence interval.

5-28 Use the information of Exercise 5-11 to obtain a confidence interval with confidence coefficient .99 for the mean breaking strength of brand F line. What is the minimum sample size required to make the length of the interval no greater than .5 of a pound?

5-5
INFERENCES ABOUT A MEAN WHEN σ^2 IS UNKNOWN

Several times we have mentioned the fact that in real-life problems the variance is generally unknown. When this is the case, the Z statistic given by formula (5-6) can no longer be used to test hypotheses and obtain confidence intervals. An intuitively reasonable thing to do is to replace σ by S, the sample standard deviation (defined in Sec. 3-5), and hope that good procedures can be developed based upon the new statistic. If we make this replacement we get

$$T_{n-1} = \frac{\bar{X} - \mu_0}{S/\sqrt{n}} \tag{5-22}$$

The sampling distribution of T_{n-1} is a t distribution with $n-1$ degrees of freedom (already discussed in Sec. 2-10) provided (a) $X_1, X_2, \ldots, X_n$ is a random sample from a distribution, (b) that distribution is normal, and (c) the mean of the distribution is μ_0. Further, it can be shown that tests and confidence intervals with good statistical properties can be based upon statistic (5-22).

We make a further comment about (b), the normality assumption. The mathematical derivation by which it is shown that statistic (5-22) has a t distribution depends upon this assumption. Fortunately, however, various investigations and empirical evidence indicate that T_{n-1} still has approximately a t distribution when the normality assumption is violated, provided n is not too small. (Again it is difficult to be specific with the term "too small." The closer the original distribution is to being normal, the smaller the n required. For the kinds of problems which we will consider, perhaps $n = 10$ or 15 is sufficiently large.) That is, probabilities based upon T_{n-1} will still be approximately correct, even if the X distribution is not normal.

Except for minor changes the testing procedures described in Sec. 5-2 are the same. Again we are usually interested in one of the standard hypothesis testing situations given by (5-1) to (5-5). Critical regions are obtained with the assistance of Appendix B6 which is entered with degrees of freedom equal to one less than the sample size (instead of infinity with

FIGURE 5-6

Density of T_{n-1} if $\mu = \mu_0$, showing significance level α
and critical region $t_{n-1} < t_{n-1;\alpha}$.

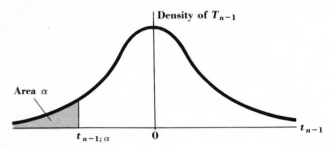

the Z statistic). When testing (5-1) or (5-4) small $\bar{x}$ and small t_{n-1} support
the alternative H_1. With a significance level α the critical region is

$$t_{n-1} = \frac{\bar{x} - \mu_0}{s/\sqrt{n}} < t_{n-1;\alpha} \tag{5-23}$$

where $\Pr(T_{n-1} < t_{n-1;\alpha}) = \alpha$ if $\mu = \mu_0$ (see Fig. 5-6). Similarly, only
large $\bar{x}$ and large t_{n-1} support the alternative H_1 of (5-2) and (5-5). A
test with significance level α is obtained by rejecting H_0 when

$$t_{n-1} = \frac{\bar{x} - \mu_0}{s/\sqrt{n}} > t_{n-1;1-\alpha} \tag{5-24}$$

where $\Pr(T_{n-1} > t_{n-1;1-\alpha}) = \alpha$ if $\mu = \mu_0$ (see Fig. 5-7). Finally, when
testing (5-3) for which the alternative is two-sided, both small and large
$\bar{x}$, and small and large t_{n-1}, provide support for H_1. The critical region
usually chosen is

$$t_{n-1} = \frac{\bar{x} - \mu_0}{s/\sqrt{n}} < t_{n-1;\alpha/2} \quad \text{and} \quad t_{n-1} = \frac{\bar{x} - \mu_0}{s/\sqrt{n}} > t_{n-1;1-\alpha/2} \tag{5-25}$$

FIGURE 5-7

Density of T_{n-1} if $\mu = \mu_0$, showing significance level α
and critical region $t_{n-1} > t_{n-1;1-\alpha}$.

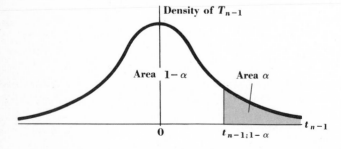

FIGURE 5-8

Density of T_{n-1} if $\mu = \mu_0$, showing significance level α and critical region $t_{n-1} < t_{n-1;\alpha/2}$ and $t_{n-1} > t_{n-1;1-\alpha/2}$.

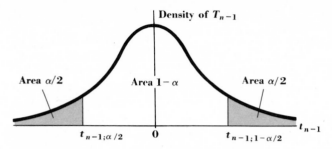

Density of T_{n-1}

Area $\alpha/2$ Area $1 - \alpha$ Area $\alpha/2$

$t_{n-1;\alpha/2}$ 0 $t_{n-1;1-\alpha/2}$ t_{n-1}

where $\Pr(T_{n-1} < t_{n-1;\alpha/2}) + \Pr(T_{n-1} > t_{n-1;1-\alpha/2}) = \alpha/2 + \alpha/2 = \alpha$ when $\mu = \mu_0$ (see Fig. 5-8). Because the t distribution is symmetric about 0, we have $t_{n-1;\alpha/2} = -t_{n-1;1-\alpha/2}$.

EXAMPLE 5-9

Suppose that we reconsider Example 5-1. Let us assume that the problem is unchanged except that the variance is unknown. Suppose that now a random sample of size 25 is observed yielding $\bar{x} = 985$ and $s = 90.2$. What conclusions should be drawn and what decision should be made?

Solution

As before we will follow the outline. Necessary changes will be made.

1 No change is necessary. We have the same information regarding the hypothesis and the alternative.
2 Again the significance level is .05, but the sample size has been chosen as $n = 25$.
3 The statistic we use is $T_{n-1} = (\bar{X} - 1{,}000)/(S/\sqrt{n})$, which has a t distribution provided
   (a) The sample of light bulbs tested was randomly selected.
   (b) The length of life for new-brand light bulbs is normally distributed.
   (c) The distribution has mean $\mu_0 = 1{,}000$ hours.
   A sample of size 25 is probably large enough to make (b) an unnecessary assumption.
4 The critical region given by formula (5-23) is $t_{24} = (\bar{x} - 1{,}000)/(s/\sqrt{n}) < t_{24;\,.05} = -1.711$. (Figure 5-6 applies with $\alpha = .05$.)
5 We perform the experiment and observe $\bar{x} = 985$, $s = 90.2$. Thus

$$t_{24} = \frac{\bar{x} - 1{,}000}{s/\sqrt{n}} = \frac{985 - 1{,}000}{90.2/5} = -.831$$

6 Since the observed value of the statistic does not lie in the critical region, again $H_0 : \mu = 1{,}000$ (or greater) is not rejected and the company will switch to the new brand.

For tests based upon the Z statistic we can compute power from the standard normal table or, more conveniently, from the graphs of Appendix B10 using the $f = \infty$ curve. If $\mu \neq \mu_0$, then T_{n-1} given by formula (5-22) no longer has a t distribution so that special tables or graphs must be used for power calculations. (When $\mu \neq \mu_0$, T_{n-1} has what is called a noncentral t distribution. The graph of the density is similar to a t density, but the center is no longer at 0 nor is the curve symmetric about a value.) Such graphs yielding power of the t test are contained in Appendix B10 with f being the degrees of freedom. Again we need α to find the correct page. Unfortunately, $\delta = |\mu - \mu_0|\sqrt{n}/\sigma$ contains σ which is now unknown. One way out of the difficulty is to choose $|\mu - \mu_0|$ as a multiple of σ, say $a\sigma$, so that $|\mu - \mu_0|/\sigma = a$ and $\delta = a\sqrt{n}$. (A second way requires two-stage sampling and is discussed in the next section.) If we make such a choice, then the graphs can be used as described in Sec. 5-3, except that now we need to find the curve labeled with the appropriate degrees of freedom.

EXAMPLE 5-10

Find the power of the test used in Example 5-9 if the actual mean μ is one-half of a standard deviation below the hypothesized value $\mu_0 = 1{,}000$.

Solution

We are given $|\mu - \mu_0| = \sigma/2$ so that

$$\delta = \frac{|\mu - \mu_0|\sqrt{n}}{\sigma} = \frac{(\sigma/2)\sqrt{25}}{\sigma} = 2.5$$

The degrees of freedom is $f = n - 1 = 24$. Since the test is one-sided with significance level .05, turn to the page of Appendix B10 labeled $\alpha = .05$. The vertical line through $\delta = 2.5$ crosses the curve for $f = 24$ opposite .80. Hence the required power is .80.

EXAMPLE 5-11

Rework Example 5-10 for $n = 16$ (instead of $n = 25$).

Solution

Now

$$\delta = \frac{(\sigma/2)\sqrt{16}}{\sigma} = 2 \qquad \text{and} \qquad f = n - 1 = 15$$

Curves are available for $f = 12$ and $f = 24$, but not for $f = 15$. With $\delta = 2$ the graphs yield

Power $= .60$ if $f = 12$

Power $= .63$ if $f = 24$

The power for $f = 15$ is obviously between $.60$ and $.63$. Perhaps this information is sufficient for an actual application. If we use linear interpolation (see Appendix A2-1), we get that the power is approximately $.61$ when $f = 15$.

The type of sample size problem demonstrated in Examples 5-5 and 5-6 (with σ known) can still be solved with the graphs of Appendix B10 and formula (5-14), provided we choose $|\mu - \mu_0|$ as a multiple of σ. With the δ which yields power $1 - \beta$ for $f = \infty$, use formula (5-14) to compute an n. Since this δ is smaller than would be required to achieve power $1 - \beta$ with $f = n - 1$ degrees of freedom, the computed n will be smaller than the minimum actually required. In other words, this computed n is a lower bound for the value we seek. On the other hand, if we use a δ which yields power $1 - \beta$ for an f that is too small, this δ will be larger than necessary and formula (5-14) will produce a computed n which may be larger than is needed to meet the requirements. In most of our problems this will be achieved with the δ which gives power $1 - \beta$ for $f = 24$. The actual minimum n is thus usually cornered in a relatively small interval. (Actually f should be decreased from ∞ to 24, to 12, to 6, etc., until the computed n is greater than the chosen $n = f + 1$. Then, the latter computed n is an upper bound for the minimum n and the computed n obtained just previously is a lower bound.) Let us demonstrate with an example.

EXAMPLE 5-12

For the situation of Example 5-10 how large does n have to be to raise the power to $.99$ if $|\mu - \mu_0| = \sigma/2$?

Solution

Since $|\mu - \mu_0|/\sigma = 1/2$, formula (5-14) becomes $n = (2\delta)^2$. We turn to the page of Appendix B10 headed by $\alpha = .05$, the significance level of the test. The smallest δ yielding power $= .99$ is associated with $f = \infty$ and is $\delta = 3.97$. This δ yields $n = [2(3.97)]^2 = 63.04$, which will be too small to guarantee a power of at least $.99$ when $|\mu - \mu_0|/\sigma = 1/2$. On the other hand if we use $\delta = 4.10$ associated with $f = 24$, $n = 25$, this δ will be larger than necessary (since we already know that $n > 63$) and may produce an n a little larger than necessary. This δ gives $n = [2(4.10)]^2 = 67.24$, which is larger than the chosen $n = 25$. Hence, we have shown that $63.04 \leqq n \leqq 67.24$. To be conservative we could take $n = 67$ or 68. Obviously, from a practical standpoint any n from 64 to 68 would approximately fulfill our requirements.

(Owen [2, pp. 37–40] has prepared graphs from which n can be read directly provided n does not exceed 100. The graphs give $n = 65$ as the solution to our problem. If we really want to be precise in determining the minimum n, we can use a table prepared by Owen [3] which gives δ to six decimal places for all combinations of power = .10(.10).90, .95, .99, $f = 1(1)30(5)100(10)200$, ∞, $\alpha = .05, .025, .01, .005$. To use these tables this way increase n (and $f = n - 1$) until the chosen n just exceeds the n computed from formula (5-14). In the above example we find that with chosen $n = 64$ the computed $n = 64.48$, while with chosen $n = 65$ the computed $n = 64.45$. Thus the minimum n is 65.)

To obtain a confidence interval for μ with confidence coefficient $1 - \alpha$ we start with

$$\Pr\left(-t_{n-1;\,1-\alpha/2} < \frac{\overline{X} - \mu}{S/\sqrt{n}} < t_{n-1;\,1-\alpha/2}\right) = 1 - \alpha \tag{5-26}$$

(dropping the subscript 0 on μ in the statistic T_{n-1}). We note that formula (5-26) is almost the same as formula (5-17), the only differences being that $t_{n-1;\,1-\alpha/2}$ has replaced $z_{1-\alpha/2}$ and S has replaced σ. Consequently, we can write immediately

$$\Pr\left(\overline{X} - t_{n-1;\,1-\alpha/2}\frac{S}{\sqrt{n}} < \mu < \overline{X} + t_{n-1;\,1-\alpha/2}\frac{S}{\sqrt{n}}\right) = 1 - \alpha \tag{5-27}$$

the counterpart of formula (5-18). Hence a confidence interval for μ with confidence coefficient $1 - \alpha$ is

$$\left(\bar{x} - t_{n-1;\,1-\alpha/2}\frac{s}{\sqrt{n}}, \ \bar{x} + t_{n-1;\,1-\alpha/2}\frac{s}{\sqrt{n}}\right) \tag{5-28}$$

The interval (5-28) has the same desirable properties mentioned earlier for the interval (5-19). (It is the shortest interval based upon T_{n-1}, and the probability that the random interval contains any constant other than μ is less than $1 - \alpha$.) As with the interval based upon the Z statistic, the interval (5-28) covers a value μ_0 if, and only if, the test of $H_0 : \mu = \mu_0$ against $H_1 : \mu \neq \mu_0$ accepts H_0 when the same α (and the same sample) is used. Thus, if we prefer to do so, to test H_0 we can compute the interval rather than the observed value of the t statistic.

EXAMPLE 5-13

Find a confidence interval with confidence coefficient .95 for the mean of the distribution using the sample results obtained in Example 5-9 (the light bulb problem).

Solution

We had $\bar{x} = 985$, $s = 90.2$, $n = 25$. Since $1 - \alpha = .95$, $\alpha = .05$, $\alpha/2 = .025$, $1 - \alpha/2 = .975$, we need $t_{24;.975} = 2.064$ (from Appendix B6). Thus the required interval is

$$\left(985 - 2.064 \frac{90.2}{\sqrt{25}}, \; 985 + 2.064 \frac{90.2}{\sqrt{25}}\right)$$

which reduces to (947.8, 1022.2).

The length of the interval (5-28) is

$$2t_{n-1;1-\alpha/2} \frac{s}{\sqrt{n}}$$

and the length of the corresponding random interval is

$$2t_{n-1;1-\alpha/2} \frac{S}{\sqrt{n}} \tag{5-29}$$

Unlike (5-20), this latter quantity is a random variable. When σ was known, the length (5-20) could be made as short as we please for given α by making n large enough. Now the best we can do is to make probability statements about the length. For a given n, α, and L we could seek

$$\Pr\left(2t_{n-1;1-\alpha/2} \frac{S}{\sqrt{n}} < L\right) \tag{5-30}$$

A more interesting problem requires the minimum n to make

$$\Pr\left(2t_{n-1;1-\alpha/2} \frac{S}{\sqrt{n}} < L\right) = \gamma \tag{5-31}$$

for given α, L, and γ (another sample size problem). As we shall demonstrate, both types of problems are easily solved using the graphs of Appendix B11.

The probability (5-30) can be rewritten as

$$\Pr\left[\frac{(n-1)S^2}{\sigma^2} < \frac{n(n-1)L^2}{4\sigma^2 t_{n-1;1-\alpha/2}^2}\right] \tag{5-32}$$

From (5-32) we observe two things. First, $Y_{n-1} = (n-1)S^2/\sigma^2$, a random variable discussed in Sec. 3-6, has a chi-square distribution provided (a) $X_1, X_2, \ldots, X_n$ is a random sample from a distribution and (b) the distribution is normal. Since the distribution of Y_{n-1} depends heavily on the normality assumption, so does the probability (5-30). Second, since σ^2 appears on the right-hand side of the less than sign in (5-32), these probabilities, like power, depend upon the unknown quantity σ^2. To get around this difficulty, we will select L in σ units so that L/σ will be a known number (as we did with δ).

To use the graphs of Appendix B11 we need the confidence coefficient (either .90 or .95) to determine the page and L (in σ units) to enter the horizontal scale. If n is given, enter the vertical scale with that n and read the probability (5-30) at the intersection of the lines through L and n. If L and γ are given and we seek n, find the intersection of the vertical line through L and the curve labeled $\Pr = \gamma$; then read n from the vertical scale.

EXAMPLE 5-14

In Example 5-13 we found a confidence interval for the mean with confidence coefficient .95 using a sample of size 25. Find the probability that a random sample of size 25 drawn from a normal distribution produces a confidence interval with confidence coefficient .95 shorter than 1 standard deviation.

Solution

We have $1 - \alpha = .95$, $n = 25$, $L = \sigma$ (or $L/\sigma = 1$). From the first page of Appendix B11 we find that the vertical line through $L = 1\sigma$ intersects the horizontal line through $n = 25$ between $\Pr = .90$ and $\Pr = .99$. The intersection is close to the $\Pr = .90$ curve, and so we might estimate the probability to be about .92.

EXAMPLE 5-15

Find the minimum n required if it is desired to obtain a confidence interval with confidence coefficient .95 for the mean of a normal distribution and to have the probability be .90 that the random interval be shorter than $.5\sigma$.

Solution

We have $1 - \alpha = .95$, $L = .5\sigma$ (or $L/\sigma = .5$), and $\Pr = \gamma = .90$. From the first page of Appendix B11 we find that the vertical line through $L = .5\sigma$ crosses the curve labeled $\Pr = .90$ opposite $n = 77$, approximately. Thus the probability is .90 that a random sample of size 77 drawn from a normal distribution will produce a confidence interval with confidence coefficient .95 for μ that is shorter than $.5\sigma$.

EXERCISES

◈ 5-29   Consider again the frozen shrimp problem of Exercise 5-1. Assume that the hypothesis testing situation is unchanged but now the variance is unknown. If an observed random sample of size $n = 25$ yields $\bar{x} = 12.24$ ounces and $s = .60$ of an ounce, what conclusion should be drawn using $\alpha = .05$? Follow the recommended outline.

◈ 5-30   Find the power of the test used in Exercise 5-29 if the mean weight μ differs from the hypothesized value μ_0 by one-half of a standard deviation. Approximately how large would n have to be to raise this to .99?

◈ 5-31   Use the information given in Exercise 5-29 to find a confidence interval for μ with confidence coefficient .95. How does the interval relate to the test of Exercise 5-29?

◈ 5-32   In Exercise 5-31 a confidence interval was found using a confidence coefficient of .95 and a sample size $n = 25$. Find the probability that a random sample of size 25 drawn from a normal distribution produces a confidence interval (with confidence coefficient .95) shorter than $.75\sigma$. How large would n have to be to raise this probability to .99?

5-33   A famous bridge player has found that his average score for a session of tournament play is 60.0 percent when he uses a standard bidding system. When he uses a new system for 16 sessions, his scores yield an observed sample mean of 62.1 and an observed sample standard deviation of 3.0 percent. Unless the player is convinced at the .01 level of significance that the new system produces a higher average, he will discontinue using it. What decision should he make? Follow the recommended outline.

5-34   Find the power of the test used in Exercise 5-33 if the mean score μ is actually one-half of a standard deviation above 60. Approximately what is the minimum n required to raise this power to .90?

5-35   Use the sample results given in Exercise 5-33 to find a confidence interval with confidence coefficient .90 for the μ of the new system.

5-36   In Exercise 5-35 a confidence interval was requested using a confidence coefficient .90 and a sample size $n = 16$. Find the probability that a random sample of size 16 drawn from a normal distribution produces a confidence interval (with confidence coefficient .90) shorter than σ. Shorter than $.5\sigma$. How large would n have to be to raise the probability to .90 when $L = .5\sigma$?

◈ 5-37   Reconsider Exercise 5-2 and assume that the hypothesis and alternative of that exercise are still of interest but that the standard deviation is unknown. With $\alpha = .01$ find the power of the test if the mean life μ is two-fifths of a standard deviation above 30 hours. What is the minimum sample size required to raise the power for this μ to .90? (Compare the answer to the one obtained in Exercise 5-13.)

◈ 5-38 Consider again the electricity consumption problem of Exercise 5-3. Assume that the same hypothesis and alternative are still of interest but that now the standard deviation is unknown. Again use $\alpha = .05$, $n = 49$ and assume that an observed random sample yields $\bar{x} = 48.9$ units, $s = 3.5$ units. What action should the chain take? Follow the recommended outline.

◈ 5-39 For the problem of Exercise 5-38, find the probability that the merchandising chain will buy the 100-watt bulbs if the average amount of electricity consumed is $\sigma/4$ units below 50. Approximately what is the minimum n required to raise this power to .80?

◈ 5-40 Use the sample results given in Exercise 5-38 to find a confidence interval with confidence coefficient .90 for the mean amount of consumption.

◈ 5-41 In Exercise 5-40 a confidence interval was required using a confidence coefficient of .90 and a sample of size $n = 49$. Find the probability that a random sample of size 49 drawn from a normal distribution produces a confidence interval (with confidence coefficient .90) shorter than $.5\sigma$. How large would n have to be to raise this probability to .99?

5-42 A standard variety of wheat produces on the average 30 bushels per acre in a certain region. A new imported variety is planted on nine randomly selected acre plots. The observed sample average for the new variety is 33.4 bushels per acre. If the observed sample standard deviation is 5.1 bushels, what conclusion can be drawn by using $\alpha = .05$? Follow the recommended outline.

5-43 Find the power of the test used in Exercise 5-42 if the average yield for the new variety differs from 30 by one-half of a standard deviation. How many acres should have been planted if it is desired that this probability should be .80?

5-44 For the new variety of Exercise 5-42 find a confidence interval with confidence coefficient .95 for the mean yield per acre.

◈ 5-45 Consider again the psychologist of Exercise 5-5. Assume that the same hypothesis and alternative are still of interest but that now the standard deviation is unknown. Again use $n = 16$, $\alpha = .10$, and assume that the observed random sample yields $\bar{x} = 229$ days, $s = 24$ days. What should the psychologist do? Follow the recommended outline.

◈ 5-46 Use the sample results given in Exercise 5-45 to find a confidence interval with confidence coefficient .80 for the mean age of maturity for the new variety of rats.

5-47 Consider again the problem of the statistics department discussed in Exercise 5-8. Assume that the same hypothesis and the same alternative are still of interest but that now the standard deviation is unknown. Again use $n = 64$, $\alpha = .05$ and assume that the observed random sample yields $\bar{x} = 61.7$, $s = 7.2$. What action should be taken? Follow the recommended outline.

5-48 Find the power of the test used in Exercise 5-47 if the average score produced by the new method is one-fourth of a standard deviation larger than 60. How many students should have been used if it is desired that this probability should be at least .95?

5-49 Use the sample results given in Exercise 5-47 to find a confidence interval with confidence coefficient .95 for the mean score produced by the new method.

5-50 In Exercise 5-49 a confidence interval was required using a confidence coefficient of .95 and a sample of size $n = 64$. Find the probability that a random sample of size 64 drawn from a normal distribution produces a confidence interval (with confidence coefficient .95) shorter tha $.4\sigma$. How large would n have to be to raise this probability to .99?

5-6
INFERENCES ABOUT A MEAN WITH A TWO-STAGE PROCEDURE

In Sec. 5-5 we found that both power calculations and probability statements concerning the length of a confidence interval depend upon the unknown quantity σ. To overcome this obstacle we chose $|\mu - \mu_0|$ and L in σ units, a choice that enabled us to use the graphs of Appendixes B10 and B11. In this section we will discuss another procedure for eliminating the unknown standard deviation.

Two types of problems will be considered. In the first type we will obtain tests about the mean μ such that two power conditions are met (as indicated in Fig. 5-5). In the second type we will obtain confidence intervals of fixed length with given confidence coefficient. [The interval (5-19), found with σ known, can be so determined, but (5-28) will have length depending on s.] A two-stage sampling scheme consisting of the following steps will be used:

1 Specify a first (or preliminary) sample size n_0. Usually n_0 is arbitrarily selected to be a small number (like 8, 10, or perhaps 15).

2 Compute a value of k which is determined by the conditions of the problem.

3 Observe a random sample of size n_0 getting $x_1, x_2, \ldots, x_{n_0}$.

4 Compute

$$s^2 = \frac{\sum_{i=1}^{n_0}(x_i - \bar{x})^2}{n_0 - 1}$$

for this preliminary sample.

5   If $ks^2 \leqq n_0$, take no more observations, and the total sample size is n_0. If $ks^2 > n_0$, continue sampling until the total sample size (including the first n_0), say n, is the smallest integer such that $n \geqq ks^2$.

6   Conduct the appropriate test or form the confidence interval.

For the first type of problem suppose that we specify significance level α (power when $\mu = \mu_0$) and power at least $1 - \beta$ if $\mu = \mu_1$. Then the k mentioned above is

$$k = \left(\frac{t_{n_0-1;\,1-\alpha} + t_{n_0-1;\,1-\beta}}{\mu_1 - \mu_0} \right)^2 \qquad (5\text{-}33)$$

for one-sided testing situations (5-1), (5-2), (5-4), (5-5). (μ_1 must be a value of μ in the alternative H_1.) When the alternative is two-sided replace α by $\alpha/2$ in formula (5-33). The test statistic is

$$T_{n_0-1} = \frac{\overline{X} - \mu_0}{S/\sqrt{n}} \qquad (5\text{-}34)$$

where s is obtained from the preliminary sample (using n_0 observations) and $\bar{x}$ is obtained from all n observations. If $\mu = \mu_0$, the sampling distribution of T_{n_0-1} is a t distribution with $n_0 - 1$ degrees of freedom provided (a) the samples are selected at random from a distribution and (b) that distribution is normal. Critical regions are again given by formulas (5-23) to (5-25) with n being n_0.

EXAMPLE 5-16

For the tomato juice problem of Sec. 4-5 we tested $H_0 : \mu = 46$ against $H_1 : \mu \neq 46$ with $\alpha = .05$. Suppose that σ is unknown and we desire that the power be .99 if the mean is really 45.5 or 46.5 so that $|\mu_1 - \mu_0| = .5$ of an ounce. If a preliminary sample of size 10 is used, find k. If the preliminary sample yields $s = .61$, how large should n be? Suppose that $\bar{x}$ based upon all n observations is 46.18. What conclusions should be reached?

Solution

As before we can follow the outline.

1   The hypothesis and alternative suggested by the problem are $H_0 : \mu = 46$ and $H_1 : \mu \neq 46$.

2   It is desired to have the power be $\alpha = .05$ if $\mu = 46$ and be at least .99 if $\mu = 45.5$ or 46.5. To compute k from formula (5-33) we identify $n_0 = 10$, $1 - \beta = .99$, $|\mu_1 - \mu_0| = .5$. Since the alternative is two-sided α is replaced by $\alpha/2 = .025$ and $1 - \alpha/2 = .975$. Appendix B6 gives $t_{9;\,.975} = 2.262$, $t_{9;\,.99} = 2.821$. Thus

$$k = \left(\frac{2.262 + 2.821}{.50} \right)^2 = (10.166)^2 = 103.35$$

3 The statistic we use is $T_9 = (\bar{X} - 46)/(S/\sqrt{n})$ where $\bar{x}$ is based on all n observations, s on the first 10 observations. The random variable T_9 has a t distribution provided

   (a) The cans of tomato juice are selected at random.

   (b) The distribution of weights is normal.

   (c) The mean of the distribution is $\mu_0 = 46$.

4 The critical region given by formula (5-25) with degrees of freedom $n_0 - 1 = 9$ is

$$t_9 = \frac{\bar{x} - 46}{s/\sqrt{n}} < -2.262 \text{ and } t_9 = \frac{\bar{x} - 46}{s/\sqrt{n}} > 2.262$$

(Fig. 5-8 applies.)

5 At this time the sampling is performed. The first sample of size 10 yields $s = .61$. Next $ks^2 = (103.35)(.61)^2 = 38.46$ so that we need $n > 38.46$, or $n = 39$, observations. This means that the second sample must contain 29 cans. From all 39 weights we compute $\bar{x} = 46.18$, and the observed value of the statistic is

$$t_9 = \frac{46.18 - 46}{.61/\sqrt{39}} = 1.84$$

6 Since the observed value of the statistic does not lie in the critical region, the hypothesis $H_0 : \mu = 46$ is accepted and no action is taken.

Next we consider the second type of problem mentioned in the beginning of the section. Suppose that we desire a confidence interval with confidence coefficient at least $1 - \alpha$ and would also like to have the interval be of length L. It can be shown that the appropriate value of k is

$$k = \left(\frac{2t_{n_0-1;\,1-\alpha/2}}{L}\right)^2 \tag{5-35}$$

and this choice makes

$$\left(\bar{x} - \frac{L}{2}, \bar{x} + \frac{L}{2}\right) \tag{5-36}$$

a confidence interval with the desired properties. Again $\bar{x}$ is based upon all observations in both the first and second sample.

EXAMPLE 5-17

Use the information from the preliminary sample in Example 5-16 to compute k and then the minimum n which yields a confidence interval of length .5 of an ounce for μ with confidence coefficient at least .95. If the results from both samples yield $\bar{x} = 46.18$, exhibit the interval.

Solution

For formula (5-35) we need $n_0 = 10$, $\alpha = .05$, $1 - \alpha/2 = .975$, $t_{9;\,.975} = 2.262$, $L = .5$. Then

$$k = \left[\frac{2(2.262)}{.5}\right]^2 = (9.048)^2 = 81.87$$

Since $s = .61$, we get $ks^2 = 81.87(.61)^2 = 30.46$. We need $n \geq 30.46$ or $n = 31$ observations (or 21 more). If the $\bar{x}$ based on 31 observations is 46.18, then the desired interval is $(46.18 - .25, 46.18 + .25)$ or $(45.93, 46.43)$.

EXERCISES

◈ 5-51   In the light bulb problem discussed in Example 5-1 we tested $H_0 : \mu = 1,000$ against $H_1 : \mu < 1,000$ with $\alpha = .05$. Now assume that σ is unknown, and suppose that we also desire to have the power be at least .90 if the mean is actually $\mu = 975$. Find k if a preliminary sample of size 20 is to be used. If the preliminary sample yields $s = 90.2$, what is the minimum total sample size? Suppose that $\bar{x}$ based upon all n observations turns out to be 970. What conclusion is reached? Follow the recommended six-step outline.

◈ 5-52   Use the information from the preliminary sample in Exercise 5-51 to compute k and the minimum n necessary to ensure that a confidence interval for μ of length 50 hours has a confidence coefficient of at least .95. Exhibit the interval if $\bar{x}$ based upon all n observations is 985 hours.

5-53   Consider again the frozen shrimp problem discussed in Exercises 5-1 and 5-29 and assume that σ is unknown. Suppose we again use $\alpha = .05$ and would like to have the power be at least .95 if the mean is actually 11.75 or 12.25 so that $|\mu - \mu_0| = .25$. If a preliminary sample of size 25 is to be used, find k. If the first sample then yields $s = .60$, how large must n be? Suppose that $\bar{x}$ based upon all n observations turns out to be 12.24. What conclusion is reached? Follow the six-step outline.

5-54   Use the information from the preliminary sample in Exercise 5-53 to compute k and the minimum n necessary to ensure that a confidence interval for μ of length .25 of an ounce has a confidence coefficient of at least .95. Exhibit the interval if $\bar{x}$ based upon all n observations is 12.24 ounces.

◈ 5-55   Consider again the electricity consumption problem of Exercises 5-3 and 5-38 and assume that σ is unknown. Suppose now that the merchandising chain would like to be quite sure of making the correct decision if $\mu = 50$ units and selects $\alpha = .01$. In addition if $\mu = 48$ units it would like to have the probability be at least .90

that it makes the purchase. A preliminary sample of size 14 yields $s^2 = 15.9$. Find k and determine how many additional observations are required. If $\bar{x}$ based upon all n observations is 49.5 units, what action should be taken? Follow the six-step outline.

◈ 5-56 Use the information from the preliminary sample in Exercise 5-55 to compute k and the minimum n necessary to ensure that a confidence interval for μ of length 1 unit has a confidence coefficient of at least .90. Exhibit the interval if $\bar{x}$ based upon all n observations is 49.5 units.

5-7
INFERENCES ABOUT A VARIANCE

The normal model, a probability model which we have used frequently, depends upon the parameter σ^2. As we have already learned, the variance is a measure of variability, and in Sec. 3-5 we encountered some problems in which the variability of the distribution associated with the model is an important consideration. In this section we will consider statistical inferences about the variance of a normal distribution. (Point estimation was discussed in Sec. 3-5.)

The standard hypothesis testing situations for a variance are similar to those already encountered with the mean of a distribution. Suppose that we have been using an old machine to make bolts and have established that the diameters of these bolts are normally distributed with variance σ_0^2 (to be specific we could take $\sigma_0^2 = .00042$). A salesman is trying to sell us a new machine which he claims produces bolts with smaller variance, say σ^2. If we decide to buy the new machine only if its variance is smaller, we will test

$$H_0 : \sigma^2 = \sigma_0^2 \qquad \text{against} \qquad H_1 : \sigma^2 < \sigma_0^2 \qquad (5\text{-}37)$$

(with $\sigma_0^2 = .00042$) and replace our present equipment if we conclude that H_1 is true. If the new machine is capable of producing many more parts than the old, we may decide to switch unless the variance associated with the new machine is larger than that associated with the old. Under the latter circumstances we would test

$$H_0 : \sigma^2 = \sigma_0^2 \qquad \text{against} \qquad H_1 : \sigma^2 > \sigma_0^2 \qquad (5\text{-}38)$$

and replace our present machine unless we conclude H_1 is true. Sometimes a problem will lead to a two-sided alternative. For example, suppose it is known that a standard method for the teaching of reading produces reading rates that are normally distributed with variance σ_0^2.

A new method is tried, and it is desired to know whether the new method has the same variance as the old. In this instance it would be appropriate to test

$$H_0 : \sigma^2 = \sigma_0^2 \qquad \text{against} \qquad H_1 : \sigma^2 \neq \sigma_0^2 \qquad (5\text{-}39)$$

Sometimes it may be preferable to replace (5-37) by

$$H_0 : \sigma^2 \geqq \sigma_0^2 \qquad \text{against} \qquad H_1 : \sigma^2 < \sigma_0^2 \qquad (5\text{-}40)$$

or (5-38) by

$$H_0 : \sigma^2 \leqq \sigma_0^2 \qquad \text{against} \qquad H_1 : \sigma^2 > \sigma_0^2 \qquad (5\text{-}41)$$

putting "satisfactory" values of σ^2 in one set, "unsatisfactory" values in another set. The testing procedure will be unchanged but again we must regard α to be the largest value of the power when H_0 is true (which will occur when $\sigma^2 = \sigma_0^2$).

It seems reasonable to base inferences about σ^2 on S^2, the sample variance. In the above hypothesis testing situations it is more convenient to use the statistic

$$Y_{n-1} = \frac{(n-1)S^2}{\sigma_0^2} \qquad (5\text{-}42)$$

The sampling distribution of Y_{n-1} is chi-square with $n-1$ degrees of freedom provided (a) $X_1, X_2, \ldots, X_n$ is a random sample from a distribution (the requirement used in the definition of S^2), (b) that distribution is normal, and (c) the distribution has variance σ_0^2. Unfortunately (as we commented in Sec. 3-5), the distribution of Y_{n-1} depends heavily upon the normality assumption. If the model which governs the behavior of the experimental values is not normal or, nearly so, probability statements can be seriously in error so that inferences may not be much better than those based upon a guess. [Instead of statistic (5-42) we can use $Y_{n-1}/(n-1) = S^2/\sigma_0^2$ whose sampling distribution is chi-square divided by degrees of freedom.]

When testing (5-37) or (5-40) only small values of s^2, and hence small y_{n-1}, support H_1. If the significance level is α, then the critical region is

$$y_{n-1} = \frac{(n-1)s^2}{\sigma_0^2} < \chi_{n-1;\,\alpha}^2 \qquad (5\text{-}43)$$

where $\Pr(Y_{n-1} < \chi_{n-1;\,\alpha}^2) = \alpha$ if $\sigma^2 = \sigma_0^2$ (see Fig. 5-9). Similarly only large s^2 and large y_{n-1} support H_1 of (5-38) and (5-41). A test of significance level α is obtained by rejecting H_0 when

$$y_{n-1} = \frac{(n-1)s^2}{\sigma_0^2} > \chi_{n-1;\,1-\alpha}^2 \qquad (5\text{-}44)$$

FIGURE 5-9

Density of Y_{n-1} (chi-square) if $\sigma^2 = \sigma_0^2$, showing significance level α and critical region $y_{n-1} < \chi_{n-1;\alpha}^2$.

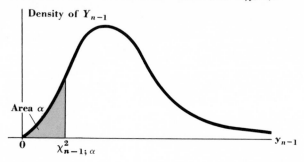

where $\Pr(Y_{n-1} > \chi_{n-1;1-\alpha}^2) = \alpha$ if $\sigma^2 = \sigma_0^2$ (see Fig. 5-10). Finally, when testing (5-39) both small and large values of s^2, and hence y_{n-1}, provide support for H_1. The intuitively appealing critical region which we shall use is

$$y_{n-1} = \frac{(n-1)s^2}{\sigma_0^2} < \chi_{n-1;\alpha/2}^2 \quad \text{and} \quad y_{n-1} = \frac{(n-1)s^2}{\sigma_0^2} > \chi_{n-1;1-\alpha/2}^2$$

(5-45)

again an "equal-tails" choice. If $\sigma^2 = \sigma_0^2$, then $\Pr(Y_{n-1} < \chi_{n-1;\alpha/2}^2) + \Pr(Y_{n-1} > \chi_{n-1;1-\alpha/2}^2) = \alpha/2 + \alpha/2 = \alpha$ (see Fig. 5-11).

EXAMPLE 5-18

A standard machine produces 1/2-inch bolts whose diameters are normally distributed with a variance of .00042. A new machine is offered to us

FIGURE 5-10

Density of Y_{n-1} (chi-square) if $\sigma^2 = \sigma_0^2$, showing significance level α and critical region $y_{n-1} > \chi_{n-1;1-\alpha}^2$.

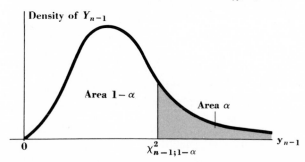

FIGURE 5-11

Density of Y_{n-1} (chi-square) if $\sigma^2 = \sigma_0^2$, showing significance level α and critical region $y_{n-1} < \chi^2_{n-1;\alpha/2}$ and $y_{n-1} > \chi^2_{n-1;1-\alpha/2}$.

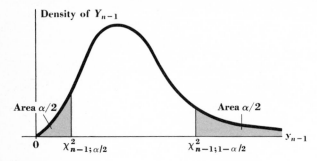

with the claim that it produces 1/2-inch bolts with smaller variance, say σ^2. Suppose that we are willing to buy the new machine if we can "prove" (with significance level $\alpha = .05$) that it produces bolts with smaller variance. A random sample of 25 bolts produced by the new machine yields an observed sample variance of .00028. What decision should be made?

Solution

We shall follow the six-step outline.

1 Letting σ^2 be the variance of diameters associated with the new machine, the problem suggests that we test $H_0 : \sigma^2 = .00042$ against $H_1 : \sigma^2 < .00042$ and purchase the new machine if H_0 is rejected.
2 The significance level $\alpha = .05$ and the sample size $n = 25$ have already been chosen.
3 The statistic we will use is $Y_{n-1} = (n-1)S^2/.00042$, which has a chi-square distribution with $n - 1$ degrees of freedom provided
   (a) The sample of bolts is randomly selected.
   (b) The diameters produced by the new machine are normally distributed. Since the normal was appropriate for the old machine, the same model seems reasonable for the new machine.
   (c) The distribution of diameters produced by the new machine has variance $\sigma_0^2 = .00042$.
4 The critical region given by formula (5-43) is $y_{24} = 24s^2/.00042 < \chi^2_{24; .05} = 13.85$. Figure 5-9 applies with $\alpha = .05$, $n = 25$.
5 The sample would now be examined. It yields $s^2 = .00028$, and we calculate $y_{24} = 24(.00028)/.00042 = 16$.
6 Since the observed value of the statistic does not lie in the critical region, H_0 is accepted and the new machine is not purchased.

EXAMPLE 5-19

With the currently used method of teaching statistics, scores on a standard-ized final examination are normally distributed with a standard deviation of 10. A new method involving the use of a new textbook is being considered as a replacement. If the standard deviation σ associated with final examinations scores produced by the new method is too small or too large, then the new method is considered unsatisfactory and will not be adopted. It is decided to adopt the new method unless there is evidence to indicate that the standard deviation differs from 10, the figure associated with the current method. An experiment is conducted with 81 students taught by the new method. Their final examination scores produce an observed sample standard deviation of 12. What decision should be made if we take $\alpha = .05$?

Solution

We shall follow the six-step outline.

1 Letting σ^2 be the variance associated with final-examination scores produced by the new method, the problem suggests that we test $H_0 : \sigma^2 = 100$ against $H_1 : \sigma^2 \neq 100$ and adopt the new method if H_0 is accepted.

2 The significance level $\alpha = .05$ and the sample size $n = 81$ have already been selected.

3 The statistic we will use is $Y_{n-1} = (n-1)S^2/100$, which has a chi-square distribution with $n-1$ degrees of freedom provided

   (a) The sample of students (and their scores) can be regarded as a random sample.

   (b) The scores produced by the new method are normally distributed. Since the normal was appropriate for the old method scores, this is a reasonable assumption.

   (c) The distribution of scores produced by the new method has variance $\sigma_0^2 = 100$.

4 The critical region given by (5-45) is $y_{80} = 80s^2/100 < \chi^2_{80; .025} = 57.15$ and $y_{80} = 80s^2/100 > \chi^2_{80; .975} = 106.6$ obtained from Appendix B7. Figure 5-11 applies with $\alpha = .05$, $n = 81$.

5 The sample results yield $s^2 = (12)^2 = 144$, and we calculate

$$y_{80} = 80(144)/100 = 115.2$$

6 The observed value of the statistic falls in the critical region and H_0 is rejected. We would probably conclude that the new method produces scores which are more variable than those produced by the old method. At any rate, based upon the experiment, the new method will not be adopted.

If $\sigma^2 \neq \sigma_0^2$, then statistic (5-42) does not have a chi-square distribution since the wrong constant appears in the denominator. It is, however, easy to correct for this in power calculations. For example, the power of the test of situation (5-37) with critical region (5-43) is

$$\Pr\left[\frac{(n-1)S^2}{\sigma_0^2} < \chi^2_{n-1;\,\alpha}\right] = \Pr\left[\frac{(n-1)S^2}{\sigma^2} < \frac{\sigma_0^2}{\sigma^2}\chi^2_{n-1;\,\alpha}\right]$$

where σ^2 is the true variance of the normal distribution. Hence $Y_{n-1} = (n-1)S^2/\sigma^2$ has a chi-square distribution with $n-1$ degrees of freedom, and we can write

$$\text{Power} = \Pr\left(Y_{n-1} < \frac{\sigma_0^2}{\sigma^2}\chi^2_{n-1;\,\alpha}\right) \tag{5-46}$$

$$= \Pr\left(\frac{Y_{n-1}}{n-1} < \frac{\sigma_0^2}{\sigma^2}\frac{\chi^2_{n-1;\,\alpha}}{n-1}\right) \tag{5-47}$$

The probabilities (5-46) and (5-47) can be evaluated from Appendix B7 and Appendix B8 respectively (subject to the limitations of the table).

EXAMPLE 5-20

Find the power of test used in Example 5-18 if the true variance is $\sigma^2 = .00028$ and not $\sigma_0^2 = .00042$.

Solution

Since $n = 25$, $\alpha = .05$, $\chi^2_{24;\,.05} = 13.85$, formula (5-47) becomes

$$\Pr\left(\frac{Y_{24}}{24} < \frac{.00042}{.00028}\frac{13.85}{24}\right) = \Pr\left(\frac{Y_{24}}{24} < .866\right)$$

From Appendix B8 we find with 24 degrees of freedom that

$$\Pr\left(\frac{Y_{24}}{24} < .831\right) = .30 \qquad \Pr\left(\frac{Y_{24}}{24} < .902\right) = .40$$

Hence

$$\Pr\left(\frac{Y_{24}}{24} < .866\right) \cong .35$$

Rather than doing calculations with formulas (5-46) and (5-47), as in Example 5-20, it is somewhat easier to work power problems (including sample size problems) using the graphs of Appendix B12. The first set of three give power for testing $H_0 : \sigma^2 = \sigma_0^2$ against $H_1 : \sigma^2 < \sigma_0^2$ for

$\alpha = .01, .025, .05$. The second set of three are for testing $H_0 : \sigma^2 = \sigma_0^2$ against $H_1 : \sigma^2 > \sigma_0^2$ with the same α's. All six can be used to approximate the power of the test of $H_0 : \sigma^2 = \sigma_0^2$ against $H_1 : \sigma^2 \neq \sigma_0^2$ with maximum error $\alpha/2$ (negligible error if the power is greater than .30). When using the graphs for a two-tailed test, use the ones labeled one-half the significance level of the test. In order to find power, enter the horizontal scale with $R = \sigma/\sigma_0$, follow the vertical line through this value up to the curve with the appropriate degrees of freedom, and read off power on the vertical scale.

EXAMPLE 5-21

Rework Example 5-20 using the graphs of Appendix B12.

Solution

We have $\alpha = .05$, $n = 25$, $\sigma^2 = .00028$, $\sigma_0^2 = .00042$, $R^2 = .00028/.00042 = 2/3$ so that $R = \sqrt{2/3} = .816$. Turning to the page headed by $\alpha = .05$, we find no curve for 24 degrees of freedom. However, for 20 degrees of freedom the power is about .30, for 30 degrees of freedom it is about .41, and so for 24 degrees of freedom it is about

$$.30 + \frac{24 - 20}{30 - 20}(.41 - .30) = .34$$

The reading obtained with visual interpolation is probably sufficiently close for practical purposes.

EXAMPLE 5-22

Suppose a sample of size $n = 101$ had been used in Example 5-18. Now what is the power if $\sigma^2 = .00028$? How large must n be to raise this to .95?

Solution

Again $R = .816$ (from Example 5-21) and $\alpha = .05$. There is a curve for 100 degrees of freedom and we read power $= .88$ approximately.

To answer the second question, we must obviously have n larger than 101. The graphs show that with $n = 151$ ($v = 150$) the power is approximately .97. We take

$$n = 101 + \frac{.95 - .88}{.97 - .88}(151 - 101) = 101 + \frac{7}{9}(50) \cong 140$$

Again, a quick visual guess is probably sufficiently accurate.

(With a good chi-square table such as Harter's [1], it is easy to get an exact answer here. In order to have Power $\geq 1 - \beta$ we need from formula (5-46)

$$\frac{\sigma_0^2}{\sigma^2} \chi_{n-1;\alpha}^2 \geq \chi_{n-1;1-\beta}^2$$

or

$$\frac{\chi_{n-1;1-\beta}^2}{\chi_{n-1;\alpha}^2} \leq \frac{\sigma_0^2}{\sigma^2}$$

Since in our problem $\sigma_0^2/\sigma^2 = 3/2 = 1.5$, $1 - \beta = .95$, $\alpha = .05$, we need the smallest n to satisfy

$$\frac{\chi_{n-1;.95}^2}{\chi_{n-1;.05}^2} \leq 1.5$$

With a little trial we find

$$\frac{\chi_{132;.95}^2}{\chi_{132;.05}^2} = \frac{159.81}{106.46} = 1.5011$$

$$\frac{\chi_{133;.95}^2}{\chi_{133;.05}^2} = \frac{160.92}{107.36} = 1.4989$$

Hence $n - 1 = 133$, $n = 134$ is the minimum n which will meet the requirements. It can be shown that the corresponding inequality for (5-38) is

$$\frac{\chi_{n-1;\beta}^2}{\chi_{n-1;1-\alpha}^2} \geq \frac{\sigma_0^2}{\sigma^2}$$

and both inequalities can be used for the two-sided case with α replaced by $\alpha/2$.)

EXAMPLE 5-23

Find the approximate power of the test used in Example 5-19 if $\sigma = 12.5$.

Solution

We have $n = 81$, $\alpha = .05$, $\sigma_0 = 10$ and desire power for $R = 12.5/10 = 1.25$, $v = 80$. Since the alternative was two-sided, we use the graphs labeled $\alpha/2 = .025$ (the fifth page of Appendix B12). For $v = 75$ the power is about .82, for $v = 100$ it is about .92. Hence for $v = 80$ the power is about .85.

To obtain a confidence interval for σ^2 (or σ) with confidence coefficient $1 - \alpha$ we start with

$$\Pr\left[\chi^2_{n-1;\,\alpha/2} < \frac{(n-1)S^2}{\sigma^2} < \chi^2_{n-1;\,1-\alpha/2}\right] = 1 - \alpha \tag{5-48}$$

Easy manipulation with the inequalities yields

$$\Pr\left[\frac{(n-1)S^2}{\chi^2_{n-1;\,1-\alpha/2}} < \sigma^2 < \frac{(n-1)S^2}{\chi^2_{n-1;\,\alpha/2}}\right] = 1 - \alpha \tag{5-49}$$

Then, using s^2, an observed value produced by experimentation, we can calculate a specific interval

$$\left[\frac{(n-1)s^2}{\chi^2_{n-1;\,1-\alpha/2}}, \frac{(n-1)s^2}{\chi^2_{n-1;\,\alpha/2}}\right] \tag{5-50}$$

a confidence interval for σ^2 with confidence coefficient $1 - \alpha$. Taking square roots of the end points of the interval gives

$$\left(\sqrt{\frac{(n-1)s^2}{\chi^2_{n-1;\,1-\alpha/2}}}, \sqrt{\frac{(n-1)s^2}{\chi^2_{n-1;\,\alpha/2}}}\right) \tag{5-51}$$

a confidence interval for σ, also with confidence coefficient $1 - \alpha$. The intervals (5-50) and (5-51) cover a value σ_0^2 (or σ_0) if, and only if, the test of $H_0 : \sigma^2 = \sigma_0^2$ against $H_1 : \sigma^2 \neq \sigma_0^2$ accepts when using the same α (and the same sample). Thus again to test H_0 we could compute the interval rather than the observed value of the statistic Y_{n-1}.

EXAMPLE 5-24

Find a confidence interval with confidence coefficient .95 for σ^2 and σ using the sample results of Example 5-19.

Solution

First we have $s = 12$, $n = 81$, $\alpha = .05$, $\alpha/2 = .025$. Appendix B7 yields $\chi^2_{80;\,.025} = 57.15$, $\chi^2_{80;\,.975} = 106.6$. Thus using interval (5-50) the confidence interval for σ^2 is

$$\left[\frac{80(144)}{106.6}, \frac{80(144)}{57.15}\right]$$

which reduces to $(108.1, 201.6)$. (This interval does not include $\sigma^2 = 100$, and the H_0 of Example 5-19 was rejected.) Taking square roots yields $(\sqrt{108.1}, \sqrt{201.6})$ or $(10.4, 14.2)$ as a confidence interval for σ with confidence coefficient .95.

The length of interval (5-51) for σ depends upon the particular value of s obtained from the experimental values. Let

$$I = \sqrt{\frac{(n-1)S^2}{\chi^2_{n-1;\,\alpha/2}}} - \sqrt{\frac{(n-1)S^2}{\chi^2_{n-1;\,1-\alpha/2}}}$$

Then for given n, α, and L we could seek

$$\Pr(I < L) \tag{5-52}$$

A more interesting problem requires the minimum n needed to make

$$\Pr(I < L) = \gamma \tag{5-53}$$

for given α, L, and γ. Formulas (5-52) and (5-53) are the counterparts of formulas (5-30) and (5-31), which apply to a confidence interval for μ. Problems of interest are essentially the same as those illustrated in Examples 5-14 and 5-15 except that now Appendix B13 is used (also entered with L in σ units).

EXAMPLE 5-25

Find the probability that a random sample of size 15 drawn from a normal distribution produces a confidence interval for the standard deviation with confidence coefficient .95 which is shorter than σ.

Solution

We have $1 - \alpha = .95$, $n = 15$, $L = \sigma$ (or $L/\sigma = 1$). From the first page of Appendix B13 we find that the vertical line through $L = 1\sigma$ intersects the horizontal line through $n = 15$ between $\Pr = .70$ and $\Pr = .90$. The intersection is close to $\Pr = .90$, and we estimate the probability to be about .86.

EXAMPLE 5-26

Find the minimum n required if it is desired to obtain a confidence interval with confidence coefficient .90 for the standard deviation σ of a normal distribution and to have the probability be .99 that the random interval be shorter than $.5\sigma$.

Solution

We have $1 - \alpha = .90$, $L = .5\sigma$, and $\Pr = \gamma = .99$. From the second page of Appendix B13 we find that the vertical line through $L = .5\sigma$ crosses the curve labeled $\Pr = .99$ opposite $n = 39$. Thus, the probability is .99 that a random sample of size 39 drawn from a normal distribution will produce a confidence interval with confidence coefficient .90 for σ that is shorter than $.5\sigma$.

Some other tables which can be used to determine sample size associated with confidence intervals for the standard deviation of a normal distribution are discussed in Sec. 3.3 of Owen's "Handbook of Statistical Tables [2]."

EXERCISES

◈ 5-57 Assume that the frozen shrimp company of Exercise 5-1 does not know the standard deviation of package weights. Suppose that the company desires to maintain a standard deviation of .50 ounce or less. A random sample of 25 packages yields an observed sample standard deviation of .60 ounce. Assuming that package weights are normally distributed, can we conclude that the desired standard deviation is not being maintained? Use $\alpha = .05$ and follow the six-step outline.

◈ 5-58 Find the power of test used in Exercise 5-57 if the actual value of σ is .60. If $\sigma = .75$.

◈ 5-59 Approximately how large does n have to be in Exercise 5-57 if a power of at least .95 is desired when $\sigma = .75$?

◈ 5-60 Using the sample information given in Exercise 5-57, find a confidence interval for σ with confidence coefficient .95.

◈ 5-61 Find the probability that a random sample of size 25 drawn from a normal distribution produces a confidence interval for σ with confidence coefficient .95 which is shorter than $.7\sigma$. How large would n have to be to raise this probability to at least .90?

5-62 Grade-point averages for students in the College of Arts and Sciences at a certain university over a number of years have been normally distributed with standard deviation .5 of a unit. A random sample of 20 grade-point averages is selected from last semester's reports from the College of Education and yields an observed sample standard deviation of .35. Does this indicate that last semester's College of Education grades have a standard deviation that is different from .5? Follow the six-step outline using $\alpha = .05$. Find the approximate power of the test if $\sigma = .40$.

5-63 Exercise 5-33 deals with a famous bridge player and a new system. Suppose that a player with somewhat less talent finds that he averages about 52.0 percent whether he uses the old or the new bidding system. He would like to use the new system, even though it entails more memory work, if he can "prove" that it yields scores with a larger standard deviation, since that would increase his chances of occasionally obtaining a really high score and thus winning. Using the old system his scores are normally distributed with a standard deviation of 5.0 percent. Using the new system for 10 sessions, his observed sample standard deviation is 8.0 percent. If the significance level is $\alpha = .01$, should he adopt the new system? Follow the six-step outline.

5-64   In Exercise 5-63 find the probability of adopting the new system if its scores have an actual σ of 7.0 percent. If $n = 36$ (not 10), what is the probability of adopting the new system if $\sigma = 7.0$ percent?

5-65   Using the sample information given in Exercise 5-63, find a confidence interval for σ, the standard deviation associated with the new system, with confidence coefficient .90.

◆ 5-66   A pill for upset stomach is being manufactured for a drugstore chain. The weights are normally distributed with mean 2 grams and standard deviation 50 milligrams. A new method of producing the pill has been discovered in the manufacturer's laboratories. Since there are some costs associated with conversion to the new method of production, it is decided that the new method will be adopted only if the standard deviation of weights is "proved" to be less than 50 milligrams using $\alpha = .01$. A random sample of 31 pills produced by the new method has an observed sample standard deviation of 35.0 milligrams. What action should be taken? Follow the six-step outline.

◆ 5-67   In Exercise 5-66 find the probability of converting to the new method of production if it produces pills with a standard deviation of 40 milligrams. How large a sample size should be taken to raise this probability to at least .60? To at least .95?

◆ 5-68   Using the sample information given in Exercise 5-66, find a confidence interval for σ, the standard deviation associated with the new method of production, with confidence coefficient .95.

◆ 5-69   Find the probability that a random sample of size 31 drawn from a normal distribution produces a confidence interval for σ with confidence coefficient .95 which is shorter than $.5\sigma$. How large would n have to be to raise this probability to at least .99?

5-70   The statistics department at the state university has had several complaints from other academic departments concerning the difficulty of the current introductory course. In response to these complaints the statistics department is considering the adoption of a new text entitled "Statistics for Those Who Hate to Think." It is decided that any text cannot be worthwhile unless it can be "proved" that it produces final examination scores which have a standard deviation in excess of 5 units (percent). An experimental class containing 21 randomly selected students is taught with the new text, and the final examinations yield $s = 6.0$ percent. Using $\alpha = .05$, should the new text be adopted? Follow the six-step outline.

5-71   With a sample of size 21 and $\alpha = .05$, used in Exercise 5-70, what is the probability that the new text will be adopted if it produces scores with standard deviation 7.5? What minimum sample size should have been used to make this probability be at least .95?

5-72 Find a table of the chi-square distribution with more entries for degrees of freedom than is found in Appendix B7. Then, use the last inequality which appears in Example 5-22 to find the exact minimum sample size requested in Exercise 5-71.

5-8
INFERENCES ABOUT TWO VARIANCES

In the preceding section we encountered some problems in which we tested hypotheses about a variance associated with a new process or procedure. In both Examples 5-18 and 5-19 the σ_0^2 used in step 1 was the variance, assumed to be known, arising from a standard process. Situations in which an experimenter may want to test hypotheses about two variances, neither of which is known, may also be encountered. Instead of having a standard and a new machine, as in Example 5-18, we may be choosing between two machines on the basis of variance. Instead of the situation of Example 5-19, two new statistics textbooks may be under consideration with both variances unknown. The company which sells frozen shrimp (Exercise 5-1) may have two methods of packaging which yield an average content of 12 ounces and wishes to choose one on the basis of variance.

Suppose that two machines both produce 1/2-inch bolts. Machine 1 is somewhat more expensive than machine 2. We would like to buy machine 2 unless we can "prove" that machine 1 makes bolts with lower variance of diameters. Let σ_1^2 and σ_2^2 be the unknown variances associated with machines 1 and 2, respectively. Then we would test

$$H_0 : \sigma_1^2 = \sigma_2^2 \qquad \text{against} \qquad H_1 : \sigma_1^2 < \sigma_2^2$$

$$\left(\text{or } H_0 : \frac{\sigma_1^2}{\sigma_2^2} = 1 \qquad \text{against} \qquad H_1 : \frac{\sigma_1^2}{\sigma_2^2} < 1 \right) \tag{5-54}$$

and buy the cheaper machine if H_0 is accepted. If the two machines are about the same price and machine 1 turns out more parts than machine 2, we may decide to buy machine 1 unless the variance associated with machine 1 can be "proved" to be larger than the variance for machine 2. Under the latter circumstances we would test

$$H_0 : \sigma_1^2 = \sigma_2^2 \qquad \text{against} \qquad H_1 : \sigma_1^2 > \sigma_2^2$$

$$\left(\text{or } H_0 : \frac{\sigma_1^2}{\sigma_2^2} = 1 \qquad \text{against} \qquad H_1 : \frac{\sigma_1^2}{\sigma_2^2} > 1 \right) \tag{5-55}$$

and buy machine 1 unless H_0 is rejected. Note that the alternative of (5-55) can be written $H_1 : \sigma_2^2 < \sigma_1^2$. If we interchange the names of the machines, calling the one that turns out more parts machine 2, then

(5-54) and (5-55) are exactly the same and need not be considered separate cases. We can always use (5-54) and associate the smaller variance under H_1 with the subscript 1. Some problems lead to two-sided alternatives. A test for equality of two means, to be encountered in the next section, depends upon the assumption that two unknown variances are equal. The assumption could be checked by testing

$$H_0 : \sigma_1^2 = \sigma_2^2 \quad \text{against} \quad H_1 : \sigma_1^2 \neq \sigma_2^2 \tag{5-56}$$

$$\left(\text{or } H_0 : \frac{\sigma_1^2}{\sigma_2^2} = 1 \quad \text{against} \quad H_1 : \frac{\sigma_1^2}{\sigma_2^2} \neq 1\right)$$

In some problems we may prefer to replace (5-54) by

$$H_0 : \sigma_1^2 \geq \sigma_2^2 \quad \text{against} \quad H_1 : \sigma_1^2 < \sigma_2^2 \tag{5-57}$$

or (5-55) by

$$H_0 : \sigma_1^2 \leq \sigma_2^2 \quad \text{against} \quad H_1 : \sigma_1^2 > \sigma_2^2 \tag{5-58}$$

putting "satisfactory" values of σ_1^2/σ_2^2 in one set, "unsatisfactory" values in another set. As in previous situations the testing procedure is unchanged, but again we must regard α to be the largest value of the power when H_0 is true (which will occur when $\sigma_1^2 = \sigma_2^2$).

Let $X_{11}, X_{21}, \ldots, X_{n_1 1}$ and $X_{12}, X_{22}, \ldots, X_{n_2 2}$ denote random samples of size n_1 and n_2 from distributions having variances σ_1^2 and σ_2^2, respectively. The sample variances are

$$S_1^2 = \frac{\sum_{i=1}^{n_1} (X_{i1} - \overline{X}_1)^2}{n_1 - 1} \quad \text{and} \quad S_2^2 = \frac{\sum_{i=1}^{n_2} (X_{i2} - \overline{X}_2)^2}{n_2 - 1} \tag{5-59}$$

where $\overline{X}_1$ and $\overline{X}_2$ are the sample means of the first and second samples, respectively. Intuitively, it seems reasonable to base inferences about σ_1^2 and σ_2^2 on S_1^2 and S_2^2 and inferences about σ_1^2/σ_2^2 on S_1^2/S_2^2. It can be shown that

$$F_{n_1-1, n_2-1} = \frac{S_1^2/\sigma_1^2}{S_2^2/\sigma_2^2} = \frac{S_1^2}{S_2^2}\left(\frac{\sigma_2^2}{\sigma_1^2}\right) = \frac{S_1^2}{S_2^2}\left(\frac{1}{R_0^2}\right) \tag{5-60}$$

is a good statistic to use to test hypotheses about σ_1^2/σ_2^2. Although other ratios could be used in tests, the H_0's of (5-54) to (5-56) specify $\sigma_1^2/\sigma_2^2 = 1$ so that (5-60) becomes

$$F_{n_1-1, n_2-1} = \frac{S_1^2}{S_2^2} \tag{5-61}$$

The statistic (5-61) has an F distribution with $n_1 - 1$, $n_2 - 1$ degrees of freedom provided (a) each sample is a random sample from a distribution, (b) both sampled distributions are normal, and (c) the variances are equal, that is, $\sigma_1^2 = \sigma_2^2$. [If (a) and (b) are satisfied but (c) is not, then (5-60) has an F distribution.] As in the case with one variance, failure to satisfy

FIGURE 5-12

Density of F_{n_1-1,n_2-1} if $\sigma_1^2 = \sigma_2^2$, showing significance level α and critical region $f_{n_1-1,n_2-1} < f_{n_1-1,n_2-1;\alpha}$.

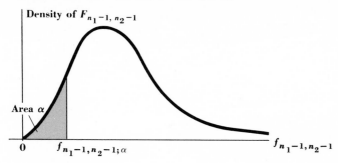

the normality assumption may produce serious errors in probability statements.

When testing (5-54) or (5-57), H_1 is supported only if s_1^2 is small by comparison to s_2^2. Consequently, H_0 is rejected only if f_{n_1-1,n_2-1} is small. If the significance level is α, then the critical region is

$$f_{n_1-1,n_2-1} = \frac{s_1^2}{s_2^2} < f_{n_1-1,n_2-1;\alpha} \tag{5-62}$$

where $\Pr(F_{n_1-1,n_2-1} < f_{n_1-1,n_2-1;\alpha}) = \alpha$ if $\sigma_1^2 = \sigma_2^2$ (see Fig. 5-12). Similarly, when testing (5-55) or (5-58), only large values of s_1^2 (when compared to s_2^2) support H_1, and a test with significance level α is obtained by rejecting H_0 when

$$f_{n_1-1,n_2-1} = \frac{s_1^2}{s_2^2} > f_{n_1-1,n_2-1;1-\alpha} \tag{5-63}$$

where $\Pr(F_{n_1-1,n_2-1} > f_{n_1-1,n_2-1;1-\alpha}) = \alpha$ if $\sigma_1^2 = \sigma_2^2$ (see Fig. 5-13).

FIGURE 5-13

Density of F_{n_1-1,n_2-1} if $\sigma_1^2 = \sigma_2^2$, showing significance level α and critical region $f_{n_1-1,n_2-1} > f_{n_1-1,n_2-1;1-\alpha}$.

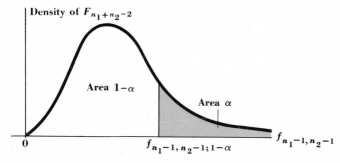

Finally, when testing (5-56), H_0 is rejected when s_1^2/s_2^2 is either too small or too large. Thus, with significance level α an intuitively appealing critical region is

$$f_{n_1-1,\,n_2-1} = \frac{s_1^2}{s_2^2} < f_{n_1-1,\,n_2-1;\,\alpha/2}$$

and   (5-64)

$$f_{n_1-1,\,n_2-1} = \frac{s_1^2}{s_2^2} > f_{n_1-1,\,n_2-1;\,1-\alpha/2}$$

the "equal-tails" choice. If $\sigma_1^2 = \sigma_2^2$ then

$$\Pr(F_{n_1-1,\,n_2-1} < f_{n_1-1,\,n_2-1;\,\alpha/2}) + \Pr(F_{n_1-1,\,n_2-1} > f_{n_1-1,\,n_2-1;\,1-\alpha/2})$$

$$= \alpha/2 + \alpha/2 = \alpha$$

(see Fig. 5-14).

FIGURE 5-14

Density of F_{n_1-1,n_2-1} if $\sigma_1^2 = \sigma_2^2$, showing significance level α and critical region $f_{n_1-1,n_2-1} < f_{n_1-1,n_2-1;\alpha/2}$ and $f_{n_1-1,n_2-1} > f_{n_1-1,n_2-1;\,1-\alpha/2}$.

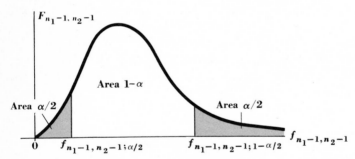

EXAMPLE 5-27

Consider again the two machines used to illustrate (5-54). Suppose that a random sample of 25 bolts produced by each machine is observed and we find $s_1^2 = .00028$, $s_2^2 = .00045$ where the subscripts indicate results from machines 1 and 2, respectively. If we use $\alpha = .05$, what decision should be made?

Solution

We shall follow the six-step outline.

1 In the previous discussion we have already indicated that we wish to test $H_0 : \sigma_1^2 = \sigma_2^2$ against $H_1 : \sigma_1^2 < \sigma_2^2$ and buy machine 2 unless H_0 is rejected.

2 The significance level $\alpha = .05$ and the sample sizes $n_1 = n_2 = 25$ have already been chosen.

3 The statistic we will use is $F_{n_1-1, n_2-1} = S_1^2/S_2^2$ which has an F distribution with $n_1 - 1$ and $n_2 - 1$ degrees of freedom provided

(a) Both samples of bolts are selected randomly.

(b) The diameters associated with each machine are normally distributed, probably a reasonable assumption.

(c) The distributions have equal variances.

4 The critical region given by inequality (5-62) is $f_{24, 24} = s_1^2/s_2^2 < f_{24, 24; .05} = .504$. Figure 5-12 applies with $n_1 = n_2 = 25$, $\alpha = .05$.

5 From the samples we compute $s_1^2 = .00028$, $s_2^2 = .00045$, and the observed value of the statistic is $f_{24, 24} = .00028/.00045 = .62$.

6 Since the observed value of the statistic does not lie in the critical region, H_0 is accepted. Consequently, machine 2 is purchased.

EXAMPLE 5-28

A standard and a new method of teaching statistics are being compared. It is desired to know whether or not there is a difference in variability, as measured by variance, of final examination scores produced by the two methods. One class is taught by each method, and both classes take the same final examination. The observed sample variances are $s_1^2 = 100$, $s_2^2 = 144$, where subscript 1 is used for the standard method and 2 is used for the new. If the classes contain $n_1 = 101$ and $n_2 = 61$ students, what conclusions would be drawn with a significance level $\alpha = .05$?

Solution

We shall follow the six-step outline.

1 The problem suggests that we test $H_0 : \sigma_1^2 = \sigma_2^2$ against $H_1 : \sigma_1^2 \neq \sigma_2^2$.

2 The significance level $\alpha = .05$ and the sample sizes $n_1 = 101$, $n_2 = 61$ have already been chosen.

3 We use the statistic $F_{n_1-1, n_2-1} = S_1^2/S_2^2$ which has an F distribution with $n_1 - 1$, $n_2 - 1$ degrees of freedom provided

(a) Both samples of students (classes) can be regarded as randomly selected.

(b) The final examination grades produced by each method are normally distributed. This is probably a reasonable assumption.

(c) Both normal distributions have the same variance, that is, $\sigma_1^2 = \sigma_2^2$.

4 The critical region given by (5-64) is $f_{100, 60} = s_1^2/s_2^2 < f_{100, 60; .025} = .641$ and $f_{100, 60} = s_1^2/s_2^2 > f_{100, 60; .975} = 1.60$. Figure 5-14 applies with $n_1 = 101$, $n_2 = 61$, $\alpha = .05$.

5 From the observed random samples we compute $s_1^2 = 100$, $s_2^2 = 144$, and the observed value of the statistic is

$$f_{100, 60} = \frac{s_1^2}{s_2^2} = \frac{100}{144} = .694$$

6 We accept H_0. The observed sample variances are such that it is reasonable that distributions with equal variances could have produced them.

If $\sigma_1^2 \neq \sigma_2^2$, then we use (5-60) to compute probabilities (power). For example, when we test (5-54) with the critical region (5-62), the power is

$$\Pr\left(\frac{S_1^2}{S_2^2} < f_{n_1 - 1, n_2 - 1; \alpha}\right) = \Pr\left(\frac{S_1^2}{S_2^2} \frac{1}{R_0^2} < \frac{1}{R_0^2} f_{n_1 - 1, n_2 - 1; \alpha}\right)$$

$$= \Pr\left(F_{n_1 - 1, n_2 - 1} < \frac{1}{R_0^2} f_{n_1 - 1, n_2 - 1; \alpha}\right) \qquad (5\text{-}65)$$

Subject to the limitations of the tables, (5-65) can be evaluated from Appendix B9.

EXAMPLE 5-29

Suppose that we test the hypothesis of Example 5-27 with $n_1 = n_2 = 25$, $\alpha = .05$. What is the probability that H_0 is rejected if $\sigma_1^2/\sigma_2^2 = (.7)^2 = .49$?

Solution

Since $f_{24, 24; .05} = .504$, $R_0^2 = .49$, we have $f_{24, 24; .05}/R_0^2 = .504/.49 = 1.03$, and (5-65) is $\Pr(F_{24, 24} < 1.03)$. From Appendix B9 we get $\Pr(F_{24, 24} < 1.00) = .50$, $\Pr(F_{24, 24} < 1.32) = .75$. Thus the probability we seek is slightly greater than .50, and interpolation gives

$$\Pr(F_{24, 24} < 1.03) = .52$$

If both sample sizes are the same, as in Example 5-29, then power can be read from the graphs of Appendix B14 when $\alpha = .01, .025, .05$. These graphs were constructed for testing $H_0 : \sigma_1^2 = \sigma_2^2$ against $H_1 : \sigma_1^2 < \sigma_2^2$ but can be used for $H_0 : \sigma_1^2 = \sigma_2^2$ against $H_1 : \sigma_1^2 > \sigma_2^2$ by interchanging the roles of the 1 and 2 subscripts (as we indicated previously). In order to find power, enter the horizontal scale with $R_0 = \sigma_1/\sigma_2$, follow the vertical line through this number up to the curve with the appropriate degrees of freedom, $v = n_1 - 1 = n_2 - 1$, and read power from the vertical scale. The graphs also give approximate power for $H_0 : \sigma_1^2 = \sigma_2^2$ against

$H_1 : \sigma_1^2 \neq \sigma_2^2$ when entered with $\alpha/2$ (instead of α).   The error is negligible when the power is greater than .30.   With a two-tailed test and $R_0 > 1$, enter the horizontal scale at $1/R_0$.

EXAMPLE 5-30

Rework Example 5-29 by using the graphs of Appendix B14.   Suppose samples of size $n_1 = n_2 = 41$ had been used.   Now what is power if $R_0 = .7$?   How large would $n_1 = n_2$ have to be to raise this to at least .95?

Solution

We have $\alpha = .05$, $R_0 = .7$.   With $n_1 = n_2 = 25$, $v = 24$, no curve is available.   However, with $R_0 = .7$, $v = 20$ we read power $= .47$, and with $R_0 = .7$, $v = 30$ we read power $= .60$.   Since $\frac{4}{10}$ of $.60 - .47 = .13$ is about .05, interpolation yields power $= .52$.

If $n_1 = n_2 = 41$, $v = 40$, $\alpha = .05$, $R_0 = .7$, with no difficulty we find power $= .72$.

Next we seek v that makes the power at least .95 when $\alpha = .05$, $R_0 = .7$. We see that $v = 90$ produces a power just greater than .95.   If we interpolate we may get $v = 88$ or 89.   In an actual problem we would probably take $n_1 = n_2 = 90$, realizing that sample sizes within 2 or 3 of an actual minimum are usually sufficiently accurate.

To obtain a confidence interval for σ_1^2/σ_2^2 (or σ_1/σ_2) with confidence coefficient $1 - \alpha$, we start with

$$\Pr\left(f_{n_1-1,\,n_2-1;\,\alpha/2} < \frac{S_1^2}{S_2^2}\frac{\sigma_2^2}{\sigma_1^2} < f_{n_1-1,\,n_2-1;\,1-\alpha/2} \right) = 1 - \alpha \qquad (5\text{-}66)$$

After manipulating the inequalities in formula (5-66), we get

$$\Pr\left(\frac{S_1^2}{S_2^2}\frac{1}{f_{n_1-1,\,n_2-1;\,1-\alpha/2}} < \frac{\sigma_1^2}{\sigma_2^2} < \frac{S_1^2}{S_2^2}\frac{1}{f_{n_1-1,\,n_2-1;\,\alpha/2}} \right) = 1 - \alpha \qquad (5\text{-}67)$$

Then using s_1^2, s_2^2, observed sample variances, we can calculate a specific interval

$$\left(\frac{s_1^2}{s_2^2}\frac{1}{f_{n_1-1,\,n_2-1;\,1-\alpha/2}} , \frac{s_1^2}{s_2^2}\frac{1}{f_{n_1-1,\,n_2-1;\,\alpha/2}} \right) \qquad (5\text{-}68)$$

which is a confidence interval for σ_1^2/σ_2^2 with confidence coefficient $1 - \alpha$. Taking square roots of the end points of the interval gives

$$\left(\frac{s_1}{s_2}\frac{1}{\sqrt{f_{n_1-1,\,n_2-1;\,1-\alpha/2}}} , \frac{s_1}{s_2}\frac{1}{\sqrt{f_{n_1-1,\,n_2-1;\,\alpha/2}}} \right) \qquad (5\text{-}69)$$

a confidence interval for σ_1/σ_2, also with confidence coefficient $1 - \alpha$. The intervals (5-68) and (5-69) cover 1 if, and only if, the test of $H_0 : \sigma_1^2 = \sigma_2^2$ against $H_1 : \sigma_1^2 \neq \sigma_2^2$ accepts H_0 when using the same α (and the same samples).

EXAMPLE 5-31

Using the sample information given in Example 5-28, find a confidence interval with confidence coefficient .95 for σ_1^2/σ_2^2 and σ_1/σ_2.

Solution

We had $s_1^2 = 100$, $s_2^2 = 144$, $n_1 = 101$, $n_2 = 61$. Since the confidence coefficient is $1 - \alpha = .95$, $\alpha = .05$, $\alpha/2 = .025$ and we need $f_{100, 60; .025} = .641$, $f_{100, 60; .975} = 1.60$. The interval (5-68) is

$$\left(\frac{100}{144}\frac{1}{1.60}, \frac{100}{144}\frac{1}{.641}\right) = (.434, 1.08)$$

Taking square roots yields $(\sqrt{.434}, \sqrt{1.08}) = (.659, 1.04)$ as a confidence interval for σ_1/σ_2 with confidence coefficient .95. We note that both intervals cover 1 as they should since H_0 was accepted in Example 5-28.

EXERCISES

◈ 5-73 Two methods of packaging frozen shrimp yield about the same average value. However, method 2 is somewhat faster, and the company would like to use it unless the variance of weights can be shown to be larger than the variance of weights for method 1 using the .05 level of significance. A random sample of 41 packages from those produced by each method is examined. The observed sample standard deviations are $s_1 = .50$ of an ounce for method 1 and $s_2 = .62$ of an ounce for method 2. What decision should be made? Follow the six-step outline.

◈ 5-74 Find the power of the test used in Exercise 5-73 if $\sigma_1/\sigma_2 = .5$. If $\sigma_1/\sigma_2 = .8$. If $\sigma_1/\sigma_2 = .8$, how large does $n_1 = n_2$ have to be to raise the power to .75?

◈ 5-75 Find a confidence interval for σ_1/σ_2 in Exercise 5-73 with confidence coefficient .95.

5-76 It is known that two brands of tires have an average life of 25,000 miles. However, there may be some difference in the variance of mileages associated with the two brands. An experiment is conducted in which 16 tires of each brand are used. It is found that the two observed sample standard deviations are $s_1 = 4,200$ miles, $s_2 = 2,800$ miles. Using $\alpha = .05$, what conclusions should be drawn? Follow the six-step outline.

5-77 Find the power of the test used in Exercise 5-76 if $\sigma_1/\sigma_2 = 1/2$. If $\sigma_1/\sigma_2 = 2$. If 41 of each kind of tires had been used (instead of 16), what is the power if $\sigma_1/\sigma_2 = 1/2$?

5-78 Find a confidence interval for σ_1/σ_2 in Exercise 5-76 with confidence coefficient .95.

◈ 5-79 Two classes are taught statistics, each with a different text. Both classes take the same final examination. The class sizes are $n_1 = 61$, $n_2 = 121$. If the observed sample standard deviations for the final examination grades are $s_1 = 10$, $s_2 = 15$, what conclusion can be drawn concerning the variances of final examination scores associated with the two textbooks? Follow the six-step outline.

◈ 5-80 Find a confidence interval for σ_1/σ_2 in Exercise 5-79 with confidence coefficient .90.

5-81 In Exercise 5-63 we considered a bridge player of modest talents and his experimentation with a new bidding system. Suppose now that he is considering two new systems both of which seem to produce about the same average score for him. He would prefer to use system 1 unless there is very good evidence that system 2 produces higher variance of scores. Using each method for 21 sessions he finds observed sample standard deviations $s_1 = 5.5$ percent, $s_2 = 10.0$ percent. With $\alpha = .01$ what decision should be made? Follow the six-step outline.

5-82 In Exercise 5-81 find the probability that system 2 will be adapted if $\sigma_1/\sigma_2 = .5$. How many sessions should have been used to raise this probability to at least .90?

5-9
INFERENCES ABOUT TWO MEANS
WITH TWO RANDOM SAMPLES

In Secs. 5-2 and 5-5 we considered a number of hypothesis testing problems concerning one mean. Sometimes an investigator will want to test hypotheses involving two means. The large corporation of Example 5-1 may be choosing between two brands of light bulbs and not know either average life. The smoker discussed in Sec. 4-4 may be interested in comparing the average nicotine content of two brands of cigarettes. A wheat grower may desire information about the average yields of two different varieties which he could plant. A teacher may be interested in comparing average results achieved by two teaching methods.

Suppose that two brands of light bulbs are being considered by a large corporation. Brand 1 is slightly less expensive than brand 2. The company would like to buy brand 1 unless there is good evidence that the average life of brand 1 is less than the average life of brand 2. If we let

μ_1 and μ_2 be the average length of life for brands 1 and 2, respectively, then we will be interested in testing

$$H_0 : \mu_1 = \mu_2 \qquad \text{against} \qquad H_1 : \mu_1 < \mu_2 \tag{5-70}$$

$$(\text{or } H_0 : \mu_1 - \mu_2 = 0 \qquad \text{against} \qquad H_1 : \mu_1 - \mu_2 < 0)$$

If H_0 is accepted, the cheaper brand will be purchased. If the two brands sell for about the same price and brand 2 is more attractive than brand 1 (perhaps because the company provides better service), we may want to buy brand 2 unless brand 1 has a longer average life. Under the latter circumstances we will test

$$H_0 : \mu_1 = \mu_2 \qquad \text{against} \qquad H_1 : \mu_1 > \mu_2 \tag{5-71}$$

$$(\text{or } H_0 : \mu_1 - \mu_2 = 0 \qquad \text{against} \qquad H_1 : \mu_1 - \mu_2 > 0)$$

and buy brand 2 unless H_0 is rejected. Actually, (5-70) and (5-71) need not be considered separate cases, because they can be reduced to the same problem merely by interchanging the roles of the subscripts. Some problems lead to two-sided alternatives. In comparing two varieties of wheat for average yield, it may be that there is no reason to prefer one variety over the other. In such a situation it would be logical to test

$$H_0 : \mu_1 = \mu_2 \qquad \text{against} \qquad H_1 : \mu_1 \neq \mu_2 \tag{5-72}$$

$$(\text{or } H_0 : \mu_1 - \mu_2 = 0 \qquad \text{against} \qquad H_1 : \mu_1 - \mu_2 \neq 0)$$

We may prefer to replace (5-70) by

$$H_0 : \mu_1 \geq \mu_2 \qquad \text{against} \qquad H_1 : \mu_1 < \mu_2 \tag{5-73}$$

or (5-71) by

$$H_0 : \mu_1 \leq \mu_2 \qquad \text{against} \qquad H_1 : \mu_1 > \mu_2 \tag{5-74}$$

putting "satisfactory" values of $\mu_1 - \mu_2$ in one set, "unsatisfactory" values in another set. As in previous situations the testing procedure is unchanged but again we must regard α to be the largest value of the power when H_0 is true (which will occur when $\mu_1 = \mu_2$).

Let $X_{11}, X_{21}, \ldots, X_{n_1 1}$ and $X_{12}, X_{22}, \ldots, X_{n_2 2}$ denote random samples of size n_1 and n_2 from distributions having means μ_1 and μ_2 and variances σ_1^2 and σ_2^2, respectively. We will denote the sample means by $\bar{X}_1$ and $\bar{X}_2$ and the sample variances by S_1^2 and S_2^2 [defined by (5-59)]. Intuitively, it seems reasonable to base inferences about μ_1 and μ_2 on $\bar{X}_1$ and $\bar{X}_2$ and inferences about $\mu_1 - \mu_2$ on $\bar{X}_1 - \bar{X}_2$. It can be shown that

$$T_{n_1 + n_2 - 2} = \frac{\bar{X}_1 - \bar{X}_2 - (\mu_1 - \mu_2)}{S_p \sqrt{1/n_1 + 1/n_2}} \tag{5-75}$$

is a good statistic to use to test hypotheses about $\mu_1 - \mu_2$. Here

$$S_p^2 = \frac{(n_1 - 1)S_1^2 + (n_2 - 1)S_2^2}{n_1 + n_2 - 2}$$

[which reduces to $S_p^2 = (S_1^2 + S_2^2)/2$ when the sample sizes are equal, that is, $n_1 = n_2 = n$]. Although other differences could be used in tests, the H_0's of (5-70) to (5-72) specify $\mu_1 - \mu_2 = 0$ so that (5-75) becomes

$$T_{n_1+n_2-2} = \frac{\bar{X}_1 - \bar{X}_2}{S_p\sqrt{1/n_1 + 1/n_2}} \tag{5-76}$$

The statistic (5-76) has a t distribution with $n_1 + n_2 - 2$ degrees of freedom provided (a) each sample is a random sample from a distribution, (b) both sampled distributions are normal, (c) the variances are equal, that is, $\sigma_1^2 = \sigma_2^2 = \sigma^2$, and (d) the two means are equal so that $\mu_1 - \mu_2 = 0$. [If (a), (b), (c) are satisfied but (d) is not, then statistic (5-75) still has a t distribution and could be used to test $H_0 : \mu_1 - \mu_2 = \Delta_0 \neq 0$ against various alternatives.]

Experimental evidence indicates that the above normality assumption is not essential provided the sample sizes are not too small. Such experiments have also indicated that the consequences of not satisfying the equal-variance assumption can be minimized by taking equal sample sizes. Provided sample sizes are equal (or nearly so), and the sample sizes are not too small, we still get fairly accurate probability statements by assuming $T_{n_1+n_2-2}$ has a t distribution.

When testing (5-70) or (5-73), only small values of $\bar{x}_1 - \bar{x}_2$, and hence small values of $t_{n_1+n_2-2}$, support H_1 and lead to rejection of H_0. If the significance level is α, then the critical region is

$$t_{n_1+n_2-2} = \frac{\bar{x}_1 - \bar{x}_2}{s_p\sqrt{1/n_1 + 1/n_2}} < t_{n_1+n_2-2;\alpha} \tag{5-77}$$

where $\Pr(T_{n_1+n_2-2} < t_{n_1+n_2-2;\alpha}) = \alpha$ if $\mu_1 - \mu_2 = 0$ (see Fig. 5-15).

FIGURE 5-15

Density of $T_{n_1+n_2-2}$, showing significance level α and critical region $t_{n_1+n_2-2} < t_{n_1+n_2-2;\alpha}$.

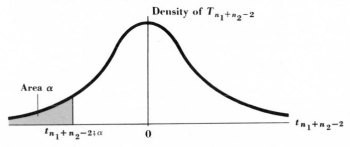

FIGURE 5-16

Density of $T_{n_1+n_2-2}$, showing significance level α and critical region $t_{n_1+n_2-2} > t_{n_1+n_2-2;1-\alpha}$.

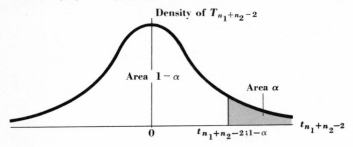

Density of $T_{n_1+n_2-2}$

Area $1-\alpha$

Area α

0 $\quad t_{n_1+n_2-2;1-\alpha}$ $\quad t_{n_1+n_2-2}$

Similarly, when testing (5-71) or (5-74) only large values of $\bar{x}_1 - \bar{x}_2$ support H_1, and a test with significance level α is obtained by rejecting H_0 when

$$t_{n_1+n_2-2} = \frac{\bar{x}_1 - \bar{x}_2}{s_p\sqrt{1/n_1 + 1/n_2}} > t_{n_1+n_2-2;1-\alpha} \tag{5-78}$$

where $\Pr(T_{n_1+n_2-2} > t_{n_1+n_2-2;1-\alpha}) = \alpha$ if $\mu_1 - \mu_2 = 0$ (see Fig. 5-16). Finally, when testing (5-72) H_0 is rejected if $\bar{x}_1 - \bar{x}_2$ is too small or too large. Thus with significance level α an intuitively appealing critical region is

$$t_{n_1+n_2-2} < t_{n_1+n_2-2;\alpha/2} \quad \text{and} \quad t_{n_1+n_2-2} > t_{n_1+n_2-2;1-\alpha/2} \tag{5-79}$$

where $t_{n_1+n_2-2} = (\bar{x}_1 - \bar{x}_2)/s_p\sqrt{1/n_1 + 1/n_2}$. If $\mu_1 - \mu_2 = 0$, then

$$\Pr(T_{n_1+n_2-2} < t_{n_1+n_2-2;\alpha/2})$$
$$+ \Pr(T_{n_1+n_2-2} > t_{n_1+n_2-2;1-\alpha/2}) = \alpha/2 + \alpha/2 = \alpha$$

(see Fig. 5-17).

FIGURE 5-17

Density of $T_{n_1+n_2-2}$ showing significance level α and critical region $t_{n_1+n_2-2} < t_{n_1+n_2-2;\alpha/2}$ and $t_{n_1+n_2-2} > t_{n_1+n_2-2;1-\alpha/2}$.

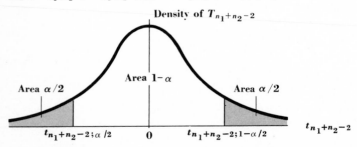

Density of $T_{n_1+n_2-2}$

Area $\alpha/2$

Area $1-\alpha$

Area $\alpha/2$

$t_{n_1+n_2-2;\alpha/2}$ $\quad 0 \quad$ $t_{n_1+n_2-2;1-\alpha/2}$ $\quad t_{n_1+n_2-2}$

If the variances are not equal, then only approximate tests are available for the hypothesis testing situations we have discussed in this section. A statistic which can be used in this case is

$$T_v' = \frac{\bar{X}_1 - \bar{X}_2 - (\mu_1 - \mu_2)}{\sqrt{S_1^2/n_1 + S_2^2/n_2}} \qquad (5\text{-}80)$$

(where $\mu_1 - \mu_2 = 0$ in the problems we have proposed) which is approximately distributed as T_v with degrees of freedom for an observed value t_v' being

$$v = \frac{(s_1^2/n_1 + s_2^2/n_2)^2}{\dfrac{(s_1^2/n_1)^2}{n_1 - 1} + \dfrac{(s_2^2/n_2)^2}{n_2 - 1}}$$

if the two distributions are normal (again, probably not a necessary assumption in practice). Since v will usually not be an integer, round it off to the nearest whole number. If $n_1 = n_2$ statistics (5-75) and (5-80) reduce to the same formula except for the degrees of freedom. It is quite likely that statistic (5-80) would be used only if the sample sizes differ greatly and we are quite sure that the variances are unequal. (For a good discussion of this problem, see a recent paper by Scheffé [4].)

EXAMPLE 5-32

The large corporation of Example 5-1 wants to choose between two brands of light bulbs on the basis of average life. Brand 1 is slightly less expensive than brand 2. The company would like to buy brand 1 unless the average life for brand 2 is shown to be greater using the .05 level of significance. A random sample of 100 bulbs of each brand is tested, and it is found that $\bar{x}_1 = 985$ hours, $\bar{x}_2 = 1{,}003$ hours, $s_1 = 80$ hours, $s_2 = 60$ hours. What conclusion should be drawn?

Solution

We shall follow the six-step outline.

1　The problem suggests that we test $H_0 : \mu_1 = \mu_2$ against $H_1 : \mu_1 < \mu_2$ and buy brand 1 if H_0 is accepted.
2　The significance level $\alpha = .05$ and the sample sizes $n_1 = n_2 = 100$ have already been chosen.
3　The statistic we use is (5-76). This statistic will have a t distribution with $100 + 100 - 2 = 198$ degrees of freedom provided (a) the samples are randomly selected, (b) the lengths of life associated with each brand are normally distributed, (c) both distributions have the same variance, and (d) $\mu_1 = \mu_2$ as hypothesized. With equal sample sizes this large we do not have to worry about (b) and (c).

4 The critical region given by (5-77) is $t_{198} < t_{198;.05} = -1.653$.
5 From the samples we compute $\bar{x}_1 = 985$, $\bar{x}_2 = 1,003$, $s_1 = 80$, $s_2 = 60$. Then we find

$$s_p^2 = \frac{99(80)^2 + 99(60)^2}{198} = \frac{6,400 + 3,600}{2} = \frac{10,000}{2}$$

The denominator of the observed value of the statistic is

$$s_p \sqrt{\frac{1}{n_1} + \frac{1}{n_2}} = \frac{100}{\sqrt{2}} \sqrt{\frac{1}{100} + \frac{1}{100}} = \frac{100}{\sqrt{2}} \sqrt{\frac{2}{100}} = 10$$

Thus

$$t_{198} = \frac{985 - 1,003}{10} = -1.8$$

6 Since the observed value of the statistic lies in the critical region, H_0 is rejected. Consequently, brand 2 should be purchased.

EXAMPLE 5-33

In Example 5-32 suppose that both brands sell for the same price and that the company has no reason to prefer one over the other. Now, how should the outline be changed, and what is the appropriate decision?

Solution

1 Now we would test $H_0 : \mu_1 = \mu_2$ against $H_1 : \mu_1 \neq \mu_2$.
4 The critical region given by inequalities (5-79) is

$$t_{198} < t_{198;.025} = -1.972 \quad \text{and} \quad t_{198} > t_{198;.975} = 1.972$$

6 H_0 is now accepted. The difference in sample means is not sufficient to demonstrate that brand 2 is superior. Lacking any other information or considerations, we would no doubt still choose brand 2.

If $\mu_1 \neq \mu_2$ then the statistic (5-76) no longer has a t distribution, so that special tables or graphs must be used for power calculations. Fortunately, the graphs of Appendix B10 still apply. Now

$$\delta = \frac{|\mu_1 - \mu_2|}{\sigma \sqrt{1/n_1 + 1/n_2}} \tag{5-81}$$

where $\mu_1 - \mu_2$ must be in the range designated by the alternative H_1. Again δ depends upon the unknown σ and to use the graphs we choose $\mu_1 - \mu_2$ in σ units.

EXAMPLE 5-34

Find the power of the test used in Example 5-32 if $\mu_2 = \mu_1 + .5\sigma$; that is, the average life for brand 2 is .5 of a standard deviation larger than for brand 1.

Solution

We are given $n_1 = n_2 = 100$, $|\mu_1 - \mu_2| = .5\sigma$. Thus

$$\delta = \frac{.5\sigma}{\sigma\sqrt{1/100 + 1/100}} = .5\sqrt{50} = 3.54$$

The degrees of freedom is $f = 198$. Turning to the page of Appendix B10 labeled $\alpha = .05$, the significance level, we find that the vertical line through $\delta = 3.54$ crosses the curves for both $f = 24$ and $f = \infty$ at about .97, which is the required power.

EXAMPLE 5-35

Find the power of test used in Example 5-33 if $|\mu_1 - \mu_2| = .5\sigma$.

Solution

Again $n_1 = n_2 = 100$, $f = 198$, $\delta = 3.54$. Since the test is two-sided, we look at the graph labeled $\alpha/2 = .05/2 = .025$. With $f = 24$ the power is .925, and with $f = \infty$ the power $= .945$. We can guess that for $f = 198$ the power is about .94. (This can also be obtained by inverse linear interpolation as described in Appendix A2-2.)

With equal sample sizes $n_1 = n_2 = n$, the standard sample size problem (power $= \alpha$ if H_0 is true, power $\geq 1 - \beta$ for a specific alternative) is handled as in Example 5-12. For this case formula (5-81) reduces to

$$\delta = \frac{|\mu_1 - \mu_2|\sqrt{n}}{\sigma\sqrt{2}} \tag{5-82}$$

and, solving for n, the counterpart of formula (5-14) is

$$n = 2\left(\frac{\sigma\delta}{\mu_1 - \mu_2}\right)^2 \tag{5-83}$$

This time $f = n + n - 2 = 2(n - 1)$ and the chosen $n = (f/2) + 1$.

EXAMPLE 5-36

In Example 5-34 we found that with $n_1 = n_2 = 100$, $\alpha = .05$ that the power is .97 if $\mu_2 = \mu_1 + .5\sigma$. Suppose that we are satisfied if the power is at least .90 at that alternative. What sample sizes should be taken?

Solution

Again we have $\alpha = .05$, $|\mu_1 - \mu_2|/\sigma = .5$. Formula (5-83) becomes $n = 2(2\delta)^2 = 8\delta^2$. As in Example 5-12, if we choose the δ that yields power $= .90$ for $f = \infty$ the computed n will be too small. The graphs of Appendix B10 give $\delta = 2.90$, which yields $n = 8(2.90)^2 = 67.28$. On the other hand, if we use $\delta = 3.00$, the value of δ which produces power $= .90$ for $f = 24$, the computed n may be larger than necessary. This time we get $n = 8(3.00)^2 = 72.00$, which is larger than the chosen $n = (f/2) + 1 = (24/2) + 1 = 13$. Hence, we have shown $67.28 \leqq n \leqq 72.00$, and any n in this interval, say $n = 70$, will be satisfactory.

(Using the Owen [3] table mentioned in Example 5-12, we find that with chosen $n = 69$ the computed $n = 69.20$, while with chosen $n = 70$ the computed $n = 69.20$. Thus the minimum n is 70.)

We observe that if we take $n = 70$ we need only 140 bulbs for the experiment as opposed to the 200 which were used. Without some preliminary power and sample size calculations, one may be inclined to use more observations than are necessary.

To obtain a confidence interval for $\mu_1 - \mu_2$ we start with

$$\Pr\left(-t_{n_1+n_2-2;\,1-\alpha/2} < \frac{\bar{X}_1 - \bar{X}_2 - (\mu_1 - \mu_2)}{S_p\sqrt{1/n_1 + 1/n_2}} < t_{n_1+n_2-2;\,1-\alpha/2}\right) = 1 - \alpha$$

(5-84)

If we compare (5-84) with (5-26), used in the case of one mean, the two probability statements are the same except for minor differences in symbols. We have μ replaced by $\mu_1 - \mu_2$, $\bar{X}$ replaced by $\bar{X}_1 - \bar{X}_2$, $S/\sqrt{n}$ replaced by $S_p\sqrt{1/n_1 + 1/n_2}$, and degrees of freedom $n - 1$ replaced by $n_1 + n_2 - 2$. Hence, looking at (5-27) and making these replacements, it is easy to see that the usual manipulations with inequalities will yield

$$\Pr\left(\bar{X}_1 - \bar{X}_2 - t_{n_1+n_2-2;\,1-\alpha/2}\,S_p\sqrt{\frac{1}{n_1} + \frac{1}{n_2}} < \mu_1 - \mu_2\right.$$

$$\left. < \bar{X}_1 - \bar{X}_2 + t_{n_1+n_2-2;\,1-\alpha/2}\,S_p\sqrt{\frac{1}{n_1} + \frac{1}{n_2}}\right) = 1 - \alpha \quad (5\text{-}85)$$

The desired confidence interval for $\mu_1 - \mu_2$ with confidence coefficient $1 - \alpha$ is

$$\left(\bar{x}_1 - \bar{x}_2 - t_{n_1+n_2-2;\,1-\alpha/2}\,S_p\sqrt{\frac{1}{n_1} + \frac{1}{n_2}},\right.$$

$$\left. \bar{x}_1 - \bar{x}_2 + t_{n_1+n_2-2;\,1-\alpha/2}\,S_p\sqrt{\frac{1}{n_1} + \frac{1}{n_2}}\right) \quad (5\text{-}86)$$

The interval (5-86) has the usual relationship to tests. In other words (5-86) will cover 0 if, and only if, the corresponding test of $H_0 : \mu_1 - \mu_2 = 0$ against $H_1 : \mu_1 - \mu_2 \neq 0$ accepts using the same α (and the same samples).

EXAMPLE 5-37

Using the experimental results of Example 5-32, find a confidence interval for $\mu_1 - \mu_2$ with confidence coefficient .95.

Solution

We had $\bar{x}_1 - \bar{x}_2 = 985 - 1{,}003 = -18$ hours and $s_p\sqrt{1/n_1 + 1/n_2} = 10$. Also $t_{198;\,.975} = 1.972$. Thus the required interval is $[-18 - (1.972)10, -18 + (1.972)10]$, which reduces to $(-37.7, 1.7)$. We note that this interval includes 0, as it should, since in Example 5-33 we accepted $H_0 : \mu_1 - \mu_2 = 0$ when tested against $H_1 : \mu_1 - \mu_2 \neq 0$.

EXERCISES

◈ 5-83 Two different types of wheat are being considered, and it is desired to know which (if either) produces the greater average yield. Thirty-two acres of each kind are planted with fairly uniform growing conditions. The results are: variety A, observed sample mean 33 bushels per acre, observed sample variance 5.9; variety B, observed sample mean 35.7 bushels per acre, observed sample variance 12.1. Analyze these results by using the six-step outline with $\alpha = .01$.

◈ 5-84 Find the power of the test used in Exercise 5-83 if the average yield for variety B is σ bushels per acre more than for variety A. About how large should $n = n_1 = n_2$ be to raise the power to .99?

◈ 5-85 With the sample results of Exercise 5-83 find a confidence interval for $\mu_1 - \mu_2$ with confidence coefficient .99. How does the interval relate to the test of Exercise 5-83?

5-86 A company is trying to decide which of two brands of tires to buy for its trucks. It would like to adopt brand G unless there is some evidence that brand F is better, in which case it will adopt the latter brand. An experiment in which 16 tires from each brand are used is conducted. The results are

Brand F: $\bar{x}_1 = 26{,}000$ miles $\quad$ $s_1 = 4{,}200$ miles
Brand G: $\bar{x}_2 = 25{,}000$ miles $\quad$ $s_2 = 2{,}800$ miles

where measurements were the length of life in miles. What conclusions can be drawn about the means of the two distributions? Follow the six-step outline using $\alpha = .05$.

5-87 Find the power of the test used in Exercise 5-86 if brand F has an average life that is $\sigma/2$ miles greater than that for brand G. How many tires of each brand should have been used to make this probability at least .80?

5-88 With the sample results of Exercise 5-86 find a confidence interval for $\mu_1 - \mu_2$ with confidence coefficient .90.

◇ 5-89 Two methods of teaching statistics are being tried by a professor. A class of 64 students is taught by method 1 and a class of 36 students by method 2. The two classes are given the same final examination. These scores yield $\bar{x}_1 = 67$, $\bar{x}_2 = 63$, $s_1 = 10$, $s_2 = 8$. Using $\alpha = .05$, can we conclude that the average final examination scores produced by the two methods are different? Follow the six-step outline.

◇ 5-90 Find the power of the test used in Exercise 5-89 if the two means differ by $(1/2)\sigma$. By $(3/4)\sigma$.

◇ 5-91 With the sample results of Exercise 5-89 find a confidence interval for the difference of the two means using confidence coefficient .95. How does this interval relate to the test of Exercise 5-89?

5-92 A psychology professor believes that IQ can be increased by studying courses specifically designed to give experience in types of skills which questions on the IQ test seem to emphasize. To test his theory two randomly selected groups of tenth graders are used for one semester. Sixteen in one group take the standard high school schedule of courses while sixteen in the second group study courses designed by the professor. At the end of the term all 32 students are given the same IQ test with the following results:

Ordinary schedule: $\bar{x}_1 = 102.5$ $s_1^2 = 100$
Special schedule: $\bar{x}_2 = 111.5$ $s_2^2 = 125$

If $\alpha = .05$, has the professor proved his theory? Follow the six-step outline.

5-93 Find the power of the test used in Exercise 5-92 if the average IQ is increased by 1 standard deviation. About how large should $n = n_1 = n_2$ be to raise this to probability .95?

5-94 Suppose that the merchandising chain mentioned in Exercise 5-3 is going to buy a large order of 100-watt bulbs. It has a choice between brand 1 and brand 2. The chain would rather buy brand 2, since it knows that the product is dependable, but will buy brand 1 if it can be proved that the average amount of electricity consumed over a 10-minute period is less for that brand. Twenty randomly selected bulbs of each brand are tested, and the results are $\bar{x}_1 = 48.0$, $s_1^2 = 18$, $\bar{x}_2 = 51.9$, $s_2^2 = 27$. If the significance level $\alpha = .01$ is used, which brand should be purchased? Use the six-step outline.

5-95 Find the power of the test used in Exercise 5-94 if the average amount of consumption for brand 1 is 1 standard deviation below that of brand 2. About how large should $n_1 = n_2 = n$ be to raise this probability to .90?

◈ 5-96 A cattle feeder would like to know whether or not to give his animals a diet supplement which is supposed to help them gain weight. He has 20 animals about the same size available for the experiment which are divided at random into two groups of 10 each. For a period of time group 1 is fed the standard diet while group 2 is fed the standard diet plus the supplement. The gain in weight for each animal is recorded and calculations yield: $\bar{x}_1 = 250$ pounds, $s_1 = 27$ pounds, $\bar{x}_2 = 265$ pounds, $s_2 = 36$ pounds. What conclusion should be drawn with a significance level of $\alpha = .01$? Follow the six-step outline.

◈ 5-97 Find the power of the test used in Exercise 5-96 if the supplement produces an average gain that is 1 standard deviation above the average achieved using only the standard diet. About how large should $n_1 = n_2 = n$ be to raise this probability to .80?

◈ 5-98 With the sample results of Exercise 5-96 find a confidence interval for the difference between the two means using confidence coefficient .95.

5-10
INFERENCES ABOUT TWO MEANS
WITH PAIRED OBSERVATIONS

There are some situations in which we may wish to draw inferences about two means, yet find that observations occur in pairs. Instead of obtaining two random samples, we will be able to get one random sample of pairs, and the two measurements associated with a pair will be related to each other. Under such circumstances the procedures of Sec. 5-9 no longer apply. This kind of a problem arises from before-and-after type experiments. For example, suppose that 20 adults are given a strenuous physical training program. Their weights are recorded before they begin and after they complete the course. We desire to know if the program produces a change in the average weight of individuals. Obviously the measurements are related. That is, if an individual weighs 200 pounds before beginning, he will probably weigh about 200 pounds at the end. Similarly, a person who begins at 100 pounds will not differ too greatly from that weight after completing the course. The final weight depends to a large extent upon the initial weight.

In some experimental situations it is advantageous to observe random variables in pairs. Suppose that two teaching methods are to be compared on the basis of average scores by using 50 students (who we hope constitute a random sample of those students for whom results would be applicable) divided into two equal-size classes, one taught by each method. When experimental results (perhaps final examination scores) are available, we plan to test the hypothesis that the means associated with the two methods are equal ($H_0 : \mu_1 = \mu_2$). Students could be assigned to the two classes in various ways. One possibility would be to select at random 25 for method 1 with the remainder being taught by method 2. With this assignment we have available two random samples and can use the t test of Sec. 5-9. However, scores so obtained will reflect not only differences in methods, but also differences in ability. As a consequence one of two undesirable things is more apt to happen: (1) We may conclude that methods produce a difference in means when actually the only difference is overall class ability (due to the luck of the draw), or (2) we may fail to detect a difference in methods when a real difference exists. To overcome this difficulty we can attempt to eliminate the variable (here, ability) in which we have no interest. To do this we first pair students according to ability (as measured by some test or by IQ), selecting the pair so that they are as nearly alike as possible, and then assign at random one of each pair to each class. The two top IQs could form the first pair, the next highest two IQs the next pair, etc. From each pair's scores we obtain a difference, and these differences are used to conduct our test.

Suppose that we are interested in one of the standard hypothesis testing situations for two means given by (5-70) to (5-74). However, either because of unavoidable circumstances (as in the before-after problem) or due to deliberate design, the experimental situation produces a random sample of n pairs of measurements $(X_{11}, X_{12}), (X_{21}, X_{22}), \ldots, (X_{n1}, X_{n2})$. From each pair a difference can be computed. Denote these by D_1, $D_2, \ldots, D_n$ where $D_i = X_{i1} - X_{i2}$, $i = 1, 2, \ldots, n$. Let

$$\bar{D} = \frac{\sum_{i=1}^{n} D_i}{n} \qquad S_D^2 = \frac{\sum_{i=1}^{n} (D_i - \bar{D})^2}{n - 1}$$

be the sample mean and sample variance of the D_i. It can be shown that

$$T_{n-1} = \frac{\bar{D} - \mu_D}{S_D / \sqrt{n}} \tag{5-87}$$

is a good statistic to use to test hypotheses about $\mu_D = \mu_1 - \mu_2$. (We note that $\bar{D} = \bar{X}_1 - \bar{X}_2$, where $\bar{X}_1, \bar{X}_2$ are the sample means for the first and second samples.) To test (5-70) to (5-74) we use $\mu_D = 0$ so that (5-87) becomes

$$T_{n-1} = \frac{\bar{D}}{S_D / \sqrt{n}} \tag{5-88}$$

The statistic (5-88) has a t distribution with $n - 1$ degrees of freedom provided (a) $D_1, D_2, \ldots, D_n$ is a random sample from a distribution,

(b) that distribution is normal, and (c) the two means μ_1, μ_2 are equal. [If (a) and (b) are satisfied but not (c), (5-87) still has a t distribution and could be used to test $H_0 : \mu_1 - \mu_2 = \Delta_0 \neq 0$ against various alternatives.] Except for the fact that differences D_i, rather than X_i, are used, (5-87) and (5-22) are exactly the same. Consequently, all the remarks made in Secs. 5-5 and 5-6 about tests, power, and confidence intervals still apply.

When testing (5-70) or (5-73), only small $\bar{d}$, and hence small t_{n-1}, support the alternative H_1 and lead to the rejection of H_0. If the significance level is α, then the critical region is

$$t_{n-1} = \frac{\bar{d}}{s_D/\sqrt{n}} < t_{n-1;\alpha} \qquad (5\text{-}89)$$

where $\Pr(T_{n-1} < t_{n-1;\alpha}) = \alpha$ if $\mu_1 = \mu_2$ (see Fig. 5-6). Similarly, when testing (5-71) or (5-74) only large values of $\bar{d}$ support H_1, and a test of significance level α is obtained by rejecting H_0 when

$$t_{n-1} = \frac{\bar{d}}{s_D/\sqrt{n}} > t_{n-1;1-\alpha} \qquad (5\text{-}90)$$

where $\Pr(T_{n-1} > t_{n-1;1-\alpha}) = \alpha$ when $\mu_1 = \mu_2$ (see Fig. 5-7). Finally when testing (5-72), H_0 is rejected when the observed value of the statistic is too small or too large. The critical region usually selected to give significance level α is

$$t_{n-1} = \frac{\bar{d}}{s_D/\sqrt{n}} < t_{n-1;\alpha/2} \qquad \text{and} \qquad t_{n-1} = \frac{\bar{d}}{s_D/\sqrt{n}} > t_{n-1;1-\alpha/2} \qquad (5\text{-}91)$$

where

$$\Pr(T_{n-1} < t_{n-1;\alpha/2}) + \Pr(T_{n-1} > t_{n-1;1-\alpha/2}) = \alpha/2 + \alpha/2 = \alpha$$

when $\mu_1 = \mu_2$ (see Fig. 5-8).

The teaching methods example illustrates (as is frequently the case) that two or more procedures can sometimes be used for the same problem. We note that the paired t test has only $n - 1$ degrees of freedom as compared to $n + n - 2 = 2(n - 1)$ for the unpaired test of Sec. 5-9. If experimental units (such as students) selected at random from the population of interest display considerable differences with regard to some variable which is not of interest, then it usually pays to use this variable to form pairs. On the other hand, if the units are quite similar (i.e., all IQs between 100 and 105), then we lose by pairing because of the loss of degrees of freedom. Perhaps the best advice that can be given is that we should prefer the paired procedure whenever we suspect that it would be appropriate.

EXAMPLE 5-38

Ten young recruits were put through a strenuous physical training program by the Army. Their weights were recorded before and after with the following results:

Recruit	1	2	3	4	5	6	7	8	9	10
Weight before	127	195	162	170	143	205	168	175	197	136
Weight after	135	200	160	182	147	200	172	186	194	141
Difference	−8	−5	2	−12	−4	5	−4	−11	3	−5

Using $\alpha = .05$ should we conclude that the program affects the average weight of young recruits?

Solution

We shall follow the six-step outline.

1   The problem suggests that we test $H_0 : \mu_1 = \mu_2$ against $H_1 : \mu_1 \neq \mu_2$ (or $H_0 : \mu_D = 0$ against $H_1 : \mu_D \neq 0$).

2   The significance level $\alpha = .05$ and the sample size $n = 10$ have already been chosen.

3   Since measurements occur as pairs we will use the statistic $T_{n-1} = \bar{D}/(S_D/\sqrt{n})$ which has a t distribution provided

   (a)   The sample of young recruits is selected at random from the group for which conclusions are to apply.

   (b)   The distribution of differences is normal (probably not a necessary assumption for n as large as 10).

   (c)   The means are equal as hypothesized.

4   The critical region is $t_9 = \bar{d}/(s_D/\sqrt{10}) < t_{9;\,.025} = -2.262$ and $t_9 > t_{9;\,.975} = 2.262$.

5   At this time it would be appropriate to observe the sample results. We calculate

$$\bar{d} = \frac{-8 - 5 + 2 - 12 - 4 + 5 - 4 - 11 + 3 - 5}{10} = \frac{-39}{10} = -3.9$$

$$\sum_{i=1}^{10} d_i^2 = (-8)^2 + (-5)^2 + 2^2 + \cdots + (-5)^2 = 449$$

$$s_D^2 = \frac{n \sum_{i=1}^{n} d_i^2 - (\sum_{i=1}^{n} d_i)^2}{n(n-1)} = \frac{10(449) - (-39)^2}{10(9)} = \frac{4,490 - 1,521}{90}$$

$$= 32.99$$

$$s_D = \sqrt{32.99} = 5.74$$

Finally

$$t_9 = \frac{\bar{d}}{s_D/\sqrt{10}} = \frac{-3.9(3.16)}{5.74} = -2.13$$

6 Since the observed value of the statistic does not fall in the critical region, we accept H_0. The experiment does not provide sufficient evidence to conclude that the program changes the average weight.

EXAMPLE 5-39

Suppose that two teaching methods are being compared by using 50 students divided into two equal classes. The students are paired on the basis of IQ, and one of each pair is assigned at random to each group. At the end of the course an examination is given and differences in paired scores are computed by subtracting the scores produced by method 2 from those produced by method 1. Method 2 requires less work and is preferred unless it can be demonstrated that method 1 is superior. The sample results are $\bar{d} = 5.6$, $s_D = 9.6$. Using the significance level $\alpha = .01$, what decision should be made?

Solution

We shall follow the six-step outline.

1 The problem suggests that we test $H_0 : \mu_1 = \mu_2$ against $H_1 : \mu_1 > \mu_2$ (or $H_0 : \mu_D = 0$ against $H_1 : \mu_D > 0$). If H_0 is rejected we will conclude that method 1 produces superior results.
2 The significance level $\alpha = .01$ and the sample size $n = 25$ pairs have already been selected.
3 As in the last example we will use the statistic $T_{n-1} = \bar{D}/(S_D/\sqrt{n})$ which has a t distribution provided
  (a) $D_1, D_2, \ldots, D_{25}$ is a random sample of pairs from a distribution for which the conclusions are to apply.
  (b) The distribution of differences is normal (not really necessary with $n = 25$).
  (c) The mean scores associated with both methods are equal (as hypothesized).
4 The critical region is $t_{24} = \bar{d}/(s_D/\sqrt{25}) > t_{24; .99} = 2.492$.
5 The experiment is performed and we compute $\bar{d} = 5.6$, $s_D = 9.6$, and

$$t_{24} = \frac{5.6}{9.6/\sqrt{25}} = \frac{5.6(5)}{9.6} = \frac{28.0}{9.6} = 2.92$$

6 The statistic falls in the critical region and we reject H_0. As a consequence method 1 will be adopted.

Except for minor changes in notation, formulas associated with power calculations and confidence intervals are the same as those used in Sec. 5-5. For power we need

$$\delta = \frac{|\mu_D| \sqrt{n}}{\sigma_D} \tag{5-92}$$

so that μ_D replaces $\mu - \mu_0$, σ_D replaces σ. A confidence interval for $\mu_D = \mu_1 - \mu_2$ with confidence coefficient $1 - \alpha$ is

$$\left(\bar{d} - t_{n-1;1-\alpha/2} \frac{s_D}{\sqrt{n}}, \bar{d} + t_{n-1;1-\alpha/2} \frac{s_D}{\sqrt{n}} \right) \tag{5-93}$$

which is formula (5-28) with $\bar{x}$ replaced by $\bar{d}$, s (or s_X) replaced by s_D.

EXAMPLE 5-40

Find the power of the test used in Example 5-39 if $\mu_1 - \mu_2 = \mu_D = \sigma_D/2$. How many pairs of students should have been used to make this probability at least .90?

Solution

With $\mu_D = \sigma_D/2$, $n = 25$ we get

$$\delta = \frac{|\mu_D| \sqrt{n}}{\sigma_D} = \frac{(\sigma_D/2)\sqrt{25}}{\sigma_D} = \frac{5}{2} = 2.5$$

With this δ and $f = n - 1 = 24$, $\alpha = .01$ we find from Appendix B10 that the power is .51.

The sample size required to achieve power $= .90$ is obtained as in Example 5-12. Since $|\mu_D|/\sigma_D = 1/2$, formula (5-14) yields $n = (2\delta)^2$. With $\alpha = .01$, $f = \infty$ we find $\delta = 3.60$ yields power $= .90$. Then $n = [2(3.60)]^2 = 51.84$ is too small. With $\alpha = .01$, $f = 24$ we find $\delta = 3.80$ yields power .90 and this δ gives $n = [2(3.80)]^2 = 57.76$ which may be slightly larger than necessary. Hence, we have shown that the minimum n is such that $51.84 \leq n \leq 57.76$. We could take $n = 55$ (or 110 students) to satisfy the given requirements. (Here it may be shown that the exact minimum n is 55.)

EXAMPLE 5-41

Using the sample results of Example 5-38, find a confidence interval for $\mu_1 - \mu_2$ with confidence coefficient .95.

Solution

We had $\bar{d} = -3.9$, $s_D = 5.74$, $n = 10$, $t_{9;.975} = 2.262$. Using (5-93) the required interval is

$$\left(-3.9 - 2.262 \frac{5.74}{\sqrt{10}}, -3.9 + 2.262 \frac{5.74}{\sqrt{10}}\right)$$

which reduces to $(-8.0, .2)$. We note that the interval covers 0 as it should since in Example 5-38 we accepted $H_0 : \mu_1 - \mu_2 = 0$ when tested against $H_1 : \mu_1 - \mu_2 \neq 0$ with the same α and the same sample.

EXERCISES

◈ 5-99   Each of two hybrid seed corn companies claims that its product is superior to the other's. Three scientists are hired to settle the dispute. They plant 1 acre of each kind in eight different localities which represent eight different soil and climate conditions. When the corn is harvested, they record the yields in bushels per acre. These are

Locality	1	2	3	4	5	6	7	8
Long Ear	114	86	93	75	102	89	64	95
Fat Kernel	107	94	86	70	90	82	73	81

Using $\alpha = .05$ test the appropriate hypothesis following the six-step outline. What conclusion should be expressed in the scientists' report?

◈ 5-100 Find the power of the test used in Exercise 5-99 if the average yield for Long Ear is σ_D bushels per acre more than for Fat Kernel. What would be the power for this alternative if 25 acres (instead of 8) had been planted with each brand? Suppose it had been desired to make this probability at least .90. What is the minimum number of acres that should have been planted with each brand? (*Hint:* Since the δ obtained with $f = 24$, $n = 25$ produces a computed n smaller than 25, δ is too small. Similarly the δ obtained with $f = 12$, $n = 13$ produces a computed n smaller than 13. Hence bounds are obtained using the δ's for $f = 6$ and $f = 12$.)

◈ 5-101 Using the sample results given in Exercise 5-99, find a confidence interval for $\mu_D = \mu_1 - \mu_2$ with confidence coefficient .95. How is the interval related to the test?

5-102   Consider once more the bridge player of Exercise 5-33. Suppose that he wishes to compare two bidding systems, say, method 1 and method 2. He prefers method 1 because it is simpler, but he will use method 2 if he can prove at the .01 level of significance that it produces a higher average. In order to reach a decision, he plays two sessions, one using each method, with seven of his favorite partners. The results in percentages are as tabulated below. Which method should he use? Follow the six-step outline.

Method	1	2	3	4	5	6	7
1	59	61	58	59	64	60	61
2	63	67	55	65	67	58	68

5-103 Find the power of the test used in Exercise 5-102 if $\mu_2 = \mu_1 + \sigma_D$. If $\mu_2 = \mu_1 + 2\sigma_D$. To what is the power increased when $\mu_2 = \mu_1 + \sigma_D$ if 25 partners (instead of 7) are used? Suppose that the player desires that the power be at least .80 when $\mu_2 = \mu_1 + \sigma_D$. How many partners should have been used? (The hint of Exercise 5-100 is useful here.)

5-104 Using the sample results given in Exercise 5-102 find a confidence interval for $\mu_D = \mu_1 - \mu_2$ with confidence coefficient .99.

◈ 5-105 A medical researcher would like to know if on the average blood pressure measurements of humans are the same when recorded with the person in a standing position as when the person is lying on his back. Using 12 randomly selected subjects, he takes one measurement by each method (randomizing the order by a toss of a coin). The results for the lying position are subtracted from those obtained standing with the following results: $-7, 1, -7, -4, 1, -9, 2, -3, -6, -5, 1, -4$. What conclusion can be drawn about μ_1, μ_2, the means for standing and lying positions, respectively? Use $\alpha = .05$ and follow the six-step outline.

◈ 5-106 Find the power of the test used in Exercise 5-105 if the means differ by $\sigma_D/2$. How many subjects should have been used (instead of 12) if this probability is to be at least .80?

◈ 5-107 Using the sample results given in Exercise 5-105, find a confidence interval for $\mu_D = \mu_1 - \mu_2$ with confidence coefficient .95.

5-108 A drug company sells an ointment which helps speed the recovery from the rash produced by poison oak. Its research laboratory has produced an improved version which is supposed to have on the average a shorter recovery time than the current product. To test the new ointment, 10 individuals are used. An area of each person's body is infected with the poison and 2 days are allowed for the rash to develop. Then half of the area on each individual receives treatment from the new ointment, the other half being treated with the old. The total length of elapsed time in days to complete recovery was recorded yielding the following results:

Individual	Old ointment	New ointment
1	14.6	14.2
2	9.8	9.4
3	15.0	14.5
4	17.8	17.0
5	11.4	11.0
6	8.7	8.3
7	12.6	12.8
8	13.2	12.7
9	10.4	9.8
10	8.9	8.7

Unless it can be proved at the .01 level of significance that the new ointment produces a shorter average recovery time, the company will not adopt it. What decision should be made? Follow the six-step outline.

5-109 Find the power of the test used in Exercise 5-108 if the average recovery time associated with the new ointment is $\sigma_D/2$ days less than the average associated with the old. Suppose that it had been desired to make this probability at least .95. What is the minimum number of individuals required for the experiment?

5-110 Using the sample results given in Exercise 5-108, find a confidence interval for μ_D = the old average minus the new using confidence coefficient .95.

REFERENCES

1 Harter, H. Leon: "New Tables of the Incomplete Gamma-function Ratio and of Percentage Points of the Chi-square and Beta Distributions," Superintendent of Documents, U.S. Government Printing Office, Washington, D.C., 1964.

2 Owen, D. B.: "Handbook of Statistical Tables," Addison-Wesley Publishing Company, Inc., Reading, Mass., 1962.

3 Owen, D. B.: The Power of Student's t-Test, Journal of the American Statistical Association, vol. 60, pp. 320–333, 1965.

4 Scheffé, Henry: Practical Solutions of the Behrens-Fisher Problem, Journal of the American Statistical Association, vol. 65, pp. 1501–1508, 1970.

Summary of Results

I. Inferences concerning one mean, variance σ^2 known
   A. Standard hypotheses, alternatives, statistics, and critical regions
   (Sec. 5-2) (significance level α)

Hypothesis	Alternative	Statistic†	Critical region
$H_0: \mu = \mu_0$	$H_1: \mu < \mu_0$	$Z = \dfrac{\bar{X} - \mu_0}{\sigma/\sqrt{n}}$	$z < z_\alpha$
$H_0: \mu \geqq \mu_0$	$H_1: \mu < \mu_0$		$z < z_\alpha$
$H_0: \mu = \mu_0$	$H_1: \mu > \mu_0$		$z > z_{1-\alpha}$
$H_0: \mu \leqq \mu_0$	$H_1: \mu > \mu_0$		$z > z_{1-\alpha}$
$H_0: \mu = \mu_0$	$H_1: \mu \neq \mu_0$		$z < z_{\alpha/2} = -z_{1-\alpha/2}$,
			$z > z_{1-\alpha/2}$

† Note that whenever a statistic is given in this summary, certain assumptions are made. These assumptions, giving the statistic the required distribution, may be found in the text material.

   B. Power calculations (Sec. 5-3)
      1. Use Appendix B10 with entries α ($\alpha/2$ with two-sided alternatives)

$$\delta = \frac{|\mu - \mu_0|}{\sigma/\sqrt{n}} \qquad f = \infty$$

      2. μ must be in the range specified by the alternative.
      3. Given μ_0, μ_1, σ, α, the minimum n required to yield power at least $1 - \beta$ when $\mu = \mu_1$ satisfies $n \geqq (\sigma\delta/|\mu_1 - \mu_0|)^2$.
   C. Two-sided confidence interval with confidence coefficient $1 - \alpha$ (Sec. 5-4)
      1. The interval:

$$\left(\bar{x} - z_{1-\alpha/2} \frac{\sigma}{\sqrt{n}}, \ \bar{x} + z_{1-\alpha/2} \frac{\sigma}{\sqrt{n}} \right)$$

2. Given length L and α, the n required to achieve this length (or shorter) is given by $n \geqq (2\sigma z_{1-\alpha/2}/L)^2$.

II. Inferences concerning one mean, variance σ^2 unknown (single sample) (Sec. 5-5)

A. Standard hypotheses, alternatives, statistics, and critical regions (significance level α)

Hypothesis	Alternative	Statistic	Critical region
$H_0: \mu = \mu_0$	$H_1: \mu < \mu_0$	$T_{n-1} = \dfrac{\bar{X} - \mu_0}{S/\sqrt{n}}$	$t_{n-1} < t_{n-1;\alpha}$
$H_0: \mu \geqq \mu_0$	$H_1: \mu < \mu_0$		$t_{n-1} < t_{n-1;\alpha}$
$H_0: \mu = \mu_0$	$H_1: \mu > \mu_0$		$t_{n-1} > t_{n-1;1-\alpha}$
$H_0: \mu \leqq \mu_0$	$H_1: \mu > \mu_0$		$t_{n-1} > t_{n-1;1-\alpha}$
$H_0: \mu = \mu_0$	$H_1: \mu \neq \mu_0$		$t_{n-1} < t_{n-1;\alpha/2}$ $= -t_{n-1;1-\alpha/2},$ $t_{n-1} > t_{n-1;1-\alpha/2}$

B. Power calculations

1. Use Appendix B10 with entries α ($\alpha/2$ with two-sided alternatives), $\delta = |\mu - \mu_0|/(\sigma/\sqrt{n})$, $f = n - 1$.

2. μ must be in the range specified by the alternative.

3. Given α, the minimum n required to yield power at least $1 - \beta$ for given $|\mu - \mu_0|/\sigma$ can be approximated with (5-14) and Appendix B10. Starting with δ which gives power $1 - \beta$ for $f = \infty$, decrease f until $n = f + 1$ is less than the computed n. The latter computed n is an upper bound for the minimum n; the computed n obtained just previously is a lower bound.

C. Two-sided confidence interval with confidence coefficient $1 - \alpha$

1. The interval:

$$\left(\bar{x} - t_{n-1;1-\alpha/2} \frac{s}{\sqrt{n}}, \bar{x} + t_{n-1;1-\alpha/2} \frac{s}{\sqrt{n}}\right)$$

2. The relationship between n and the probability that the length of the random interval is less than L is given in Appendix B11. In particular, for given L/σ we can find n such that the probability is γ that the length is less than L.

III. Inferences concerning one mean, variance σ^2 unknown (two-stage sample) (Sec. 5-6)

A. Standard hypotheses, alternatives, statistics, and critical regions (significance level α)

Hypothesis	Alternative	Statistic	Critical region
$H_0: \mu = \mu_0$	$H_1: \mu < \mu_0$	$T_{n_0-1} = \dfrac{\bar{X} - \mu_0}{S/\sqrt{n}}$ where $n_0 =$ preliminary sample size $n =$ total sample size $\bar{X} =$ sample mean of all observations $S =$ sample standard deviation of preliminary sample	$t_{n_0-1} < t_{n_0-1;\alpha}$
$H_0: \mu \geqq \mu_0$	$H_1: \mu < \mu_0$		$t_{n_0-1} < t_{n_0-1;\alpha}$
$H_0: \mu = \mu_0$	$H_1: \mu > \mu_0$		$t_{n_0-1} > t_{n_0-1;1-\alpha}$
$H_0: \mu \leqq \mu_0$	$H_1: \mu > \mu_0$		$t_{n_0-1} > t_{n_0-1;1-\alpha}$
$H_0: \mu = \mu_0$	$H_1: \mu \neq \mu_0$		$t_{n_0-1} < t_{n_0-1;\alpha/2}$, $t_{n_0-1} > t_{n_0-1;1-\alpha/2}$

B. Calculation of n required to give a specified power independent of σ
1. With given α, μ, μ_0, n_0, and power $1 - \beta$ calculate k as given by (5-33).
2. Compute ks^2. If $ks^2 \leqq n_0$, no more observations are necessary. If $ks^2 > n_0$, continue sampling until the total sample size just exceeds ks^2.
3. For two-sided tests replace α by $\alpha/2$ in (5-33).

C. Two-sided confidence interval of specified length L with confidence coefficient at least $1 - \alpha$
1. The interval: $(\bar{x} - L/2, \bar{x} + L/2)$.
2. With given α, n_0, and L, calculate k from (5-35). Then compute ks^2 and determine n as in III-B-2.

IV. Inferences concerning one variance (Sec. 5-7)

A. Standard hypotheses, alternatives, statistics, and critical regions (significance level α)

Hypothesis	Alternative	Statistic	Critical region
$H_0: \sigma^2 = \sigma_0^2$	$H_1: \sigma^2 < \sigma_0^2$	$Y_{n-1} = \dfrac{(n-1)S^2}{\sigma_0^2}$ $\left(\text{or } \dfrac{Y_{n-1}}{n-1} = \dfrac{S^2}{\sigma_0^2}\right)$	$y_{n-1} < \chi_{n-1;\alpha}^2$
$H_0: \sigma^2 \geqq \sigma_0^2$	$H_1: \sigma^2 < \sigma_0^2$		$y_{n-1} < \chi_{n-1;\alpha}^2$
$H_0: \sigma^2 = \sigma_0^2$	$H_1: \sigma^2 > \sigma_0^2$		$y_{n-1} > \chi_{n-1;1-\alpha}^2$
$H_0: \sigma^2 \leqq \sigma_0^2$	$H_1: \sigma^2 > \sigma_0^2$		$y_{n-1} > \chi_{n-1;1-\alpha}^2$
$H_0: \sigma^2 = \sigma_0^2$	$H_1: \sigma^2 \neq \sigma_0^2$		$y_{n-1} < \chi_{n-1;\alpha/2}^2$, $y_{n-1} > \chi_{n-1;1-\alpha/2}^2$

B. Power calculations
1. Use Appendix B12 with entries α ($\alpha/2$ with two-sided alternatives), $R = \sigma/\sigma_0$, and $v = n - 1$.
2. σ must be in the range specified by the alternative.
3. Given α, the minimum n required to yield power at least $1 - \beta$ for given R can be read approximately from Appendix B12.
C. Two-sided confidence interval with confidence coefficient $1 - \alpha$
1. The interval:

$$\left[\frac{(n-1)s^2}{\chi^2_{n-1;\,1-\alpha/2}}, \frac{(n-1)s^2}{\chi^2_{n-1;\,\alpha/2}} \right]$$

2. The relationship between n and the probability that the length of the random interval is less than L is given in Appendix B13. In particular, for given L/σ we can find n such that the probability is γ that the length is less than L.

V. Inferences concerning two variances (Sec. 5-8)
A. Standard hypotheses, alternatives, statistics, and critical regions (significance level α)

Hypothesis	Alternative	Statistic	Critical region
$H_0: \sigma_1^2 = \sigma_2^2$	$H_1: \sigma_1^2 < \sigma_2^2$	$F_{n_1-1,\,n_2-1}$	$f_{n_1-1,\,n_2-1} < f_{n_1-1,\,n_2-1;\,\alpha}$
$H_0: \sigma_1^2 \geqq \sigma_2^2$	$H_1: \sigma_1^2 < \sigma_2^2$	$= \dfrac{S_1^2}{S_2^2}$	$f_{n_1-1,\,n_2-1} < f_{n_1-1,\,n_2-1;\,\alpha}$
$H_0: \sigma_1^2 = \sigma_2^2$	$H_1: \sigma_1^2 > \sigma_2^2$		$f_{n_1-1,\,n_2-1} > f_{n_1-1,\,n_2-1;\,1-\alpha}$
$H_0: \sigma_1^2 \leqq \sigma_2^2$	$H_1: \sigma_1^2 > \sigma_2^2$	where S_1^2, S_2^2	$f_{n_1-1,\,n_2-1} > f_{n_1-1,\,n_2-1;\,1-\alpha}$
$H_0: \sigma_1^2 = \sigma_2^2$	$H_1: \sigma_1^2 \neq \sigma_2^2$	are the sample variances	$f_{n_1-1,\,n_2-1} < f_{n_1-1,\,n_2-1;\,\alpha/2},$ $f_{n_1-1,\,n_2-1} > f_{n_1-1,\,n_2-1;\,1-\alpha/2}$

B. Power calculations
1. Use Appendix B14 with entries α ($\alpha/2$ with two-sided alternatives), $R_0 = \sigma_1/\sigma_2$, and $v = n - 1$, if $n_1 = n_2 = n$. (If the sample sizes are not equal, use (5-65) for (5-54) or (5-57).)
2. σ_1/σ_2 must be in the range specified by the alternative.
3. For $n_1 = n_2 = n$ (equal sample sizes) and given α, the minimum n required to yield power at least $1 - \beta$ for given R_0 can be read approximately from Appendix B14.
C. Two-sided confidence interval with confidence coefficient $1 - \alpha$ for σ_1^2/σ_2^2
1. The interval:

$$\left[\frac{s_1^2}{s_2^2} \frac{1}{f_{n_1-1,\,n_2-1;\,1-\alpha/2}}, \frac{s_1^2}{s_2^2} \frac{1}{f_{n_1-1,\,n_2-1;\,\alpha/2}} \right]$$

2. The interval for σ_1/σ_2 is obtained by taking the square roots of the numbers forming the interval in V-C-1.

VI. Inferences concerning two means (with two random samples) (Sec. 5-9)

    A. Standard hypotheses, alternatives, statistics, and critical regions (significance level α)

Hypothesis	Alternative	Statistic	Critical region
$H_0: \mu_1 = \mu_2$	$H_1: \mu_1 < \mu_2$	$T_{n_1+n_2-2}$	$t_{n_1+n_2-2} < t_{n_1+n_2-2;\,\alpha}$
$H_0: \mu_1 \geqq \mu_2$	$H_1: \mu_1 < \mu_2$		$t_{n_1+n_2-2} < t_{n_1+n_2-2;\,\alpha}$
$H_0: \mu_1 = \mu_2$	$H_1: \mu_1 > \mu_2$	$= \dfrac{\bar{X}_1 - \bar{X}_2}{S_p\sqrt{1/n_1 + 1/n_2}}$	$t_{n_1+n_2-2} > t_{n_1+n_2-2;\,1-\alpha}$
$H_0: \mu_1 \leqq \mu_2$	$H_1: \mu_1 > \mu_2$		$t_{n_1+n_2-2} > t_{n_1+n_2-2;\,1-\alpha}$
$H_0: \mu_1 = \mu_2$	$H_1: \mu_1 \neq \mu_2$	where $S_p^2 =$	$t_{n_1+n_2-2} < t_{n_1+n_2-2;\,\alpha/2}$,
		$\dfrac{(n_1-1)S_1^2 + (n_2-1)S_2^2}{n_1+n_2-2}$	$t_{n_1+n_2-2} > t_{n_1+n_2-2;\,1-\alpha/2}$

    B. Power calculations

      1. Use Appendix B10 with entries α ($\alpha/2$ with two-sided alternatives)

$$\delta = \frac{|\mu_1 - \mu_2|}{\sigma\sqrt{1/n_1 + 1/n_2}} \qquad f = n_1 + n_2 - 2$$

      2. $\mu_1 - \mu_2$ must be in the range specified by the alternative.

      3. If $n_1 = n_2 = n$ (equal sample sizes) and α is given, then the minimum n required to yield power at least $1 - \beta$ for given $|\mu_1 - \mu_2|/\sigma$ can be approximated with (5-83) and Appendix B10. Starting with δ which gives power $1 - \beta$ for $f = \infty$, decrease f until $n = (f/2) + 1$ is less than the computed n. The latter computed n is an upper bound for the minimum n; the computed n obtained just previously is a lower bound.

    C. Two-sided confidence interval with confidence coefficient $1 - \alpha$ for $\mu_1 - \mu_2$

      1. The interval:

$$\left(\bar{x}_1 - \bar{x}_2 - t_{n_1+n_2-2;\,1-\alpha/2}\, S_p\sqrt{1/n_1 + 1/n_2}\,,\; \bar{x}_1 - \bar{x}_2 \right.$$
$$\left. + t_{n_1+n_2-2;\,1-\alpha/2}\, S_p\sqrt{1/n_1 + 1/n_2}\right)$$

VII. Inferences concerning two means (paired observations) (Sec. 5-10)

    A. Standard hypotheses, alternatives, statistics, and critical regions (significance level α)

Hypothesis	Alternative	Statistic	Critical region
$H_0: \mu_1 = \mu_2$	$H_1: \mu_1 < \mu_2$	$T_{n-1} = \dfrac{\bar{D}}{S_D/\sqrt{n}}$	$t_{n-1} < t_{n-1;\alpha}$
$H_0: \mu_1 \geqq \mu_2$	$H_1: \mu_1 < \mu_2$		$t_{n-1} < t_{n-1;\alpha}$
$H_0: \mu_1 = \mu_2$	$H_1: \mu_1 > \mu_2$		$t_{n-1} > t_{n-1;1-\alpha}$
$H_0: \mu_1 \leqq \mu_2$	$H_1: \mu_1 > \mu_2$	where	$t_{n-1} > t_{n-1;1-\alpha}$
$H_0: \mu_1 = \mu_2$	$H_1: \mu_1 \neq \mu_2$	$\bar{D} =$ sample mean of differences	$t_{n-1} < t_{n-1;\alpha/2},$
		$S_D^2 =$ sample variance of differences	$t_{n-1} > t_{n-1;1-\alpha/2}$

B. Power calculations

1. Use Appendix B10 with entries α ($\alpha/2$ with two-sided alternatives), $\delta = |\mu_1 - \mu_2|/(\sigma_D/\sqrt{n}), f = n - 1$.
2. $\mu_1 - \mu_2$ must be in the range specified by the alternative.
3. Given α, the minimum n required to yield power at least $1 - \beta$ for given $|\mu_1 - \mu_2|/\sigma_D = |\mu_D|/\sigma_D$ can be approximated with (5-14), which is now $n = (\sigma_D \delta/\mu_D)^2$ and Appendix B10. Starting with the δ which gives power $1 - \beta$ for $f = \infty$, decrease f until $n = f + 1$ is less than the computed n. The latter computed n is an upper bound for the minimum n; the computed n obtained just previously is a lower bound.

C. Two-sided confidence interval with confidence coefficient $1 - \alpha$ for $\mu_1 - \mu_2$

1. The interval:

$$\left(\bar{d} - t_{n-1;1-\alpha/2} \frac{s_D}{\sqrt{n}}, \; \bar{d} + t_{n-1;1-\alpha/2} \frac{s_D}{\sqrt{n}} \right)$$

2. The relationship between n and the probability that the length of the random interval is less than L is given in Appendix B11. In particular, for given L/σ we can find n such that the probability is γ that the length is less than L.

HYPOTHESIS TESTING
AND INTERVAL ESTIMATION
FOR PARAMETERS OF
DISCRETE DISTRIBUTIONS;
APPROXIMATE CHI-SQUARE TESTS

6-1
INTRODUCTION

The statistical techniques which we considered in Chap. 5 were derived under the assumption that experimental values are governed by a normal distribution. We observed, however, that the normality assumption is not too crucial for drawing inferences about the mean if the sample size is moderately large. In situations where the normal model may not be appropriate we were primarily concerned with inferences about means of continuous distributions.

In this chapter our main interest is parameters which arise in discrete probability models (binomial, Poisson, hypergeometric, multinomial). Some examples which suggest hypothesis testing have already been mentioned. In Sec. 4-2 we devoted considerable space to the use of the binomial as it applied to the potato problem and observed that it led to testing $H_0: p = .10$ against $H_1: p > .10$ (where p was the fraction of bad potatoes). In Example 2-19 a secretary claimed that on the average she makes 1 error per page. The information given in that problem suggested that the Poisson is an appropriate model and that there is some interest in testing $H_0: \mu = 1$ against $H_1: \mu > 1$. If this latter hypothesis is rejected, we may seek a confidence interval for μ, the secretary's average.

As the above comments imply, the types of problems which are about to be considered are very similar to those discussed in Chap. 5.

6-2
INFERENCES ABOUT THE BINOMIAL
PARAMETER p

Problems which lead to tests of hypotheses about a proportion are encountered frequently in experimental situations. In Sec. 4-1 we mentioned the tuberculosis example (Exercise 2-7) and indicated that an experimenter might be interested in proving that the fraction of cures, say p, produced by a new method exceeds the fraction .30 achieved with an old method. A manufacturer of a mass produced product (such as camera flashbulbs) will be interested in the fraction p which are defective. The President may like to know if the fraction p of voters who support him on a controversial policy is .60 or greater. The potato problem with unknown fraction of bad potatoes has already received considerable attention.

We recall from our study of probability models that p is a parameter of the binomial distribution. If sampling is performed so that the binomial conditions (2-1) are realistic, then $X =$ the number of "successes" will have a binomial distribution with parameter p (and n). This suggests (as we illustrated with the potato problem of Sec. 4-2) that interesting probability statements can then be based on X and tests of hypotheses can be conducted using x.

The standard hypotheses testing situations which are most frequently encountered with binomial p are similar to those we have discussed for means and variances. Suppose that a manufacturing process currently in use produces bolts, .10 of which require rework. If we are interested in installing a new process, but only if there is evidence that the new p is smaller, then we would probably test

$$H_0 : p = p_0 \quad \text{against} \quad H_1 : p < p_0 \tag{6-1}$$

with $p_0 = .10$, making the change if H_0 is rejected. In the tuberculosis problem the old cure is successful .30 of the time. If the new cure will be adopted only if it is successful more often, then an experimenter would test

$$H_0 : p = p_0 \quad \text{against} \quad H_1 : p > p_0 \tag{6-2}$$

with $p_0 = .30$ and adopt the new cure if he can believe H_1 (reject H_0). If we are interested in determining whether a coin is unbiased or not, then we will want to test

$$H_0 : p = p_0 \quad \text{against} \quad H_1 : p \neq p_0 \tag{6-3}$$

with $p_0 = 1/2$.

In some problems we may prefer to replace (6-1) by

$$H_0 : p \geqq p_0 \quad \text{against} \quad H_1 : p < p_0 \tag{6-4}$$

and (6-2) by

$$H_0 : p \leqq p_0 \quad \text{against} \quad H_1 : p > p_0 \tag{6-5}$$

thus separating the p's into two sets, "satisfactory" and "unsatisfactory" values. Again the test procedure is unchanged, but the significance level α must be interpreted as the largest power under H_0 (which will occur when $p = p_0$).

As we have already indicated, the statistic used to test hypotheses about p is X, the number of successes in n trials. If conditions (2-1) are satisfied, then X has a binomial distribution. Further, if $p = p_0$ [the hypothesis of (6-1), (6-2), or (6-3) is true], then X has a binomial distribution with parameters n and p_0, a fact we use to determine critical regions. The determination of sample size n was discussed in Sec. 4-3 and will receive more attention later in this section.

When (6-1) or (6-4) is tested, small values of x support H_1 and lead to the rejection of H_0. As we observed near the end of Sec. 4-2, a specified significance level α, say α_0, is unlikely to be achieved due to the fact that X is a discrete random variable. The critical region usually chosen is $x \leq x_0$ where x_0 is the largest value of x such that $\Pr(X \leq x) \leq \alpha_0$ (determined, of course, with $p = p_0$). The actual significance level is $\Pr(X \leq x_0) = \alpha$, calculated with $p = p_0$. When we attempted to achieve significance level .05 for testing

$$H_0: p = .10 \qquad \text{against} \qquad H_1: p > .10$$

(in Sec. 4-2), we settled for the critical region $x \geq 6$ with $\alpha = .03340$ instead of .05. In other words, the convention which we are adopting leads to a significance level α as close to α_0 as possible, yet such that α is no larger than α_0. (An alternative convention would be to take the region which gives α closest α_0, thus allowing α to be larger than α_0.) Similarly, when testing (6-2) or (6-5) only large x support H_1. Now H_0 is rejected when $x \geq x_0$ where x_0 is the smallest value of x such that (when $p = p_0$) $\Pr(X \geq x) \leq \alpha_0$ and the actual significance level is $\Pr(X \geq x_0) = \alpha$. For the two-sided situation of (6-3) both small and large x contradict H_0 and lead to rejection. If we aim at a significance level α_0, the critical region usually selected is $x \leq x_1$ and $x \geq x_2$ where x_1 is the largest value of x such that $\Pr(X \leq x) \leq \alpha_0/2$ and x_2 is the smallest value of x such that $\Pr(X \geq x) \leq \alpha_0/2$. The actual significance level is $\Pr(X \leq x_1) + \Pr(X \geq x_2) = \alpha$. Of course, once more all calculations are performed with $p = p_0$.

EXAMPLE 6-1

A coin is tossed 100 times and 43 heads are recorded. Does this result indicate that the coin (or perhaps the tossing procedure) is biased? Use $\alpha_0 = .05$.

Solution

We shall follow the six-step outline.

1 Letting p be the probability that a head occurs, the problem suggests that we test $H_0: p = .5$ against $H_1: p \neq .5$. If the alternative is true the coin is biased.

2 The significance level we hope to achieve is $\alpha_0 = .05$. However, the critical region selected in step 4 will yield $\alpha < .05$. The sample size n has been selected as $n = 100$.

3 The statistic we use is X, the number of heads showing in 100 tosses. It is reasonable to use the binomial model for computing probabilities associated with X since (*a*) each toss is a head or a tail, (*b*) the probability of a head remains constant for each throw (if tossed the same way every time), (*c*) tosses are independent, and (*d*) the coin is tossed $n = 100$ times, a fixed number. Thus X has a binomial distribution with parameters $n = 100$ and p. If the hypothesis is true we have further that $p = .5$.

4 From the binomial tables we find with $n = 100$, $p = .5$ that

$$\Pr(X \leq 39) = \sum_{x=0}^{39} b(x; 100, .5) = .01760$$

$$\Pr(X \leq 40) = \sum_{x=0}^{40} b(x; 100, .5) = .02844$$

Hence, according to our convention which requires that x_1 be the largest x such that $\Pr(X \leq x) \leq \alpha_0/2 = .025$, we take $x_1 = 39$. Similarly, we find that

$$\Pr(X \geq 61) = \sum_{x=61}^{100} b(x; 100, .5) = 1 - \sum_{x=0}^{60} b(x; 100, .5)$$

$$= 1 - .98240 = .01760$$

$$\Pr(X \geq 60) = \sum_{x=60}^{100} b(x; 100, .5) = 1 - \sum_{x=0}^{59} b(x; 100, .5)$$

$$= 1 - .97156 = .02844$$

and since x_2 is the smallest x such that $\Pr(X \geq x) \leq .025$ we have $x_2 = 61$. The critical region is $x \leq 39$ and $x \geq 61$ and the actual significance level is

$$\alpha = \Pr(X \leq 39) + \Pr(X \geq 61) = .01760 + .01760 = .03520$$

5 We now perform the experiment and observe $x = 43$. No further calculations are necessary.

6 Since $x = 43$ does not fall in the critical region, we do not reject H_0. Consequently, we proceed under the assumption that the coin is unbiased.

EXAMPLE 6-2

Suppose we are considering for purchase a new machine which makes bolts. We will buy the machine unless the fraction of bolts which require rework is greater than .10. A sample of 50 bolts produced by the machine is examined, and 8 require rework. With $\alpha_0 = .05$, should we buy the machine?

Solution

We shall follow the six-step outline.

1 Letting p be the unknown fraction which require rework, the problem suggests that we test $H_0: p = .10$ against $H_1: p > .10$ and purchase the machine unless H_0 is rejected.

2 We aim at a significance level of $\alpha_0 = .05$. However, we know that the actual significance level obtained in step 4 will probably be less than .05. The sample size $n = 50$ has been selected.

3 The statistic we use is X, the number of bolts requiring rework. The sample of bolts should be obtained so as to make conditions (2-1) seem realistic. We check the assumptions of the binomial model: (*a*) A bolt requires rework or it does not. (*b*) Unless the machine shows some sign of wear or a setting is accidentally changed, it is not unreasonable to assume that p remains constant. (*c*) If the machine is working properly, it is probably reasonable to assume that the classification of one bolt is independent of the classification of others. (If the machine is inclined to produce several poor bolts in a row, the last two conditions will seem more realistic if we examine every fifth bolt, or perhaps every tenth bolt. Another possibility is to select a simple random sample from one day's production.) (*d*) The number of bolts to be examined is $n = 50$. Thus, the binomial with $n = 50$ and parameter p seems like a reasonable model to use. Further, if H_0 is true we take $p = p_0 = .10$.

4 To determine the critical region we use $n = 50$, $p = .10$. In Appendix B1 we find

$$\Pr(X \geq 9) = \sum_{x=9}^{50} b(x; 50, .10) = 1 - \sum_{x=0}^{8} b(x; 50, .10)$$

$$= 1 - .94213 = .05787$$

$$\Pr(X \geq 10) = \sum_{x=10}^{50} b(x; 50, .10) = 1 - \sum_{x=0}^{9} b(x; 50, .10)$$

$$= 1 - .97546 = .02454$$

Hence, according to our convention we take as the critical region $x \geq 10$ with actual significance level .02454.

5 We observe 50 bolts and find $x = 8$. No computations are necessary.

6 Since $x = 8$ does not fall in the critical region, H_0 is accepted and the machine is purchased.

When H_0 is not true, X still has a binomial distribution but with some other value of p. Consequently, power calculations are easy since the probability that X falls in the critical region can be evaluated from another column of the binomial table.

EXAMPLE 6-3

Find the power of the test used in Example 6-2 if $p = .20$. If $p = .30$.

Solution

Since the critical region is $x \geq 10$, we need if $p = .20$

$$\Pr(X \geq 10) = \sum_{x=10}^{50} b(x; 50, .20)$$

$$= 1 - \sum_{x=0}^{9} b(x; 50, .20)$$

$$= 1 - .44374 = .55626$$

Similarly, if $p = .30$ we get

$$\Pr(X \geq 10) = \sum_{x=10}^{50} b(x; 50, .30)$$

$$= 1 - \sum_{x=0}^{9} b(x; 50, .30)$$

$$= 1 - .04023 = .95977$$

EXAMPLE 6-4

Suppose that $n = 100$ bolts is used in Example 6-2.   Find the new critical region.   Then find the power of the test when $p = .20$.   When $p = .30$.

Solution

If $p = .10$, $n = 100$ we find that

$$\Pr(X \geq 16) = 1 - \Pr(X \leq 15) = 1 - \sum_{x=0}^{15} b(x; 100, .10)$$

$$= 1 - .96011 = .03989$$

$$\Pr(X \geq 15) = 1 - \Pr(X \leq 14) = 1 - \sum_{x=0}^{14} b(x; 100, .10)$$

$$= 1 - .92743 = .07257$$

Hence, the critical region is $x \geq 16$ with $\alpha = .03989$.

For $p = .20$, $n = 100$ we obtain

$$\Pr(X \geq 16) = 1 - \Pr(X \leq 15) = 1 - \sum_{x=0}^{15} b(x; 100, .20)$$

$$= 1 - .12851 = .87149$$

For $p = .30$, $n = 100$ we get

$$\Pr(X \geq 16) = 1 - \Pr(X \leq 15) = 1 - \sum_{x=0}^{15} b(x; 100, .30)$$

$$= 1 - .00040 = .99960$$

EXAMPLE 6-5

How large does n have to be in Example 6-2 to raise the power to .95 or greater if $p = .20$?

Solution

Our table in Appendix B1 is inadequate to solve this problem.   The calculations of Example 6-4 indicate that $n = 100$ is not large enough to meet the two power requirements.   Several possibilities are available.   We can use the formula based upon the large-sample approximation [formula (6-25) discussed in Sec. 6-4 and Example 6-17].   This yields $n = 133$.   Better yet, perhaps, we can get the Army Materiel Command table [6] and solve the problem by inspection.   A simple procedure is described in section 5 of their Introduction.   This yields $n = 135$ with critical region $x \geq 20$ and significance level .04837.

Hence, for the moment we are not prepared to solve this problem (unless we get the table mentioned above).   The best we can do is to observe that n must be larger than 100.

EXAMPLE 6-6

Find the power of the test used in Example 6-1 if $p = .40$.

Solution

The critical region is $x \leq 39$ and $x \geq 61$.   With $n = 100$, $p = .40$ we need $\Pr(X \leq 39) + \Pr(X \geq 61)$.   With Appendix B1 we find

$$\Pr(X \leq 39) = \sum_{x=0}^{39} b(x; 100, .40) = .46208$$

$$\Pr(X \geq 61) = \sum_{x=61}^{100} b(x; 100, .40) = 1 - \sum_{x=0}^{60} b(x; 100, .40)$$

$$= 1 - .99998 = .00002$$

Thus the required power is $.46208 + .00002 = .46210$. (We observe that since $.00002$ is so small, this is practically the same power as for a one-sided test with $\alpha_0 = .05$ replaced by $\alpha_0/2 = .025$. We took advantage of this fact throughout Chap. 5 to obtain power for two-sided tests from one-sided power curves.)

In Sec. 3-2 we learned that a good point estimate of p is $\hat{p} = t = x/n$. The confidence interval problem for parameters of discrete distributions is considerably more difficult than those we encountered in Chap. 5. Fortunately, graphs are available which make the problem extremely simple. These are found in Appendix B15 and can be used for confidence coefficient $1 - \alpha = .90, .95, .99$. Specifically we can find two numbers w_1, w_2 such that

$$\Pr(W_1 \leqq p \leqq W_2) \geqq 1 - \alpha \tag{6-6}$$

Then (w_1, w_2), end points included, is a confidence interval for p with confidence coefficient at least $1 - \alpha$. To use the graphs we need n, $1 - \alpha$, and the point estimate $\hat{p}$. Turn to the page headed by the desired $1 - \alpha$ and find the observed $\hat{p}$ on the horizontal axis. Draw (or imagine) a vertical line through this $\hat{p}$ which intersects the two curves labeled with the appropriate sample size. These two intersections when connected to the vertical axis (labeled p) by horizontal lines yield the desired w_1 and w_2. Let us illustrate with an example.

EXAMPLE 6-7

Using the sample results of Example 6-1, find a confidence interval with confidence coefficient at least $.95$ for the probability of obtaining a head.

Solution
We had $x = 43$, $n = 100$, so that $\hat{p} = .43$. We turn to the set of graphs in Appendix B15 headed by confidence coefficient $.95$. Next locate $\hat{p} = .43$ on the horizontal axis and imagine that a vertical line has been drawn through this point intersecting the two curves labeled 100. Connect the two intersections with horizontal lines intersecting the p axis. We read $w_1 = .33$, $w_2 = .53$ on the vertical axis, and the desired confidence interval is $(.33, .53)$.

We note that the interval found in Example 6-7 covers the hypothesized value $p = p_0 = 1/2$ of Example 6-1. In general, the interval (w_1, w_2) covers a value p_0 if, and only if, the test of $H_0: p = p_0$ against $H_1: p \neq p_0$ leads to acceptance using the same sample provided the α_0 of the test and the α of the confidence interval are the same. Hence the graphs of Appendix B15 could be used in place of the binomial tables for the two-sided hypothesis testing problem to determine whether to accept or reject H_0.

Let us again consider the hypothesis testing situation $H_0: p = p_0$ against $H_1: p > p_0$ to illustrate one further point of interest. If observations are expensive, as they may be in the tuberculosis experiment of Exercise 2-7, it may be possible to get by with fewer observations than was originally planned. If p is the fraction of cures for the new method, we may test $H_0: p = .30$ against $H_1: p > .30$ hoping to prove that H_1 is true. Suppose we select $n = 50$, $\alpha_0 = .05$. Then it is easy to verify that the critical region is $x \geq 21$ with actual significance level .04776. Now let us assume that experimentation is conducted one individual at a time. It is obvious that we could quit sampling and reject H_0 as soon as 21 cures are obtained (which might occur long before 50 patients are used). Similarly, we could terminate experimentation and accept H_0 as soon as 30 noncures are observed since it would then be impossible to reject. In other words, once the decision is obvious, there is no point in spending money and effort to continue experimentation, and there is a high probability that we will be able to make a decision with fewer than 50 observations. In general, if the above procedure is used we will

(a) Determine n and the corresponding x_0 (either select n arbitrarily or determine n from two power conditions as in Sec. 4-3) so that the critical region is $x \geq x_0$

(b) Stop sampling and   (6-7)

    (i) Reject H_0 as soon as x_0 successes are observed

    (ii) Accept H_0 as soon as $n - x_0 + 1$ failures are observed (whichever occurs first)

Procedure (6-7) is referred to as truncated (or curtailed) sampling. The number of observations required to reach a decision, say V, is now a random variable, and it can be shown that

$$E(V) = \frac{x_0}{p} [1 - B(x_0; n + 1, p)] + \frac{n - x_0 + 1}{1 - p} [B(x_0 - 1; n + 1, p)] \quad (6-8)$$

We can, of course, use truncated sampling without the result (6-8) and know that usually less sampling will be required. To illustrate the potential savings we work the following example.

EXAMPLE 6-8

For the tuberculosis problem mentioned above, find the expected number of observations for truncated sampling if $p = .40$ for the new cure. If $p = .50$.

Solution

We have $n = 50$, $x_0 = 21$. Hence, if $p = .40$

$$E(V) = \frac{21}{.40} [1 - B(21; 51, .40)] + \frac{50 - 21 + 1}{1 - .40} [B(20; 51, .40)]$$

Our binomial table in Appendix B1 does not give probabilities for $n = 51$. However, these sums are available in the Ordnance Corps table [7] and the newer Army Materiel Command table [6]. We get

$$E(V) = 52.5[.37350] + 50[.51521]$$
$$= 19.609 + 25.760 = 45.37$$

Similarly, if $p = .50$ we get

$$E(V) = \frac{21}{.50}[1 - B(21; 51, .50)] + \frac{30}{.5}[B(20; 51, .50)]$$

$$= 42[.86878] + 60[.08039]$$

$$= 36.489 + 4.823 = 41.31$$

Hence, on the average the experimenter will save $50 - 41.31 = 8.69$ observations by using truncated sampling when the real p is .50.

EXERCISES

◈ 6-1 A seed company claims that only 10 percent of its radish seeds fail to germinate. A sample of 100 seeds is planted, and 19 fail to germinate. Test a hypothesis formulated from the company's claim. Use $\alpha_0 = .05$ and follow the six-step outline.

◈ 6-2 Find the power of the test used in Exercise 6-1 if actually 20 percent fail to germinate. If 25 percent fail.

◈ 6-3 Using the sample results of Exercise 6-1, find a confidence interval with confidence coefficient at least .95 for the fraction of seeds which fail to germinate.

6-4 According to a genetic theory, 25 percent of a species have a certain characteristic. A random sample of 20 of the species contains 9 having the characteristic. Using $\alpha_0 = .05$, should we conclude that the theory is contradicted? Follow the six-step outline.

6-5 Find the power of the test used in Exercise 6-4 if 10 percent of the species have the characteristic. If 40 percent have the characteristic.

6-6 Using the sample results of Exercise 6-4, find a confidence interval with confidence coefficient at least .95 for the percent of the species which have the characteristic.

◈ 6-7 Suppose that in the bolt problem of Example 6-2 we are interested in testing $H_0: p = .20$ against $H_1: p > .20$. Let us again take $\alpha_0 = .05$. If, further, we wish the power to be at least .90 if $p = .50$, what is the smallest n of those listed in Appendix B1 which meets the requirements? What is the critical region associated with this n?

6-8 Suppose we wish to test $H_0: p = .50$ against $H_1: p < .50$ using $\alpha_0 = .01$ for a problem associated with the binomial model. If, further, we wish the power to be at least .95 if $p = .10$, what is the smallest possible n of those listed in Appendix B1? What is the critical region associated with this n?

◈ 6-9   At the state university it is known that the proportion of students who fail freshman English is .40. The English department is considering the addition of a tutorial service in order to lower this proportion. If there is good evidence that the fraction of failures can be made less than .25, they will add the service. If no such evidence is produced, the service will not be added because of the additional expense. To make a decision a randomly selected group of 50 freshmen are taught with the tutorial service available and only 8 failures are recorded. Using $\alpha_0 = .01$, what action should be taken? Follow the six-step outline.

6-10   The Chamber of Commerce of Anchorage, Alaska, claims that at least 60 percent of the city residents are male. A random sample of 100 residents contains 54 men. Using $\alpha_0 = .05$, does the sample result disprove the claim? Use the six-step outline.

6-11   Using the sample results of Exercise 6-10, find a confidence interval with confidence coefficient at least .95 for the fraction of men in Anchorage.

◈ 6-12   An entrance examination to a large university is considered satisfactory if 10 percent fail. A new examination is to be tried on a random sample of 50 applicants. If the failure rate is .10 we would like to have the probability of rejecting the examination be .05 or less. In addition, if the failure rate is .30 we would like to reject the examination with probability at least .95. Is a sample of size 50 sufficiently large to meet these requirements? If 9 out of 50 fail the examination, what decision should be made? Use the six-step outline.

◈ 6-13   Using the sample results of Exercise 6-12, find a confidence interval with confidence coefficient at least .99 for the fraction of applicants who will fail the new examination.

6-14   A fruit wholesaler inspects a hundred of each large lot of grapefruit he receives. He would like to reject the carload if the fraction of bad ones is too large. If 10 percent are bad he regards the lot as satisfactory and does not want to reject more than 1 percent of the time. On the other hand, he cannot afford to accept lots with 25 percent bad more than 10 percent of the time. Can he meet both requirements with the chosen sample size?

6-15   Suppose that the inspection process conducted by the wholesaler of Exercise 6-14 involves cutting each sampled grapefruit open. This being the case he decides to use truncated sampling. If the critical region $x \geq 19$ is used, find the expected number of grapefruit which must be cut open when 25 percent of the grapefruit are bad. [The Ordnance Corps table [7] yields $B(19; 101, .25) = .09040$, $B(18; 101, .25) = .05667$.]

6-3
INFERENCES ABOUT THE POISSON
PARAMETER μ

As we observed earlier, we have already been exposed to a hypothesis testing problem involving the Poisson parameter μ (with the secretary problem, Example 2-19). Anytime the Poisson model is appropriate to characterize the behavior of a random variable, we are apt to have an inference problem on our hands since μ will usually be unknown. In Sec. 3-3 we observed that the sample mean $\overline{X}$ was the "best" point estimator of μ. Now we turn our attention to hypothesis testing and interval estimation.

The standard hypothesis testing situations which we shall discuss are the same ones considered in Chap. 5. (At that time the normal model, or at least a continuous model, was regarded as appropriate.) If an individual decides that he will have his telephone taken out if the average number of calls is less than two per day, he will be interested in testing

$$H_0: \mu = \mu_0 \qquad \text{against} \qquad H_1: \mu < \mu_0 \tag{6-9}$$

(with $\mu_0 = 2$) and he will remove the telephone if H_0 is rejected. If an executive is willing to hire a secretary unless the secretary averages more than one error per typed page, the executive will probably test

$$H_0: \mu = \mu_0 \qquad \text{against} \qquad H_1: \mu > \mu_0 \tag{6-10}$$

(with $\mu_0 = 1$) and hire the secretary unless H_0 is rejected. If a given small volume of blood contains on the average five red cells in a healthy individual, a laboratory technician may want to test

$$H_0: \mu = \mu_0 \qquad \text{against} \qquad H_1: \mu \neq \mu_0 \tag{6-11}$$

(with $\mu_0 = 5$) if either too few or too many red cells indicate trouble. Our earlier discussion in Sec. 2-7 indicated that the "number of successes" in each of the above cases has a Poisson distribution.

In some problems we may prefer to replace (6-9) by

$$H_0: \mu \geqq \mu_0 \qquad \text{against} \qquad H_1: \mu < \mu_0 \tag{6-12}$$

and (6-10) by

$$H_0: \mu \leqq \mu_0 \qquad \text{against} \qquad H_1: \mu > \mu_0 \tag{6-13}$$

thus separating the μ's into two sets, "satisfactory" and "unsatisfactory" values. Again the test procedure will be unchanged, but the significance level α must be interpreted as the largest power under H_0 (which will occur when $\mu = \mu_0$).

If $X_1, X_2, \ldots, X_n$ is a random sample drawn from a Poisson distribution, it would seem reasonable to base tests about μ on the sample mean $\overline{X}$. It is more convenient, however, to use

$$Y = X_1 + X_2 + \cdots + X_n = \sum_{i=1}^{n} X_i \tag{6-14}$$

(the numerator of $\overline{X}$). As we have already observed in Sec. 3-3, Y has a Poisson distribution with mean $n\mu$ provided (a) $X_1, X_2, \ldots, X_n$ is a random sample from a distribution and (b) that distribution is a Poisson with mean μ. Further, if the mean of the distribution is μ_0, so that the H_0 of (6-9), (6-10), or (6-11) is true, then Y has a Poisson distribution with $n\mu_0$, a fact useful in the determination of the critical region. As in previous problems, sample size may be chosen arbitrarily or to meet two power requirements. The latter topic will receive more attention later in the section.

Since the Poisson is a discrete distribution, the problem of determining the critical region is similar to the one encountered in the binomial case. Again, if α_0 is the "aimed at" significance level, critical regions are selected so that the actual significance level α is as close to α_0 as possible, yet such that α is not larger than α_0. When (6-9) or (6-12) is tested, small values of y support H_1 and lead to the rejection of H_0. If α_0 is the significance level at which we aim, then the critical region usually chosen is $y \leq y_0$ where y_0 is the largest value of y such that $\Pr(Y \leq y) \leq \alpha_0$ (determined with $n\mu = n\mu_0$). The actual significance level is $\Pr(Y \leq y_0) = \alpha$, calculated with $n\mu = n\mu_0$. Similarly, when (6-10) or (6-13) is tested, H_0 is rejected when y is too large. Now we will take for our critical region $y \geq y_0$ where y_0 is the smallest value of y such that $\Pr(Y \geq y) \leq \alpha_0$ (when $n\mu = n\mu_0$) and the actual significance level is $\Pr(Y \geq y_0)$. For the two-sided situation of (6-11) both small and large y contradict H_0 and lead to rejection. If we aim at a significance level α_0, the critical region usually selected is $y \leq y_1$ and $y \geq y_2$ where y_1 is the largest value of y such that $\Pr(Y \leq y) \leq \alpha_0/2$ and y_2 is the smallest value of y such that $\Pr(Y \leq y) \leq \alpha_0/2$. The actual significance level is $\Pr(Y \leq y_1) + \Pr(Y \geq y_2) = \alpha$. Of course, once more all calculations are performed with $n\mu = n\mu_0$.

EXAMPLE 6-9

Suppose that an individual wants to have his telephone disconnected if his average number of calls per day is less than two. He selects five days at random and records the following number of calls on these days: 0, 2, 1, 1, 1. Using $\alpha_0 = .05$, should he have his telephone removed?

Solution

We shall follow the six-step outline.

1. Letting μ be the unknown average number of calls per day, the problem suggests that we test $H_0: \mu = 2$ against $H_1: \mu < 2$. If H_0 is rejected, the telephone will be removed.

2. The significance level we hope to achieve is $\alpha_0 = .05$. However, the critical region selected in step 4 will yield $\alpha < .05$. The sample size n has already been selected as $n = 5$.

3. The statistic we use is Y, the number of calls recorded in the 5-day period. The random variable Y has a Poisson distribution with mean $n\mu_0 = 5(2) = 10$ if (a) $X_1, X_2, \ldots, X_5$ are randomly selected observations from a distribution, (b) for each X_i that distribution is Poisson with mean μ, and (c) the value of μ is $\mu_0 = 2$, that is, H_0 is true. If (a) and (b) are satisfied but H_0 is not true, Y still has a Poisson distribution but with mean $n\mu$, not $n\mu_0$. In a practical situation it is quite possible that the distribution will be different for weekdays and weekend days, depending upon the individual's living habits. If he is a young single person who is rarely at home on weekends, then μ is undoubtedly different for Saturday and Sunday than for Monday through Friday. If this is the case, we should probably select a random sample of weekdays.

4. From the Poisson table we find with $n\mu_0 = 10$ that

$$\Pr(Y \leq 4) = \sum_{y=0}^{4} p(y; 10) = .02925$$

$$\Pr(Y \leq 5) = \sum_{y=0}^{5} p(y; 10) = .06709$$

Hence, according to our convention, the critical region is $y \leq y_0 = 4$ and the actual significance level is $.02925$.

5. The experiment is performed and we observe $y = 0 + 2 + 1 + 1 + 1 = 5$.

6. Since $y = 5$ does not fall in the critical region, H_0 is accepted. The practical consequences of this decision are that the telephone will not be removed.

EXAMPLE 6-10

During a 12-month period, the number of twin births per month recorded in a hospital are 2, 0, 1, 1, 0, 0, 3, 2, 1, 1, 0, 1. Do these results contradict the hypothesis that the average number of twin births at the hospital is .5 per month? Use $\alpha_0 = .05$.

Solution

We shall follow the six-step outline.

1. Letting μ be the average number of twin births per month at the hospital, the problem suggests that we test $H_0: \mu = .5$ against $H_1: \mu \neq .5$.

2 The significance level at which we aim is .05. However, the critical region selected in step 4 will yield $\alpha < .05$. The sample size has already been selected as $n = 12$.

3 The statistic we use is Y, the number of twin births recorded in a 12-month period. The random variable Y has a Poisson distribution with mean $n\mu_0 = 12(.5) = 6$ provided (a) $X_1, X_2, \ldots, X_{12}$ are randomly selected observations from a distribution, (b) that distribution is a Poisson, and (c) the value of μ is $\mu_0 = .5$, that is, H_0 is true. If (a) and (b) are satisfied but H_0 is not true, Y still has a Poisson distribution with mean $n\mu$, not $n\mu_0$. In discussing the Poisson in Sec. 2-7 we did not mention the number of twin births per month as a Poisson random variable. However, conditions (2-39) seem reasonable since events appear to be independent, the probability that an event occurs appears to be proportional to the length of the time interval, and the probability that two or more events occur in a small time interval is negligible.

4 From the Poisson table we find with $n\mu_0 = 6$ that

$$\Pr(Y \leq 1) = \sum_{y=0}^{1} p(y; 6) = .01735 < .025$$

$$\Pr(Y \leq 2) = \sum_{y=0}^{2} p(y; 6) = .06197 > .025$$

and

$$\Pr(Y \geq 12) = \sum_{y=12}^{\infty} p(y; 6) = 1 - \sum_{y=0}^{11} p(y; 6)$$

$$= 1 - .97991 = .02009 < .025$$

$$\Pr(Y \geq 11) = \sum_{y=11}^{\infty} p(y; 6) = 1 - \sum_{y=0}^{10} p(y; 6)$$

$$= 1 - .95738 = .04262 > .025$$

Hence, the critical region is $y \leq 1$ and $y \geq 12$ with significance level $\alpha = .01735 + .02009 = .03744$.

5 The experiment is performed and we observe $y = 2 + 0 + 1 + 1 + 0 + 0 + 3 + 2 + 1 + 1 + 0 + 1 = 12$.

6 Since $y = 12$ falls in the critical region, we reject H_0. The logical conclusion is that the average number of twin births at the hospital is larger than .5 per month.

As we have already observed, when H_0 is not true Y still has a Poisson distribution but with another mean $n\mu$, not $n\mu_0$. Consequently, power calculations are easy since the probability that Y falls in the critical region can be evaluated from another column of the Poisson table.

EXAMPLE 6-11

Find the power of the test used in Example 6-9 if $\mu = .8$.

Solution

Since the critical region is $y \leq 4$, we need with $n\mu = 5(.8) = 4$

$$\Pr(Y \leq 4) = \sum_{y=0}^{4} p(y; 4) = .62884$$

EXAMPLE 6-12

Find the power of the test used in Example 6-10 if $\mu = .75$.

Solution

The critical region is $y \leq 1$ and $y \geq 12$. With mean $n\mu = 12(.75) = 9$ we need

$\Pr(Y \leq 1) + \Pr(Y \geq 12)$. From Appendix B4 we find

$$\Pr(Y \leq 1) = \sum_{y=0}^{1} p(y; 9) = .00123$$

$$\Pr(Y \geq 12) = 1 - \Pr(Y \leq 11) = 1 - \sum_{y=0}^{11} p(y; 9)$$

$$= 1 - .80301 = .19699$$

Hence the required power is $.00123 + .19699 = .19822$.

Next let us consider the standard sample size problem first proposed in Sec. 4-3 and discussed in Chap. 5 for various situations. For the Poisson case we desire to have the power be no more than α_0 if $\mu = \mu_0$ and at least $1 - \beta_1$ if $\mu = \mu_1$, a value in the alternative. As usual, we seek the minimum n which satisfies these requirements. One way to proceed (but not very efficient) is to increase n until both requirements are met. For the one-sided case (6-9), or (6-12), it can be shown that n must satisfy

$$\frac{\chi^2_{2y_0+2; 1-\alpha_0}}{2\mu_0} \leq n \leq \frac{\chi^2_{2y_0+2; \beta_1}}{2\mu_1} \tag{6-15}$$

For case (6-10), or (6-13), the corresponding formula is

$$\frac{\chi^2_{2y_0; 1-\beta_1}}{2\mu_1} \leq n \leq \frac{\chi^2_{2y_0; \alpha_0}}{2\mu_0} \tag{6-16}$$

Then y_0 and n are found by trial by increasing y_0 until the interval contains at least one integer. The smallest such n is the desired solution, and the accompanying y_0 determines the critical region.   Let us illustrate with an example.

EXAMPLE 6-13

In Example 6-11 we found that for the test of Example 6-9, a power of .62884 is achieved when $\mu = \mu_1 = .8$ and $n = 5$.   What is the minimum sample size n which will raise the power to at least .90 when $\mu = \mu_1 = .8$ if we retain $\alpha_0 = .05$?   What is the accompanying critical region?

Solution

We have $1 - \beta_1 = .90$, $\beta_1 = .10$, $\alpha_0 = .05$, $\mu_0 = 2$, $\mu_1 = .8$, and formula (6-15) becomes

$$\frac{\chi^2_{2y_0+2;\,.95}}{2(2)} \leqq n \leqq \frac{\chi^2_{2y_0+2;\,.10}}{2(.8)}$$

With $y_0 = 9$ we find $\chi^2_{20;\,.95} = 31.41$, $\chi^2_{20;\,.10} = 12.44$, and the inequality is

$$\frac{31.41}{4} \leqq n \leqq \frac{12.44}{1.6} \qquad \text{or} \qquad 7.85 \leqq n \leqq 7.78$$

The fact that the lower bound on n is smaller than the upper bound indicates that we have not chosen y_0 large enough (all $y_0 < 9$ produce a similar result).   With $y_0 = 10$ we find $\chi^2_{22;\,.95} = 33.92$, $\chi^2_{22;\,.10} = 14.04$, and the inequality is

$$\frac{33.92}{4} \leqq n \leqq \frac{14.04}{1.6} \qquad \text{or} \qquad 8.48 \leqq n \leqq 8.77$$

Although the upper bound on n is now greater than the lower bound, the interval does not contain an integer and y_0 must be further increased. Similarly, with $y_0 = 11$, we find $9.10 \leqq n \leqq 9.79$ which also does not contain an integer.   Finally, with $y_0 = 12$ we get $9.72 \leqq n \leqq 10.81$ which contains $n = 10$, the desired sample size.   The corresponding critical region is $y \leqq 12$.

With $n = 10$, $\mu_1 = .8$, $n\mu_1 = 10(.8) = 8$ the power is

$$\text{Pr}(Y \leqq 12) = \sum_{y=0}^{12} p(y;8) = .93620 > .90$$

as we required.   With $n = 10$, $\mu_0 = 2$, $n\mu_0 = 20$ the power (significance level) is

$$\text{Pr}(Y \leqq 12) = \sum_{y=0}^{12} p(y;20)$$

which is not available in our table.   The Molina table [5] gives .03901, which is less than .05 as required.

With a good chi-square table like Harter's [2] and formulas (6-15) and (6-16), nearly all the sample size problems which one is likely to encounter with the Poisson can be solved. (To determine n when the alternative is two-sided, and $\mu_1 < \mu_0$, replace α_0 by $\alpha_0/2$, y_0 by y_1, and use formula (6-15). If $\mu_1 > \mu_0$, use formula (6-16) with α_0 replaced by $\alpha_0/2$, y_0 by y_2.) A large sample formula for n will be given in the next section.

In Sec. 3-3 we learned that a good point estimate of μ is $\hat{\mu} = \bar{x} = y/n$. A confidence interval for μ can be obtained with the assistance of either Appendix B7 or B8. It can be shown that if we take

$$w_1 = \frac{\chi^2_{2y;\,\alpha/2}}{2n} = \frac{y}{n}\frac{\chi^2_{2y;\,\alpha/2}}{2y} \qquad w_2 = \frac{\chi^2_{2y+2;\,1-\alpha/2}}{2n} = \frac{y+1}{n}\frac{\chi^2_{2y+2;\,1-\alpha/2}}{2y+2} \qquad (6\text{-}17)$$

then

$$\Pr(W_1 \leq \mu \leq W_2) \geq 1 - \alpha \qquad (6\text{-}18)$$

and (w_1, w_2), end points included, is a confidence interval for μ with confidence coefficient at least $1 - \alpha$.

EXAMPLE 6-14

Using the sample information of Example 6-10, find a confidence interval for μ with confidence coefficient at least .95.

Solution

We had $n = 12$, $y = 12$. We are given $\alpha = .05$ so that $\alpha/2 = .025$. Also $2y = 24$, $2y + 2 = 26$, and we need $\chi^2_{24;\,.025} = 12.40$, $\chi^2_{26;\,.975} = 41.92$. Thus

$$w_1 = \frac{\chi^2_{24;\,.025}}{2(12)} = \frac{12.40}{24} = .52$$

$$w_2 = \frac{\chi^2_{26;\,.975}}{2(12)} = \frac{41.92}{24} = 1.75$$

and the desired interval is (.52, 1.75).

We note that the interval found in Example 6-14 does not cover $\mu_0 = .5$, the hypothesized value of Example 6-10. As in previous situations, the confidence interval (w_1, w_2) covers a value μ_0 if, and only if, $H_0: \mu = \mu_0$ is accepted when tested against $H_1: \mu \neq \mu_0$ provided the same sample is used and the α_0 of the hypothesis testing problem is equal to the α used for the confidence interval.

EXERCISES

◈ 6-16 An executive is willing to hire a secretary who has applied for a position unless she averages more than one error per typed page. A random sample of five pages is selected from some prepared by the secretary. The pages contain 3, 3, 4, 1, 2 errors. Using $\alpha_0 = .05$, what decision should be made? Follow the six-step outline.

◈ 6-17 Find the power of the test used in Exercise 6-16 if $\mu = 1.6$. If $\mu = 2$.

◈ 6-18 Using the sample results given in Exercise 6-16, find a confidence interval with confidence coefficient at least .95 for the average number of errors per page.

◈ 6-19 Suppose that we use $\alpha_0 = .05$ to test the hypothesis of Exercise 6-16. In addition, if the secretary's average is two errors per page, we would like to reject H_0 at least .90 of the time. Using only those values available to us in Appendix B, find the minimum sample size which meets both requirements. What is the accompanying critical region?

6-20 A bakery would like to have a certain kind of cookie that it makes contain two raisins on the average. A random sample of four cookies yields a total of five raisins. Test the appropriate hypothesis following the six-step outline. Use $\alpha_0 = .05$.

6-21 What is the power of the test used in Exercise 6-20 if the average number of raisins is 2.5? If the average is 1?

6-22 Using the sample results of Exercise 6-20, find a confidence interval with confidence coefficient at least .95 for the average number of raisins per cookie.

◈ 6-23 It is claimed that the distribution of the number of stoppages on a loom due to warp breakage is a Poisson. Suppose that a loom currently in use has to be stopped on the average once per hour. A new loom is being considered as a replacement, and it is decided that the purchase will be made if it can be shown at the .05 level of significance that on the average less than one stop per hour is required. During an 8-hour trial period the new loom has to be stopped three times. Use the six-step outline, and recommend the appropriate action.

◈ 6-24 What is the power of the test used in Exercise 6-23 if the average associated with the new loom is $\frac{1}{2}$ per hour? If the average is $\frac{1}{4}$ per hour?

◈ 6-25 We have tested the hypothesis of Exercise 6-23 with $\alpha_0 = .05$. If we also require that the probability of rejection be at least .99 if the average for the new loom is $\frac{1}{4}$ per hour, what is the minimum number of hours that the loom must be tested? If this period is used and eight stops are observed, what decision should be made?

◈ 6-26 Using the sample results given in Exercise 6-23, find a confidence interval with confidence coefficient at least .90 for the average number of stops associated with the new loom.

◈ 6-27 Suppose we test $H_0: \mu = .01$ against $H_1: \mu > .01$ for a Poisson random variable using $\alpha_0 = .05$. If we wish the power to be at least .90 if $\mu = .10$, is $n = 60$ large enough to meet the requirements? Use formula (6-16) to find the minimum n and the accompanying y_0.

6-28 Suppose we test $H_0: \mu = .5$ against $H_1: \mu < .5$ for a Poisson random variable using $\alpha_0 = .05$. If we wish the power to be at least .90 when $\mu = .1$, is $n = 20$ large enough to meet the requirements? Use formula (6-15) to find the minimum n and the accompanying y_0.

6-29 The number of defects per yard of cloth is known to have a Poisson distribution. A wholesaler has received a large order of cloth from a manufacturer. To determine whether or not he should accept the shipment, the wholesaler inspects 100 square yards, selected at random a yard at a time. If the average number of defects per yard is .02, then he does not want to reject such shipments more than one time in a hundred. On the other hand, if the average number of defects per yard is .10, he cannot afford to accept such shipments more than 15 percent of the time. Can this be accomplished with the sample size that is used? What decision should be made if the sample yields a total of five defects? Follow the six-step outline.

6-30 Using the sample results of Exercise 6-29, find a confidence interval with confidence coefficient at least .90 for the average number of defects per yard.

◈ 6-31 A book company is trying to decide whether or not it should engage a certain printing company for production of technical books. If the printing company's product averages 2 errors per page the output is regarded as satisfactory and the book company would like to sign a contract. On the other hand, if the average is 4 errors per page, the book company regards the expense of making corrections intolerable and will seek another printer. If we take $\alpha_0 = \beta_1 = .05$ and use only those values available to us in Appendix B, find the smallest sample size which meets both requirements. What is the accompanying critical region? If a sample of the size obtained above yields a total of 24 errors, what decision should be made?

6-4
LARGE-SAMPLE PROCEDURES FOR
BINOMIAL p AND POISSON μ

The procedures outlined in Secs. 6-2 and 6-3, used in conjunction with a good set of tables, will be sufficient to solve most problems of practical interest. However, if such tables are not available or if the range of entries of those tables is exceeded, then the approximation of this section may be used.

When existing binomial tables are of no assistance, we can use the statistic

$$Z' = \frac{(X/n) - p}{\sqrt{p(1-p)/n}} = \frac{X - np}{\sqrt{np(1-p)}} \tag{6-19}$$

to draw inferences about p. [This statistic was already used in Sec. 2-8 and is formula (2-47).] If X has a binomial distribution with parameters n and p, then (because of the Central Limit theorem) Z' has approximately a standard normal distribution.

If $p = p_0$, which we assume to obtain critical regions for the hypothesis testing situations (6-1) to (6-5), then statistic (6-19) becomes

$$Z' = \frac{X - np_0}{\sqrt{np_0(1 - p_0)}} \tag{6-20}$$

When using the approximate procedures of this section we no longer distinguish between a desired significance level α_0 and an achieved level α but use one figure designated by α for both purposes. When (6-1) or (6-4) is tested, only small values of x, and hence small z', support H_1 and lead to the rejection of H_0. If the significance level is α, then we take for the critical region

$$z' = \frac{x - np_0}{\sqrt{np_0(1 - p_0)}} < z_\alpha \tag{6-21}$$

When H_0 is true, $\Pr(Z' < z_\alpha) \cong \Pr(Z < z_\alpha) = \alpha$ where Z is our usual symbol for a standard normal random variable. Similarly, only large x and large z' support the H_1 of (6-2) and (6-5). A test with significance level approximately α is obtained by rejecting H_0 when

$$z' = \frac{x - np_0}{\sqrt{np_0(1 - p_0)}} > z_{1-\alpha} \tag{6-22}$$

If H_0 is true, then $\Pr(Z' > z_{1-\alpha}) \cong \Pr(Z > z_{1-\alpha}) = 1 - \alpha$. Finally, when the two-sided case (6-3) is tested, both small and large values of x, and hence z', provide support for H_1. Thus it is reasonable to reject H_0 with both small and large values of z', and the critical region we choose to yield a test with significance level approximately α is

$$z' = \frac{x - np_0}{\sqrt{np_0(1 - p_0)}} < z_{\alpha/2} \quad \text{and} \quad z' = \frac{x - np_0}{\sqrt{np_0(1 - p_0)}} > z_{1-\alpha/2} \tag{6-23}$$

EXAMPLE 6-15

Consider again the new machine of Example 6-2. Suppose that a sample of 2,500 bolts yields 277 which require rework. Again use significance level .05 and make the appropriate decision.

Solution

We shall follow the six-step outline, indicating the necessary changes.

1 Same as in Example 6-2.
2 The significance level is chosen as .05 and will be achieved approximately. The sample size has been selected as $n = 2,500$.
3 The statistic we use is Z' given by (6-20). The conditions that Z' has approximately a standard normal distribution are the same as for X having a binomial distribution. Hence all the comments given in Example 6-2 are still appropriate.
4 The critical region (6-22) is

$$z' = \frac{x - 2,500(.10)}{\sqrt{2,500(.10)(.90)}} > z_{.95} = 1.645$$

5 We now perform the experiment and observe $x = 277$. Then, the observed value of the statistic is

$$z' = \frac{277 - 250}{\sqrt{2,500}\sqrt{.09}} = \frac{27}{50(.3)} = \frac{27}{15} = 1.8$$

6 Since $z' = 1.8$ falls in the critical region, H_0 is rejected and the machine is not purchased.

If H_0 is not true so that $p = p_1 \neq p_0$, then the Z' of (6-19) with $p = p_1$ has approximately a standard normal distribution. With a little algebraic manipulation it can be shown that the power of the tests for the one-sided cases with critical regions (6-21) and (6-22) is approximately

$$\text{Power} = \Pr\left(Z < \frac{z_\alpha \sqrt{p_0(1 - p_0)} + \sqrt{n}|p_0 - p_1|}{\sqrt{p_1(1 - p_1)}}\right) \tag{6-24}$$

where p_1 is a value of p in the range designated by H_1. For the two-sided case with critical region (6-23), formula (6-24) still gives approximately the power if α is replaced by $\alpha/2$. If in addition to using the significance level α, we also require that the power be at least $1 - \beta$ if $p = p_1$ (our standard sample size problem), then we can show that for the one-sided cases the approximation yields

$$n \geq \left(\frac{z_\alpha \sqrt{p_0(1 - p_0)} + z_\beta \sqrt{p_1(1 - p_1)}}{p_0 - p_1}\right)^2 \tag{6-25}$$

For the two-sided case (6-3), replace α by $\alpha/2$ in inequality (6-25).

EXAMPLE 6-16

Find the power of the test used in Example 6-15 if $p = .12$.

Solution

We have a one-sided hypothesis testing situation with $p_0 = .10$, $p_1 = .12$, $\alpha = .05$ ($z_{.05} = -1.645$), $n = 2,500$. Formula (6-24) yields

$$\text{Power} = \Pr\left(Z < \frac{-1.645\sqrt{.10(.90)} + \sqrt{2,500}\,|.10 - .12|}{\sqrt{(.12)(.88)}}\right)$$

$$= \Pr\left(Z < \frac{-1.645(.30) + 50|-.02|}{.325}\right)$$

$$= \Pr(Z < 1.56) = .94$$

EXAMPLE 6-17

In Example 6-5 we were discussing the test of $H_0 : p = .10$ against $H_1 : p > .10$ (considered again in Example 6-15). We had significance level .05 and inquired how large n should be to raise the power to at least .95 if $p = p_1 = .20$. Solve this problem using inequality (6-25).

Solution

We identify $\alpha = .05$, $1 - \beta = .95$, $\beta = .05$ so that $z_\alpha = z_\beta = z_{.05} = -1.645$. Also $p_0 = .10$, $p_1 = .20$, and (6-25) becomes

$$n \geq \left(\frac{-1.645\sqrt{(.10)(.90)} - 1.645\sqrt{(.20)(.80)}}{.10 - .20}\right)^2$$

$$= \left(\frac{-1.645(.3) - 1.645(.4)}{-.10}\right)^2 = (11.515)^2 = 132.6$$

Hence, based upon our approximation, the minimum n is 133. As we commented before, the actual minimum n obtained from binomial tables is $n = 135$, and so the approximation is quite good.

To get a large-sample confidence interval for p, we start with

$$\Pr\left(-z_{1-\alpha/2} < \frac{X - np}{\sqrt{np(1 - p)}} < z_{1-\alpha/2}\right) \cong 1 - \alpha \tag{6-26}$$

To get p in the middle of the inequalities requires that a quadratic inequality be solved. If certain small terms are then discarded, probability statement (6-26) can be rewritten as

$$
\Pr\left(\frac{X}{n} - z_{1-\alpha/2}\sqrt{\frac{\frac{X}{n}\left(1 - \frac{X}{n}\right)}{n}} < p < \frac{X}{n}\right.
$$

$$
\left. + z_{1-\alpha/2}\sqrt{\frac{\frac{X}{n}\left(1 - \frac{X}{n}\right)}{n}}\right) \cong 1 - \alpha \quad (6\text{-}27)
$$

Thus, a confidence interval for p with confidence coefficient approximately $1 - \alpha$ is (w_1, w_2) where

$$
w_1 = \frac{x}{n} - z_{1-\alpha/2}\sqrt{\frac{\frac{x}{n}\left(1 - \frac{x}{n}\right)}{n}} \qquad w_2 = \frac{x}{n} + z_{1-\alpha/2}\sqrt{\frac{\frac{x}{n}\left(1 - \frac{x}{n}\right)}{n}} \quad (6\text{-}28)
$$

EXAMPLE 6-18

Using the sample results of Example 6-15, find a confidence interval with confidence coefficient approximately .95 for p, the fraction of bolts produced by the new machine which require rework.

Solution

We have $n = 2,500$, $x = 277$, $x/n = 277/2,500 = .1108$, $\alpha = .05$, $1 - \alpha/2 = .975$, $z_{.975} = 1.960$. Thus

$$
z_{1-\alpha/2}\sqrt{\frac{\frac{x}{n}\left(1 - \frac{x}{n}\right)}{n}} = 1.960\sqrt{\frac{(.1108)(.8892)}{2,500}} = \frac{1.960}{50}\sqrt{.09852}
$$

$$
= .012
$$

The interval is $(.111 - .012, .111 + .012)$, which reduces to $(.099, .123)$.

Large-sample results for Poisson μ are similar to those for binomial p. If $Y = \sum_{i=1}^{n} X_i$ has mean 100 or less, then we can proceed as in Sec. 6-3 and use the Molina table [5] described in Sec. 2-7. If that mean is greater than 100 (or the tables are not available), then we can use the statistic

$$
Z' = \frac{Y - n\mu}{\sqrt{n\mu}} \quad (6\text{-}29)
$$

to draw inferences about μ. Because of the Central Limit theorem, the random variable Z' has approximately a standard normal distribution when $X_1, X_2, \ldots, X_n$ is a random sample from a Poisson distribution with mean μ.

If $\mu = \mu_0$, then statistic (6-29) becomes

$$Z' = \frac{Y - n\mu_0}{\sqrt{n\mu_0}} \tag{6-30}$$

When (6-9) or (6-12) is tested, only small values of y, and hence z', support H_1 and lead to rejection of H_0. A significance level of approximately α is achieved by rejecting when

$$z' = \frac{y - n\mu_0}{\sqrt{n\mu_0}} < z_\alpha \tag{6-31}$$

For (6-10) and (6-13) the critical region is

$$z' = \frac{y - n\mu_0}{\sqrt{n\mu_0}} > z_{1-\alpha} \tag{6-32}$$

Finally, for the two-sided case (6-11), we reject when

$$z' = \frac{y - n\mu_0}{\sqrt{n\mu_0}} < z_{\alpha/2} \quad \text{and} \quad z' = \frac{y - n\mu_0}{\sqrt{n\mu_0}} > z_{1-\alpha/2} \tag{6-33}$$

The formula and comments concerning power calculations and confidence intervals for binomial p (given earlier in the section) apply with minor modifications. The counterparts of formulas (6-24), (6-25), and (6-28) are

$$\text{Power} = \Pr\left(Z < \frac{z_\alpha\sqrt{\mu_0} + \sqrt{n}\,|\mu_0 - \mu_1|}{\sqrt{\mu_1}} \right) \tag{6-34}$$

$$n \geq \left(\frac{z_\alpha\sqrt{\mu_0} + z_\beta\sqrt{\mu_1}}{\mu_0 - \mu_1} \right)^2 \tag{6-35}$$

and

$$w_1 = \frac{y}{n} - z_{1-\alpha/2}\sqrt{\frac{y/n}{n}} \qquad w_2 = \frac{y}{n} + z_{1-\alpha/2}\sqrt{\frac{y/n}{n}} \tag{6-36}$$

EXAMPLE 6-19

Suppose that the individual of Example 6-9 selects 200 days at random during which he received a total of 364 telephone calls. Using significance level .05, what is the appropriate decision? How large does n have to be to achieve a power of .95 if $\mu = 1.8$?

Solution

To answer the first question we shall follow the six-step outline, indicating the necessary changes.

1 Same as in Example 6-9.
2 The significance level is chosen as .05 and will be achieved approximately. The sample size $n = 200$ has been selected.
3 The statistic we use is Z' given by (6-30). The conditions that Z' has approximately a standard normal distribution are the same as for Y having a Poisson distribution already discussed in Example 6-9. [If H_0 is not true, then (6-29) with the appropriate μ has approximately a standard normal distribution.]
4 The critical region given by inequality (6-31) is

$$z' = \frac{y - 200(2)}{\sqrt{200(2)}} < z_{.05} = -1.645$$

5 The experiment yields $y = 364$. The observed value of the statistic is

$$z' = \frac{364 - 400}{\sqrt{400}} = \frac{-36}{20} = -1.8$$

6 Since $z' = -1.8$ falls in the critical region, H_0 is rejected. With this result the telephone would be removed.

To answer the second question, we identify $\alpha = .05$, $1 - \beta = .95$, $\beta = .05$, $\mu_0 = 2$, $\mu_1 = 1.8$. Since $z_{.05} = -1.645$, we get from formula (6-35)

$$n \geqq \left(\frac{-1.645\sqrt{2} - 1.645\sqrt{1.8}}{2 - 1.8} \right)^2 = 513.7$$

Thus, the minimum n is 514.

EXERCISES

⬧ 6-32 Suppose that in Exercise 6-1 the seed company's claim is investigated by using 1,600 seeds, 184 of which fail to germinate. Test the appropriate hypothesis with $\alpha = .05$. Follow the six-step outline.

⬧ 6-33 Find the power of the test used in Exercise 6-32 if actually 12 percent fail to germinate. How large would n have to be to raise this probability to .90?

⬧ 6-34 Using the sample results of Exercise 6-32, find a confidence interval with confidence coefficient approximately .90 for the fraction of seeds which fail to germinate.

6-35 According to a genetic theory, 25 percent of a species have a certain characteristic. A random sample of 1,200 of the species contains

324 having the characteristic. If $\alpha = .05$ is used, is the theory contradicted? Follow the six-step outline.

6-36 Find the power of the test used in Exercise 6-35 if 23 percent have the characteristic. If 27 percent have the characteristic.

6-37 Using the sample results of Exercise 6-35, find a confidence interval with confidence coefficient approximately .95 for the fraction of the species which have the characteristic.

◈ 6-38 Suppose that the executive of Exercise 6-16 looks at a random sample of 225 pages of the secretary's work and finds 252 errors. Use $\alpha = .05$ and recommend the appropriate action. Follow the six-step outline.

◈ 6-39 Find the power of the test used in Exercise 6-38 if $\mu = 1.21$. How large would n have to be to raise this probability to .95?

◈ 6-40 Using the sample results of Exercise 6-38, find a confidence interval with confidence coefficient approximately .95 for the average number of errors per page.

◈ 6-41 A civil rights group claims that in a certain Wyoming county blacks have been discriminated against in the selection of jurors. In the past year out of 2,600 jurors only 4 were black, while 1 percent of the county's population is black. Formulate a reasonable hypothesis and test it using $\alpha = .001$ and following the six-step outline. If prospective jurors are selected from a list of property owners (rather than from a list including everyone), what further conclusion might be drawn?

6-42 The number of defects per yard of cloth is known to have a Poisson distribution. A wholesaler will buy a large order from a cloth producer if the average number of defects is .1 per yard. However, he does not want to buy if there is evidence that the average is greater than .1. If he inspects a random sample of 1,000 yards and finds 112 defects, what decision should be made using $\alpha = .05$? Follow the six-step outline.

6-43 If the wholesaler in Exercise 6-42 wishes in addition to reject with probability at least .95 if the average is .2, how large a sample should he have taken?

6-44 Using the sample results of Exercise 6-42, find a confidence interval with confidence coefficient approximately .90 for the average number of defects per yard.

6-5
HYPOTHESIS TESTING FOR
HYPERGEOMETRIC k

In Sec. 2-6 we developed the hypergeometric probability model and observed that it depended on three parameters N, n, and k. We will assume that N (perhaps a " lot " size) is known. The sample size n may be chosen arbitrarily or determined so that two power conditions are satisfied. Frequently k (which may be the number of defectives in a lot) will be an unknown parameter about which we may wish to draw inferences. Point estimation of k was discussed in Sec. 3-2. Now we will consider some hypothesis testing problems. For the model to be appropriate, sampling must be performed so that conditions (2-30) seem realistic. We recall that those assumptions imply that a simple random sample of size n is drawn from N items, k of which have some characteristic.

The standard hypothesis testing situations follow the usual pattern. If we would purchase a lot of N items only if we were convinced that it contained less than k_0 defectives, then we would probably test

$$H_0: k = k_0 \quad \text{against} \quad H_1: k < k_0 \tag{6-37}$$

and buy the lot if H_0 is rejected. If a lot is regarded as satisfactory unless there is good evidence that it contains more than k_0 defectives, then we would test

$$H_0: k = k_0 \quad \text{against} \quad H_1: k > k_0 \tag{6-38}$$

and judge the lot satisfactory unless H_0 is rejected. Situation (6-38) is the one which is encountered most frequently in practice, being particularly important in the area of statistical quality control. Since situation (6-37) can be converted to (6-38) by interchanging the roles of "defective" and "nondefective," and since the two-sided alternative is not of much interest, we will discuss only (6-38).

In some problems we may prefer to replace (6-38) by

$$H_1: k \leq k_0 \quad \text{against} \quad H_1: k > k_0 \tag{6-39}$$

As usual, the test procedure is unchanged but the significance level α must be interpreted as the largest power under H_0 (which occurs when $k = k_0$).

Intuitively, it seems reasonable to base our decision on X, the number of defectives in a simple random sample of size n. If conditions (2-30) are satisfied, then X has a hypergeometric distribution. Further, if $k = k_0$, then X has a hypergeometric distribution with parameters N, n, and k_0, a fact we use to determine the critical region. When testing situation (6-38) or (6-39), a large value of x would support H_1 and lead to the rejection of H_0. Consequently, if α_0 is the "aimed-at" significance level, then the critical region usually chosen is $x \geq x_0$ where x_0 is the smallest value of x such that $\Pr(X \geq x) \leq \alpha_0$. In terms of a hypergeometric sum, the latter inequality is $1 - P(N, n, k_0, x - 1) \leq \alpha_0$ or

$$P(N, n, k_0, x - 1) \geq 1 - \alpha_0 \tag{6-40}$$

The actual significance level is

$$\Pr(X \geq x_0) = 1 - P(N, n, k_0, x_0 - 1)$$

EXAMPLE 6-20

A lot consisting of $N = 10$ items is satisfactory if it contains 2 defectives. If sample results indicate that it contains more than 2 defectives, the lot will be rejected and will not be purchased. If a simple random sample of size 4 contains 1 defective, what decision should be made using $\alpha_0 = .20$?

Solution

We shall follow the six-step outline.

1 The problem suggests that we test $H_0: k = 2$ against $H_1: k > 2$ and purchase the lot if H_0 is accepted.
2 The significance level we hope to achieve is $\alpha_0 = .20$. However, as usual with tests based upon discrete random variables the actual significance level may be less than .20. The sample size $n = 4$ has already been selected.
3 The statistic we use is X, the number of defectives in the sample. If a simple random sample of $n = 4$ items is selected from $N = 10$ (which is given), then X has a hypergeometric distribution. Further if H_0 is true, then $k = k_0 = 2$.
4 From the hypergeometric tables of Appendix B3 we find (with $N = 10$, $n = 4$, $k = 2$) that

$$\Pr(X \geq 1) = 1 - \Pr(X \leq 0) = 1 - P(10, 4, 2, 0)$$
$$= 1 - .333333 = .666667 > .20$$

$$\Pr(X \geq 2) = 1 - \Pr(X \leq 1) = 1 - P(10, 4, 2, 1)$$
$$= 1 - .866667 = .133333 < .20$$

Actually, it is easier to work with inequality (6-40) and find the smallest x which satisfies $P(10, 4, 2, x - 1) \geq .80$. With $x = 1$ we have $P(10, 4, 2, 0) = .333333 < .80$, and with $x = 2$ we get $P(10, 4, 2, 1) = .866667 > .80$. Hence $x_0 = 2$, the critical region is $x \geq 2$, and the actual significance level is .133333.
5 The simple random sample is observed and produces the observed result $x = 1$.
6 Since $x = 1$ does not fall in the critical region, H_0 is accepted and the lot is purchased.

When H_0 is not true, X still has a hypergeometric distribution but with some other value of k. Consequently, power can be evaluated directly from the hypergeometric table.

EXAMPLE 6-21

Find the power of the test used in Example 6-20 if $k = 3, 5, 8$.

Solution

We have $N = 10$, $n = 4$, $x_0 = 2$. Hence for any value of k the power is

$$\Pr(X \geqq 2) = 1 - \Pr(X \leqq 1) = 1 - P(10, 4, k, 1)$$

With $k = 3$ we get $P(10, 4, 3, 1) = .666667$, and the power is $1 - .666667 = .333333$. With $k = 5$ we get $P(10, 4, 5, 1) = P(10, 5, 4, 1) = .261905$, and the power is $1 - .261905 = .738095$. With $k = 8$ the sample must contain 2 or more defectives, and we reject with probability 1. Since $P(10, 4, 8, 1) = P(10, 8, 4, 1) = 0$, this value does not appear in the table.

FIGURE 6-1

Power curve for testing $H_0 : k = 2$ against $H_1 : k > 2$ with $N = 10$, $n = 4$, critical region $x \geqq 2$.

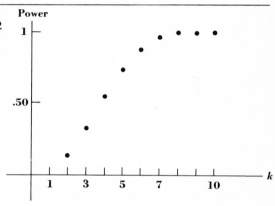

We note that if we were to plot a power "curve" for this or any other hypergeometric hypothesis testing problem, the curve would consist of a series of dots since the parameter k can assume only integer values. (See Fig. 6-1.)

The usual type of sample size problem can be solved by inspection in a hypergeometric table. For convenience let $c = x_0 - 1$. If we require power $\leqq \alpha_0$ when $k = k_0$, then from inequality (6-40) c must satisfy

$$P(N, n, k_0, c) \geqq 1 - \alpha_0 \tag{6-41}$$

If we also require power $\geqq 1 - \beta_1$ when $k = k_1 > k_0$, then c must also satisfy $1 - P(N, n, k_1, c) \geqq 1 - \beta_1$, which can be written

$$P(N, n, k_1, c) \leqq \beta_1 \tag{6-42}$$

To find the minimum n and accompanying c, first let $c = 0$ and find those n which permit inequalities (6-41) and (6-42) to be satisfied. If no n will satisfy both inequalities, repeat with $c = 1$, then $c = 2$, etc., continuing until

n which satisfy both inequalities are found. The smallest c which yields solutions also yields the minimum sample size. (This is the same procedure which was referred to in Example 6-5 for the binomial case. We did not discuss it at that time because Appendix B1 is inadequate for the problem.) We will illustrate with an example.

EXAMPLE 6-22

Find the minimum sample size required to test the hypothesis of Example 6-20 with $\alpha_0 = .20$ if we also require that power be at least .80 when $k = 4$.

Solution

We have $\alpha_0 = .20$, $1 - \alpha_0 = .80$, $1 - \beta_1 = .80$, $\beta_1 = .20$, $N = 10$, $k_0 = 2$, $k_1 = 4$. Then inequalities (6-41) and (6-42) become

$$P(10, n, 2, c) \geqq .80, \quad P(10, n, 4, c) \leqq .20$$

If we let $c = 0$, then $P(10, n, 2, 0) \geqq .80$ is satisfied only if $n = 1$ since $P(10, 1, 2, 0) = P(10, 2, 1, 0) = .800000$ while $P(10, 2, 2, 0) = .622222$. The second inequality becomes $P(10, n, 4, 0) \leqq .20$, and this is satisfied only if $n \geqq 3$ since $P(10, 2, 4, 0) = P(10, 4, 2, 0) = .333333$ while $P(10, 3, 4, 0) = P(10, 4, 3, 0) = .166667$. Since it is obviously impossible to have $n = 1$ and $n \geqq 3$ at the same time, $c = 0$ does not permit a solution.

We next try $c = 1$. The first inequality becomes $P(10, n, 2, 1) \geqq .80$. Since $P(10, 4, 2, 1) = .866667$, $P(10, 5, 2, 1) = .777778$, we must have $n \leqq 4$. The second inequality is $P(10, n, 4, 1) \leqq .20$. Since $P(10, 5, 4, 1) = .261905$, $P(10, 6, 4, 1) = .119048$, we need $n \geqq 6$. It is impossible to have $n \leqq 4$ and $n \geqq 6$ at the same time. Hence $c = 1$ does not permit a solution.

We next try $c = 2$. The first inequality becomes $P(10, n, 2, 2) \geqq .80$. This is satisfied for all $n \leqq 10$ since with only 2 defectives present, the probability of obtaining 2 or fewer is necessarily 1. The second inequality becomes $P(10, n, 4, 2) \leqq .20$. Since $P(10, 7, 4, 2) = .333333$, $P(10, 8, 4, 2) = .133333$, we need $n \geqq 8$. In other words, solutions are possible with $8 \leqq n \leqq 10$, and so we take $n = 8$, $c = 2$, $x_0 = 3$. The critical region is $x \geqq 3$, and the actual significance level is 0.

The result found in Example 6-22, sampling 8 out of 10 items, would not be a very appealing one in a practical situation. Probably most lots would be larger than $N = 10$, so that in general a much smaller fraction of the lot need be sampled. The tables of Lieberman and Owen [4] from which Appendix B3 was taken can be used for $N = 1(1)50(10)100$ to work problems of the type illustrated in the above examples. For other values of N we can use the binomial approximation to the hypergeometric given by formula (2-35). When we do this, we are testing hypotheses about $p = k/N$, and the procedures of Secs. 6-2 and 6-4 can be applied.

⬧ 6-45 A company assembles a certain product into lots of size 10. From a lot an inspector checks 5 items, selected at random without replacement. He classifies these items as defective or non-defective depending upon whether or not the item needs further work. Unless he is convinced that the number of defectives in the lot exceeds 3, he will pass the lot as satisfactory. Rejected lots will be held for further work. Using $\alpha_0 = .10$, what decision should be made with a lot for which the sample yields 2 defectives?

⬧ 6-46 For the test used in Exercise 6-45 what is the power if the lot contains 5 defectives? 7 defectives?

⬧ 6-47 Suppose that the inspector in Exercise 6-45 wants to reject a lot with 1 defective no oftener than .10 of the time but if the lot contains 5 defectives he would like to reject at least .80 of the time. What is the minimum sample size and the corresponding critical region?

6-48 A company receives a moderately priced piece of equipment in lots of 1,000. A simple random sample of 50 pieces is selected, and each piece is tested. If the lot contains 100 items which need repair work, the lot is considered acceptable. However, if the sample result indicates that more than 100 items need repair, the lot will be rejected. Using $\alpha_0 = .10$, what decision should be made if 8 defectives are found in the sample? Follow the six-step outline.

6-49 Find the power of the test used in Exercise 6-48 if the lot contains 200 items which require repair work. What is the minimum sample size required to raise this power to at least .90 for a lot that contains 200 items which require repair work?

⬧ 6-50 A lot of 10 items is checked by inspecting a random sample of 3 items. If the lot contains 1 defective item, it is regarded as satisfactory. We do not want to reject satisfactory lots more than .30 of the time. If the sample contains 1 defective, what decision should be made? Follow the six-step outline.

⬧ 6-51 Find the power of the test used in Exercise 6-50 if the lot actually contains 3 defectives. 5 defectives.

⬧ 6-52 Obtain a copy of the Lieberman-Owen table [4] and find the minimum n and accompanying c which satisfies (6-41) and (6-42) with $N = 100$, $k_0 = 5$, $k_1 = 20$, $\alpha_0 = .05$, $\beta_1 = .10$.

6-53 Obtain a copy of the Lieberman-Owen table [4], and find the minimum n and accompanying c which satisfies formulas (6-41) and (6-42) with $N = 50$, $k_0 = 2$, $k_1 = 10$, $\alpha_0 = .10$, $\beta_1 = .20$.

6-6
INFERENCES ABOUT TWO POISSON MEANS

Sometimes an experimenter will want to test a hypothesis about the means of two Poisson distributions. The executive of Exercise 6-16 may be considering two secretaries instead of one for his position. He has a theory that women who are high school graduates make better secretaries than college-trained women. If μ_1 is the average number of errors per page associated with a college-educated secretary and μ_2 is the average for a secretary with a high school diploma, the executive may wish to test

$$H_0: \mu_1 = \mu_2 \quad \text{against} \quad H_1: \mu_1 < \mu_2 \tag{6-43}$$

and hire the less educated one unless H_0 is rejected. If a second applicant has a higher wage rate than the first, the executive will probably test

$$H_0: \mu_1 = \mu_2 \quad \text{against} \quad H_1: \mu_1 > \mu_2 \tag{6-44}$$

and hire the first woman unless H_0 is rejected. [Situation (6-44) can be converted to (6-43) by interchanging the roles of the subscripts.] If a grocery chain is considering two bakeries as possible suppliers for its cookies and will make the choice on the basis of the average number of raisins per cookie, then the chain's statistician is likely to test

$$H_0: \mu_1 = \mu_2 \quad \text{against} \quad H_1: \mu_1 \neq \mu_2 \tag{6-45}$$

provided there is no reason to favor either firm.

As usual, if (6-43) is replaced by

$$H_0: \mu_1 \geqq \mu_2 \quad \text{against} \quad H_1: \mu_1 < \mu_2 \tag{6-46}$$

or (6-44) by

$$H_0: \mu_1 \leqq \mu_2 \quad \text{against} \quad H_1: \mu_1 > \mu_2 \tag{6-47}$$

the testing procedure will be unchanged. More comments about significance level will be made later.

Let $X_{11}, X_{21}, \ldots, X_{n1}$ and $X_{12}, X_{22}, \ldots, X_{n2}$ denote random samples of size n drawn from Poisson distributions with means μ_1 and μ_2, respectively. From our study of one Poisson it would seem reasonable to base tests of hypotheses on $Y_1 = \sum_{i=1}^{n} X_{i1}$ and $Y_2 = \sum_{i=1}^{n} X_{i2}$. It can be shown that for a given value of $Y_1 + Y_2 = T$, say t, Y_1 has a binomial distribution with parameters t (replacing n in our usual binomial notation) and $p = \mu_1/(\mu_1 + \mu_2)$. If we let $\mu_1/\mu_2 = a$, then $p = \mu_1/(\mu_1 + \mu_2) = a/(a + 1)$. In particular, if $\mu_1 = \mu_2$ so that $a = 1$, then $p = 1/2$. This means that (6-43) to (6-45) are equivalent to

$$H_0 : p = 1/2 \quad \text{against} \quad H_1 : p < 1/2 \qquad (6\text{-}48)$$

$$H_0 : p = 1/2 \quad \text{against} \quad H_1 : p > 1/2 \qquad (6\text{-}49)$$

$$H_0 : p = 1/2 \quad \text{against} \quad H_1 : p \neq 1/2 \qquad (6\text{-}50)$$

The latter three hypotheses (and hence our original hypotheses) can be tested using the random variable Y_1 given t and the procedures discussed in Sec. 6-2 for binomial p. For (6-48) we reject if y_1 given t is too small, for (6-49) we reject if y_1 given t is too large, and for (6-50) we reject if y_1 given t is either too small or too large. With desired significance level α_0 critical regions based on Y_1 given t are selected in exactly the same way as in Sec. 6-2 using $p = p_0 = 1/2$. It can be shown that the test procedure which we have described has very good statistical properties and that the significance level is no greater than α_0.

EXAMPLE 6-23

An executive is considering two secretaries for a position. One is a college graduate and the other has a high-school education. The executive decides that he will hire the less educated secretary unless there is evidence to indicate that the other woman makes on the average fewer errors per page of typing. From each person's work 10 pages are selected at random. If 9 errors are found in the material prepared by the more educated secretary and 16 are recorded for the less educated secretary, which should he hire if he uses $\alpha_0 = .05$?

Solution

We shall follow the six-step outline.

1 The problem suggests that we test $H_0 : \mu_1 = \mu_2$ against $H_1 : \mu_1 < \mu_2$, where μ_1, μ_2 are the averages of the college-educated and high-school–educated secretaries respectively. As we have indicated above, we test instead $H_0 : p = 1/2$ against $H_1 : p < 1/2$ with the observed value of a binomial random variable.

2 The significance level we hope to achieve is $\alpha_0 = .05$. However, the actual significance α will be no greater than .05. The sample sizes $n = 10$ have been selected arbitrarily. (In the testing procedure t plays the role of sample size and will not be known until the experiment is performed.)

3 The statistic we use is Y_1, the total number of errors made by the college-educated secretary. If (a) each sample is a random sample from a distribution, (b) both of the distributions are Poisson, and (c) $\mu_1 = \mu_2$ so that H_0 is true, then Y_1 given $t =$ the observed total number of errors made by both women has a binomial distribution with $p = 1/2$.

4   The critical region depends on the results of the experiment.   Hence, we now observe and find $t = 9 + 16 = 25$.   From Appendix B1 we find for the given value of t

$$\Pr(Y_1 \leqq 7) = \sum_{y_1 = 0}^{7} b(y_1; 25, .50) = .02164$$

$$\Pr(Y_1 \leqq 8) = \sum_{y_1 = 0}^{8} b(y_1; 25, .50) = .05388$$

and the critical region is $y_1 \leqq 7$ with actual significance level given $t = 25$ being .02164.   (As we shall point out, independent of which t is obtained, the significance level is no greater than .05).

5   No computations are necessary.   We note that we have observed $y_1 = 9$.

6   Since $y_1 = 9$ does not fall in the critical region, H_0 is accepted and the secretary with the high school background is hired.

The testing procedure which we have just described and illustrated is called a conditional test.   The name arises from the fact that

$$\Pr(Y_1 = y_1 | T = t) = b(y_1; t, p)$$

is a conditional probability.   One important difference between a conditional test and those tests considered previously is that in the former the critical region depends upon the outcome of the experiment.   This being the case, different experimental results obviously yield different critical regions.   It can be quickly verified that if the executive of Example 6-23 had observed $t = 50$, then his critical region would have been $y_1 \leqq 18$. To compute the actual significance level of the test, we would have to find the overall probability of rejection.   That is, we would need (computed under the assumption that H_0 is true)

Power $= \Pr(T = 0$ and H_0 is rejected$) + \Pr(T = 1$ and H_0 is rejected$)$

$$+ \Pr(T = 2 \text{ and } H_0 \text{ is rejected}) + \cdots \qquad (6\text{-}51)$$

a sum of probabilities associated with a series of mutually exclusive events. Using the multiplication theorem for two events [formula (1-13)] on each term of (6-51) yields

Power $= \Pr(H_0$ is rejected$| T = 0) \Pr(T = 0)$

$$+ \Pr(H_0 \text{ is rejected} | T = 1) \Pr(T = 1)$$

$$+ \Pr(H_0 \text{ is rejected} | T = 2) \Pr(T = 2) + \cdots \qquad (6\text{-}52)$$

Since according to our test procedure we make $\Pr(H_0$ is rejected when true $|T = t) \leqq \alpha_0$ for all t, we have from (6-52)

Significance level $\leqq \alpha_0 \Pr(T = 0) + \alpha_0 \Pr(T = 1) + \alpha_0 \Pr(T = 2) + \cdots$

$$= \alpha_0[\Pr(T = 0) + \Pr(T = 1) + \Pr(T = 2) + \cdots]$$

$$= \alpha_0[1] = \alpha_0$$

where the quantity in the brackets is the sum of all the probabilities in a probability distribution. In other words, the significance level is less than or equal to α_0, as desired.

To find power for any value of $\mu_1/\mu_2 = a$ we would have to evaluate (6-52) under the assumption that $p = a/(a + 1)$. (Unfortunately the result depends not only on a but also on $\mu_1 + \mu_2$.) This kind of calculation need not concern us and probably would not be done under any circumstances. We can, however, still solve our usual "sample size" problem which requires that the power be $\leqq \alpha_0$ if H_0 is true and $\geqq 1 - \beta_1$ if a specific alternative is true. To do this determine t as in the binomial case using $p = 1/2$ when H_0 is true, $p = a/(a + 1)$ when $\mu_1/\mu_2 = a \neq 1$. Then, observe pairs (x_{11}, x_{12}), (x_{21}, x_{22}), ... one at a time, terminating when $y_1 + y_2 \geqq t$. By an argument similar to the one used in the previous paragraph, it can be shown that the test satisfies the given power conditions.

EXAMPLE 6-24

Consider again the secretary problem of Example 6-23. Suppose that if $\mu_1 = \mu_2$ as hypothesized, the executive wants the test procedure to reject H_0 (so that the college-educated secretary will be hired) no oftener than $\alpha_0 = .05$ of the time. In addition, if the more educated applicant makes on the average $1/3$ as many errors as the less educated one, the executive would like to have the probability be at least .90 that he rejects the less educated woman. Make the appropriate changes in the outline.

Solution

1   The comments given in Example 6-23 still apply.
2   The significance level is to be no greater than .05. In addition if $\mu_1/\mu_2 = 1/3$ so that $p = (1/3)/[1/3 + 1] = .25$, it is desired that the power be at least .90. To calculate t we use formula (6-25) with $\alpha = .05$, $\beta = .10$, $p_0 = 1/2 = .50$, $p_1 = .25$, and get

$$t \geqq \left[\frac{-1.645\sqrt{(1/2)(1/2)} - 1.282\sqrt{(1/4)(3/4)}}{1/2 - 1/4}\right]^2$$

$$= [-1.645(2) - 1.282\sqrt{3}]^2 = [5.510]^2 = 30.4$$

Hence, we sample pairs until $y_1 + y_2 = t \geqq 31$. (With binomial tables it can be verified that we should have $y_1 + y_2 \geqq 33$.)

3 Again we use Y_1 if binomial tables are available. Without the table we use statistic (6-20), which becomes

$$Z' = \frac{Y_1 - t(1/2)}{\sqrt{t(1/2)(1/2)}} = \frac{2Y_1 - t}{\sqrt{t}}$$

The other comments of Example 6-23 still apply.

4 The critical region depends upon the results of the experiment if binomial tables are used. If we use the normal approximation mentioned in step 3, then the critical region is

$$z' = \frac{2y_1 - t}{\sqrt{t}} < z_{.05} = -1.645$$

5 In Example 6-23 10 pairs of observations produced $y_1 = 9$, $t = 25$. Suppose we use these results and an eleventh pair yields 1 error for the first secretary, 4 for the second. Now $y_1 = 10$, $t = 30$, but t is not large enough. If the twelfth pair yields 1 error for the first woman, 5 for the second, then $y_1 = 11$, $t = 36 > 31$, and the experimentation terminates. We calculate

$$z' = \frac{2(11) - 36}{\sqrt{36}} = \frac{22 - 36}{6} = -2.33$$

(With binomial tables it can be verified that with $t = 36$ we should reject if $y_1 \leqq 12$.)

6 Since $z' = -2.33$ falls in the critical region, H_0 is rejected and the college-educated secretary is hired.

In the discussion accompanying probability statement (6-6) we learned that the interval (w_1, w_2), obtainable from the graphs of Appendix B15 [or from (6-28)], is a confidence interval for p with confidence coefficient at least $1 - \alpha$. Thus, given t we can find a confidence interval for p using $\hat{p} = y_1/t$ and t to enter the graphs. Then, replacing p by $a/(a + 1)$, where $a = \mu_1/\mu_2$, the interval $w_1 \leqq p \leqq w_2$ is easily converted to $w_1/(1 - w_1) \leqq \mu_1/\mu_2 \leqq w_2/(1 - w_2)$. With an argument similar to that used to show that the significance level of the test is no greater than α_0, it can be shown that

$$\left(\frac{w_1}{1 - w_1}, \frac{w_2}{1 - w_2} \right) \tag{6-53}$$

is a confidence interval for μ_1/μ_2 with confidence coefficient at least $1 - \alpha$.

EXAMPLE 6-25

Using the sample results of Example 6-23, find a confidence interval for μ_1/μ_2 with confidence coefficient at least .90.

Solution

We had $y_1 = 9$, $t = 25$ so that $\hat{p} = 9/25 = .36$. From the graph of Appendix B15 headed by confidence coefficient .90, we read with $\hat{p} = .36$, $t = 25$ ($n = 25$) that $w_1 = .20$, $w_2 = .55$ (approximately). Then, according to interval (6-53) we need $.20/(1 - .20) = .25$ and $.55/(1 - .55) = 1.22$. The desired confidence interval for μ_1/μ_2 is $(.25, 1.22)$.

EXERCISES

◈ 6-54 A wholesaler is considering the purchase of a large order of cloth from either firm 1 or firm 2. He prefers to deal with firm 2 but will buy from firm 1 if convinced that the average number of defects per yard for its product is less than the average for firm 2. From each firm's stock a random sample of 100 yards is inspected. A total of 20 defects is counted, 5 appearing in cloth produced by firm 1, 15 in cloth produced by firm 2. With $\alpha_0 = .05$, what is the appropriate decision? Follow the six-step outline.

◈ 6-55 If in addition to having $\alpha_0 = .05$ in Exercise 6-54, it is desirable to have the probability be at least .90 of rejecting the product of firm 2 in favor of firm 1 when $\mu_1 = \mu_2/4$, how large does t have to be? Describe how the sampling would be conducted.

◈ 6-56 Using the sample results given in Exercise 6-54, find a confidence interval for μ_1/μ_2 with confidence coefficient at least .95.

6-57 A book company can engage one of two printers to do technical books. To help the company make a decision each printer submits 50 pages of galley proofs (covering the same material). If the first printer's work contains 43 errors while the second contains 57 errors, can we conclude that there is a difference in means using $\alpha_0 = .05$? Follow the six-step outline.

6-58 Using the sample results of Exercise 6-57, find a confidence interval for μ_1/μ_2 with confidence coefficient at least .95. What is the relationship between the confidence interval and the test of Exercise 6-57?

◈ 6-59 A city government is interested in reducing the average number of traffic deaths per time period. Each month is divided at random into two parts. During one part there is heavy patrolling by the police department, while during the other part there is light patrolling. The heavy patrolling is expensive and will not be adopted permanently unless it can be proved that the average number of deaths per half-month period is reduced. After one year 15 deaths were recorded during heavy patrolling, 35 deaths during light patrolling. Using $\alpha_0 = .01$ what decision should be made? Follow the six-step outline.

◆ 6-60 Using the sample results given in Exercise 6-59, find a confidence interval for the ratio of the heavy patrolling average per half month to the light patrolling average with confidence coefficient at least .95.

◆ 6-61 If in addition to having $\alpha_0 = .01$ in Exercise 6-59, it is desired to have the probability be at least .80 of adopting heavy patrolling when the average for heavy patrolling is two-thirds that for light patrolling, how large does t have to be? Describe how sampling would be conducted.

6-62 An appliance repair shop is interested in knowing if the average number of requests for service that it receives via the telephone is increased by advertising in the local newspaper. (The shop is listed in the yellow pages of the telephone directory.) During a week without advertising 46 requests were received, while during a week with advertising 54 requests were received. Assuming that the Poisson is the appropriate model, test the hypothesis suggested by the problem. Use $\alpha_0 = .05$ and follow the six-step outline.

6-63 Using the sample results given in Exercise 6-62, find a confidence interval for the ratio of the average number of requests per week received via telephone without advertising to the average received with advertising with confidence coefficient at least .95.

6-64 Suppose that in Exercise 6-62 it is desired to have the probability be at least .80 of concluding that the average associated with advertising is higher when actually it is twice as high as the average with no advertising. Can both requirements be met with the observed value of t?

6-7
HYPOTHESIS TESTING FOR TWO BINOMIALS

Sometimes an experimenter will want to test a hypothesis concerning the p's of two binomial distributions. Instead of considering one machine to make bolts (as in Example 6-2), we may have a choice between two machines each producing an unknown fraction of defectives. Let p_1 and p_2 be the fractions associated with machines 1 and 2 respectively. If machine 1 is more expensive, we may prefer to buy machine 2 unless there is evidence to indicate that the first machine produces a lower fraction of defectives. Under such circumstances we would probably test

$$H_0: p_1 = p_2 \quad \text{against} \quad H_1: p_1 < p_2 \quad (6\text{-}54)$$

and buy the cheaper machine unless H_0 is rejected. If p_1 is the fraction of successes associated with a new cure for tuberculosis and p_2 is the fraction of successes associated with an older less expensive cure, we would probably be interested in testing

$$H_0: p_1 = p_2 \quad \text{against} \quad H_1: p_1 > p_2 \tag{6-55}$$

and in adopting the new cure only if H_0 is rejected. [Situation (6-55) can be converted to (6-54) by interchanging the roles of the subscripts.] If both cures are new and we have no reason to prefer one over the other, then we would probably test

$$H_0: p_1 = p_2 \quad \text{against} \quad H_1: p_1 \neq p_2 \tag{6-56}$$

As usual, if situation (6-54) is replaced by

$$H_0: p_1 \geq p_2 \quad \text{against} \quad H_1: p_1 < p_2 \tag{6-57}$$

or (6-55) by

$$H_0: p_1 \leq p_2 \quad \text{against} \quad H_1: p_1 > p_2 \tag{6-58}$$

the testing procedure will be unchanged.

If X has a binomial distribution with parameters n_1, p_1 and independently Y has a binomial distribution with parameters n_2, p_2, then under the condition that $p_1 = p_2$ [in other words, the H_0 of situation (6-54), (6-55), or (6-56) is true] it can be shown that the distribution of X, given $X + Y = T = t$, has a hypergeometric distribution with parameters $N = n_1 + n_2$, $n = t$, $k = n_1$. When testing situation (6-54), or (6-57), the alternative is supported by small values of x given t. If for every value of t we use the critical region $x \leq x_0$, where x_0 is the largest value of x which satisfies

$$\Pr(X \leq x \,|\, t) = P(n_1 + n_2, t, n_1, x) \leq \alpha_0 \tag{6-59}$$

then the test will have a significance level $\leq \alpha_0$ (shown by the same type of argument used for the Poisson case in Sec. 6-6). When testing situation (6-55) or (6-58), interchange the roles of the subscripts and again use the critical region $x \leq x_0$ as determined above. Finally, when testing situation (6-56) the alternative is supported by both small and large values of x given t. To obtain a test with significance level $\leq \alpha_0$, reject H_0 if $x \leq x_1$ or $x \geq x_2$ where x_1 is the largest value of x that satisfies

$$\Pr(X \leq x \,|\, t) = P(n_1 + n_2, t, n_1, x) \leq \frac{\alpha_0}{2} \tag{6-60}$$

and x_2 is the smallest value of x which satisfies

$$\Pr(X \geq x \,|\, t) = 1 - \Pr(X \leq x - 1 \,|\, t)$$

$$= 1 - P(n_1 + n_2, t, n_1, x - 1) \leq \frac{\alpha_0}{2} \tag{6-61}$$

It can be shown that the tests described in this paragraph have good statistical properties.

When H_0 is not true, the distribution of X given t is not a standard distribution, so that special tables would be required for the computation of power. In addition power depends not only on p_1/p_2 but also on p_1 and p_2. Because of these complications we will not consider power and sample size type problems.

EXAMPLE 6-26

A company can purchase either machine 1 or machine 2 to make bolts. Machine 1 is more expensive, and the company prefers to buy machine 2 unless machine 1 produces a lower fraction of defectives. A sample of 5 bolts is selected from those produced by each machine. A total of 4 defectives are found, 1 produced by machine 1, 3 produced by machine 2. Using $\alpha_0 = .10$, what decision should be made?

Solution

We shall follow the six-step outline.

1 Letting p_1, p_2 be the fraction defective produced by machines 1, 2, respectively, the problem suggests that we test

$$H_0: p_1 = p_2 \quad \text{against} \quad H_1: p_1 < p_2$$

and buy the cheaper machine if H_0 is accepted.

2 The significance level given in the problem is $\alpha_0 = .10$. The significance level we will actually obtain will be no greater than .05. The sample sizes $n_1 = n_2 = 5$ have already been chosen arbitrarily.

3 The statistic we use is X, the number of defectives produced by machine 1. If the binomial conditions (2-1) are satisfied when determining observed values of X and Y, then X given $t =$ the observed total number of defectives contained in both samples has a hypergeometric distribution with parameters $N = n_1 + n_2 = 5 + 5 = 10$, $t = 1 + 3 = 4$ (playing the role of n), $n_1 = 5$ (playing the role of k). In Example 6-2 we discussed procedures for selecting samples which make the binomial conditions seem realistic.

4 The critical region depends upon the results of the experiment (specifically it depends on t). Hence, if we had not already done so, we would now observe $t = 4$. From Appendix B3 we find

$$\Pr(X \leq 0 \,|\, t = 4) = P(10, 4, 5, 0) = P(10, 5, 4, 0) = .023810$$

$$\Pr(X \leq 1 \,|\, t = 4) = P(10, 4, 5, 1) = P(10, 5, 4, 1) = .261905$$

and the critical region is $x \leq 0$; that is, $x = 0$.

5 No computations are necessary. We have already observed $x = 1$.

6 Since $x = 1$ does not fall in the critical region, H_0 is accepted and we buy the less expensive machine.

EXAMPLE 6-27

A corn producer can purchase the same kind of seed from two companies. Since both charge the same price, he has no reason to prefer one over the other. In order to make a decision, he plants 50 seeds furnished by each company. Of those furnished by company 1 a total of 2 fail to germinate, while 8 of those belonging to company 2 fail to germinate. Using $\alpha_0 = .10$, test a hypothesis suggested by the problem and recommend the appropriate action.

Solution

We shall follow the six-step outline.

1 Letting p_1, p_2 be the fraction of seeds associated with company 1 and 2, the problem suggests that we test $H_0 : p_1 = p_2$ against $H_1 : p_1 \neq p_2$.
2 The sample sizes $n_1 = n_2 = 50$ and $\alpha_0 = .10$ have already been selected.
3 The statistic we use is X, the number of seeds associated with company 1 which fail to germinate. Both samples of seeds should be selected so that the binomial conditions seem realistic. Then X given t, the total number of seeds failing to germinate, has a hypergeometric distribution with parameters $N = 50 + 50 = 100$, $t = 2 + 8 = 10$, $n_1 = 50$.
4 Again we need t to determine the critical region. Here $t = 10$. Since $N = 100$ our hypergeometric table in Appendix B3 cannot be used. Perhaps the best course of action is to obtain a copy of the Lieberman-Owen table [4] for the evaluation of probabilities. We need x_1, x_2 which satisfy inequalities (6-60) and (6-61). Since

$$\Pr(X \leq 2 \,|\, t = 10) = P(100, 10, 50, 2) = P(100, 50, 10, 2) = .045824$$
$$\Pr(X \leq 3 \,|\, t = 10) = P(100, 10, 50, 3) = P(100, 50, 10, 3) = .158920$$

We see that $x_1 = 2$. Similarly

$$\Pr(X \leq 6 \,|\, t = 10) = P(100, 10, 50, 6) = P(100, 50, 10, 6) = .841080$$
$$\Pr(X \leq 7 \,|\, t = 10) = P(100, 10, 50, 7) = P(100, 50, 10, 7) = .954176$$

so that $x_2 - 1 = 7$, $x_2 = 8$. The critical region is $x \leq 2$ and $x \geq 8$ when $t = 10$.

Without the hypergeometric tables we could calculate the required hypergeometric sums by doing a little arithmetic. Alternatively, we might try the binomial approximation to the hypergeometric given by formula (2-35). Then, to find x_1 given by inequality (6-60) we need the largest x to satisfy $P(100, 10, 50, x) \cong B(x; 10, \frac{50}{100}) \leq .05$. Since Appendix B1 gives $B(1; 10, .50) = .01074$, $B(2; 10, .50) = .05469$, we have $x_1 = 1$. Similarly, $x_2 = 9$. The critical region based on the approximation is $x \leq 1$ and $x \geq 9$. Obviously, a more extreme value of x than necessary is required for rejection.

5 We note that we have observed $x = 2$.
6 Since $x = 2$ falls in the critical region, H_0 is rejected. The recommended action is to buy seeds from company 1.

We observe that binomial sums used to approximate hypergeometric sums can in turn be approximated by using the standard normal distribution. It may be shown, however, that the statistic

$$Z' = \frac{\dfrac{X}{n_1} - \dfrac{Y}{n_2}}{\sqrt{\left(\dfrac{1}{n_1} + \dfrac{1}{n_2}\right)\left(\dfrac{t}{n_1 + n_2}\right)\left(1 - \dfrac{t}{n_1 + n_2}\right)}} \tag{6-62}$$

is a reasonable one to use when the larger of t and n_1 is not too small. For situation (6-54), or (6-57), reject when x/n_1 is small compared to y/n_2 or, in other words, when z' is small. Hence to obtain a test with significance level no larger than $\alpha_0 = \alpha$ (approximately), reject when

$$z' < z_\alpha \tag{6-63}$$

When testing situation (6-56) we should reject H_0 if x/n_1 is either small or large compared to y/n_2. A critical region no larger than $\alpha_0 = \alpha$ (approximately) is obtained by rejecting when

$$z' < z_{\alpha/2} \quad \text{and} \quad z' > z_{1-\alpha/2} \tag{6-64}$$

EXAMPLE 6-28

Two instructors have been teaching beginning statistics over the past several years. During this period one failed 28 out of 400 while the second failed 72 out of 600. Does this indicate that the two instructors fail a different fraction of students? Use the statistic (6-62) and an approximate significance level of .05.

Solution

We shall follow the six-step outline.

1 If we let p_1 be the fraction of students failed by the first instructor and p_2 the fraction failed by the second, the problem suggests that we test $H_0: p_1 = p_2$ against $H_1: p_1 \neq p_2$.
2 The sample sizes $n_1 = 400$, $n_2 = 600$ and the significance level $\alpha_0 = .05$ have already been chosen.
3 The statistic we will use is Z' given by (6-62). Let X and Y be the number failed by the first and second instructors, respectively. If X and Y have independent binomial distributions, then Z' has approximately a standard normal distribution. Instructors probably grade independently of one another. The binomial conditions seem reasonable.

4   The critical region is $z' < z_{.025} = -1.960$ and $z' > z_{.975} = 1.960$ as given by (6-64).

5   We have $x = 28, n_1 = 400, y = 72, n_2 = 600, t = 28 + 72 = 100$.   Then

$$z' = \frac{\dfrac{28}{400} - \dfrac{72}{600}}{\sqrt{\left(\dfrac{1}{400} + \dfrac{1}{600}\right)\dfrac{100}{1,000}\left(1 - \dfrac{100}{1,000}\right)}} = -2.58$$

6   Since $z' = -2.58$ falls in the critical region we reject H_0.   Undoubtedly, we will conclude that the first instructor fails a smaller fraction of students.

EXERCISES

⬥ 6-65   It is claimed that of those who die from natural causes a smaller fraction smoke than of those who die from heart disease.   A random sample of 4 cards drawn from the "natural causes" file yields 0 smokers, while a random sample of 6 cards drawn from the "heart disease" file yields 5 smokers.   Test the appropriate hypothesis with $\alpha_0 = .05$.   Follow the six-step outline.

⬥ 6-66   Repeat Exercise 6-65 if 80 cards drawn from the "natural causes" file yields 3 smokers while 120 cards drawn from the heart disease file yields 12 smokers.   Use the appropriate binomial approximation to the hypergeometric.

⬥ 6-67   In a certain large city it has been proposed that property taxes be increased to meet the rising expenditures of the city government and the school system.   It is anticipated that the fraction of property owners who favor the proposal will be less than the corresponding fraction of non-property owners.   A random sample of 800 belonging to each category is polled.   Of these 370 property owners favor the proposal while 430 non-property owners are in favor.   Does this support the anticipated result?   Use $\alpha_0 = .01$ and follow the six-step outline.   [*Hint:* Use (6-62).]

6-68   A product called Ambitol is advertised with the claim that it makes a person more ambitious if he drinks 1 ounce three times a day.   An investigating scientist doubts the claim, believing it may even be possible that Ambitol lessens ambition.   To test the product 50 subjects are given Ambitol for a week and 50 subjects are given plain water flavored to taste like Ambitol.   After a week 15 of the first group and 10 of the second group felt that their ambition had been increased.   Test the appropriate hypothesis with $\alpha_0 = .10$ using a hypergeometric table if one is available.   Follow the six-step outline.

6-69   A violent storm in East Pakistan was responsible for a typhoid
epidemic.   In a community of 12,000 people it was found that 20 of
the 2,000 who had typhoid shots were attacked by the sickness while
280 out of the 10,000 with no shots became ill.   Do these results
indicate that during an epidemic people with shots have a smaller
chance of getting typhoid than those with no shots?   Use $\alpha_0 = .01$
and follow the six-step outline.

6-70   A fruit wholesaler can buy his grapefruit from either a dealer in
Texas or from one in Arizona.   Sometimes he buys from one,
sometimes from the other, depending upon the quality of year's
crop.   This year the Arizona crop has a superior appearance and is
the one he prefers to buy unless there is good evidence that the
Texas crop has a smaller fraction of bad ones.   To make a decision
he visits both states and in each inspects a random sample of 100
grapefruit.   The Texas sample yields 7 bad grapefruit while the
Arizona sample yields 13.   Use $\alpha_0 = .05$ and recommend the appropriate course of action.   (*Hint:* Use the binomial approximation
to the hypergeometric.)

6-8
TESTING HYPOTHESES CONCERNING
$p_1, p_2, \ldots, p_k$ OF THE MULTINOMIAL
DISTRIBUTION

In Sec. 2-4 we derived formula (2-18), which enables us to compute
probabilities associated with a series of experiments satisfying the multinomial conditions (2-16).   As we have already discovered, if the parameters $p_1, p_2, \ldots, p_k$, n are known, it is easy to write down, though tedious
to evaluate, a probability associated with a specific outcome $x_1, x_2, \ldots, x_k$.
We now turn our attention to some hypothesis testing problems in which
the multinomial model is appropriate but the p's are unknown.

Only the simplest type of situation will be discussed in this section.
Here we will consider only the case in which the p's are completely specified by the problem.   That is, once the problem information is available,
the hypothesized values of the p's will be known numbers.   For example,
suppose that we would like to know whether or not a six-sided die is
unbiased (or symmetric).   A reasonable hypothesis to test is

$$H_0: p_1 = 1/6, p_2 = 1/6, p_3 = 1/6, p_4 = 1/6, p_5 = 1/6, p_6 = 1/6$$

against   (6-65)

$$H_1: \text{not all } p\text{'s are } 1/6$$

The choice presented by situation (6-65) is a special case of

$$H_0: p_1 = p_{10}, p_2 = p_{20}, \ldots, p_k = p_{k0}$$

against (6-66)

$H_1:$ not all p's are as specified by H_0

where $p_{10}, p_{20}, \ldots, p_{k0}$ are known probabilities suggested by the problem. As we know from our previous study of the multinomial model, we must have $\sum_{i=1}^{k} p_i = \sum_{i=1}^{k} p_{i0} = 1$ since the result of each experiment in the series must be the occurrence of one of the k possible categories.

It is possible to get an exact test for situation (6-66). To do so, however, is very tedious and time-consuming even for small n. A critical region of "unusual" values of $x_1, x_2, \ldots, x_k$, such that a known probability no greater than α is associated with the region, would have to be determined. Both the enumeration of points belonging to the critical region and the computation of probabilities get rather involved. Fortunately, an approximate test that is applicable in most real-life situations is available.

If $X_1, X_2, \ldots, X_k$ are random variables associated with a series of experiments satisfying the multinomial conditions with $p_1 = p_{10}, p_2 = p_{20}, \ldots, p_k = p_{k0}$ [that is, the H_0 of situation (6-66) is true], then it is known that

$$Y'_{k-1} = \sum_{i=1}^{k} \frac{(X_i - np_{i0})^2}{np_{i0}} \qquad (6\text{-}67)$$

has approximately a chi-square distribution with $k - 1$ degrees of freedom. Recall from the section on the multinomial that $E(X_i) = np_i \, (= np_{i0}$ here). If the observed values $x_1, x_2, \ldots, x_k$ are close to the expected values $np_{10}, np_{20}, \ldots, np_{k0}$, then the observed value of statistic (6-67), say y'_{k-1}, will tend to be small. Such a result might be expected if the H_0 of (6-66) is true. On the other hand if one or more of x_i differ considerably from np_{i0}, then y'_{k-1} will be large. Intuitively, one would feel that large deviations between observed and expected values lend support to the alternative of situation (6-66). In other words, if observed values differ considerably from expected values, the reason may be that the expected values (and hence the p_{i0}'s) are incorrect. Hence it is reasonable to reject if y'_{k-1} is large. A test with significance level approximately equal to α is obtained by using the critical region

$$y'_{k-1} = \sum_{i=1}^{k} \frac{(x_i - np_{i0})^2}{np_{i0}} > \chi^2_{k-1; 1-\alpha} \qquad (6\text{-}68)$$

Because Y'_{k-1} has approximately a chi-square distribution we have

$$\Pr(Y'_{k-1} > \chi^2_{k-1; 1-\alpha}) \cong \Pr(Y_{k-1} > \chi^2_{k-1; 1-\alpha}) = \alpha \qquad (6\text{-}69)$$

where Y_{k-1} has a chi-square distribution with $k - 1$ degrees of freedom and $\chi^2_{k-1; 1-\alpha}$ is defined by formula (2-54). (See also Fig. 6-2.)

FIGURE 6-2

Density of Y_{k-1} (chi-square), showing significance level α and $\chi^2_{k-1;1-\alpha}$.

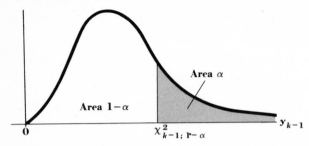

EXAMPLE 6-29

Suppose that we have purchased a new six-sided die. It is of interest to know whether or not the die is unbiased (or symmetric). The die is rolled 120 times and the following results are obtained:

Spots showing	1	2	3	4	5	6
Observed number	11	21	29	30	19	10

What conclusion should be drawn if we take $\alpha = .05$?

Solution

We shall follow the six-step outline.

1 The problem suggests that we test

$$H_0: p_1 = 1/6, p_2 = 1/6, p_3 = 1/6, p_4 = 1/6, p_5 = 1/6, p_6 = 1/6$$

against

H_1: not all p's are $1/6$

If H_0 is rejected, we will conclude that the die is biased.

2 The significance level $\alpha = .05$ and the sample size $n = 120$ have already been selected.

3 The statistic we use is $Y'_{k-1} = \sum_{i=1}^{k}(X_i - np_{i0})^2/np_{i0}$ with $n = 120, k = 6$, and each $p_{i0} = 1/6$. Here $X_1, X_2, \ldots, X_6$ are the number of times $1, 2, \ldots, 6$ show, respectively, in 120 rolls. The statistic has approximately a chi-square distribution provided (1) the hypothesis is true and (2) the multinomial conditions are satisfied. The multinomial seems reasonable, since (a) the result of each roll can be classified into one of six categories, 1, 2, 3, 4, 5, 6, (b) the probabilities of falling into these categories remain constant for each roll (all $1/6$ if H_0 is true), (c) each roll is independent of all the others, and (d) $n = 120$ is a fixed number of rolls.

4 The critical region given by inequality (6-68) is

$$y'_5 > \chi^2_{5;\,.95} = 11.07.$$

5 We perform the experiment, observe

$$x_1 = 11,\ x_2 = 21,\ x_3 = 29,\ x_4 = 30,\ x_5 = 19,\ x_6 = 10 \text{ and compute}$$

$$y'_5 = \frac{(11-20)^2}{20} + \frac{(21-20)^2}{20} + \frac{(29-20)^2}{20} + \frac{(30-20)^2}{20} + \frac{(19-20)^2}{20}$$

$$+ \frac{(10-20)^2}{20}$$

$$= \frac{81}{20} + \frac{1}{20} + \frac{81}{20} + \frac{100}{20} + \frac{1}{20} + \frac{100}{20} = \frac{364}{20}$$

$$= 18.2$$

6 Since the observed value of the statistic is $y'_5 = 18.2$ which falls in the critical region, H_0 is rejected. We conclude that the die is not symmetric. Since $\chi^2_{5;\,.995} = 16.75$, we even reject at the .005 level of significance.

EXAMPLE 6-30

A coin is tossed 1,600 times, and 840 heads are recorded. Test the hypothesis that the coin is unbiased. Use both the normal approximation of Sec. 6-4 and the approximate chi-square test of this section and compare the results. Take $\alpha = .05$.

Solution

We shall follow the six-step outline.

1 Letting p be the probability of obtaining a head on a single toss, the problem suggests that we test $H_0: p = 1/2$ against $H_1: p \neq 1/2$. In multinomial notation we would let $p_1,\ p_2$ be the probabilities of obtaining heads and tails, respectively, on a single toss and test H_0: $p_1 = 1/2,\ p_2 = 1/2$ against H_1: both p's are not 1/2.

2 The significance level $\alpha = .05$ and the sample size $n = 120$ have already been selected.

3 With the normal approximation we use the statistic $Z' = (X - np_0)/\sqrt{np_0 q_0}$, where X is the number of heads, $n = 120$, $p_0 = 1/2$. With the chi-square approximation we use

$$Y'_1 = \frac{(X_1 - np_{10})^2}{np_{10}} + \frac{(X_2 - np_{20})^2}{np_{20}}$$

where X_1, X_2 are the number of heads and tails, respectively, $n = 120$, $p_{10} = p_{20} = 1/2$. The statistic Z' has approximately a standard normal distribution if (1) X has a binomial distribution and (2) the hypothesis is true. The statistic Y_1' has approximately a chi-square distribution if (1) X_1, X_2 have a multinomial distribution and (2) the hypothesis is true. Since the multinomial with $k = 2$ is the binomial, these are equivalent conditions. We have already discussed the reasonableness of the binomial conditions for coin-tossing problems.

4   The critical region for the normal approximation is $z' < z_{\alpha/2} = z_{.025} = -1.960$ and $z' > z_{1-\alpha/2} = z_{.975} = 1.960$. For the approximate chi-square test we reject when $y_1' > \chi^2_{1;1-\alpha} = \chi^2_{1;.95} = 3.84$. (We note that $1.960^2 = 3.84$.)

5   We perform the experiment and observe $x = 840$. Then

$$z' = \frac{840 - 1{,}600(1/2)}{\sqrt{1{,}600(1/2)(1/2)}} = \frac{40}{20} = 2$$

Alternatively, we observe $x_1 = 840$, $x_2 = 760$, and compute

$$y_1' = \frac{[840 - 1{,}600(1/2)]^2}{1{,}600(1/2)} + \frac{[760 - 1{,}600(1/2)]^2}{1{,}600(1/2)} = 2 + 2 = 4$$

(We note $z'^2 = y_1'$.)

6   With either procedure we reject H_0 since the observed value of the statistic falls in the critical region. The conclusion is that the coin is biased.

Example 6-30 demonstrates that the approximate chi-square test and the approximate normal test are equivalent when $k = 2$. It is easy to show that we always have $z'^2 = y_1'$. Further, it can be verified that if one statistic falls in the critical region, the other does, and vice versa. Consequently, the approximate chi-square test can be regarded as a generalization of the approximate normal test for the case $H_0: p = p_0$ against $H_1: p \neq p_0$. With two categories, either procedure can be used. With more than two categories, we use the approximate chi-square test.

In order to get reasonable accuracy for critical regions, each expected number (each np_{i0}) should be at least 5. The sample size n should be no less than 20 and preferably somewhat larger.

Power calculations for the test of situation (6-66) are not difficult provided suitable tables are available. When H_0 is not true Y_{k-1}' has approximately a distribution called the noncentral chi-square. Those interested should see volume 1 of tables sponsored by the Institute of Mathematical Statistics [3].

EXERCISES

Use the six-step outline.

◆ 6-71 The number of male births occurring in a random sample of 120 families each containing three children is

Male births	0	1	2	3
Number of families	21	37	44	18

According to theory, the probabilities of 0, 1, 2, 3 male births are respectively 1/8, 3/8, 3/8, 1/8. Does this sample contradict the theory? Use $\alpha = .05$.

6-72 Use the one-digit numbers in the first four rows of Appendix B2 to test the hypothesis that the numbers in the table are randomly distributed. By randomly distributed we mean that $X =$ a number drawn from the table has probability function $f(x) = 1/10$, $x = 0, 1, 2, \ldots, 9$.

◆ 6-73 In a classical genetic experiment involving the crossing of two types of peas, Mendel found that the seeds from the plants could be classified as follows:

Round and yellow	315
Round and green	108
Wrinkled and yellow	101
Wrinkled and green	32

According to his theory, the frequencies should be in the ratio of $9:3:3:1$, or the probabilities associated with the four classes should be respectively 9/16, 3/16, 3/16, 1/16. Does this sample contradict his theory?

6-74 Consider again the blindfolded individual of Exercise 1-59. Recall that with no discriminatory ability the probabilities of making 0, 1, 2, 4 correct identifications are respectively 9/24, 8/24, 6/24, 1/24. Suppose that in 24 performances of the experiment, the individual gets 0 correct identifications 3 times, 1 correct identification 10 times, 2 correct identifications 9 times, and 4 correct identifications 2 times. Test the hypothesis that the individual possesses no discriminatory ability. Use $\alpha = .05$.

◆ 6-75 According to long-term university records regarding beginning-level courses in various fields, 40 percent of the students receive D's and F's, 40 percent receive C's, and 20 percent receive A's and B's. After many complaints that a certain chemistry professor gives too many low grades, the academic dean examined a random sample of 100 grades given by the professor and found 60 D's and F's, 30 C's, and 10 A's and B's. Do these results contradict the assumption that the professor conforms to overall university percentages in his grading? Use $\alpha = .01$.

6-76 A professor frequently gives five-answer multiple-choice examinations. The answers are designated by A, B, C, D, E. Supposedly he assigns at random the correct answer to one of the five choices. A random selection of 100 questions taken from the professor's file of old examinations yields 14 A answers, 20 B answers, 27 C answers, 27 D answers, and 12 E answers. If we use $\alpha = .05$, what conclusion can be drawn about the ability of the professor to make random choices?

◆ 6-77 Suppose that it is hypothesized that twice as many automobile accidents resulting in deaths occur on Saturday and Sunday as on other days of the week. That is, the probability that such accidents occur on Saturday is 2/9, on Sunday is 2/9, and on each other day of the week is 1/9. From the national record file, cards for 90 accidents are selected at random. These yield the following distribution of accidents according to the days of the week:

Sun.	Mon.	Tues.	Wed.	Thurs.	Fri.	Sat.
30	6	8	11	7	10	18

Using $\alpha = .05$, do these results contradict the hypothesis?

6-78 Mortality records for a recent period give the following information for American males:

Age bracket	Under 40	40 to 50	50 to 60	60 to 70	70 to 80	over 80
Probability of dying	.10	.10	.20	.30	.20	.10

From records of heavy smokers selected at random it is found that 15 died under the age of 40, 24 between the ages of 40 and 50, 32 between the ages of 50 and 60, 23 between the ages of 60 and 70, 5 between the ages of 70 and 80, and 1 over 80 years of age. Does the mortality table for heavy smokers conform to the one for all American males?

6-9
APPROXIMATE CHI-SQUARE TEST FOR POISSON AND NORMAL MODELS

In Sec. 2-7 we gave conditions (2-39), under which the Poisson probability function can be derived. Although in given situations we may be able to argue that the conditions seem reasonable or unreasonable, we usually justified the use of the model on the basis of experience. The tests of Chap. 5 are derived by assuming that random samples are drawn from normal distributions. Although the normality assumption is not too critical in drawing inferences about the mean, nonnormality of the model can seriously affect inferences concerning variances. Experience shows that many types of random variables have normal (or near normal) distributions. Sometimes an experimenter may have reason to believe that the model he has chosen is not appropriate. On such occasions it would be helpful to have a statistical procedure to assist in passing judgment on the model. We shall now discuss approximate chi-square tests which apply to the Poisson and normal cases.

We seek a test of the hypothesis that the model is appropriate against the alternative that the model is inappropriate. In terms of symbols for the Poisson case we test

$$H_0: p_1 = p(0; \mu), \, p_2 = p(1; \mu), \, \ldots, \, p_{k-1} = p(k - 2; \mu), \, p_k = 1 - \sum_{i=1}^{k-1} p_i$$

against

$$H_1: \text{the } p\text{'s are not given by the Poisson}$$

$$(6\text{-}70)$$

where $p(x; \mu)$ is defined by formula (2-40). The kth category includes all values which can be assumed that are greater than or equal to $k - 1$. (As we shall illustrate in Example 6-31, k should be chosen so that all estimated expected numbers are at least 5. Sometimes this can be accomplished in several ways, each of which may yield a different value of k.) Now suppose that we consider a random sample of size n (supposedly a random sample from a Poisson distribution). If we let X_1 be the number of occurrences of 0 successes, X_2 be the number of occurrences of 1 success, $\ldots$, X_k be the number of occurrences of $k - 1$ or more successes, then $X_1, X_2, \ldots, X_k$ have a multinomial distribution with probabilities $p_1, p_2, \ldots, p_k$ (where $x_1 + x_2 + \cdots + x_k = n$). Now situation (6-70) sounds like (6-66), and it would appear that we should be able to use the statistic (6-67) for our test. There is, however, one major difference between our current problem and the one of the previous section. The p_i of (6-70) are not completely specified but depend upon μ, an unknown parameter. We could estimate μ by $\hat{\mu} = \bar{x}$ (as decided in Sec. 3-3) and then estimate the p_i by $\hat{p}_i$ obtained using $\bar{x}$ in place of μ. If we do this the test of the last section can be used if we replace the former degrees of freedom $k - 1$ by $k - 2$. (One degree of freedom is lost due to the fact one parameter is estimated from the observed sample.) Thus the observed value of the statistic we now use is

$$y'_{k-2} = \sum_{i=1}^{k} \frac{(x_i - n\hat{p}_i)^2}{n\hat{p}_i}$$

$$(6\text{-}71)$$

and a critical region with significance level approximately α is obtained by rejecting when

$$y'_{k-2} > \chi^2_{k-2;1-\alpha} \tag{6-72}$$

If H_0 is true, Y'_{k-2} has approximately a chi-square distribution with $k-2$ degrees of freedom.

EXAMPLE 6-31

Suppose that we have some doubts concerning the adequacy of the Poisson model for the secretary problem of Example 2-19. We examine 100 pages, selected at random, of the secretary's work and find the following results:

Errors	0	1	2	3	4	5	6
Pages	36	40	19	2	0	2	1

Do these results justify our doubts if we use $\alpha = .05$?

Solution

We shall follow the six-step outline.

1 The hypotheses and the alternative are given by (6-70).
2 The significance level $\alpha = .05$ and the sample size $n = 100$ have already been chosen.
3 The statistic we use is Y'_{k-2} with observed value given by (6-71). If the hypothesis is true and the multinomial conditions are satisfied, then the statistic has approximately a chi-square distribution with $k-2$ degrees of freedom. Checking the multinomial conditions we observe (a) the result from each page can be classified into one of k categories, namely 0 errors, 1 error, 2 errors, ..., $k-2$ errors, and $k-1$ or more errors, (b) $p_1, p_2, \ldots, p_k$ remain constant provided the number of errors on a page has the same distribution for each page, (c) the results on one page can be regarded as independent of the results on any other, and (d) a fixed number of pages, $n = 100$, is used.
4 The critical region is $y'_{k-2} > \chi^2_{k-2;.95}$. The sample results suggest that we may use $k = 7$. However, after some calculations in step 5 we decide to use $k = 4$, and at that time we list the critical region as $y'_2 > \chi^2_{2;.95} = 5.99$.
5 After observing the sample we compute

$$\hat{\mu} = \bar{x} = \frac{0(36) + 1(40) + 2(19) + 3(2) + 4(0) + 5(2) + 6(1)}{36 + 40 + 19 + 2 + 0 + 2 + 1}$$

$$= \frac{100}{100} = 1$$

From Appendix B4 with $\mu = 1$ we find $p(0;1) = .36788, p(1;1) = .73576$ $-.36788 = .36788, p(2;1) = .91970 - .73576 = .18394, p(3;1) = .98101$ $-.91970 = .06131, p(4;1) = .99634 - .98101 = .01533, p(5;1) = .99941$ $-.99634 = .00307, 1 - P(5;1) = 1 - .99941 = .00059$. The estimates of the expected numbers are obtained by multiplying the above probabilities by 100. Since the estimated expected numbers associated with 4 errors, 5 errors, and 6 or more are respectively 1.533, .307, .059 (all less than 5), we use one category for 3 or more errors, thus making all estimated expected numbers at least 5. Summarized in table form, the calculation $y'_{k-2} = y'_2$ is

Errors	x_i	$\hat{p}_i$	$n\hat{p}_i$	$(x_i - n\hat{p}_i)^2/n\hat{p}_i$
0	36	.36788	36.788	.017
1	40	.36788	36.788	.280
2	19	.18394	18.394	.020
3 or more	5	.08030	8.030	1.143
	100	1.00000	100.00	$1.460 = y'_2$

6 Since the critical region is $y'_2 > \chi^2_{2;.95} = 5.99$, we accept H_0 and conclude that the Poisson is a reasonable model for giving probabilities associated with the number of errors per page.

In the normal case we test

H_0: the density function is given by formula (2-49) (the distribution is normal)

against   (6-73)

H_1: the density function is not given by formula (2-49) (the distribution is not normal)

For situation (6-70), the Poisson case, natural choices were available for $p_1, p_2, \ldots, p_k$ in terms of the $p(x;\mu)$ even though there was some leeway in grouping categories. In the normal case there is even more leeway since a set of p's can be chosen in many ways such that $p_1 + p_2 + \cdots + p_k = 1$. All we have to do is divide the range of the normal random variable into k mutually exclusive intervals say, $(-\infty, a_1), (a_1, a_2), \ldots, (a_{k-2}, a_{k-1})$, (a_{k-1}, ∞), and let the probabilities of a normal random variable falling in these intervals be $p_1, p_2, \ldots, p_k$, respectively. Now the p's depend upon two unknown parameters, μ and σ^2, both unknown. As in the Poisson case, unknown parameters are replaced by estimates. Hence if we use $\bar{x}$ for μ, s^2 for σ^2, we can estimate the p's getting $\hat{p}_1, \hat{p}_2, \ldots, \hat{p}_k$. Now, letting X_1 be the number of occurrences in $(-\infty, a_1)$, X_2 be the number of occurrences in (a_1, a_2), ..., X_k be the number of occurrences in (a_{k-1}, ∞), the observed value of the statistic is

$$y'_{k-3} = \sum_{i=1}^{k} \frac{(x_i - n\hat{p}_i)^2}{n\hat{p}_i} \qquad (6\text{-}74)$$

and Y'_{k-3} has approximately a chi-square distribution with $k - 3$ degrees of freedom if H_0 is true. (Two degrees of freedom are lost since two parameters are estimated from the observed sample.) To obtain a critical region with significance level approximately α, we reject when

$$y'_{k-3} > \chi^2_{k-3;1-\alpha} \qquad (6\text{-}75)$$

The number of intervals, k, and the way the intervals are selected can be chosen in a variety of ways by the investigator. One reasonable choice is to make all the estimated expected numbers the same. Thus each $\hat{p}_i = 1/k$ and $n\hat{p}_i = n/k$, where k is chosen so that the estimated expected numbers are at least 5 and preferably at least 10. The interval boundaries $a_1, a_2, \ldots, a_{k-1}$ can then be determined, after which $x_1, x_2, \ldots, x_k$ (the observed numbers in each interval or category) can be counted. Let us illustrate by an example.

EXAMPLE 6-32

The following 40 numbers represent an observed random sample of corn yields per acre for a given variety: 32, 40, 93, 53, 93, 15, 92, 44, 33, 83, 61, 51, 46, 41, 100, 26, 46, 28, 88, 36, 66, 22, 82, 39, 81, 69, 78, 13, 74, 48, 61, 96, 108, 46, 74, 90, 53, 59, 79, 85. Do these results contradict the assumption that corn yields are normally distributed if we use $\alpha = .05$?

Solution

We shall follow the six-step outline.

1 The hypothesis and the alternative are given by (6-73).
2 The significance level $\alpha = .05$ and the sample size $n = 40$ have already been chosen.
3 The statistic we use is Y'_{k-3} with observed value given by formula (6-74). If the hypothesis is true and the multinomial conditions are satisfied, then the statistic has approximately a chi-square distribution with $k - 3$ degrees of freedom. After dividing yields into k intervals, we observe (a) each yield falls into one of k intervals, (b) the probabilities of obtaining yields in the various intervals remain constant for each yield if each yield comes from the same distribution, (c) yields are independent if sampling is random, and (d) a fixed number of yields $n = 40$ is used.
4 The critical region is $y'_{k-3} > \chi^2_{k-3;.95}$ where k is yet to be chosen. In step 5 we use $k = 4$ so that we reject when $y'_1 > \chi^2_{1;.95} = 3.84$.

5 After observing the sample we compute $\bar{x} = 60.6$, $s^2 = 658.0$, $s = 25.7$. Suppose we decide to make each estimated expected number equal to 10. Since $n = 40$, $n\hat{p}_i = 10$, $\hat{p}_i = .25$. From the standard normal table we find that the four intervals on z meeting this requirement are $(-\infty, -.675)$, $(-.675, 0)$, $(0, .675)$, $(.675, \infty)$. Now $z = (x - \mu)/\sigma$ or $x = \sigma z + \mu$ has to be used to convert intervals on z to intervals on x. In place of σ, μ we use their estimates 25.7, 60.6. Then the four intervals on x become $(-\infty, 43.3)$, $(43.3, 60.6)$, $(60.6, 77.9)$, $(77.9, \infty)$. The number of yields falling in these intervals are 11, 9, 6, 14, respectively. The observed value of the statistic is

$$y_1' = \frac{(11 - 10)^2}{10} + \frac{(9 - 10)^2}{10} + \frac{(6 - 10)^2}{10} + \frac{(14 - 10)^2}{10} = 3.4$$

6 Since the observed value of the statistic does not fall in the critical region, H_0 is accepted. We conclude that the normal distribution is appropriate for use with corn yields.

When conducting the test illustrated in Example 6-32, it is desirable to have a sample size larger than 40. In a review-type paper published in 1952, Cochran [1] has recommended that the estimated expected numbers be about 12 with $n = 200$, 20 with $n = 400$, and 30 with $n = 1,000$.

EXERCISES

Follow the six-step outline.

◈ 6-79 An observed random sample of hills of corn inspected for corn borers produced the following results:

Number of corn borers	0	1	2	3	4	5	6
Number of hills	23	22	16	18	15	4	2

With $\alpha = .05$, use the approximate chi-square test to investigate the adequacy of the Poisson model.

6-80 Rework Example 6-32, making each estimated expected number equal to 5.

6-10
THE CHI-SQUARE TESTS FOR INDEPENDENCE AND HOMOGENEITY

Sometimes the results of multinomial-type experiments can be classified into categories according to two characteristics. Observed results for such experiments are usually presented in a two-way classification table called a contingency table. Approximate chi-square tests can be used to test a variety of hypotheses concerning the behavior of the characteristics. In this section we shall discuss two such tests.

Sometimes it is of interest to know whether or not two characteristics are independent. To illustrate let us consider an example. The tobacco companies claim that there is no relationship between smoking and lung ailments. To investigate the claim a random sample of 200 males in the 40 to 50 age bracket are given medical tests. The observed sample results appear in Table 6-1.

TABLE 6-1

Smoking versus lung ailments

	Lung ailments	No lung ailments	Totals
Smokers	40	80	120
Nonsmokers	10	70	80
Totals	50	150	200

The tobacco syndicate would like to test

H_0 : Smoking and lung ailments are independent

against   (6-76)

H_1 : There is some dependence between smoking and lung ailments

In order to get a more exact formulation of the hypothesis and alternative (6-76), for males in the 40 to 50 age bracket let

p_{11} = probability that an individual is a smoker and has a lung ailment
p_{12} = probability that an individual is a smoker and does not have a lung ailment
p_{21} = probability that an individual is a nonsmoker and has a lung ailment
p_{22} = probability that an individual is a nonsmoker and does not have a lung ailment

(The double subscript notation is used for convenience. If we insist on using one subscript, we could identify $p_1 = p_{11}$, $p_2 = p_{21}$, $p_3 = p_{12}$, $p_4 = p_{22}$. Either way there are four categories and four p's.)

$p_1. = p_{11} + p_{12}$ = probability that an individual is a smoker
$p_2. = p_{21} + p_{22}$ = probability that an individual is a nonsmoker
$p._1 = p_{11} + p_{21}$ = probability that an individual has a lung ailment
$p._2 = p_{12} + p_{22}$ = probability that an individual does not have a lung ailment

In our study of probability we learned that the probability of two independent events occurring is equal to the product of the individual probabilities [in symbols $\Pr(A_1 A_2) = \Pr(A_1)\Pr(A_2)$]. Hence, if smoking habits and lung ailments are independent characteristics, then $p_{11} = p_1 \cdot p_{\cdot 1}$, $p_{12} = p_1 \cdot p_{\cdot 2}$, $p_{21} = p_2 \cdot p_{\cdot 1}$, $p_{22} = p_2 \cdot p_{\cdot 2}$. Thus, situation (6-76) can be replaced by

$$H_0: \quad p_{11} = p_1 \cdot p_{\cdot 1} \qquad p_{12} = p_1 \cdot p_{\cdot 2} \qquad p_{21} = p_2 \cdot p_{\cdot 1} \qquad p_{22} = p_2 \cdot p_{\cdot 2}$$

against (6-77)

H_1: the equations given under H_0 are not all true

In order to test H_0, it would seem reasonable to attempt to use an approximate chi-square statistic similar to (6-67) used in the case when all parameters are known. In the problem of Sec. 6-8 the expected numbers np_{i0} were known numbers, while now the expected numbers np_{ij} ($i = 1$, 2, $j = 1$, 2) are unknown. Under the hypothesis $np_{ij} = np_i \cdot p_{\cdot j}$, but the p_i and $p_{\cdot j}$ are also unknown. However, reasonable estimates of p_i, $p_{\cdot j}$ are

$$\hat{p}_{i \cdot} = \frac{\text{total observed in row } i}{n} = \frac{t_{i \cdot}}{n}$$

$$\hat{p}_{\cdot j} = \frac{\text{total observed in column } j}{n} = \frac{t_{\cdot j}}{n}$$

Specifically we get $\hat{p}_{1 \cdot} = 120/200$, $\hat{p}_{2 \cdot} = 80/200$, $\hat{p}_{\cdot 1} = 50/200$, $\hat{p}_{\cdot 2} = 150/200$. It would seem logical to replace $np_{ij} = np_i \cdot p_{\cdot j}$ by estimates $n\hat{p}_{i \cdot} \hat{p}_{\cdot j}$ and calculate the approximate chi-square statistic as before. By mathematical arguments it can be shown that this is a legitimate procedure provided 1 degree of freedom is subtracted for each parameter estimated. (In the smoker problem it appears that we need to estimate four parameters $p_{1 \cdot}$, $p_{2 \cdot}$, $p_{\cdot 1}$, $p_{\cdot 2}$. Since $p_{1 \cdot} + p_{2 \cdot} = 1$, an estimate of $p_{1 \cdot}$ gives an estimate of $p_{2 \cdot}$. Similarly, since $p_{\cdot 1} + p_{\cdot 2} = 1$, an estimate of $p_{\cdot 1}$ produces an estimate of $p_{\cdot 2}$. Hence, only two parameters are estimated, and the degrees of freedom is $4 - 1 - 2 = 1$.) If we let X_{ij} be the number of occurrences in row i and column j, then the observed value of the statistic to use for the smoker problem is

$$y_1' = \sum_{i=1}^{2} \sum_{j=1}^{2} \frac{(x_{ij} - n\hat{p}_{i \cdot} \hat{p}_{\cdot j})^2}{n\hat{p}_{i \cdot} \hat{p}_{\cdot j}} \tag{6-78}$$

The associated random variable Y_1' has an approximate chi-square distribution with 1 degree of freedom. To obtain a critical region with significance level approximately α we reject when

$$y_1' > \chi_{1;1-\alpha}^2 \tag{6-79}$$

EXAMPLE 6-33

Write out the six-step outline for testing independence in the smoker problem. Use $\alpha = .05$.

Solution

1　The hypothesis and the alternative are given by (6-77).
2　The significance level $\alpha = .05$ and the sample size $n = 200$ have already been chosen.
3　The statistic we use has observed value given by (6-78). If H_0 is true and the multinomial conditions are satisfied, then this statistic has approximately a chi-square distribution with 1 degree of freedom. Let us check the multinomial conditions. We have: (a) There are four categories. (b) For males in the 40 to 50 age bracket the probabilities $p_{11}, p_{21}, p_{12}, p_{22}$ remain essentially constant for each selection since there is a very large number of subjects available. (c) The independence condition is practically satisfied when drawing is made from a large number of subjects. (With a small number of subjects, draw with replacement when it is possible to do so. Here it might result in the same individual being counted twice.) (d) The experiment is repeated a fixed number of times.
4　The critical region is $y_1' > \chi^2_{1;.95} = 3.84$.
5　After performing the experiment and making Table 6-1 we first compute the estimated expected numbers. These are

$$n\hat{p}_1.\hat{p}._1 = 200\,\frac{120}{200}\frac{50}{200} = \frac{(120)(50)}{200} = 30$$

$$n\hat{p}_1.\hat{p}._2 = 200\,\frac{120}{200}\frac{150}{200} = \frac{(120)(150)}{200} = 90$$

$$n\hat{p}_2.\hat{p}._1 = 200\,\frac{80}{200}\frac{50}{200} = \frac{(80)(50)}{200} = 20$$

$$n\hat{p}_2.\hat{p}._2 = 200\,\frac{80}{200}\frac{150}{200} = \frac{(80)(150)}{200} = 60$$

Since $x_{11} = 40$, $x_{12} = 80$, $x_{21} = 10$, $x_{22} = 70$, the observed value of the statistic is

$$y_1' = \frac{(40-30)^2}{30} + \frac{(80-90)^2}{90} + \frac{(10-20)^2}{20} + \frac{(70-60)^2}{60}$$

$$= 3.33 + 1.11 + 5.00 + 1.67 = 11.1$$

6　Since the observed value of the statistic falls in the critical region, we reject the hypothesis of independence. In fact, since $\chi^2_{1;.995} = 7.88$, H_0 would be rejected with $\alpha = .005$. Examination of our results given in Table 6-1 leads to the conclusion that H_0 is rejected because lung ailments occur more frequently with smokers than nonsmokers for males in the 40 to 50 age group.

The smoker example was made as simple as possible in order to facilitate the discussion. In a real problem we might include other age brackets. Possibly we would include women smokers and nonsmokers as additional categories. Maybe lung ailments could be classified as serious and not serious. The cigarette companies would undoubtedly prefer to have only a 20 to 25 age bracket since such individuals are unlikely to have developed ailments as yet.

Next we shall generalize the preceding discussion. Suppose that characteristic A is divided into r categories and characteristic B into c categories. Let the multinomial-type experiment be repeated n times with the observed results appearing in Table 6-2. The random variable

TABLE 6-2

Contingency table

Characteristic A	1	2	Characteristic B j		c	Totals
1	x_{11}	x_{12}	$\cdots$	x_{1j}	$\cdots$ x_{1c}	$t_{1.}$
2	x_{21}	x_{22}	$\cdots$	x_{2j}	$\cdots$ x_{2c}	$t_{2.}$
i	x_{i1}	x_{i2}	$\cdots$	x_{ij}	$\cdots$ x_{ic}	$t_{i.}$
r	x_{r1}	x_{r2}	$\cdots$	x_{rj}	$\cdots$ x_{rc}	$t_{r.}$
Totals	$t_{.1}$	$t_{.2}$		$t_{.j}$	$t_{.c}$	$t_{..} = n$

X_{ij} is the number of occurrences in category i of characteristic A and category j of characteristic B. Let p_{ij} be the probability that an object (or subject) falls in category i of characteristic A and category j of characteristic B, $p_{i.}$ be the probability that an object falls in category i of characteristic A, and $p_{.j}$ be the probability that an object falls in category j of characteristic B. Then the generalization of (6-77) is

$$H_0 : p_{ij} = p_{i.} p_{.j} \qquad i = 1, 2, \ldots, r \qquad j = 1, 2, \ldots, c$$

against $\hspace{10cm}$ (6-80)

H_1: the equations given under H_0 are not all true

The observed value of the statistic we use to test H_0 is

$$y'_{(r-1)(c-1)} = \sum_{i=1}^{r} \sum_{j=1}^{c} \frac{(x_{ij} - n\hat{p}_{i\cdot}\hat{p}_{\cdot j})^2}{n\hat{p}_{i\cdot}\hat{p}_{\cdot j}} \tag{6-81}$$

[Since there are $r - 1$ of the $p_{i\cdot}$'s and $c - 1$ of the $p_{\cdot j}$'s to estimate, a total of $r - 1 + c - 1$ parameters are estimated under H_0. The total number of categories is $k = rc$. Hence, associated with the statistic is $rc - 1 - (r - 1) - (c - 1) = (r - 1)(c - 1)$ degrees of freedom.] Replacing the estimates by their equivalents in terms of totals and n, that is, $\hat{p}_{i\cdot} = t_{i\cdot}/n$, $\hat{p}_{\cdot j} = t_{\cdot j}/n$, (6-81) reduces to the more convenient form

$$y'_{(r-1)(c-1)} = \sum_{i=1}^{r} \sum_{j=1}^{c} \frac{(x_{ij} - t_{i\cdot}t_{\cdot j}/n)^2}{t_{i\cdot}t_{\cdot j}/n} \tag{6-82}$$

A little further algebra reduces (6-82) to

$$y'_{(r-1)(c-1)} = n\left(\sum_{i=1}^{r} \sum_{j=1}^{c} \frac{x_{ij}^2}{t_{i\cdot}t_{\cdot j}} - 1 \right) \tag{6-83}$$

a computational form which is sometimes used. The statistic $Y'_{(r-1)(c-1)}$ has approximately a chi-square distribution with $(r - 1)(c - 1)$ degrees of freedom (provided, of course, the X_{ij} have a multinomial distribution and H_0 is true). Consequently, to obtain a critical region with significance level approximately equal to α, we reject when

$$y'_{(r-1)(c-1)} > \chi^2_{(r-1)(c-1);1-\alpha} \tag{6-84}$$

A second hypothesis of interest which sometimes arises in the study of contingency tables is known as the hypothesis of homogeneity. To illustrate the concept, let us consider an example. Suppose that it is of some interest to know whether or not university students of various levels of classification feel the same in regard to the amount of work required by their professors. In order to make a decision, three random samples of university students are selected and interviewed. The first sample contains 300 lowerclassmen, the second sample contains 200 upperclassmen, and the third sample contains 100 graduate students. Each student is then asked to select one of the following three categories according to which best characterizes his feeling: (1) Professors expect too much work from their students. (2) The amount of work professors require is about what it should be. (3) Professors do not demand enough work from their students. The results of the questioning are presented in Table 6-3. In line with our objectives it is reasonable to choose between

H_0 : the proportion of students falling into each work category is the same irrespective of classification

against $\hspace{8cm}$ (6-85)

H_1 : the proportions in the work categories are not the same for each classification

If we let p_{ij} be the probability that those in row i fall into category j, then we can rewrite the hypothesis and alternative of (6-85) as

$$H_0 : p_{11} = p_{21} = p_{31} \qquad p_{12} = p_{22} = p_{32} \qquad p_{13} = p_{23} = p_{33}$$

against (6-86)

H_1: the equations given under H_0 are not all satisfied

We observe that the organization of the experiment indicates that we are concerned with 3 multinomial distributions (each with 3 categories and 3 p's). The hypothesis states that all three of these multinomials have exactly the same set of p's. Instead of one sample as we used with the independence problem, we now have 3 samples, one for each of the three

TABLE 6-3

Professors' workload versus student classification

Classification	Too much work	Right amount of work	Too little work	Totals
Lowerclassmen	182	33	85	300
Upperclassmen	68	72	60	200
Graduate	32	15	53	100
Totals	282	120	198	600

distributions. (Alternatively, we could take *one* sample of 600 students and conduct a test of independence. If this were done, any student could fall into any one of the 9 categories, and the row totals would not be fixed numbers. One advantage of fixing the row totals is that we can take sample sizes approximately proportional to the number of subjects falling in each row classification. As we shall see, the test statistic is the same for both the independence and the homogeneity problems. Consequently, if we incorrectly regarded the workload-classification problem as an independence problem, we would probably still make the right practical decision.)

To test the hypothesis of (6-86), we again compute the usual-type sum for an approximate chi-square test. If H_0 is true, then a reasonable estimate of $p_{11} = p_{21} = p_{31} = p_1$ is

$$\hat{p}_1 = \frac{\text{total of column 1}}{n} = \frac{t_{.1}}{n} = \frac{282}{600}$$

Similarly, an estimate of $p_{12} = p_{22} = p_{32} = p_2$ is

$$\hat{p}_2 = \frac{\text{total of column 2}}{n} = \frac{t_{.2}}{n} = \frac{120}{600}$$

and an estimate of $p_{13} = p_{23} = p_{33} = p_3$ is

$$\hat{p}_3 = \frac{\text{total of column 3}}{n} = \frac{t_{.3}}{n} = \frac{198}{600}$$

Hence of the 300 lowerclassmen we "expect" $(300)(282)/600 = 141$, $(300)(120)/600 = 60$, $(300)(198)/600 = 99$, respectively, in the three categories. Similar calculations (as demonstrated in Example 6-34) yield the estimated expected numbers associated with the other cells of the table. We note that under H_0 the estimated expected numbers are of the form $t_i.\hat{p}_{ij} = t_i.(t_{.j}/n) = t_i.t_{.j}/n$, the same as obtained under the hypothesis of independence. Hence, the observed value of the statistic must again reduce to (6-82) or (6-83). (To justify retaining the same number of degrees of freedom, we argue as follows. For an r by c table like Table 6-2 there are $c - 1$ degrees of freedom associated with each of the r rows, a total of $r(c - 1)$. If the hypothesis is true, then $c - 1$ of the parameters $p_1, p_2, \ldots, p_c$ need to be estimated. If $c - 1$ is subtracted from $r(c - 1)$ we again get $(r - 1)(c - 1)$.) Once more a critical region with significance level approximately equal to α is given by inequality (6-84).

The generalization of (6-86) is fairly obvious. With r rows and c columns it becomes

$$H_0 : p_{1j} = p_{2j} = \cdots = p_{rj} \qquad j = 1, 2, \ldots, c$$

against   (6-87)

H_1: the equations given under H_0 are not all satisfied

As we have already commented, $Y'_{(r-1)(c-1)}$ with observed value given by formula (6-82) is the statistic we use, and the critical region is given by inequality (6-84). The row totals are *fixed* chosen numbers. In summary, we compute the same observed statistic for both the hypothesis of independence and the hypothesis of homogeneity and use the same critical region. In the homogeneity case the X_{ij} in *each row* must have a multinomial distribution (and the hypothesis must be true) in order that $Y'_{(r-1)(c-1)}$ be approximately distributed as chi-square.

EXAMPLE 6-34

Write out the six-step outline for testing homogeneity in the workload-classification problem. Use $\alpha = .05$.

Solution

1 The hypotheses and alternative are given by (6-85).
2 The significance level $\alpha = .05$ and the three sample sizes 300, 200, 100 have already been chosen.
3 The statistic we use has observed value given by (6-82) with $r = c = 3$. If H_0 is true and each of the three samples are selected in accordance with the multinomial conditions, then the statistic has approximately a chi-square distribution with $(r - 1)(c - 1) = (2)(2) = 4$ degrees of freedom. Let us check the multinomial conditions. We have: (*a*) Under each classification there are three categories. (*b*) In each classification it is reasonable to regard the probabilities of falling into the respective categories as constants if a large number of students are available or if sampling is performed with replacement. (*c*) If the number of students in each classification is large, draws can be regarded as independent, since probabilities will change very little from draw to draw. If samples are drawn with replacement, the independence condition will be satisfied. Of course, in either case an effort should be made to obtain students at random. (*d*) Each of the three experiments is repeated a fixed number of times (300, 200, 100).
4 The critical region is $y_4' > \chi^2_{4;.95} = 9.49$.
5 After performing the three experiments and obtaining Table 6-3, we compute the observed value of the statistic. The estimated expected numbers are given by $t_{i.}t_{.j}/n$ for each cell. Arranging these in the same order as the observed numbers given in Table 6-3, we get

$$\frac{(300)(282)}{600} = 141 \qquad \frac{(300)(120)}{600} = 60 \qquad \frac{(300)(198)}{600} = 99$$

$$\frac{(200)(282)}{600} = 94 \qquad \frac{(200)(120)}{600} = 40 \qquad \frac{(200)(198)}{600} = 66$$

$$\frac{(100)(282)}{600} = 47 \qquad \frac{(100)(120)}{600} = 20 \qquad \frac{(100)(198)}{600} = 33$$

Hence

$$y_4' = \frac{(182 - 141)^2}{141} + \frac{(33 - 60)^2}{60} + \frac{(85 - 99)^2}{99}$$

$$+ \frac{(68 - 94)^2}{94} + \frac{(72 - 40)^2}{40} + \frac{(60 - 66)^2}{66}$$

$$+ \frac{(32 - 47)^2}{47} + \frac{(15 - 20)^2}{20} + \frac{(53 - 33)^2}{33}$$

$$= 11.92 + 12.15 + 1.98 + 7.19 + 25.60 + .55 + 4.79 + 1.25 + 12.12$$

$$= 77.55$$

6 Since the observed value of the statistic falls in the critical region, we reject the hypothesis of homogeneity. In fact, since $\chi^2_{4;.995} = 14.86$, H_0 would be rejected with $\alpha = .005$ (and much smaller α's). Examination of our results given in Table 6-3 lead to the conclusion that we reject primarily because a large fraction of lowerclassmen believe that professors expect too much work.

As with previous approximate chi-square tests, estimated expected numbers smaller than 5 should be avoided. This can frequently be accomplished by regrouping the categories.

EXERCISES

Follow the six-step outline.

◆ 6-81 A random sample of 200 students is selected from each of four undergraduate classes. Each is asked to express his preference as a Republican, Democrat, or Independent voter. The results are

Class	Republican	Preference Democrat	Independent	Totals
Freshman	51	110	39	200
Sophomore	74	106	20	200
Junior	25	124	51	200
Senior	90	60	50	200
Totals	240	400	160	800

Are the classes alike in their preference?

6-82 A random sample of 600 grades is selected from those earned in the colleges of engineering, agriculture, and Arts and Sciences. The results are

Grades	Engineering	Colleges Agriculture	A and S	Totals
A	25	15	30	70
B	41	20	79	140
C	82	70	138	290
Below C	18	30	52	100
Totals	166	135	299	600

What conclusion can be drawn?

◆ 6-83 In a certain large city 1,000 people were selected at random and questioned about their smoking and drinking habits. The results of the survey were

Drink	Never	Smoke Occasionally	Moderately	Heavily
Never	85	23	56	36
Occasionally	153	44	128	75
Moderately	128	26	101	45
Heavily	34	7	15	44

What conclusion can be drawn concerning the smoking and drinking habits of people in the city?

6-84 In order to compare a new machining process with a standard one, 500 parts are selected from those produced by each process. The results are

Process	Defective parts	Nondefective parts
New	52	448
Old	77	423

What conclusions can be drawn?

◈ 6-85 The prerequisite for elementary statistics is the completion (with a passing grade) of one college mathematics course. For 500 students who have taken elementary statistics in the past several years, the following results were tabulated:

Grade in statistics	Grade in prerequisite math course A or B	C	D
A or B	33	71	21
C	67	121	62
D or F	20	48	57

Use $\alpha = .01$ and draw the appropriate conclusions.

6-86 A new disease which attacks horses is carried by mosquitoes. Authorities are interested in knowing if a horse's ability to survive is influenced by his age. At the beginning of the mosquito season a random sample of 100 horses is selected in each of four age categories. At the end of the season the following results were tabulated:

Age in years as of January 1	Survived	Died
5 and under	77	23
6 to 10	65	35
11 to 15	74	26
16 and over	64	36

With $\alpha = .05$ what conclusion can be drawn?

◆ 6-87   A product called Ambitol is advertised with the claim that it makes a person more ambitious if he drinks 1 ounce three times a day. To test the product 100 subjects are given Ambitol for a week and 100 subjects are given water flavored to taste like Ambitol. After a week the subjects are quizzed, and the following results are obtained:

	Subject felt		
Subjects	More ambitious	About the same	Less ambitious
Ambitol	50	35	15
Water	30	45	25

Use $\alpha = .01$ to draw the appropriate conclusions.

6-88   A psychologist has constructed an examination which supposedly measures aptitude for high school teaching. He is interested in learning whether or not the attitude of the individual administering the examination influences the results. Three random samples of size 100 are selected from those in the process of seeking their first position, and each group is given the examination. The first group is handled by an individual who has an arrogant attitude and treats the subjects with obvious contempt. The second group is examined by an individual who conducts the proceedings in an efficient manner much like a robot. The third group is led by a very friendly and cooperative individual. The results are classified as follows:

	Results on examination		
Attitude of examiner	Good	Intermediate	Poor
Contemptuous	18	54	28
Indifferent	23	53	24
Friendly	28	61	11

If the psychologist uses $\alpha = .05$, what will he conclude?

REFERENCES

1   Cochran, William G.: The χ^2 Test of Goodness of Fit, *Annals of Mathematical Statistics*, vol. 23, no. 3, pp. 315–345, 1952.

2   Harter, H. Leon: "New Tables of the Incomplete Gamma-function Ratio and of Percentage Points of the Chi-square and Beta Distributions," Superintendent of Documents, U.S. Government Printing Office, Washington, D.C., 1964.

3 Harter, H. L., and D. B. Owen (eds.): "Selected Tables in Mathematical Statistics," Markham Publishing Company, Chicago, 1970.

4 Lieberman, G. J., and D. B. Owen: "Tables of the Hypergeometric Probability Distribution," Stanford University Press, Stanford, Calif., 1961.

5 Molina, E. C.: "Poisson's Exponential Binomial Limit," D. Van Nostrand Company, Inc., Princeton, N.J., 1949.

6 U.S. Army Materiel Command: "Tables of the Cumulative Binomial Probabilities," AMC Pamphlet AMCP 706-109, January 1971.

7 U.S. Army Ordnance Corps: "Tables of the Cumulative Binomial Probabilities," Ordnance Corps Pamphlet ORDP20-1, September 1952.

Summary of Results
DRAWING INFERENCES ABOUT PARAMETERS OF DISCRETE DISTRIBUTIONS

I. Inferences concerning binomial p (Secs. 6-2 and 6-4)

A. Standard hypothesis, alternatives, statistics, and critical regions [significance level α less than or equal to an aimed-at significance level α_0 (disregard difference between α, α_0 with large sample)]

Hypothesis	Alternative	Statistic†	Critical region
$H_0: p = p_0$ $H_0: p \geq p_0$	$H_1: p < p_0$ $H_1: p < p_0$		All calculations performed with $p = p_0$ Small sample: reject if $x \leq x_0$, where x_0 is the largest value of x such that $\Pr(X \leq x) = \alpha \leq \alpha_0$. Large sample: reject if $z' < z_\alpha$.
$H_0: p = p_0$ $H_0: p \leq p_0$	$H_1: p > p_0$ $H_1: p > p_0$	Small sample: $X =$ number of successes in n trials if p is the probability of a success on each trial Large sample: $$Z' = \frac{X - np_0}{\sqrt{np_0(1 - p_0)}}$$	Small sample: reject if $x \geq x_0$, where x_0 is the smallest value of x such that $\Pr(X \geq x) = \alpha \leq \alpha_0$. Large sample: reject if $z' > z_{1-\alpha}$.
$H_0: p = p_0$	$H_1: p \neq p_0$		Small sample: reject if $x \leq x_1, x \geq x_2$, where x_1 is the largest value of x such that $\Pr(X \leq x) = \alpha_1 \leq \alpha_0/2$ and x_2 is the smallest value of x such that $\Pr(X \geq x) = \alpha_2 \leq \alpha_0/2$ (the significance level is then $\alpha_1 + \alpha_2$). Large sample: reject if $z' < z_{\alpha/2}$, $z' > z_{1-\alpha/2}$.

† Note that whenever a statistic is given in this summary, certain assumptions are made. These assumptions, giving the statistic the required distribution, may be found in the text material.

B. Power calculations
   1. With small sample use binomial tables with appropriate value of p, say p_1, and find the probability that X falls in the critical region if $p = p_1$.
   2. With large sample and one-sided alternative the power is approximately

$$\Pr\left(Z < \frac{z_\alpha\sqrt{p_0(1 - p_0)} + \sqrt{n}|p_0 - p_1|}{\sqrt{p_1(1 - p_1)}}\right)$$

where p_1 is a value of p in the range designated under H_1. The n required to yield power at least $1 - \beta$ when $p = p_1$ is

$$n \geqq \left[\frac{z_\alpha\sqrt{p_0(1 - p_0)} + z_\beta\sqrt{p_1(1 - p_1)}}{p_0 - p_1}\right]^2$$

(With a two-sided alternative, replace α by $\alpha/2$.)
C. Two-sided confidence interval with confidence coefficient at least $1 - \alpha$
   1. With small sample find the interval (w_1, w_2) from Appendix B15. Entries required are α, n, $\hat{p} = x/n$.
   2. With large sample the interval is

$$\left(\hat{p} - z_{1-\alpha/2}\sqrt{\frac{\hat{p}(1 - \hat{p})}{n}}, \hat{p} + z_{1-\alpha/2}\sqrt{\frac{\hat{p}(1 - \hat{p})}{n}}\right)$$

II. Inferences concerning Poisson μ (Secs. 6-3 and 6-4)
   A. Standard hypothesis, alternatives, statistics, and critical regions [significance level α less than or equal to an aimed-at significance level α_0 (disregard difference between α, α_0 with large sample)]

Hypothesis	Alternative	Statistic	Critical region
$H_0: \mu = \mu_0$	$H_1: \mu < \mu_0$		Small sample: reject if $y \leqq y_0$, where y_0 is the largest value y which can be assumed by a Poisson random variable with parameter $n\mu_0$ such that $\Pr(Y \leqq y) = \alpha \leqq \alpha_0$. Large sample: reject if $z' < z_\alpha$.
$H_0: \mu \geqq \mu_0$	$H_1: \mu < \mu_0$		

Continued

Hypothesis	Alternative	Statistic	Critical region
$H_0: \mu = \mu_0$ $H_0: \mu \leq \mu_0$	$H_1: \mu > \mu_0$ $H_1: \mu > \mu_0$	Small sample: $Y = \sum_{i=1}^{n} X_i$ where $X_1, \ldots, X_n$ is a random sample from a Poisson distribution with mean μ Large sample:	Small sample: reject if $y \geq y_0$, where y_0 is the smallest value y which can be assumed by a Poisson random variable with parameter $n\mu_0$ such that $\Pr(Y \geq y) = \alpha \leq \alpha_0$. Large sample: reject if $z' > z_{1-\alpha}$.
$H_0: \mu = \mu_0$	$H_1: \mu \neq \mu_0$	$Z' = \dfrac{Y - n\mu_0}{\sqrt{n\mu_0}}$	Small sample: reject if $y \leq y_1$, $y \geq y_2$, where y_1 is the largest value of y such that $\Pr(Y \leq y) = \alpha_1 < \alpha_0/2$ and y_2 is the smallest value of y such that $\Pr(Y \geq y) = \alpha_2 < \alpha_0/2$ (the significance level is $\alpha_1 + \alpha_2$), where Y is a Poisson random variable with parameter $n\mu_0$. Large sample: reject if $z' < z_{\alpha/2}$, $z' > z_{1-\alpha/2}$.

B. Power calculations
   1. With small sample use Poisson tables with parameter $n\mu_1$ and
      find the probability that Y falls in the critical region.   For
      the one-sided alternative $\mu < \mu_0$ the n required to yield power
      at least $1 - \beta_1$ when $\mu = \mu_1$ must satisfy

$$\frac{\chi^2_{2y_0+2;1-\alpha_0}}{2\mu_0} \leq n \leq \frac{\chi^2_{2y_0+2;\beta_1}}{2\mu_1}$$

If the alternative is $\mu > \mu_0$, then n must satisfy

$$\frac{\chi^2_{2y_0;1-\beta_1}}{2\mu_1} \leq n \leq \frac{\chi^2_{2y_0;\alpha_0}}{2\mu_0}$$

2. With large sample and one-sided alternative the power is approximately

$$\Pr\left(Z < \frac{z_\alpha\sqrt{\mu_0} + \sqrt{n}|\mu_0 - \mu_1|}{\sqrt{\mu_1}}\right)$$

where μ_1 is a value of μ in the range designated under H_1. The n required to yield power at least $1 - \beta$ when $\mu = \mu_1$ is

$$n \geq \left[\frac{z_\alpha\sqrt{\mu_0} + z_\beta\sqrt{\mu_1}}{\mu_0 - \mu_1}\right]^2$$

(With a two-sided alternative, replace α by $\alpha/2$.)
C. Two-sided confidence interval with confidence coefficient at least $1 - \alpha$
   1. With small sample compute the interval (w_1, w_2) where

$$w_1 = \frac{\chi^2_{2y;\alpha/2}}{2n} \qquad w_2 = \frac{\chi^2_{2y+2;1-\alpha/2}}{2n}$$

   2. With large sample the interval is

$$\left(\hat{\mu} - z_{1-\alpha/2}\sqrt{\frac{\hat{\mu}}{n}}, \hat{\mu} + z_{1-\alpha/2}\sqrt{\frac{\hat{\mu}}{n}}\right) \qquad \text{where } \hat{\mu} = \frac{y}{n}$$

III. Hypothesis testing for hypergeometric k (Sec. 6-5)
   A. The standard hypothesis testing situations considered are

$$H_0 : k = k_0 \qquad \text{against} \qquad H_1 : k > k_0$$

and

$$H_0 : k \leq k_0 \qquad \text{against} \qquad H_1 : k > k_0$$

The statistic used is X, the number of "defectives" in a simple random sample of size n. The critical region is $x \geq x_0$ where x_0 is the smallest value of x such that $\Pr(X \geq x) \leq \alpha_0$, the aimed-at significance level. Here X has a hypergeometric distribution with parameters N, n, k ($k = k_0$ is used to find the critical region).
   B. Power calculations
      1. With small N use hypergeometric tables with appropriate value k, say k_1, and find the probability that X falls in the critical region if $k = k_1$. The n required to yield power at least $1 - \beta_1$ when $k = k_1$ is found by trial as described in Example 6-22.
      2. With large N, base probability statements on the binomial approximation to the hypergeometric with $p = k/N$.

IV. Inferences about two Poisson means (Sec. 6-6)
  A. Standard hypotheses, alternatives, statistics, and critical regions
     [significance level α less than or equal to an aimed-at significance
     level α_0 (disregard difference between α, α_0 with large samples)]

Hypothesis	Alternative	Statistic	Critical region
$H_0: \mu_1 = \mu_2$ $H_0: \mu_1 \geqq \mu_2$	$H_1: \mu_1 < \mu_2$ $H_1: \mu_1 < \mu_2$	Small samples: Y_1 given $Y_1 + Y_2$ $= t$ where Y_1, Y_2 are the sample sums based upon random samples of size n from each distribution. The distribution of Y_1 given t is binomial with $n =$ $t, p = 1/2$	Small samples: reject if y_1 is too small, deter- mined as in I-A. with $n = t, p = 1/2$. Large samples: reject if $z' < z_\alpha$.
$H_0: \mu_1 = \mu_2$	$H_1: \mu_1 > \mu_2$	Large samples:	Interchange roles of 1 and 2 subscripts and proceed as with alter-
$H_0: \mu_1 \leqq \mu_2$	$H_1: \mu_1 > \mu_2$	$Z' = \dfrac{2Y_1 - t}{\sqrt{t}}$	native $\mu_1 < \mu_2$.
$H_0: \mu_1 = \mu_2$	$H_1: \mu_1 \neq \mu_2$		Small samples: reject if y_1 is too small or too large, determined as in I-A with $n = t, p =$ $1/2$. Large sample: reject if $z' < z_{\alpha/2}$, $z' > z_{1-\alpha/2}$.

  B. Power calculations
     1. Except for sample size problems, power calculations are
        generally avoided.
     2. To obtain a test with power $\leqq \alpha_0$ if H_0 is true and $\geqq 1 - \beta_1$
        when $\mu_1/\mu_2 = a \neq 1$, determine t as in binomial case. Then
        sample, one pair at a time, until $y_1 + y_2 \geqq t$. (See Example
        6-24.)
  C. Two-sided confidence interval with confidence coefficient at
     least $1 - \alpha$ for μ_1/μ_2
     1. Determine w_1, w_2 as in binomial case using $n = t$, $\hat{p} = y_1/t$.
        Then interval for μ_1/μ_2 is

$$\left(\frac{w_1}{1 - w_1}, \frac{w_2}{1 - w_2} \right)$$

V. Hypothesis testing for two binomials (Sec. 6-7)

A. Standard hypothesis, alternatives, statistics, and critical regions [significance level α less than or equal to an aimed-at significance level α_0 (disregard difference between α, α_0 with large samples)]

Hypothesis	Alternative	Statistic	Critical region
$H_0: p_1 = p_2$ $H_0: p_1 \geq p_2$	$H_1: p_1 < p_2$ $H_1: p_1 < p_2$	Small samples: X given $X + Y = t$ where X is binomial n_1, p_1 and Y is binomial n_2, p_2. The distribution of Y given t is hypergeometric with $N = n_1 + n_2$, $n = t$, $k = n_1$ if $p_1 = p_2$. Large samples: $$Z' = \dfrac{\dfrac{X}{n_1} - \dfrac{Y}{n_2}}{\sqrt{\left(\dfrac{1}{n_1} + \dfrac{1}{n_2}\right)\Delta}}$$ where $$\Delta = \left(\dfrac{t}{n_1 + n_2}\right)$$ $$\cdot \left(1 - \dfrac{t}{n_1 + n_2}\right)$$	Small samples: reject if $x \leq x_0$ where x_0 is the largest value of x such that $\Pr(X \leq x \mid t) \leq \alpha_0$ Large samples: reject if $z' < z_\alpha$.
$H_0: p_1 = p_2$ $H_0: p_1 \leq p_2$	$H_1: p_1 > p_2$ $H_1: p_1 > p_2$		Interchange roles of 1 and 2 subscripts and proceed as with alternative $p_1 < p_2$.
$H_0: p_1 = p_2$	$H_1: p_1 \neq p_2$		Small samples: reject if $x \leq x_1$ or $x \geq x_2$ where x_1 is the largest value of x such that $\Pr(X \leq x \mid t) \leq \alpha_0/2$ and x_2 is the smallest value of x such that $\Pr(X \geq x \mid t) \leq \alpha_0/2$. Large samples: reject if $z' < z_{\alpha/2}$, $z' > z_{1-\alpha/2}$.

VI. Testing hypotheses concerning multinomial $p_1, p_2, \ldots, p_k$ (all p's specified by the hypothesis) (Sec. 6-8)

A. The standard hypothesis and alternative are

$$H_0: p_1 = p_{10}, \ p_2 = p_{20}, \ \ldots, \ p_k = p_{k0}$$

against

H_1: not all p's are as specified by H_0

The observed value of the statistic is

$$y'_{k-1} = \sum_{i=1}^{k} \frac{(x_i - np_{i0})^2}{np_{i0}}$$

A critical region with significance level approximately α is

$$y'_{k-1} > \chi^2_{k-1;1-\alpha}$$

B. Power calculations require special tables and have been omitted.

VII. Tests for verification of the adequacy of the Poisson and normal models (Sec. 6-9)

A. For the Poisson case we test

H_0 : The Poisson model is appropriate

against

H_1 : The Poisson model is not appropriate

Let X_i be the number of occurrences of $i - 1$ successes, $i = 1, 2, \ldots, k - 1$ and X_k be the number of occurrences of $k - 1$ or more successes when a Poisson experiment is repeated n times. Then, the observed value of the statistic is

$$y'_{k-2} = \sum_{i=1}^{k} \frac{(x_i - n\hat{p}_i)^2}{n\hat{p}_i}$$

where the $\hat{p}_i$ are obtained using $\hat{\mu} = \bar{x}$ in the Poisson distribution and k is chosen so that each $n\hat{p}_i \geq 5$. (See Example 6-31.) The critical region is $y'_{k-2} > \chi^2_{k-2;1-\alpha}$.

B. For the normal case we test

H_0 : The normal model is appropriate

against

H_1 : The normal model is not appropriate

Let X_i be the number of occurrences in the interval (a_{i-1}, a_i), where $a_0 = -\infty$, $a_k = \infty$. Then, the observed value of the statistic is

$$y'_{k-3} = \sum_{i=1}^{k} \frac{(x_i - n\hat{p}_i)^2}{n\hat{p}_i}$$

where the $\hat{p}_i$ are obtained using $\hat{\mu} = \bar{x}$, $\hat{\sigma} = s$ in the normal distribution. One way to choose the a_i is so that each $n\hat{p}_i \geq 5$ is the same. (See Example 6-32.) The critical region is $y'_{k-3} > \chi^2_{k-3;1-\alpha}$.

VIII. Hypothesis of independence and hypothesis of homogeneity for multinomial p's (Sec. 6-10)

A. When investigating the independence of two characteristics we choose between

$$H_0 : p_{ij} = p_{i.}p_{.j} \qquad i = 1, 2, \ldots, r \qquad j = 1, 2, \ldots, c$$

against

H_1 : The equations given under H_0 are not all true

The observed value of the statistic is

$$y'_{(r-1)(c-1)} = \sum_{i=1}^{r} \sum_{j=1}^{c} \frac{(x_{ij} - t_i.t._j/n)^2}{t_i.t._j/n}$$

The critical region is $y'_{(r-1)(c-1)} > \chi^2_{(r-1)(c-1);1-\alpha}$.

B. For the homogeneity case we test

$$H_0 : p_{1j} = p_{2j} = \cdots = p_{rj} \qquad j = 1, 2, \ldots, c$$

against

H_1: The equations given under H_0 are not all true

The observed value of the statistic and the critical region are the same as given in A.

C. Power calculations have been omitted.

7

ANALYSIS OF VARIANCE

7-1
INTRODUCTION

In Sec. 5-9 we considered some problems associated with drawing inferences about the means of two distributions by using a random sample from each distribution. In particular we discussed the testing of

$$H_0 : \mu_1 = \mu_2 \qquad \text{against} \qquad H_1 : \mu_1 \neq \mu_2 \tag{7-1}$$

power calculations, and confidence intervals. Now we would like to generalize those procedures so that we can draw inferences about the means of $r(r \geq 2)$ distributions based upon r random samples, one drawn from each distribution. The generalization of (7-1) is

$$H_0 : \mu_1 = \mu_2 = \cdots = \mu_r \qquad \text{against} \qquad H_1 : \text{not all } r \text{ means are equal} \tag{7-2}$$

where $\mu_1, \mu_2, \ldots, \mu_r$ are the means of the r distributions.

It is not difficult to imagine situations in which an experimenter may want to draw inferences about several means. The large corporation of Example 5-32 may have five brands of light bulbs from which to choose. The smoker discussed in Sec. 4-4 may be interested in comparing the average nicotine contents of four brands of cigarettes. A wheat grower may want to obtain information about average yields for six varieties of wheat. A teacher may be interested in comparing average results achieved by three teaching methods.

The topics considered in this chapter (and also in Sec. 5-9) are a small part of an important branch of statistics which is called *analysis of variance*. It is moderately difficult to give a good meaningful definition of analysis of variance, and one will not be presented here. Our objectives have already been listed, and for our purposes it will be sufficient to consider analysis of variance as a body of statistical techniques which is useful in drawing inferences about the means of r distributions.

7-2
SOME NOTATION

Before discussing the test procedure for situation (7-2), we shall define some notation which will be used throughout the chapter.

For convenience we will refer to the distributions as distribution 1, distribution 2, ..., distribution r. As we have already indicated the means will be denoted by μ_1, μ_2, ..., μ_r, respectively. In addition each distribution will have a variance. These will be denoted by $\sigma_1^2, \sigma_2^2, \ldots, \sigma_r^2$, respectively.

When experimentation is performed we will observe a random sample of size n_1 from distribution 1, a random sample of size n_2 from distribution 2, ..., a random sample of size n_r from distribution r. Let x_{ij} be the ith sample value observed from distribution j. These observed values can be exhibited as in Table 7-1. Thus the second subscript identifies the

TABLE 7-1

r observed random samples with totals and means

	Distribution				
	1	2	j	r	
	x_{11}	x_{12}	$\cdots$	x_{1j}	$\cdots$	x_{1r}
	x_{21}	x_{22}	$\cdots$	x_{2j}	$\cdots$	x_{2r}
	$\cdots$					
	x_{n_11}	x_{n_22}		x_{n_jj}		x_{n_rr}
Totals	$t_{.1}$	$t_{.2}$		$t_{.j}$	$t_{.r}$	$t_{..}$
Means	$\bar{x}_{.1}$	$\bar{x}_{.2}$		$\bar{x}_{.j}$	$\bar{x}_{.r}$	$\bar{x}_{..}$

distribution from which the observation comes, while the first subscript indicates the order in which observations are made. The totals and means are defined in the obvious way. We have

$$t_{.1} = \sum_{i=1}^{n_1} x_{i1} = \text{sum of the observed sample values drawn from distribution 1}$$

$$t_{.2} = \sum_{i=1}^{n_2} x_{i2} = \text{sum of the observed sample values drawn from distribution 2}$$

$$t_{.j} = \sum_{i=1}^{n_j} x_{ij} = \text{sum of the observed sample values drawn from distribution } j$$

$$t._r = \sum_{i=1}^{n_r} x_{ir} = \text{sum of the observed sample values drawn from distribution } r$$

$$t.. = \sum_{j=1}^{r} t._j = \sum_{j=1}^{r} \sum_{i=1}^{n_j} x_{ij} = \text{total of all observations}$$

$$\bar{x}._1 = \frac{t._1}{n_1} = \text{observed sample mean for distribution 1}$$

$$\bar{x}._2 = \frac{t._2}{n_2} = \text{observed sample mean for distribution 2}$$

$$\bar{x}._j = \frac{t._j}{n_j} = \text{observed sample mean for distribution } j$$

$$\bar{x}._r = \frac{t._r}{n_r} = \text{observed sample mean for distribution } r$$

$$\bar{x}.. = \frac{t..}{N} = \text{observed sample mean of all observations}$$

where

$$N = \sum_{j=1}^{r} n_j = \text{total number of observations}$$

To illustrate Table 7-1 with a numerical example, suppose that final-examination scores for three statistics classes are the numbers appearing in Table 7-2. All three classes were taught by the same instructor, but

TABLE 7-2

Final-examination scores

		Textbook		
	1	**2**	**3**	
	56	77	70	
	53	73	61	
	68	63	67	
		74	62	
		68		
Totals	$t._1 = 177$	$t._2 = 355$	$t._3 = 260$	$t.. = 792$
Sample means	$\bar{x}._1 = 59$	$\bar{x}._2 = 71$	$\bar{x}._3 = 65$	$\bar{x}.. = 66$
Sample sizes	$n_1 = 3$	$n_2 = 5$	$n_3 = 4$	$N = 12$

each had a different text. We might be interested in testing the hypothesis that all three texts produce the same average score. In other words, we would probably test

$$H_0 : \mu_1 = \mu_2 = \mu_3 \quad \text{against} \quad H_1 : \text{not all means are equal}$$

for the distributions of final-examination scores. The sample sizes are small to make the example simple.

7-3
TESTING THE HYPOTHESIS OF EQUAL MEANS

We recall from Sec. 5-9 that to test the hypothesis of (7-1), the statistic $T_{n_1 + n_2 - 2} = (\overline{X}_1 - \overline{X}_2)/S_p \sqrt{1/n_1 + 1/n_2}$ was used [formula (5-76)]. When certain assumptions are satisfied, then $T_{n_1 + n_2 - 2}$ has a t distribution with $n_1 + n_2 - 2$ degrees of freedom. The assumptions we now make are an obvious generalization of those needed for the t statistic. If we assume

(a) Each of the r samples is a random sample from a distribution
(b) Each of the r sampled distributions is normal
(c) Each of the r distributions has the same variance (that is, $\quad$ (7-3) $\sigma_1^2 = \sigma_2^2 = \cdots = \sigma_r^2 = \sigma^2$)
(d) Each of the r means is the same [H_0 of situation (7-2) is true]

then it can be shown that the statistic

$$F_{r-1, N-r} = \frac{\sum_{j=1}^{r} \sum_{i=1}^{n_j} (\overline{X}._j - \overline{X}..)^2/(r - 1)}{\sum_{j=1}^{r} \sum_{i=1}^{n_j} (X_{ij} - \overline{X}._j)^2/(N - r)} \tag{7-4}$$

has an F distribution with $r - 1$ and $N - r$ degrees of freedom. It also can be shown that the statistic (7-4) is a good statistic to use to test the hypothesis of equal means and that the critical region contains only large values of $f_{r-1, N-r}$. To obtain a test with significance level α we reject the hypothesis of equal means if

$$f_{r-1, N-r} > f_{r-1, N-r; 1-\alpha} \tag{7-5}$$

The critical region (7-5) can be justified on an intuitive basis. The numerator of $f_{r-1, N-r}$ contains N terms of the type $(\overline{x}._j - \overline{x}..)^2$, one for each cell of Table 7-1. If all the observed sample means are equal (that is, $\overline{x}._1 = \overline{x}._2 = \cdots = \overline{x}._r$, an event which occurs with probability 0), then each observed sample mean is equal to $\overline{x}..$ and every term of the type $(\overline{x}._j - \overline{x}..)^2$ is zero making $f_{r-1, N-r}$ zero. Such an unlikely result would support the hypothesis of equal means. If the observed sample means are nearly equal, then the terms $(\overline{x}._j - \overline{x}..)^2$ are small and $f_{r-1, N-r}$ is small. The more the observed sample means differ, the larger $f_{r-1, N-r}$ becomes. Consequently, it seems reasonable to reject H_0 only if $f_{r-1, N-r}$ is large.

Concerning the mathematical assumptions (7-3), we can make essentially the same comments given for the t statistic. Provided the sample sizes are not too small, failure to satisfy the normality assumption does not seriously affect probability statements based on statistic (7-4). The consequences of not satisfying the equal-variance assumption can be minimized by taking equal sample sizes. Consequently, if sample sizes are equal (or nearly so) and the sample sizes are not too small, assumptions (b) and (c) are not too crucial.

EXAMPLE 7-1

Evaluate the observed value of statistic (7-4) for Table 7-2.

Solution

Consider the denominator first. We need to evaluate $(x_{ij} - \bar{x}._j)^2$ for all 12 observations in the table, sum the results, and divide by $N - r = 12 - 3 = 9$. We get

$$\sum_{j=1}^{3} \sum_{i=1}^{n_j} (x_{ij} - \bar{x}._j)^2/9 = [(56 - 59)^2 + (53 - 59)^2 + (68 - 59)^2$$
$$+ (77 - 71)^2 + (73 - 71)^2 + (63 - 71)^2$$
$$+ (74 - 71)^2 + (68 - 71)^2 + (70 - 65)^2$$
$$+ (61 - 65)^2 + (67 - 65)^2 + (62 - 65)^2]/9$$
$$= [9 + 36 + 81 + 36 + 4 + 64 + 9 + 9 + 25 + 16$$
$$+ 4 + 9]/9$$
$$= 302/9 = 33.56$$

In evaluating the numerator terms will be repeated, the number of repetitions being equal to the number of observations in the sample. We get

$$\sum_{j=1}^{3} \sum_{i=1}^{n_j} (\bar{x}._j - \bar{x}..)^2/(3 - 1) = [(59 - 66)^2 + (59 - 66)^2 + (59 - 66)^2$$
$$+ (71 - 66)^2 + (71 - 66)^2 + (71 - 66)^2$$
$$+ (71 - 66)^2 + (71 - 66)^2 + (65 - 66)^2$$
$$+ (65 - 66)^2 + (65 - 66)^2 + (65 - 66)^2]/2$$
$$= [49 + 49 + 49 + 25 + 25 + 25 + 25 + 25$$
$$+ 1 + 1 + 1 + 1]/2$$
$$= 276/2 = 138$$

Finally, we calculate

$$f_{2,9} = \frac{138}{33.56} = 4.11$$

EXAMPLE 7-2

Follow the six-step outline for testing the hypothesis of equal means with the samples given in Table 7-2. Use $\alpha = .05$.

Solution

1 We test $H_0 : \mu_1 = \mu_2 = \mu_3$ against H_1: not all three means are equal.
2 The significance level $\alpha = .05$ and the sample sizes $n_1 = 3$, $n_2 = 5$, $n_3 = 4$ have already been chosen.
3 The statistic we use is (7-4) with $r = 3$, $N = 12$, $N - r = 9$. This statistic will have an F distribution provided (a) the samples are randomly selected, (b) the final-examination grades associated with each text have a normal distribution, (c) all three distributions have the same variance, and (d) the hypothesis is true. Samples can be selected in a manner to make (a) seem realistic. Perhaps (b) and (c) are reasonable for final examinations, assumptions we would not need to worry about with larger sample sizes that are nearly equal.
4 The critical region is $f_{2,9} > f_{2,9;.95} = 4.26$.
5 From Example 7-1 the observed value of the statistic is $f_{2,9} = 4.11$.
6 We accept H_0. If the means are not equal, the samples of Table 7-2 fail to provide sufficient proof.

It is not too difficult to show that when $r = 2$ the F test of this section and the t test of Sec. 5-9 produce the same decision. When there are only two distributions, statistic (7-4) is the square of the t statistic. That is $F_{1, n_1 + n_2 - 2} = T^2_{n_1 + n_2 - 2}$. (This is easily verified with a little algebra.) In addition, $f_{r-1, N-r} = f_{1, n_1 + n_2 - 2}$ falls in the critical region if, and only if, $t_{n_1 + n_2 - 2}$ falls in its critical region. Hence the F test for the testing situation (7-2) can be regarded as a generalization of the t test for (7-1).

In the next three sections we discuss some additional topics that are usually helpful in analyzing and presenting the results of an analysis of variance. As we have demonstrated in Example 7-2, we can already solve the hypothesis testing problem proposed at the beginning of the chapter. Consequently, we could work problems like those appearing at the end of Sec. 7-6 without reading these sections, but it is generally more difficult to do so.

7-4
AN IMPORTANT IDENTITY

Consider the identity

$$\sum_{j=1}^{r} \sum_{i=1}^{n_j} (x_{ij} - \bar{x}..)^2 = \sum_{j=1}^{r} \sum_{i=1}^{n_j} [(x_{ij} - \bar{x}._j) + (\bar{x}._j - \bar{x}..)]^2 \tag{7-6}$$

Obviously the right-hand side of (7-6) reduces to the left-hand side since $\bar{x}._j$ has been subtracted and added to each term in the sum. The left-hand side is obtained by subtracting the mean of all N observations from each sample value, squaring each difference, and summing all these values. The sum is nothing more than the numerator of the observed sample variance if all N observations are regarded as one sample. [Except for the double subscript notation, the sum is exactly the same as the numerator of s^2, given by formula (3-25).] In analysis of variance problems we call the left-hand side of (7-6) the total sum of squares and designate it by SS_T. In other words

$$SS_T = \sum_{j=1}^{r} \sum_{i=1}^{n_j} (x_{ij} - \bar{x}..)^2 \tag{7-7}$$

If each bracket on the right-hand side of identity (7-6) is squared, it becomes

$$\sum_{j=1}^{r} \sum_{i=1}^{n_j} [(x_{ij} - \bar{x}._j)^2 + (\bar{x}._j - \bar{x}..)^2 + 2(x_{ij} - \bar{x}._j)(\bar{x}._j - \bar{x}..)]$$

$$= \sum_{j=1}^{r} \sum_{i=1}^{n_j} (x_{ij} - \bar{x}._j)^2 + \sum_{j=1}^{r} \sum_{i=1}^{n_j} (\bar{x}._j - \bar{x}..)^2$$

$$+ 2 \sum_{j=1}^{r} \sum_{i=1}^{n_j} (x_{ij} - \bar{x}._j)(\bar{x}._j - \bar{x}..) \tag{7-8}$$

It is fairly easy to show (Exercise 7-1) that

$$2 \sum_{j=1}^{r} \sum_{i=1}^{n_j} (x_{ij} - \bar{x}._j)(\bar{x}._j - \bar{x}..) = 0 \tag{7-9}$$

The first term on the right-hand side of (7-8) will be called the within-samples sum of squares, and it will be designated by SS_W. Thus, we have

$$SS_W = \sum_{j=1}^{r} \sum_{i=1}^{n_j} (x_{ij} - \bar{x}._j)^2 \tag{7-10}$$

The second term on the right-hand side of (7-8) will be called the among-samples sum of squares, denoted by SS_A. Hence

$$SS_A = \sum_{j=1}^{r} \sum_{i=1}^{n_j} (\bar{x}._j - \bar{x}..)^2 \tag{7-11}$$

Putting together the shorthand notation and the results of squaring the right-hand side of (7-6) yields the identity

$$SS_T = SS_A + SS_W \tag{7-12}$$

The observed value of statistic (7-4) can now be rewritten as

$$f_{r-1,\,N-r} = \frac{SS_A/(r-1)}{SS_W/(N-r)} \tag{7-13}$$

If we let $SS_A/(r - 1) = MS_A$, called the mean square for among samples, and $SS_W/(N - r) = MS_W$, called the mean square for within samples, then the observed value of statistic (7-4) can be written even more simply as

$$f_{r-1, N-r} = \frac{MS_A}{MS_W} \tag{7-14}$$

EXAMPLE 7-3

Verify that the identity (7-12) holds for the observed samples given in Table 7-2.

Solution

In Example 7-1 we found $SS_A = 276$, $SS_W = 302$. We need to show that $SS_T = 276 + 302 = 578$. We have

$$
\begin{aligned}
SS_T = \sum_{j=1}^{3} \sum_{i=1}^{n_j} (x_{ij} - \bar{x}..)^2 &= (56 - 66)^2 + (53 - 66)^2 + (68 - 66)^2 \\
&\quad + (77 - 66)^2 + (73 - 66)^2 + (63 - 66)^2 + (74 - 66)^2 + (68 - 66)^2 \\
&\quad + (70 - 66)^2 + (61 - 66)^2 + (67 - 66)^2 + (62 - 66)^2 \\
&= 100 + 169 + 4 + 121 + 49 + 9 + 64 + 4 + 16 + 25 + 1 + 16 \\
&= 578
\end{aligned}
$$

as required.

EXAMPLE 7-4

Verify that (7-9) holds for the observed samples given in Table 7-2.

Solution

We have

$$
\begin{aligned}
\sum_{j=1}^{3} \sum_{i=1}^{n_j} (x_{ij} &- \bar{x}._j)(\bar{x}._j - \bar{x}..) \\
&= (56 - 59)(59 - 66) + (53 - 59)(59 - 66) + (68 - 59)(59 - 66) \\
&\quad + (77 - 71)(71 - 66) + (73 - 71)(71 - 66) + (63 - 71)(71 - 66) \\
&\quad + (74 - 71)(71 - 66) + (68 - 71)(71 - 66) + (70 - 65)(65 - 66) \\
&\quad + (61 - 65)(65 - 66) + (67 - 65)(65 - 66) + (62 - 65)(65 - 66) \\
&= (59 - 66)[56 + 53 + 68 - 3(59)] \\
&\quad + (71 - 66)[77 + 73 + 63 + 74 + 68 - 5(71)] \\
&\quad + (65 - 66)[70 + 61 + 67 + 62 - 4(65)] \\
&= (-7)[0] + (5)[0] + (-1)[0] = 0
\end{aligned}
$$

The proof that equation (7-9) holds in general follows the scheme we have used in the above numerical example.

EXERCISE

◈ 7-1   Follow the scheme used in Example 7-4 to prove (7-9).

7-5
COMPUTING FORMULAS

In Example 7-1 the numerator and denominator of $f_{r-1, N-r}$ were evaluated directly from the definitions of MS_A and MS_W by first calculating SS_A and SS_W from their definitional forms. As in the case of s^2, the observed sample variance, we can obtain computing formulas that are more adaptable to a desk calculator or hand calculations.

Consider first

$$SS_T = \sum_{j=1}^{r} \sum_{i=1}^{n_j} (x_{ij} - \bar{x}..)^2 \tag{7-15}$$

Except for the fact that we are using double-summation notation, (7-15) is exactly the same as the numerator of

$$s^2 = \frac{\sum_{i=1}^{n} (x_i - \bar{x})^2}{n - 1}$$

By squaring out each individual term we showed [formula (3-27)] that

$$\sum_{i=1}^{n} (x_i - \bar{x})^2 = \sum_{i=1}^{n} x_i^2 - n\bar{x}^2 \tag{7-16}$$

Upon conversion to the notation of this chapter, (7-16) becomes

$$\sum_{j=1}^{r} \sum_{i=1}^{n_j} (x_{ij} - \bar{x}..)^2 = \sum_{j=1}^{r} \sum_{i=1}^{n_j} x_{ij}^2 - N\bar{x}..^2. \tag{7-17}$$

Since $\bar{x}.. = t../N$, we see that $N\bar{x}..^2 = N(t../N)^2 = t..^2/N$. Substituting this result in the right-hand side of (7-17) yields

$$SS_T = \sum_{j=1}^{r} \sum_{i=1}^{n_j} x_{ij}^2 - \frac{t..^2}{N} \tag{7-18}$$

the desired computing formula.

Next consider

$$SS_A = \sum_{j=1}^{r} \sum_{i=1}^{n_j} (\bar{x}._j - \bar{x}..)^2 = \sum_{j=1}^{r} n_j (\bar{x}._j - \bar{x}..)^2$$

Upon squaring and using properties of summation (discussed in Appendix A1), we get

$$SS_A = \sum_{j=1}^{r} n_j (\bar{x}._j^2 - 2\bar{x}..\,\bar{x}._j + \bar{x}..^2)$$

$$= \sum_{j=1}^{r} n_j \bar{x}._j^2 - 2\bar{x}.. \sum_{j=1}^{r} n_j \bar{x}._j + \bar{x}..^2 \sum_{j=1}^{r} n_j$$

Recalling that $\bar{x}._j = t._j/n_j$, $\sum_{j=1}^r n_j = N$, $\bar{x}.. = t../N$, we can write

$$SS_A = \sum_{j=1}^r n_j \frac{t^2._j}{n_j^2} - 2\bar{x}.. \sum_{j=1}^r n_j \frac{t._j}{n_j} + N\bar{x}^2..$$

$$= \sum_{j=1}^r \frac{t^2._j}{n_j} - 2\frac{t..}{N} t.. + N\frac{t^2..}{N^2}$$

$$= \sum_{j=1}^r \frac{t^2._j}{n_j} - 2\frac{t^2..}{N} + \frac{t^2..}{N}$$

Combining the last two terms yields the computing formula

$$SS_A = \sum_{j=1}^r \frac{t^2._j}{n_j} - \frac{t^2..}{N} \tag{7-19}$$

Finally, because of identity (7-12), SS_W can be obtained by subtraction. We have

$$SS_W = SS_T - SS_A \tag{7-20}$$

EXAMPLE 7-5

Use formulas (7-18) to (7-20) on the observed samples given in Table 7-2.

Solution

We have

$$SS_T = 56^2 + 53^2 + 68^2 + 77^2 + 73^2 + 63^2 + 74^2 + 68^2 + 70^2 + 61^2$$

$$+ 67^2 + 62^2 - \frac{792^2}{12}$$

$$= 52,850 - 52,272 = 578$$

$$SS_A = \frac{177^2}{3} + \frac{355^2}{5} + \frac{260^2}{4} - \frac{792^2}{12}$$

$$= 52,548 - 52,272 = 276$$

$$SS_W = 578 - 276 = 302$$

as before.

7-6
THE ANALYSIS OF VARIANCE TABLE

Frequently the results from an analysis of variance are presented in a table. The form we shall use when testing the hypothesis of (7-2) with assumptions (7-3) is given in Table 7-3. Here, *SS*, *d.f.*, *MS*, *f* are

TABLE 7-3

An analysis of variance table

Source of variation	SS	$d.f.$	MS	f
Among samples	SS_A	$r-1$	MS_A	$\dfrac{MS_A}{MS_W}$
Within samples	SS_W	$N-r$	MS_W	
Total	SS_T	$N-1$		

abbreviations for sum of squares, degrees of freedom, mean square, and $f_{r-1,\,N-r}$, respectively. The analysis of variance table for the observed samples of Table 7-2 is given in Table 7-4. In the six-step outline the table should be given under step 5.

TABLE 7-4

Analysis of variance for Table 7-2

Source of variation	SS	$d.f.$	MS	f
Among samples	276	2	138	4.11
Within samples	302	9	33.56	
Total	578	11		

Quite often the notation 4.11* and 4.11** is used to indicate that 4.11 is significant at the .05 and .01 levels, respectively. Of course, in the Table 7-4 example, it is not significant at either level.

EXERCISES

◈ 7-2 A 33-acre field is divided into acre plots. Eleven plots are selected at random and planted with variety 1 corn. Eleven more are selected at random from the remaining 22 and planted with variety 2 corn. The other 11 acres are planted with variety 3 corn. When the corn is harvested, the yields in bushels per acre are as recorded in Table 7-5. Test the hypothesis that the average yield for each variety is the same. Use $\alpha = .05$.

TABLE 7-5

Corn yields

Variety 1	Variety 2	Variety 3
89	122	95
97	120	80
115	115	83
82	87	100
105	105	95
108	90	100
110	105	113
80	130	105
99	130	80
110	110	80
110	111	92

7-3 A professor teaches four classes of elementary statistics. Each class is conducted in the same way except that a different text is used for each class. Assume that the students are selected at random for the four classes. After the course is over, all students take the same final examination. The scores are recorded in Table 7-6. Do the results contradict the hypothesis that all four distributions have the same mean? Use $\alpha = .01$.

TABLE 7-6

Final-examination scores

"Statistics Cookbook"	"Statistics with Humor"	"Statistics Made Useful"	"Statistics in Story Form"
78	51	64	54
78	57	54	61
79	64	61	79
70	75	66	69
83	42	57	69
74	83	71	65
81	54	69	76
77	58	59	
85	74	51	
	64	71	
		83	

◈ 7-4 A psychologist has devised an examination in such a way that the score achieved depends almost entirely on the ability to follow instructions. The examination is administered to 40 students who have been divided into four equal-size groups at random. The instructions are given in the following ways: to group I, written and brief; to group II, oral and brief; to group III, written and detailed; to group IV, oral and detailed. Computations with the scores yield $SS_T = 870$, $SS_A = 150$. If we use $\alpha = .05$, what conclusion should be drawn concerning the hypothesis of equal means?

7-5 Suppose that a merchandising chain has a choice of purchasing brand 1, brand 2, or brand 3 100-watt bulbs. Twenty-one randomly selected bulbs of each brand are tested by recording the amount of electricity consumed in a 10-minute period. If the results yield $\bar{x}_{.1} = 48.0$, $\bar{x}_{.2} = 49.0$, $\bar{x}_{.3} = 51.0$, $SS_A = 98$, $SS_T = 1,598$, can we conclude that the three sampled distribution do not have the same means? Use $\alpha = .05$.

◈ 7-6 A company packages frozen shrimp at four different locations. Six packages selected at random from each of the four locations yield the following weights in ounces:

Location 1—12.00, 12.05, 12.04, 12.07, 12.05, 12.03

Location 2—11.96, 11.98, 11.96, 11.97, 12.00, 11.95

Location 3—12.04, 12.03, 11.96, 11.98, 12.03, 12.02

Location 4—11.98, 12.05, 11.98, 12.06, 12.02, 12.03

Calculations yield $\bar{x}_{.1} = 12.04$, $\bar{x}_{.2} = 11.97$, $\bar{x}_{.3} = 12.01$, $\bar{x}_{.4} = 12.02$, $SS_A = .0156$, $SS_T = .0310$. With $\alpha = .01$ can we conclude that the mean weights associated with the four locations are not all the same?

7-7 Worms are classified into three groups by a structural characteristic (small, medium, or large ventral flap in the region of the vulva). From each of the three groups a random sample of 11 worms is taken, and the length in centimeters of each worm is measured and recorded, giving the results in Table 7-7. Use $\alpha = .01$ to test the

TABLE 7-7

Lengths of worms

Group 1	Group 2	Group 3
10.5	9.2	9.9
13.0	8.0	8.0
13.0	8.0	11.0
11.0	10.5	10.8
11.1	11.3	10.5
9.0	9.5	8.2
10.5	8.0	11.5
8.7	8.3	9.7
11.5	10.0	8.9
12.0	9.5	11.0
12.2	10.0	11.0

hypothesis that the mean length of each group is the same. Calculations yield $SS_A = 18.77$, $SS_T = 66.30$.

◊ 7-8   Each of 5 major automobile manufacturers has provided 6 engines for a government air pollution study. All 30 engines are rated at the same horsepower. After testing, the 30 engines are scored and the lower the score, the less the pollution. The results yield $\bar{x}_{.1} = 37.4$, $\bar{x}_{.2} = 35.3$, $\bar{x}_{.3} = 41.4$, $\bar{x}_{.4} = 45.2$, $\bar{x}_{.5} = 40.8$, $SS_A = 202$, $SS_T = 452$. With $\alpha = .01$ can we conclude that the mean score for each type of engine is the same?

7-7

POWER OF THE F TEST

If $r = 2$, the t test can be used to test the hypothesis of equal means. For this case computation of power has already been discussed in Sec. 5-9. If $r \geq 3$, then additional graphs or tables are needed to obtain power. Part of one such set of graphs appears in Appendix B16. In order to simplify the problems, we will consider only the case of equal sample sizes. That is, we will require that $n_1 = n_2 = \cdots = n_r = n$.

In order to obtain power from the graphs of Appendix B16, we need certain information. First, we must have $v_1 = r - 1$ to find the correct page of the table. This number appears in the upper left-hand corner of each page and runs from 2 to 5 inclusive. (The journal article from which these graphs were taken gives curves for $v_1 = 1$ to $v_1 = 8$.) Second, we need α to determine which set of graphs on the page are to be used. Each page contains two sets, one for $\alpha = .01$ and one for $\alpha = .05$, each clearly labeled. Third, we need $v_2 = N - r$ to determine which curve of a particular set we are to use. On each page there are curves for $v_2 = 6, 7, 8, 9, 10, 12, 15, 20, 30, 60, \infty$. For other values of v_2 it is necessary to interpolate. Finally, to read power from the vertical scale, we enter the horizontal scale with ϕ, a parameter which must be computed for the particular alternative hypothesis for which power is sought. This parameter has the value

$$\phi = \frac{\sqrt{\dfrac{n}{r} \displaystyle\sum_{j=1}^{r} (\mu_j - \mu)^2}}{\sigma} \tag{7-21}$$

where $\mu = \sum_{j=1}^{r} \mu_j / r$, the average of the r means. (When the n's are not equal, the expression for ϕ is slightly more complicated.)

Unfortunately ϕ, as given by formula (7-21), contains the unknown common standard deviation σ. We also had this problem with δ, needed for power of the t test. In Sec. 5-9 we overcame the difficulty by choosing alternatives for which $\mu_1 - \mu_2$ is a multiple of σ. Now, in order to make ϕ a pure number (not involving σ), we consider alternatives for which several differences of means are multiples of σ. As an alternative we might be tempted to replace σ by its estimate s. However, the latter procedure can lead to large errors in power if s differs from σ by a relatively small amount.

We shall now consider some examples which illustrate the use of the graphs.

EXAMPLE 7-6

The psychologist of Exercise 7-4 would like to know the probability of rejecting the hypothesis of equal means if three of the means are equal and the fourth exceeds the other three by σ.

Solution

The psychologist sampled four distributions, one each with group I, group II, group III, and group IV. Let the means be μ_1, μ_2, μ_3, and μ_4. We had $r = 4$ and $n_1 = n_2 = n_3 = n_4 = n = 10$.

Suppose we consider the case for which the distribution sampled with groups I, II, and III have equal means. Then, the alternative of interest has $\mu_1 = \mu_2 = \mu_3$, and the fourth mean is $\mu_4 = \mu_1 + \sigma$. We find

$$\mu = \frac{\mu_1 + \mu_2 + \mu_3 + \mu_4}{4} = \frac{\mu_1 + \mu_1 + \mu_1 + \mu_1 + \sigma}{4} = \mu_1 + \frac{\sigma}{4}$$

and

$$\sum_{j=1}^{4} (\mu_j - \mu)^2 = \left(\mu_1 - \mu_1 - \frac{\sigma}{4}\right)^2 + \left(\mu_1 - \mu_1 - \frac{\sigma}{4}\right)^2$$

$$+ \left(\mu_1 - \mu_1 - \frac{\sigma}{4}\right)^2 + \left(\mu_1 + \sigma - \mu_1 - \frac{\sigma}{4}\right)^2$$

$$= \frac{\sigma^2}{16} + \frac{\sigma^2}{16} + \frac{\sigma^2}{16} + \frac{9\sigma^2}{16} = \frac{3\sigma^2}{4}$$

Then, since $n = 10$, $r = 4$, we get

$$\phi = \frac{\sqrt{(10/4)(3\sigma^2/4)}}{\sigma} = \sqrt{\frac{30}{16}} = \frac{\sqrt{30}}{4} = \frac{5.477}{4} = 1.37$$

For this example $v_1 = r - 1 = 3$, $v_2 = N - r = 40 - 4 = 36$, $\alpha = .05$. Turning to the second page of Appendix B16, we find

Power $= .60$ if $v_2 = 60$

Power $= .57$ if $v_2 = 30$

For $v_2 = 36$ we take power $= .57 + (1/6)(.60 - .57) = .575$ or we have approximately power $= .58$. For practical purposes, anything reasonably close to .58 would be satisfactory.

Had we used $\alpha = .01$, then we can verify power $\cong .32$ from the right-hand set of graphs.

The standard sample size problem (power $= \alpha$ if H_0 is true, power $\geqq 1 - \beta$ for a specified alternative) can be solved for the minimum n by using the graphs of Appendix B16. One way to proceed is to increase n until both requirements are just satisfied. Alternatively, the problem can be solved by the same procedure we used with t tests (demonstrated in Examples 5-12 and 5-36). To use the latter method we first solve equation (7-21) for n, getting

$$n = \left[\frac{\sigma^2}{\sum_{j=1}^{r}(\mu_j - \mu)^2} \right] r\phi^2 \tag{7-22}$$

which we use to compute n for various ϕ for which power $= 1 - \beta$. If we choose the ϕ which makes power $= 1 - \beta$ for $v_2 = \infty$ ($n = \infty$), then again the computed n will be too small and this computed n will be a lower bound for the minimum n. On the other hand, if we use a ϕ which yields power $1 - \beta$ for a v_2 that is too small, this ϕ will be larger than necessary, and (7-22) computed from this ϕ will yield an n which may be larger than is needed to meet the requirements. Hence, once more we can corner the actual minimum n in a relatively small interval. Since $v_2 = N - r = nr - r = r(n - 1)$, the chosen n is $n = (v_2/r) + 1$. Decrease v_2 from ∞, to 60, to 30, etc., until the chosen n is smaller than the computed n. Then, the latter computed n is an upper bound on the minimum n, while the computed n obtained just previously is a lower bound. We will demonstrate both methods of solution in the following example.

EXAMPLE 7-7

For the situation described in Example 7-6 find the minimum n such that the power is $\alpha = .05$ when the means are equal and the power is at least .95 when three of the means are equal and the fourth exceeds the other three by σ.

Solution

First, we find the minimum n by trial. Since $n = 10$ yields a power of about .58, we need a larger n.

For $n = 20$ we have

$$\phi = \frac{\sqrt{(20/4)(3\sigma^2/4)}}{\sigma} = \sqrt{60/4} = \frac{\sqrt{60}}{4} = \frac{7.746}{4} = 1.94$$

$v_1 = 3$, $v_2 = 80 - 4 = 76$.

Even with $v_2 = \infty$ the power is less than .95, and so $n = 20$ is not large enough. Actually the graphs yield power $\cong .91$.

Suppose we next try $n = 25$. Then

$$\phi = \frac{\sqrt{(25/4)(3\sigma^2/4)}}{\sigma} = \sqrt{75/16} = \frac{\sqrt{75}}{4} = 2.17$$

$$v_1 = 3 \qquad v_2 = 100 - 4 = 96$$

Since the power is greater than .95 for $v_2 = 60$, it is greater than .95 for $v_2 = 96$, and $n = 25$ satisfies the requirements.

To determine whether or not a smaller n satisfies the requirements, we try $n = 24$. Now

$$\phi = \sqrt{72/16} = 2.12 \qquad v_1 = 3 \qquad v_2 = 92$$

and we read power = .95 for $v_2 = 60$. Hence again power > .95 with $v_2 = 92$, and $n = 24$ also works.

With $n = 23$ we have

$$\phi = \sqrt{69/16} = 2.08 \qquad v_1 = 3 \qquad v_2 = 88$$

and power $\cong .95$ with $v_2 = \infty$. Hence, it appears that with $v_2 = 88$, power < .95 and n is too small. Thus $n = 24$ is the exact minimum n.

To use the alternative method, we first evaluate (7-22). Since $\sum_{j=1}^{4} (\mu_j - \mu)^2 = 3\sigma^2/4$, we get $n = (4/3)4\phi^2 = 5.33\phi^2$. With $v_2 = \infty$ ($n = \infty$), $v_1 = 3$, $\alpha = .05$, we find power = .95 if $\phi = 2.04$. The computed n is $(5.33)(2.04)^2 = 22.18$, which is a lower bound for the minimum n. With $v_2 = 60$ [$n = (60/4) + 1 = 16$], $v_1 = 3$, $\alpha = .05$, we find power = .95 if $\phi = 2.13$. The computed n is $(5.33)(2.13)^2 = 24.18$. Since this exceeds the chosen $n = 16$, we have found an upper bound for the minimum n. Thus, we know that the minimum n is in the interval $22.18 \leq n \leq 24.18$. We would probably take $n = 23$ or 24 or even 25. For practical purposes anything close to 24 would be satisfactory.

Another interesting type of problem involving power calculations arises when we seek information about the power if the maximum difference between two means is a given amount, say $a\sigma$. A little reflection reveals that many different choices of the μ's will yield a maximum difference of $a\sigma$ and that a power value will be associated with each choice. For an illustration consider again Example 7-6. In that problem the maximum difference between two means was 1σ. For the alternative $\mu_1 = \mu_2$, $\mu_3 = \mu_1 - (1/2)\sigma$, $\mu_4 = \mu_1 + (1/2)\sigma$ the maximum difference is also 1σ. Similarly, the alternative $\mu_1 = \mu_2$, $\mu_3 = \mu_1 - .2\sigma$, $\mu_4 = \mu_1 + .8\sigma$ also has a maximum difference of 1σ between two means. For the choice of Example 7-6 we found power $\cong .58$. For the second choice given above we find in Example 7-8 that power = .40. For the third choice still another value would be obtained. It can be shown that the minimum value of power for all such choices is obtained by taking one μ_j, say μ_1, equal to $\mu + a\sigma/2$, another μ_j, say μ_2, equal to $\mu - a\sigma/2$, and all other μ_j's equal to each other (equal to μ). Then

$$\sum_{j=1}^{r} (\mu_j - \mu)^2 = \left(\mu + \frac{a\sigma}{2} - \mu\right)^2 + \left(\mu - \frac{a\sigma}{2} - \mu\right)^2 + 0 + \cdots + 0$$

$$= \frac{a^2\sigma^2}{2}$$

and

$$\phi = \frac{\sqrt{(n/r)(a^2\sigma^2/2)}}{\sigma} = a\sqrt{\frac{n}{2r}} \tag{7-23}$$

EXAMPLE 7-8

Find the minimum value of the power if the maximum difference between two means in Exercise 7-4 is σ.

Solution

As we have commented above, the minimum occurs when $\mu_1 = \mu_2$, $\mu_3 = \mu_1 - (1/2)\sigma$, $\mu_4 = \mu_1 + (1/2)\sigma$; or when $\mu_3 = \mu_4$, $\mu_1 = \mu_3 - (1/2)\sigma$, $\mu_2 = \mu_4 + (1/2)\sigma$; or whenever two means are equal (to say, μ), another is $\mu - (1/2)\sigma$, the last is $\mu + (1/2)\sigma$. For all these cases ϕ is given by formula (7-23) with $a = 1$, $n = 10$, $r = 4$, so that $\phi = \sqrt{10/8} = \sqrt{5/4} = 2.236/2 = 1.12$. Again $v_1 = 3$, $v_2 = 36$. Now the graphs of Appendix B16 yield a power of approximately .40.

A table which can be used to advantage for power calculations with analysis of variance tests has been published by Tiku [3]. He has given $1 -$ power to four decimal places for all combinations of

$\alpha = .005, .01, .025, .05$

$v_1 = 1(1)10, 12$

$v_2 = 2(2)30, 40, 60, 120, \infty$

$\phi = .5, 1.0(.2)2.2(.4)3.0$

EXERCISES

◈ 7-9   If $\alpha = .05$, find the probability of rejecting the hypothesis of equal means in Exercise 7-2 when two means are equal and the third exceeds these two by 1.2σ.

◈ 7-10   If $\alpha = .05$, find the minimum n such that the probability of rejecting the hypothesis of equal means for the problem of Exercise 7-2 will be at least .95 when two means are equal and the third exceeds these two by 1.2σ.

◈ 7-11   If $\alpha = .05$, find the least value for the probability of rejecting the hypothesis of equal means in Exercise 7-2 if the largest difference between two means is 1.2σ.

7-12   Consider again the problem of Exercise 7-5. If two of the means are equal and the third exceeds the other two by 1 standard deviation, find the power of the test.

7-13   The power in Exercise 7-12 was obtained with $n = 21$. For the same alternative find the minimum value of n such that the power of the test is at least .80.

7-14   For the test used in Exercise 7-5 find the least value for the power if the largest difference between two means is 1σ.

◈ 7-15   Consider again the problem of Exercise 7-6. If the first two means are equal and the last two means are equal, but the last two are 1.5σ larger than the first two, find the power of the test. If we had used $n = 16$ instead of $n = 6$, what is the power at the above alternative? How large would n have to be to raise this power to at least .90?

7-16   Consider again the problem of Exercise 7-8. If the first two means are equal, the fifth mean exceeds the first two by 1σ, and the third and fourth mean each exceed the first two by 2σ, find the power of the test. Suppose it is desired to have the power at the above alternative be at least .80. Could this be accomplished with $\alpha = .01$, using 5 engines from each manufacturer?

7-17   Suppose we had conducted the test of Exercise 7-8 with $\alpha = .05$ and $n = 5$ engines from each manufacturer. Is the power at the alternative considered in Exercise 7-16 at least .90?

7-8
CONTRASTS AND THE METHOD OF SCHEFFÉ

If the hypothesis of equal means (7-2) is accepted, the analysis of the observed sample results is usually terminated. However, if H_0 is rejected, we may not be content to conclude only that the r means are not all the same. After rejection it is logical to inquire as to what further inferences can be drawn about the unknown means. In particular, we may like to know whether the observed samples indicate that certain means are larger than certain other means. Investigation of further inferences about the means leads us to a study of contrasts.

If $c_1, c_2, \ldots, c_r$ are known constants such that

$$\sum_{j=1}^{r} c_j = c_1 + c_2 + \cdots + c_r = 0$$

then

$$L = c_1\mu_1 + c_2\mu_2 + \cdots + c_r\mu_r \qquad (7\text{-}24)$$

is called a contrast. Actually we have already considered the contrast $\mu_1 - \mu_2$ in Sec. 5-9 when we tested hypotheses and obtained confidence intervals for this difference. For this simple case involving only two means μ_1, μ_2 we have $r = 2$, and there are only two constants c_1, c_2 which we identify as $c_1 = 1$, $c_2 = -1$. Of course, $\mu_2 - \mu_1$, or $3\mu_1 - 3\mu_2$, or any multiple of $\mu_1 - \mu_2$ is also a contrast. Since any conclusion about $\mu_1 - \mu_2$ immediately implies a conclusion about a multiple of $\mu_1 - \mu_2$, there is essentially only one contrast involving two means. With $r = 3$ and three means μ_1, μ_2, μ_3

$\mu_1 - \mu_2$ is a contrast with $c_1 = 1$, $c_2 = -1$, $c_3 = 0$

$\mu_1 - \mu_3$ is a contrast with $c_1 = 1$, $c_2 = 0$, $c_3 = -1$

$\mu_2 - \mu_3$ is a contrast with $c_1 = 0$, $c_2 = 1$, $c_3 = -1$

$2\mu_1 - \mu_2 - \mu_3$ is a contrast with $c_1 = 2$, $c_2 = -1$, $c_3 = -1$ $\qquad (7\text{-}25)$

$\mu_1 - 2\mu_2 + \mu_3$ is a contrast with $c_1 = 1$, $c_2 = -2$, $c_3 = 1$

$\mu_1 + \mu_2 - 2\mu_3$ is a contrast with $c_1 = 1$, $c_2 = 1$, $c_3 = -2$

$5\mu_1 + 2\mu_2 - 7\mu_3$ is a contrast with $c_1 = 5$, $c_2 = 2$, $c_3 = -7$

etc.

Thus, we can write down infinitely many contrasts involving three means. However, there are essentially only two contrasts since any other one is a linear combination of these two. [To illustrate consider the first two contrasts of (7-25). Then, any other contrast is of the form $a_1(\mu_1 - \mu_2)$ $+ a_2(\mu_1 - \mu_3)$ where a_1 and a_2 are constants. For example, if we take $a_1 = -1$, $a_2 = 1$ we get $\mu_2 - \mu_3$, the third contrast of (7-25). If we take $a_1 = -2$, $a_2 = 7$, we get the last contrast of (7-25).] In general, with r means we have essentially $r - 1$ contrasts, any others being a linear combination of these $r - 1$. Hence, when we use the expression "all possible contrasts," there are not as many as there might seem to be.

We observe that if all means are equal, then every contrast is zero. On the other hand, we can show that if every contrast is zero, all means are equal. If not all the means are equal, then one or more contrasts must be different from zero. When the H_0 of situation (7-2) is rejected, we attempt to determine which contrasts are nonzero. As we shall see, this attempt leads us to certain conclusions about the means.

In Sec. 5-9 we found a confidence interval for $\mu_1 - \mu_2$. The interval was determined so that we had a high degree of confidence that the interval captured $\mu_1 - \mu_2$. We would now like to generalize that procedure so that it applies to all possible contrasts when r means are under consideration (and reduces to our former result where $r = 2$). The interval for every L in the desired set is given by

$$(\hat{L} - S\hat{\sigma}_L, \ \hat{L} + S\hat{\sigma}_L) \tag{7-26}$$

where

$$\hat{L} = c_1 \bar{x}._1 + c_2 \bar{x}._2 + \cdots + c_r \bar{x}._r \tag{7-27}$$

(an unbiased estimate of L)

$$S^2 = (r-1)f_{r-1, \, N-r; \, 1-\alpha} \tag{7-28}$$

(not to be confused with the sample variance)

and

$$\hat{\sigma}_L^2 = MS_W \sum_{j=1}^{r} \frac{c_j^2}{n_j} \tag{7-29}$$

Scheffé [2] has shown that the probability is $1 - \alpha$ that the set of random intervals associated with (7-26) captures all the corresponding contrasts.

When we first discussed confidence intervals for one parameter in Sec. 5-4, we pointed out that the computation of a confidence interval is somewhat analogous to throwing an interval in one-dimensional space at a point which represents the parameter. With r means and essentially $r - 1$ contrasts we can compare the computation of the intervals to throwing an $r - 1$ dimensional box (or parallelepiped) at an $r - 1$ dimensional point whose coordinates are the corresponding $r - 1$ contrasts, with the throwing procedure being such that the point is captured by the box about $1 - \alpha$ of the time in the long run.

EXAMPLE 7-9

Using $1 - \alpha = .95$ and the observed samples given in Table 7-2, compute the Scheffé intervals given by (7-26) for the first six contrasts of (7-25).

Solution

We have $\bar{x}._1 = 59$, $\bar{x}._2 = 71$, $\bar{x}._3 = 65$, $n_1 = 3$, $n_2 = 5$, $n_3 = 4$, $r = 3$, $N - r = 9$, $f_{2,9;.95} = 4.26$, $MS_W = 33.56$. Thus $S^2 = 2f_{2,9;.95} = 2(4.26) = 8.52$, $S = 2.92$.

For the contrast $\mu_1 - \mu_2$ the unbiased estimate is

$$\hat{L} = \bar{x}._1 - \bar{x}._2 = 59 - 71 = -12$$

Also

$$\hat{\sigma}_L^2 = MS_W\left(\frac{1}{n_1} + \frac{1}{n_2}\right) = 33.56\left(\frac{1}{3} + \frac{1}{5}\right) = 17.90$$

$$S^2\hat{\sigma}_L^2 = (8.52)(17.90) = 152.5 \qquad S\hat{\sigma}_L = 12.35$$

The interval is $(-12 - 12.35, \ -12 + 12.35)$ or $(-24.35, .35)$.

For the contrast $\mu_1 - \mu_3$ the unbiased estimate is

$$\hat{L} = \bar{x}_{.1} - \bar{x}_{.3} = 59 - 65 = -6$$

Then

$$\hat{\sigma}_L^2 = MS_W\left(\frac{1}{n_1} + \frac{1}{n_3}\right) = 33.56\left(\frac{1}{3} + \frac{1}{4}\right) = 19.58$$

$$S^2\hat{\sigma}_L^2 = (8.52)(19.58) = 166.7 \qquad S\hat{\sigma}_L = 12.91$$

The interval is $(-6 - 12.91, -6 + 12.91)$ or $(-18.91, 6.91)$.

For the contrast $\mu_2 - \mu_3$ the unbiased estimate is

$$\hat{L} = \bar{x}_{.2} - \bar{x}_{.3} = 71 - 65 = 6$$

Then

$$\hat{\sigma}_L^2 = MS_W\left(\frac{1}{n_2} + \frac{1}{n_3}\right) = 33.56\left(\frac{1}{5} + \frac{1}{4}\right) = 15.10$$

$$S^2\hat{\sigma}_L^2 = (8.52)(15.10) = 128.7 \qquad S\hat{\sigma}_L = 11.34$$

The interval is $(6 - 11.34, 6 + 11.34)$ or $(-5.34, 17.34)$.

For the contrast $2\mu_1 - \mu_2 - \mu_3$ the unbiased estimate is

$$\hat{L} = 2\bar{x}_{.1} - \bar{x}_{.2} - \bar{x}_{.3} = 2(59) - 71 - 65 = -18$$

Then

$$\hat{\sigma}_L^2 = MS_W\left(\frac{4}{n_1} + \frac{1}{n_2} + \frac{1}{n_3}\right) = 33.56\left(\frac{4}{3} + \frac{1}{5} + \frac{1}{4}\right) = 59.85$$

$$S^2\hat{\sigma}_L^2 = (8.52)(59.85) = 509.9 \qquad S\hat{\sigma}_L = 22.58$$

The interval is $(-18 - 22.58, -18 + 22.58)$ or $(-40.58, 4.58)$.

For the contrast $\mu_1 - 2\mu_2 + \mu_3$ the unbiased estimate is

$$\hat{L} = \bar{x}_{.1} - 2\bar{x}_{.2} + \bar{x}_{.3} = 59 - 2(71) + 65 = -18$$

Then

$$\hat{\sigma}_L^2 = MS_W\left(\frac{1}{n_1} + \frac{4}{n_2} + \frac{1}{n_3}\right) = 33.56\left(\frac{1}{3} + \frac{4}{5} + \frac{1}{4}\right) = 46.42$$

$$S^2\hat{\sigma}_L^2 = (8.52)(46.42) = 395.5 \qquad S\hat{\sigma}_L = 19.89$$

The interval is $(-18 - 19.89, -18 + 19.89)$ or $(-37.89, 1.89)$.

Finally, for the contrast $\mu_1 + \mu_2 - 2\mu_3$ the unbiased estimate is

$$\hat{L} = \bar{x}_{.1} + \bar{x}_{.2} - 2\bar{x}_{.3} = 59 + 71 - 2(65) = 0$$

$$\hat{\sigma}_L^2 = MS_W\left(\frac{1}{n_1} + \frac{1}{n_2} + \frac{4}{n_3}\right) = 33.56\left(\frac{1}{3} + \frac{1}{5} + \frac{4}{4}\right) = 51.46$$

$$S^2\hat{\sigma}_L^2 = (8.52)(51.46) = 438.4 \qquad S\hat{\sigma}_L = 20.94$$

The interval is $(0 - 20.94, 0 + 20.94)$ or $(-20.94, 20.94)$.

If the experimentation whose results are summarized in Table 7-2 were to be repeated many times and the above six intervals computed each time, then about .95 of the time the set of computed intervals will capture all six contrasts.

We note that each of the six intervals obtained in Example 7-9 contains 0. For example, for $\mu_2 - \mu_3$ we found $(-5.34, 17.34)$. In other words, this implies

$$-5.34 \leqq \mu_2 - \mu_3 \leqq 17.34$$

When the interval computed for a contrast L contains 0, the interpretation is that the experiment has failed to demonstrate that L is different from 0, and we proceed as though $L = 0$. For the above contrast $\mu_2 - \mu_3 = 0$ is equivalent to $\mu_2 = \mu_3$. Now suppose that calculations yield an interval which does not contain 0. Then the estimated contrast $\hat{L}$ is said to be significantly different from 0, and the conclusion is that either $L > 0$ or $L < 0$. Had we obtained $(6.52, 17.36)$ for $\mu_2 - \mu_3$, implying

$$6.52 \leqq \mu_2 - \mu_3 \leqq 17.36$$

we would conclude that $\mu_2 - \mu_3 > 0$ or $\mu_2 > \mu_3$. Similarly, had we obtained $(-12.34, -3.07)$ for $2\mu_1 - \mu_2 - \mu_3$, implying

$$-12.34 \leqq 2\mu_1 - \mu_2 - \mu_3 \leqq -3.07$$

we would conclude that $2\mu_1 - \mu_2 - \mu_3 < 0$ or $\mu_1 < (\mu_2 + \mu_3)/2$ (the mean for distribution 1 is less than the average of the means for distributions 2 and 3).

Probably the most interesting contrasts are those which contain only two means, such as the first three considered in Example 7-9. For these simple-type contrasts the Scheffé intervals enable us to make inferences about which distribution means are greater than which others. If, for example, we find that the computed intervals imply

$$-3.47 \leqq \mu_1 - \mu_2 \leqq 8.62$$

$$4.05 \leqq \mu_1 - \mu_3 \leqq 11.71$$

$$6.52 \leqq \mu_2 - \mu_3 \leqq 17.36$$

then we conclude that the means for distributions 1 and 2 are higher than the mean for distribution 3 and that no difference between the means of distributions 1 and 2 has been demonstrated.

The fact that all the intervals computed in Example 7-9 contained 0 was no accident. This will always be the case when H_0 is accepted and the intervals are computed using the same α. (We used $\alpha = .05$ for both the test and the intervals.) On the other hand, when the hypothesis of equal means is rejected, some intervals computed with the same α will not contain 0. Thus, the Scheffé procedure has that property that one or more of the intervals will not cover 0 if, and only if, the F test rejects the hypothesis of equal means when the same α is used. This is the same relationship which we previously observed exists between the t test and the confidence interval for $\mu_1 - \mu_2$.

When conducting an analysis of variance, we first perform the F test, then look at intervals for contrasts only if the hypothesis of equal means is rejected. We can regard the Scheffé intervals as something thrown in free with the F test. That is, no additional assumptions are needed and there is no problem concerning significance levels or probability statements. Further, contrasts of interest can be selected *after* the experimentation is performed, contrary to usual statistical procedures.

EXERCISES

◈ 7-18   In Exercise 7-2 we rejected H_0 with $\alpha = .05$. Use the Scheffé procedure with $\alpha = .05$ to find intervals for the first six contrasts of (7-25). Then draw the appropriate conclusions.

7-19   In Exercise 7-3 we rejected H_0 with $\alpha = .01$. Use the Scheffé procedure with $\alpha = .01$ to find intervals for the six simple-type contrasts involving two means (that is $\mu_1 - \mu_2$, $\mu_1 - \mu_3$, $\mu_1 - \mu_4$, $\mu_2 - \mu_3$, $\mu_2 - \mu_4$, $\mu_3 - \mu_4$). Then draw the appropriate conclusions.

◈ 7-20   In Exercise 7-6 we rejected H_0 with $\alpha = .01$. Use the Scheffé procedure with $\alpha = .01$ to find intervals for the six simple-type contrasts involving two means (as given in Exercise 7-19). Then draw the appropriate conclusions.

◈ 7-21   In Exercise 7-6 suppose that locations 2 and 3 are on the East Coast, while locations 1 and 4 are on the Gulf Coast. Using $\alpha = .01$, investigate a contrast comparing the averages for the East Coast and the Gulf Coast. [*Hint:* The averages are equal if $(\mu_2 + \mu_3)/2 = (\mu_1 + \mu_4)/2$ or if $L = \mu_1 - \mu_2 - \mu_3 + \mu_4 = 0$. Determine whether or not the interval for L covers 0.]

7-22   In Exercise 7-8 we rejected H_0 with $\alpha = .01$. If we use the Scheffé procedure with $\alpha = .01$, can we conclude that the mean score associated with manufacturer number 2 is less than the mean score associated with manufacturer number 1? With manufacturer number 3? With manufacturer number 4? With manufacturer number 5?

7-23   Five statistics classes were taught in the following ways: class 1, by correspondence with textbook A; class 2, by Professor Jones with textbook A; class 3, by Professor Jones with textbook B; class 4, by Professor Smith with textbook A; class 5, by Professor Smith with textbook B. Fifty students were assigned randomly, 10 to each class. All classes took a standard final examination, with 150 representing a perfect score. The totals were $t_{\cdot 1} = 500$, $t_{\cdot 2} = 700$, $t_{\cdot 3} = 1{,}000$, $t_{\cdot 4} = 600$, $t_{\cdot 5} = 1{,}200$, and the total sum of squares was 56,500. It is easy to verify that the hypothesis of equal means is rejected. Using $\alpha = .01$ with the Scheffé procedure, investigate a contrast comparing textbook A and textbook B. Investigate a contrast comparing Professor Jones and Professor Smith. [*Hint:* The averages for textbook A and textbook B are the same if $(\mu_1 + \mu_2 + \mu_4)/3 = (\mu_3 + \mu_5)/2$ or if

$$L = 2\mu_1 + 2\mu_2 - 3\mu_3 + 2\mu_4 - 3\mu_5 = 0$$

Determine whether or not the interval for L covers 0.]

7-9
A TEST FOR THE EQUALITY OF VARIANCES

One of the assumptions given in the list (7-3) required for the use of the F test (and the Scheffé procedure) is that the r distributions have equal variances. Several procedures are available to test that assumption. We shall discuss one, called Cochran's test, which requires that all sample sizes be equal. That is, we must have $n_1 = n_2 = \cdots = n_r = n$. (For other tests see Guenther [1] and Scheffé [2].)

The hypothesis and alternative of interest are

$$H_0: \sigma_1^2 = \sigma_2^2 = \cdots = \sigma_r^2 \qquad \text{against} \qquad H_1: \text{not all variances are equal}$$
$$(7\text{-}30)$$

Let $s_1^2, s_2^2, \ldots, s_r^2$ be the observed sample variances. If we assume

(*a*) Each of the r samples is a random sample from a distribution
(*b*) Each of the r sampled distributions is normal   (7-31)
(*c*) Each of the r variances is the same (H_0 is true)

then the statistic whose observed value is

$$R_{n,r} = \frac{\text{largest observed sample variance}}{s_1^2 + s_2^2 + \cdots + s_r^2} \tag{7-32}$$

can be used to test H_0 of (7-30). Obviously the largest possible value of $R_{n,r}$ is 1, which occurs when all the observed sample variances are 0 except one. It will be nearly 1 if one of the observed sample variances is very large compared to the others. Such a result supports H_1 and should lead to rejection of H_0. Consequently, it makes sense to reject if $R_{n,r}$ is large. To achieve significance level α reject when

$$R_{n,r} > R_{n,r;\,1-\alpha} \tag{7-33}$$

Values of $R_{n,r;\,1-\alpha}$ are given in Appendix B17 for $\alpha = .01$, .05.

Like previous tests concerning variances (discussed in Chap. 5), the one discussed here also appears to depend heavily upon the normality assumption.

EXAMPLE 7-10

Suppose that three samples of size 10 are selected randomly, one each from three normal distributions. The observed results yield $s_1^2 = 140$, $s_2^2 = 200$, $s_3^2 = 660$. Test the hypothesis of equal variances. Use $\alpha = .05$.

Solution

We shall follow the six-step outline.

1 We test $H_0: \sigma_1^2 = \sigma_2^2 = \sigma_3^2$ against H_1: not all three variances are equal.
2 The significance level $\alpha = .05$ and sample sizes $n_1 = n_2 = n_3 = 10$ have already been selected.
3 The statistic we use has observed value given by (7-32). If (a) the samples are selected randomly, (b) the three sampled distributions are normal, and (c) the three variances are equal, then critical regions can be obtained from Appendix B17. We are given that the first two assumptions are satisfied.
4 The critical region is $R_{10,3} > R_{10,3;.95} = .6167$, found in Appendix B17.
5 We already have $s_1^2 = 140$, $s_2^2 = 200$, $s_3^2 = 660$, the largest of which is the latter. The observed value of the statistic is

$$R_{10,3} = \frac{660}{140 + 200 + 660} = .66$$

6 Since the observed value of the statistic falls in the critical region, H_0 is rejected. (Since $R_{10,3;.99} = .6912$, H_0 is accepted with $\alpha = .01$.)

When we test the hypothesis of equal variances as a preliminary exercise before we test the hypothesis of equal means, we create some problems that deserve further comment. First, if we conclude that the variances are not equal, then it appears that the analysis of variance F test is no longer available. Second, if the hypothesis of equal variances is accepted, the significance level attached to the F test is not what it would appear to be. The probability of rejecting the hypothesis of equal means when true depends upon the distribution of $F_{r-1,N-r}$ subject to the condition that the hypothesis of equal variances has been accepted (a conditional probability). Computation of such probabilities is usually very involved, even for an expert. A partial solution to the dilemma has already been discussed earlier in the chapter. We know that provided the samples sizes are equal (or nearly so), the consequences of not satisfying the equal variance assumption can be minimized. Hence, an experimenter would rarely test for equal variances preliminary to using the F test for means.

EXERCISES

◆ 7-24 Suppose that we have four random samples, each of size $n = 8$, and one each selected from four normal distributions. If the observed results yield $s_1^2 = 190$, $s_2^2 = 527$, $s_3^2 = 66.2$, $s_4^2 = 873$, and we use $\alpha = .05$, is the hypothesis of equal variances accepted?

7-25 Use $\alpha = .05$ and test the hypothesis of equal variances for the observed sample results given in Exercise 7-2.

◆ 7-26 Suppose that the observed samples of Exercise 7-5 yield $s_1^2 = 1.61$, $s_2^2 = 5.98$, $s_3^2 = 4.43$. Use $\alpha = .05$ to test the hypothesis of equal variances. (To find the critical region, interpolation on n is necessary. It is recommended that linear interpolation on $12/\sqrt{n-1}$ be used. This is similar to inverse linear interpolation described in Appendix A2-2 where linear interpolation on $1/x$ is used.)

◆ 7-27 Use $\alpha = .01$ and test the hypothesis of equal variances for the observed sample results given in Exercise 7-6.

7-28 Suppose that the observed samples of Exercise 7-8 yield $s_1^2 = 5.27$, $s_2^2 = 8.16$, $s_3^2 = 6.82$, $s_4^2 = 25.12$, $s_5^2 = 10.05$. Use $\alpha = .05$ to test the hypothesis of equal variances.

7-10
CONCLUDING REMARKS

In this chapter we have considered only the most elementary type of analysis of variance problem. Experiments whose results can be summarized as in Table 7-1 and which satisfy conditions (7-3) are sometimes called a "one-way classification" or a "completely randomized design." Mainly we have generalized the discussion of Sec. 5-9. Another so-called design is obtained by generalizing in a similar way the type of problem considered in Sec. 5-10. Such a generalization and other more complicated analysis of variance problems are discussed in various places in the literature.

REFERENCES

1 Guenther, William C.: "Analysis of Variance," Prentice-Hall, Inc., Englewood Cliffs, N.J., 1964.

2 Scheffé, Henry: "The Analysis of Variance," John Wiley & Sons, Inc., New York, 1959.

3 Tiku, M. L.: Tables of the Power of the F-Test, *Journal of the American Statistical Association*, vol. 62, pp. 525–539, 1967.

Summary of Results

DRAWING INFERENCES IN ANALYSIS OF VARIANCE

I. Inferences in analysis of r means, $r \geq 2$

    A. Standard hypotheses, alternatives, statistics, and critical regions (significance level α)

Hypothesis	Alternative	Statistic† (observed value)	Critical region
$H_0 : \mu_1 = \mu_2$ $= \cdots = \mu_r$	H_1 : not all r means are equal	$f_{r-1,\,N-r} = \dfrac{SS_A/(r-1)}{SS_W/(N-r)}$ where $$SS_T = \sum_{j=1}^{r} \sum_{i=1}^{n_j} x_{ij}^2 - \frac{t_{..}^2}{N}$$ $$SS_A = \sum_{j=1}^{r} \frac{t_{.j}^2}{n_j} - \frac{t_{..}^2}{N}$$ $$SS_W = SS_T - SS_A$$	$f_{r-1,\,N-r} >$ $f_{r-1,\,N-r;\,1-\alpha}$

† Note that whenever a statistic is given in this summary, certain assumptions are made. These assumptions, giving the statistic the required distribution, may be found in the text material.

    B. Power calculations

        1. When $r = 2$, use Appendix B10 as outlined in VI-B of the Summary of Results, Chap. 5.

        2. If $r = 3, 4, 5, 6$ and the sample sizes are equal, use Appendix B16 with entries α, $v_1 = r - 1$, $v_2 = N - r$, and

$$\phi = \frac{\sqrt{\dfrac{n}{r} \sum_{j=1}^{r} (\mu_j - \mu)^2}}{\sigma}$$

        where $\mu = \sum_{j=1}^{r} \mu_j / r$, the average of the r distribution means.

        3. The sample size n required to guarantee a given power at a specified alternative can be found from Appendix B16 by trial and error.

    C. Confidence region with confidence coefficient $1 - \alpha$ for all possible contrasts: The intervals are given by $(\hat{L} - S\hat{\sigma}_{\hat{L}}, \hat{L} + S\hat{\sigma}_{\hat{L}})$, where

$$\hat{L} = c_1 \bar{x}_{.1} + \cdots + c_r \bar{x}_{.r}$$

$$S^2 = (r-1) F_{r-1,\,N-r;\,1-\alpha}$$

$$\hat{\sigma}_{\hat{L}}^2 = MS_W \sum_{j=1}^{r} \frac{c_j^2}{n_j}$$

369

II. Inferences in analysis of r variances, $r > 2$ (equal sample sizes)
   A. Standard hypotheses, alternatives, statistics, and critical regions
      (significance level α)

Hypothesis	Alternative	Statistic (Observed value)	Critical region
$H_0: \sigma_1^2 = \sigma_2^2$ $= \cdots = \sigma_r^2$	$H_1:$ not all variances are equal	$R_{n,r} = \dfrac{\text{largest sample variance}}{s_1^2 + s_2^2 + \cdots + s_r^2}$	$R_{n,r} > R_{n,r;\, 1-\alpha}$

REGRESSION AND CORRELATION

8-1
INTRODUCTION

In Sec. 5-10 we encountered situations in which the observations occur in pairs. One of the examples considered involved the weights of 20 adults before and after a strenuous physical training program. Thus each individual furnishes two related measurements: his weight before being subjected to the course and his weight after completing the course. It might occur to us that we could use the results associated with individuals who have completed the program to help "predict" final weights of others who are about to start. One of the objectives of this chapter will be to study some techniques which utilize the information furnished by one variable to draw inferences about the parameters which help govern the behavior of a second related variable.

As a second example, suppose we have records that give the grade-point averages from both high school and college for a number of students. We might seek methods which utilize these observed results to give a reasonable estimate of the expected college grade point for incoming freshmen whose high school grade points are available. That is, for those students with a given high school grade point, we might like to "predict" their average college grade point. Again we have two related variables, and want to use the information furnished by one to help characterize the behavior of the other.

In some types of two-variable problems we may be interested in a measure of relationship between two random variables. For example, suppose that college administrators have used in a satisfactory manner the results of a very complicated and extensive entrance examination. A new examination, much simpler to give, has been offered as a substitute. If it can be established that scores made by students taking both examinations show a sufficiently high relationship, the administrators may be willing to make the change.

Let us consider another situation in which knowledge of a measure of relationship might be of interest. Suppose that IQ has been used successfully to predict scholastic achievement. Since shoe size is far easier to obtain, we may prefer to attempt to use the latter measurement in place of IQ if we can establish some degree of relationship. Probably the most sensible hypothesis to test in this case is that there is no relationship at all.

Problems of the first type in which we are primarily interested in drawing inferences about the mean of a random variable given a second variable (which may or may not be random) are generally called *regression* problems. When a measure of relationship is the prime consideration, then the two-variable investigation is referred to as a *correlation* problem. For both types of problems we will consider the usual topics—point estimation, testing hypotheses, interval estimation, and power calculations.

8-2
LINEAR REGRESSION—ESTIMATION OF PARAMETERS

To introduce the subject of regression, let us return to the example concerning high school and college grade points. Suppose that the A, B, C, D, F grading system is used with letters being replaced by 4, 3, 2, 1, 0, respectively, to compute averages. Hence two numbers are associated with each student: his high school grade point x_i and his college grade point y_i. Assume that no student with a high school average of less than 2.00 is admitted to college, so that x_i can range from 2.00 to 4.00 and y_i can range from 0 to 4.00. From available records we select five students whose high school grade points were $x_1 = 2.00$, $x_2 = 2.50$, $x_3 = 3.00$, $x_4 = 3.50$, $x_5 = 4.00$ and find that their college grade points were $y_1 = 2.30$, $y_2 = 1.86$, $y_3 = 2.59$, $y_4 = 2.96$, $y_5 = 3.54$. The x_i's were chosen nonrandomly in order to cover the range of interest followed by a random selection of y_i for each x_i. (Thus $y_1 = 2.30$ was determined by selecting at random a card from the file containing records for those students who had a high school grade point of 2.00.) With these results we would like to estimate the average college grade point for any given value of x, the high school grade point.

In order to proceed, we first formulate a model. Actually, we have already made a start in that direction by assuming that the x_i are selected nonrandomly and y_i is a random observation of Y_i, the college grade point of a student having high school grade point x_i. To be a little more specific, let us imagine that for every value of x, $2 \leq x \leq 4$, there exists a continuous distribution of Y with $0 \leq y \leq 4$. Each of these distributions will have an unknown mean and an unknown variance. Let the mean be $\mu_{Y|x}$, called the *mean of Y given x*, and denote the variance by $\sigma^2_{Y|x}$, called the *variance of Y given x*. It is not unreasonable to assume that all the means lie on a continuous curve, called the *curve of regression*. In this chapter we shall consider only the simplest type of regression curve, namely, the straight line (actually only a line segment in our example). If we assume that all the unknown means lie on a line, this implies

$$\mu_{Y|x} = A + Bx \qquad (8\text{-}1)$$

where A and B are unknown constants. If A and B were known, then for any given x we could readily compute the mean of the Y distribution for that x. For example, suppose that $A = .1$ and $B = .9$ for the grade-point problem. Then with $x = 2.40$ we could evaluate

$$\mu_{Y|x} = .1 + .9x = .1 + .9(2.40) = 2.26$$

This would be our best guess for the college grade point of a person selected at random from those students whose high school grade point was 2.40. Unfortunately, we can only estimate $\mu_{Y|x}$ by using estimates for A and B. Denote these estimates by a, b and the corresponding estimators by $\hat{A}$, $\hat{B}$. Then, in place of the unknown (8-1) we shall use an estimated value given by

$$\hat{\mu}_{Y|x} = a + bx \qquad (8\text{-}2)$$

For the grade-point problem the true but unknown regression line and its estimate might appear as in Fig. 8-1.

Before we discuss the estimates a, b let us add one further assumption, followed by a summary of the conditions for our model. For most of our inferences we need to assume that the variance for every Y distribution is the same. With this addition we have the following assumptions for the so-called *linear regression model:*

(a)   Variables $x_1, x_2, \ldots, x_n$ are selected nonrandomly
(b) $y_1, y_2, \ldots, y_n$ are observed values of random variables $Y_1, Y_2, \ldots, Y_n$
(c)   The regression curve is linear, that is, $\mu_{Y|x} = A + Bx$
(d)   The variance of every Y distribution is the same, that is, $\sigma_{Y|x}^2 = \sigma^2$ for every x

$$(8\text{-}3)$$

FIGURE 8-1

A regression line and its estimate.

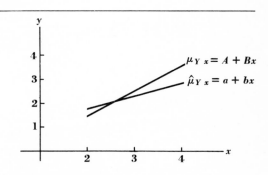

Sometimes we may prefer to consider the x's as observed values of random variables. This would be the case if students' cards were selected at random from the file without first separating the cards according to high school grade point. Then assumptions (8-3) could be changed to

(a) $(x_1, y_1), (x_2, y_2), \ldots, (x_n, y_n)$ are observed pairs of values selected at random from a joint (two-dimensional) distribution of X and Y (8-4)

(b) $\mu_{Y|x} = A + Bx$

(c) $\sigma^2_{Y|x} = \sigma^2$

(Recall that we first mentioned joint distribution in connection with the multinomial, Sec. 2-4). Although the assumptions of linear regression and equal variances seem like a lot to assume, they are often fairly reasonable. In the case of model (8-4), these assumptions are satisfied if X and Y have a two-dimensional normal distribution (which is discussed in the next section).

Without incorporating any further assumptions into our model, we can obtain estimates of a and b by a procedure known as the *method of least squares*. The main justification for using this procedure is that estimates so obtained have some good statistical properties. To follow the explanation of this new concept, consider Fig. 8-2. The five dots represent the five pairs of grade points we have previously observed. The d_i's represent the vertical distances from the points to the estimated regression line. Thus

$d_1 = 2.30 - (a + 2b)$

$d_2 = 1.86 - (a + 2.5b)$

$d_3 = 2.59 - (a + 3b)$

$d_4 = 2.96 - (a + 3.5b)$

$d_5 = 3.54 - (a + 4b)$

and, in general,

$d_i = y_i - (a + bx_i)$

FIGURE 8-2

An estimated regression line determined from five observed points by least squares.

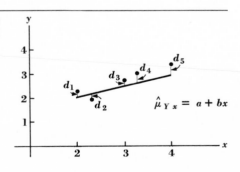

The least-squares regression line is the one obtained by choosing a and b so that

$$d_1^2 + d_2^2 + d_3^2 + d_4^2 + d_5^2 = \sum_{i=1}^{5} d_i^2$$

is a minimum.

To find the least-squares estimates a and b, we use a technique called completing the square (usually encountered in high school algebra). For example, if $u = 2v^2 + 3v + 7$, we can write

$$\begin{aligned}
u &= 2(v^2 + \tfrac{3}{2}v) + 7 \\
&= 2(v^2 + \tfrac{3}{2}v + \tfrac{9}{16} - \tfrac{9}{16}) + 7 \\
&= 2(v^2 + \tfrac{3}{2}v + \tfrac{9}{16}) - 2(\tfrac{9}{16}) + 7 \\
&= 2(v + \tfrac{3}{4})^2 + \tfrac{47}{8}
\end{aligned}$$

(8-5)

The quantity $\tfrac{9}{16} = (\tfrac{1}{2} \cdot \tfrac{3}{2})^2$ was inserted because this figure was needed to enable us to write the parentheses containing v as a perfect square. The form (8-5) makes it obvious that $v = -\tfrac{3}{4}$ gives a minimum value for u of $\tfrac{47}{8}$, since with any other choice of v, u is $\tfrac{47}{8}$ plus something. Now, in general, the sum of the vertical distances squared is

$$\begin{aligned}
\sum_{i=1}^{n} d_i^2 &= \sum_{i=1}^{n} [y_i - (a + bx_i)]^2 = \sum_{i=1}^{n} (y_i - a - bx_i)^2 \\
&= \sum_{i=1}^{n} (y_i^2 + a^2 + b^2 x_i^2 - 2ay_i - 2bx_i y_i + 2abx_i) \\
&= \sum_{i=1}^{n} y_i^2 + na^2 + b^2 \sum_{i=1}^{n} x_i^2 - 2a \sum_{i=1}^{n} y_i - 2b \sum_{i=1}^{n} x_i y_i + 2ab \sum_{i=1}^{n} x_i
\end{aligned}$$

If we complete the square, first on a, then on b, we get, after a few lines of algebra,

$$\begin{aligned}
\sum_{i=1}^{n} d_i^2 = {} & n[a - (\bar{y} - b\bar{x})]^2 \\
& + \left(\sum_{i=1}^{n} x_i^2 - n\bar{x}^2 \right) \left[b - \frac{\sum_{i=1}^{n} x_i y_i - n\bar{x}\bar{y}}{\sum_{i=1}^{n} x_i^2 - n\bar{x}^2} \right]^2 \\
& + \sum_{i=1}^{n} y_i^2 - n\bar{y}^2 - \frac{(\sum_{i=1}^{n} x_i y_i - n\bar{x}\bar{y})^2}{\sum_{i=1}^{n} x_i^2 - n\bar{x}^2}
\end{aligned}$$

(8-6)

where $\bar{x} = \sum_{i=1}^{n} x_i / n$, $\bar{y} = \sum_{i=1}^{n} y_i / n$. Obviously, $\sum_{i=1}^{n} d_i^2$ is minimized if

$$b = \frac{\sum_{i=1}^{n} x_i y_i - n\bar{x}\bar{y}}{\sum_{i=1}^{n} x_i^2 - n\bar{x}^2} \qquad a = \bar{y} - b\bar{x}$$

(8-7)

and the minimum value is

$$(n - 2)s_{Y|x}^2 = \sum_{i=1}^{n} y_i^2 - n\bar{y}^2 - \frac{(\sum_{i=1}^{n} x_i y_i - n\bar{x}\bar{y})^2}{\sum_{i=1}^{n} x_i^2 - n\bar{x}^2}$$

(8-8)

[When calculating (8-7) or (8-8), it is frequently convenient to replace $n\bar{x}^2$, $n\bar{y}^2$, $n\bar{x}\bar{y}$ by $(\sum_{i=1}^{n} x_i)^2/n$, $(\sum_{i=1}^{n} y_i)^2/n$, $(\sum_{i=1}^{n} x_i)(\sum_{i=1}^{n} y_i)/n$, respectively.] The only assumption that we needed to find the least squares estimates a and b was that the regression curve is linear. The estimated regression line we have obtained is the best in the sense of least squares whether or not we look upon the x's as nonrandom variables selected by the experimenter. It can be shown that a and b are unbiased estimates of A and B and that $\hat{\mu}_{Y|x}$ is an unbiased estimate of $\mu_{Y|x}$. If we further assume that the variances are equal, then $s_{Y|x}^2$ defined by (8-8) is an unbiased estimate of $\sigma_{Y|x}^2 = \sigma^2$.

If to assumptions (8-3) we add

(e)   The distribution of Y for each x is normal

or to (8-4) we add

(d)   The pairs (X_i, Y_i) have a two-dimensional normal distribution

then the estimates of a and b can be found by another procedure called the method of maximum likelihood (briefly mentioned in Exercise 3-14). This estimation procedure is a topic for more advanced courses. For our purposes it suffices to know that the method of maximum likelihood gives the same estimates for a and b already obtained by least squares.

EXAMPLE 8-1

Using the observed grade points given earlier in the section, find the estimated regression line and use it to estimate the mean college grade point for students with a high school average of 2.40. Assuming the equal-variance condition is satisfied, find the unbiased estimate of σ^2 and σ.

Solution

We need a and b given by (8-7). We had $x_1 = 2.00$, $y_1 = 2.30$, $x_2 = 2.50$, $y_2 = 1.86$, $x_3 = 3.00$, $y_3 = 2.59$, $x_4 = 3.50$, $y_4 = 2.96$, $x_5 = 4.00$, $y_5 = 3.54$. From these results we find

$$\sum_{i=1}^{5} x_i = 2.00 + 2.50 + 3.00 + 3.50 + 4.00 = 15.00 \qquad \bar{x} = 3.00$$

$$\sum_{i=1}^{5} y_i = 2.30 + 1.86 + 2.59 + 2.96 + 3.54 = 13.25 \qquad \bar{y} = 2.65$$

$$\sum_{i=1}^{5} x_i^2 = (2.00)^2 + (2.50)^2 + (3.00)^2 + (3.50)^2 + (4.00)^2 = 47.50$$

$$\sum_{i=1}^{5} y_i^2 = (2.30)^2 + (1.86)^2 + (2.59)^2 + (2.96)^2 + (3.54)^2 = 36.75$$

$$\sum_{i=1}^{5} x_i y_i = (2.00)(2.30) + (2.50)(1.86) + (3.00)(2.59) + (3.50)(2.96)$$
$$+ (4.00)(3.54) = 41.54$$

Thus

$$b = \frac{41.54 - 5(3.00)(2.65)}{47.50 - 5(3.00)^2} = \frac{1.79}{2.50} = .72$$

$$a = 2.65 - (.72)(3.00) = .49$$

and $\hat{\mu}_{Y|x} = .49 + .72x$

If $x = 2.40$ then

$$\hat{\mu}_{Y|2.40} = .49 + (.72)(2.40) = 2.22$$

is the estimated mean college grade point for those who have a high school grade point of 2.40.

The estimate of σ^2 is obtained from formula (8-8) by dividing by $n - 2 = 3$. We get

$$s^2_{Y|x} = \frac{1}{3}\left[36.75 - 5(2.65)^2 - \frac{(1.79)^2}{2.50} \right]$$

$$= \frac{1}{3}\left[36.75 - 5(7.02) - \frac{3.20}{2.50} \right] = \frac{1}{3}(.37) \cong .12$$

Finally, the estimate of σ is $s_{Y|x} = \sqrt{.12} = .35$.

EXERCISES

◈ 8-1 Suppose that in the grade-point problem we select for our x's, $x_1 = 2.00$, $x_2 = 2.10$, $x_3 = 2.20$, ..., $x_{20} = 3.90$, $x_{21} = 4.00$. Then we draw a random sample of size 2 (so that $n = 42$) from each of the 21 distributions of Y. Computations yield $\sum_{i=1}^{42} x_i = 126.00$, $\sum_{i=1}^{42} x_i^2 = 393.40$, $\sum_{i=1}^{42} y_i = 105.00$, $\sum_{i=1}^{42} y_i^2 = 284.30$, $\sum_{i=1}^{42} x_i y_i = 330.40$. Find (a) the estimated regression line, (b) the estimated mean college grade point for students whose high school grade point is 3.00, (c) the estimate of σ^2 (under the equal-variance assumption), (d) the estimate of σ.

8-2 A large mail order house uses the weight of incoming mail to determine how many of its employees are to be assigned to filling orders on a given day. Suppose that the observed results in Table 8-1 constitute an observed random sample of size 10 taken from the company's records. Assuming that the model with assumptions (8-4) is correct, find the estimated regression line. Use it to estimate the average number of man-hours required to fill the orders if the mail on hand at 7 A.M. weighs 673 pounds. If each employee works an 8-hour day, estimate the number of workers that should be assigned to filling orders that day if it is intended that all orders will be filled by the end of the day. Estimate σ^2 and σ.

TABLE 8-1

Weight of mail versus man-hours

Weight of mail on hand at 7 A.M. (in hundreds of pounds)	Man-hours required to fill orders (in thousands of hours)
5.21	12.6
7.16	17.3
6.34	15.2
8.41	18.5
6.94	15.8
6.52	15.0
7.33	16.8
5.87	13.8
6.61	14.9
8.03	18.0

◈ 8-3 Over a number of years the maximum snow depth has been recorded at a mountain station. In addition the water runoff in the valley below has been measured and recorded in thousands of acre feet. A random sample of 10 years yields the observed results given in Table 8-2. Assuming that model (8-4) is correct, find the estimated regression line. Suppose that this coming winter the mountain station records a maximum snow depth of 20.2 feet. What is the estimate of the expected amount of water runoff in thousands of acre feet? Estimate σ^2 and σ.

TABLE 8-2

Maximum snow depth versus water runoff

Maximum snow depth at mountain station (in feet)	Acre feet of water runoff (in thousands)
15.2	28.6
10.4	19.3
21.2	40.5
18.6	35.6
26.4	48.9
23.7	45.0
13.5	29.2
16.7	34.1
24.0	46.7
19.1	37.4

8-4 For a random sample of 12 boys heights were recorded on their fourteenth and eighteenth birthdays. The following pairs of results were recorded: (72, 74), (58, 68), (64, 70), (67, 76), (56, 66), (59, 67), (66, 70), (63, 72), (65, 70), (60, 67), (69, 70), (62, 67). Letting (x_i, y_i) be a general observed pair, calculations yield $\sum_{i=1}^{12} x_i = 761$, $\sum_{i=1}^{12} x_i^2 = 48{,}505$, $\sum_{i=1}^{12} y_i = 837$, $\sum_{i=1}^{12} y_i^2 = 58{,}483$, $\sum_{i=1}^{12} x_i y_i = 53{,}203$. Assuming that the model (8-4) is correct, find the estimated regression line. If a boy is 60 inches tall on his fourteenth birthday, what is the estimate of his expected height on his eighteenth birthday? Estimate σ^2 and σ.

◆ 8-5   A transistor is to be used in various pieces of equipment operating in temperatures ranging from 200 to 400° centigrade. It is of interest to have information concerning the average length of life of transistors when used at various temperatures. Twenty-seven transistors are tested, three each at 9 different temperature settings. The observed results are given in Table 8-3. Letting (x_i, y_i) be a general observed pair, where x_i is the temperature setting and y_i is the corresponding observed length of life, calculations yield

$$\sum_{i=1}^{27} x_i = 8{,}100 \qquad \sum_{i=1}^{27} x_i^2 = 2{,}542{,}500 \qquad \sum_{i=1}^{27} y_i = 12{,}883$$

$$\sum_{i=1}^{27} y_i^2 = 6{,}419{,}899 \qquad \sum_{i=1}^{27} x_i y_i = 3{,}697{,}425$$

Assuming that model (8-3) is correct, find the estimated regression line. If transistors are to be used at 280° centigrade, what is the estimate of the average length of life? Estimate σ^2 and σ.

TABLE 8-3

Operating temperature versus length of life

Temperature (degrees centigrade)	200	225	250	275	300	325	350	375	400
Length of life (hours)	634	551	593	530	467	455	432	396	325
	584	630	505	562	431	470	400	341	350
	670	576	528	499	502	403	385	362	302

8-3
THE BIVARIATE NORMAL PROBABILITY MODEL

In our discussion of the multinomial (Sec. 2-4), we observed that a formula which yields probabilities for two or more random variables is called a joint probability function and that the random variables are said to have a joint probability distribution. In the multinomial case we were concerned with discrete random variables. In Sec. 1-6 in connection with one continuous random variable, we introduced the concept of a density function. We learned that any function $f(x)$ which graphs into a curve and has the properties (1) it is nonnegative and (2) the total area under the curve is 1, qualifies as a density function. Partial areas under the curve are then fractions and can be interpreted as probabilities. The difficult problem we encounter in practice is the selection of the appropriate density for the problem at hand. In Chap. 2 we discussed some useful density functions, one of which was the normal density. Now we would like to generalize the concept of a density function to the two-variable case and, in particular, the one-variable normal to a two-variable normal.

Actually, we have already considered two or more continuous random variables in the same problem. In Sec. 5-9 we were concerned with drawing inferences about the means of two distributions. Each distribution was assumed to be a normal, but the two distributions were in no way related to one another. (In Chap. 7 we generalized to r such distributions.) In Sec. 5-10 we encountered problems generating a random sample of pairs such that the two random variables of a pair were related to one another. Although we did not need to do so at that time, we may seek a model (two-variable density) which gives realistic probabilities for related pairs. Since the normal has been used so extensively, it seems natural to consider a two-variable (or bivariate) normal as a possibility for such cases.

Any function $f(x, y)$ which graphs into a three-dimensional surface can be used as a density provided it has the following properties:

(a) It is nonnegative.
(b) The total volume under the surface is 1.

(To explain the phrase "under the surface," suppose some configuration which may, for example, be bell-shaped, hat-shaped, or box-shaped is sitting on a table. The volume under the surface is that volume which is under the configuration but above the table.) Partial volumes under the surface are then fractions and can be interpreted as probabilities. As in the one-variable case, we are faced with the problem of selecting a density which yields realistic probabilities for a given two-variable situation.

In this section the only joint density we will consider is the bivariate normal. It would seem reasonable to require that each variable when considered by itself should have a one-variable normal distribution. Thus, if (X, Y) represents an individual's high school and college grade points, high school grade points should have a normal distribution and college grade points should have a normal distribution. In addition, the model should reflect in some way that X and Y are related variables. The bivariate normal model has both of these properties. Its density function is

$$f(x, y) = \frac{1}{2\pi\sigma_X \sigma_Y \sqrt{1 - \rho^2}} \exp\left\{\frac{1}{2(1 - \rho^2)}\left[\frac{(x - \mu_X)^2}{\sigma_X^2} - \frac{2\rho(x - \mu_X)(y - \mu_Y)}{\sigma_X \sigma_Y} + \frac{(y - \mu_Y)^2}{\sigma_Y^2}\right]\right\}$$

$$(8\text{-}9)$$

where $\exp(u) = e^{-u}$. Fortunately, we will not have to use this complicated formula. (We did not have to work with formulas for one-variable densities either.) We do observe, however, that it depends upon the two means μ_X, μ_Y, the two variances σ_X^2, σ_Y^2, and one new parameter ρ, called the correlation coefficient. As we shall see later, ρ can be regarded as a "measure of relationship" between X and Y.

For our purposes we should know some properties of the bivariate normal distribution. The more important ones are

(a) The random variable X when considered by itself has a normal distribution with mean μ_X and variance σ_X^2. Similarly, the random variable Y when considered by itself has a normal distribution with mean μ_Y and variance σ_Y^2. If, in a given situation, we can believe that both X and Y have normal distributions, then the bivariate normal will be a reasonable model.

(b) The means of Y distributions for different fixed values of X, say x, lie on a straight line. That is, $\mu_{Y|x} = A + Bx$. (A similar comment can be made about the means of X distributions for fixed Y, say y. Thus $\mu_{X|y} = C + Dy$ where C and D are constants.)

(c) The variance of Y for any given X, say x, denoted by $\sigma_{Y|x}^2$, is the same for all x. (The same comment can be made about the variance of X for any given Y, say y.)

(d) Geometrically, (8-9) is a bell-shaped surface. Tables exist for evaluating partial volumes, but in the next two sections we have no need for such tables.

Because of properties (b) and (c), the assumptions of linear regression and equal variances (made in Sec. 8-2) are often reasonable.

We comment that it is possible to construct two-variable distributions other than the bivariate normal having property (a). However, these distributions are of mathematical, not practical, interest. Generally if we believe that both X and Y are normal random variables, the bivariate normal model is appropriate.

8-4
INFERENCES IN REGRESSION ANALYSIS

For the estimation problems of Sec. 8-2, it was not necessary to make any assumptions regarding the specific form of the sampled distribution. However, tests and confidence intervals for the parameters $\mu_{Y|x}$, A, B, and $\sigma_{Y|x}^2$ are derived under the appropriate normality assumptions. We can assume either

(a) Variables $x_1, x_2, \ldots, x_n$ are selected nonrandomly

(b) $y_1, y_2, \ldots, y_n$ are observed values of random variables
$Y_1, Y_2, \ldots, Y_n$

(c) The regression curve is linear, that is, $\mu_{Y|x} = A + Bx$ (8-10)

(d) The variance of every Y distribution is the same, that is,
$\sigma^2_{Y|x} = \sigma^2$ for every x

(e) For every x the distribution of Y is normal

[(8-10) is (8-3) plus normality] or

(a) $(x_1, y_1), (x_2, y_2), \ldots, (x_n, y_n)$ are observed pairs of random
variables selected at random from a two-dimensional distribution (8-11)

(b) The pairs (X_i, Y_i) have a bivariate normal distribution

We note that with (8-11) we need not list "the regression curve is linear"
and "the variance of every Y distribution is the same" since these are
properties of the bivariate normal distribution. Tests of hypotheses and
confidence intervals are the same for both sets of assumptions (8-10) and
(8-11).

Perhaps the most interesting tests concern $\mu_{Y|x}$ for a given value of x,
say $x = x_0$. The standard hypothesis testing situations are the same as
before with means; only the notation is different. Thus, we may be
interested in testing

$$H_0 : \mu_{Y|x_0} = \mu_0 \qquad \text{against} \qquad H_1 : \mu_{Y|x_0} < \mu_0 \qquad (8\text{-}12)$$

$$H_0 : \mu_{Y|x_0} = \mu_0 \qquad \text{against} \qquad H_1 : \mu_{Y|x_0} > \mu_0 \qquad (8\text{-}13)$$

or

$$H_0 : \mu_{Y|x_0} = \mu_0 \qquad \text{against} \qquad H_1 : \mu_{Y|x_0} \neq \mu_0 \qquad (8\text{-}14)$$

[As in earlier discussions, if we so desire we can replace the hypotheses of
(8-12) and (8-13) by $H_0 : \mu_{Y|x_0} \geqq \mu_0$ and $H_0 : \mu_{Y|x_0} \leqq \mu_0$.] If either
assumptions (8-10) or assumptions (8-11) are satisfied and if $\mu_{Y|x_0} = \mu_0$,
then the statistic T_{n-2} whose observed value is

$$t_{n-2} = \frac{a + bx_0 - \mu_0}{s_{Y|x}\sqrt{\dfrac{1}{n} + \dfrac{(x_0 - \bar{x})^2}{\sum_{i=1}^{n} x_i^2 - n\bar{x}^2}}} \qquad (8\text{-}15)$$

has a t distribution with $n - 2$ degrees of freedom. [We note that (8-15)
is very similar to the observed value of (5-22), used for tests in Sec. 5-5.
The quantity $a + bx_0$ has replaced $\bar{x}$, $s_{Y|x}$ has replaced s, the two-term
expression under the square root has replaced $1/n$, and degrees of
freedom $n - 2$ has replaced $n - 1$.] When testing (8-12) small values
of $\hat{\mu}_{Y|x_0} = a + bx_0$, and small values of t_{n-2}, support H_1 and lead to
rejection. Consequently, a reasonable critical region is

$$t_{n-2} < t_{n-2;\alpha} \qquad (8\text{-}16)$$

Similarly, the H_1 of (8-13) is supported by large $\hat{\mu}_{Y|x_0}$, and large t_{n-2}, so that a reasonable critical region is

$$t_{n-2} > t_{n-2;1-\alpha} \tag{8-17}$$

Finally, when testing (8-14) both small and large values of $\hat{\mu}_{Y|x_0}$, and t_{n-2}, support H_1. The critical region usually selected is

$$t_{n-2} < t_{n-2;\alpha/2} \quad \text{and} \quad t_{n-1} > t_{n-1;1-\alpha/2} \tag{8-18}$$

EXAMPLE 8-2

Suppose that a 2.00 college grade point is required for graduation. Using the observed results of Example 8-1, test a hypothesis which, if rejected, would indicate that the average college grade point is less than 2.00 for students whose high school grade point is 2.00. Use $\alpha = .05$.

Solution

We shall follow the six-step outline.

1 The problem suggests that we test $H_0: \mu_{Y|x_0} = 2.00$ against $H_1 : \mu_{Y|x_0} < 2.00$ where $x_0 = 2.00$.

2 The significance level $\alpha = .05$ and the sample size $n = 5$ have already been selected.

3 The statistic we use is T_{n-2}, whose observed value is given by (8-15). If H_0 is true, then $T_{n-2} = T_3$ has a t distribution with 3 degrees of freedom provided conditions (8-10) are satisfied. That is, we would like to believe that (a) the x's are selected nonrandomly (which they were), (b) for the given x's the y's were observed at random (which they were), (c) the regression curve is linear, (d) the variance of each Y distribution is the same, and (e) for every x the Y distribution is normal. Of course, we cannot be sure that conditions (c) to (e) are satisfied, but these assumptions are not too unreasonable.

4 The critical region is $t_3 < t_{3;.05} = -2.353$.

5 From Example 8-1 we have $a = .49$, $b = .72$, $s_{Y|x} = .35$, $\sum_{i=1}^{5} x_i^2 - 5\bar{x}^2 = 2.50$, $\bar{x} = 3.00$. Thus $a + bx_0 = .49 + .72(2.00) = 1.93$, and the observed value of the statistic is

$$t_3 = \frac{1.93 - 2.00}{.35\sqrt{\frac{1}{5} + (2.00 - 3.00)^2/2.50}}$$

$$= \frac{-.07}{.35\sqrt{.60}} = \frac{-.07}{(.35)(.77)} = \frac{-.07}{.27} = -.26$$

6 We do not reject H_0. Thus, on the basis of this small sample we do not conclude that the average college grade point is less than 2.00 for those whose high school grade point is 2.00.

In working a real problem like Example 8-2, we could proceed in another way. Alternatively, we could sample college grade points only from those who have a high school grade point of 2.00 and use the t test described in Sec. 5-5. One possible advantage in using the methods of this chapter is that larger sample sizes are generally available when several (instead of one) Y distribution is considered. In addition, the observed results can also be used to draw inferences for other given values of x.

Power calculations are performed in the usual manner with t tests by using

$$\delta = \frac{|\mu_0 - \mu_1|}{\sigma\sqrt{\dfrac{1}{n} + \dfrac{(x_0 - \bar{x})^2}{\sum_{i=1}^{n} x_i^2 - n\bar{x}^2}}} \tag{8-19}$$

$f = n - 2$, and Appendix B10 when we seek the probability of rejecting H_0 if $\mu_{Y|x_0} = \mu_1$, where μ_1 falls in the range of the alternative H_1. Thus, we see that power (unlike the critical region) depends on the given (or selected) set of x's, $x_1, \ldots, x_n$ unless we make $x_0 = \bar{x}$ (which we can do if we select the x's). We note that $x_0 = \bar{x}$ makes the denominator of (8-19) as small as possible which in turn makes δ as large as possible, Hence, selecting $x_0 = \bar{x}$ maximizes the power for all values of $\mu_{Y|x_0}$ in the alternative. Sample size problems are a little more complicated than with previous t tests and will be omitted.

EXAMPLE 8-3

In Example 8-2 find the power of the test if the average college grade point is σ units below 2.00 when $x_0 = 2.00$.

Solution

We have $\mu_0 = 2.00$, $\mu_1 = 2.00 - \sigma$, $|\mu_0 - \mu_1| = \sigma$. Thus $\delta = \sigma/\sigma(.77) = 1.30$. Turning to Appendix B10 with $\alpha = .05$, $f = 3$, we find that the power is .26. Of course, with a larger n we would expect to get a larger value of power.

To obtain a confidence interval for $\mu_{Y|x_0}$ with confidence coefficient $1 - \alpha$, we start with

$$\Pr(-t_{n-2;\,1-\alpha/2} < T_{n-2} < t_{n-2;\,1-\alpha/2}) = 1 - \alpha$$

where the observed value of T_{n-2} is given by (8-15) with μ_0 replaced by $\mu_{Y|x_0}$. The usual manipulation with inequalities yields the confidence interval

$$\left(a + bx_0 - t_{n-2;\,1-\alpha/2}\, s_{Y|x}\sqrt{\frac{1}{n} + \frac{(x_0 - \bar{x})^2}{\sum_{i=1}^{n} x_i^2 - n\bar{x}^2}}, \right.$$

$$\left. a + bx_0 + t_{n-2;\,1-\alpha/2}\, s_{Y|x}\sqrt{\frac{1}{n} + \frac{(x_0 - \bar{x})^2}{\sum_{i=1}^{n} x_i^2 - n\bar{x}^2}} \right) \tag{8-20}$$

The interval (8-20) can be written down immediately from (5-28) by using the similarities relating (8-15) and (5-22) (which we previously noted). The interval (8-20) covers a value μ_0 if, and only if, the test of (8-14) accepts H_0 when the same α (and the same sample) is used.

The interval (8-20) is appropriate only for one given value of $x = x_0$. Sometimes it may be desirable to have a confidence region for the entire regression line. By an argument similar to the one used to derive the Scheffé intervals for contrasts (discussed in Chap. 7), it may be shown that the desired region is given by

$$\left(a + bx - s_{Y|x} \sqrt{2f_{2,\,n-2;\,1-\alpha}\left[\frac{1}{n} + \frac{(x-\bar{x})^2}{\sum_{i=1}^{n} x_i^2 - n\bar{x}^2}\right]}, \right.$$
$$\left. a + bx + s_{Y|x} \sqrt{2f_{2,\,n-2;\,1-\alpha}\left[\frac{1}{n} + \frac{(x-\bar{x})^2}{\sum_{i=1}^{n} x_i^2 - n\bar{x}^2}\right]} \right) \tag{8-21}$$

The random interval associated with (8-21) captures the unknown regression line with probability $1 - \alpha$. While intervals computed from (8-20) contain *one* value $\mu_{Y|x_0}$ about $1 - \alpha$ of the time over the long run, intervals computed from (8-21) contain *all* values of $\mu_{Y|x}$ about $1 - \alpha$ of the time over the long run.

If we let x_0 in (8-20) take on different values, then another region is also generated. This region is the one between the two dotted curves in Fig. 8-3. The region produced by (8-21) is between the two solid curves.

FIGURE 8-3

Confidence intervals for $\mu_{Y|x_0}$ and confidence region for $\mu_{Y|x}$.

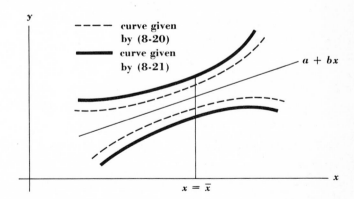

EXAMPLE 8-4

Using the observed results given for the grade-point problem of Example 8-1, find a confidence interval with confidence coefficient .95 for $\mu_{Y|x_0}$ if

$x_0 = 3.50$. Using $1 - \alpha = .95$ compute the intervals given by (8-21) if $x = 3.50$ and if $x = 3.00$.

Solution

We have $a + bx_0 = .49 + .72(3.50) = 3.01$,

$$s_{Y|x}\sqrt{\frac{1}{n} + \frac{(x_0 - \bar{x})^2}{\sum_{i=1}^{n} x_i^2 - n\bar{x}^2}} = .35\sqrt{\frac{1}{5} + \frac{(3.50 - 3.00)^2}{2.50}} = .35\sqrt{.3} = .19$$

$1 - \alpha/2 = .975$, $t_{3;.975} = 3.182$. Therefore, the interval is

$$[3.01 - 3.182(.19), 3.01 + 3.182(.19)] = (3.01 - .60, 3.01 + .60)$$

$$= (2.41, 3.61)$$

The interval is wide because n is small.

To use (8-21) to find an interval for $x = 3.50$, we need again $a + bx = .49 + .72(3.50) = 3.01$. Since $f_{2, n-2; 1-\alpha} = f_{2,3;.95} = 19.2$, we get

$$s_{Y|x}\sqrt{2f_{2, n-2; 1-\alpha}\left[\frac{1}{n} + \frac{(x - \bar{x})^2}{\sum_{i=1}^{n} x_i^2 - n\bar{x}^2}\right]}$$

$$= .35\sqrt{2(19.2)\left[\frac{1}{5} + \frac{(3.50 - 3.00)^2}{2.50}\right]} = .35\sqrt{38.4(.3)}$$

$$= .35\sqrt{11.52} = .35(3.39) = 1.19$$

Then, the interval is $(3.01 - 1.19, 3.01 + 1.19)$ or $(1.82, 4.20)$. Since grades cannot exceed 4.00 we would use the interval $(1.82, 4.00)$. We would expect to get a wider interval than by the first method since the second procedure attempts to capture all $\mu_{Y|x}$ at one time.

To use (8-21) to find an interval for $x = 3.00$, we need $a + bx = .49 + .72(3.00) = 2.65$ and

$$.35\sqrt{2(19.2)\left[\frac{1}{5} + \frac{(3.00 - 3.00)^2}{2.50}\right]} = .35\sqrt{38.4(.2)} = .97$$

Then, the interval is $(2.65 - .97, 2.65 + .97)$ or $(1.68, 3.62)$.

Sometimes inferences concerning B are of interest. This parameter is the slope of the regression line. By "slope" we mean the rate of increase of $\mu_{Y|x}$ divided by the rate of increase of x. (Hence, if $\mu_{Y|x}$ increases $\frac{1}{2}$ a unit for every unit that x increases, then $B = \frac{1}{2}$. If $\mu_{Y|x}$ decreases 2 units for every unit that x increases, then $B = -2$. If $\mu_{Y|x}$ is the same for all x, then $B = 0$ and the regression line is parallel to the x axis.) Although we can test

$$H_0 : B = B_0 \quad \text{against} \quad H_1 : B < B_0 \tag{8-22}$$

or

$$H_0 : B = B_0 \quad \text{against} \quad H_1 : B > B_0 \tag{8-23}$$

or

$$H_0 : B = B_0 \quad \text{against} \quad H_1 : B \neq B_0 \tag{8-24}$$

for any value B_0, the most useful case is (8-24) with $B_0 = 0$. If B is zero, then $\mu_{Y|x}$ is the same for all x, so that the x variable is of no assistance in drawing inferences. In other words, every Y distribution is the same. The statistic which is used to test the above hypothesis is T_{n-2} with observed value

$$t_{n-2} = \frac{b - B_0}{s_{Y|x}} \sqrt{\sum_{i=1}^{n} x_i^2 - n\bar{x}^2} \tag{8-25}$$

If either set of assumptions (8-10) or (8-11) is satisfied and $B = B_0$, then T_{n-2} has a t distribution with $n - 2$ degrees of freedom. As we might expect, we reject when t_{n-2} is small in the first situation (8-22), large in second (8-23), and either small or large in the third (8-24). Thus critical regions could again be given in the form (8-16), (8-17), and (8-18). For power calculations we use Appendix B10 in the usual way with $f = n - 2$ and

$$\delta = \frac{|B_0 - B_1| \sqrt{\sum_{i=1}^{n} x_i^2 - n\bar{x}^2}}{\sigma} \tag{8-26}$$

when we seek the probability of rejecting H_0 if $B = B_1$.

The standard argument yields

$$\left(b - \frac{t_{n-2;\,1-\alpha/2}\, s_{Y|x}}{\sqrt{\sum_{i=1}^{n} x_i^2 - n\bar{x}^2}}, \; b + \frac{t_{n-2;\,1-\alpha/2}\, s_{Y|x}}{\sqrt{\sum_{i=1}^{n} x_i^2 - n\bar{x}^2}} \right) \tag{8-27}$$

as a confidence interval for B with confidence coefficient $1 - \alpha$. The interval (8-27) has the usual relationship to the two-sided test.

EXAMPLE 8-5

Using the observed results given for the grade-point problem of Example 8-1, test the hypothesis that the regression line has 0 slope. Use $\alpha = .05$.

Solution

We shall follow the six-step outline.

1 We test $H_0 : B = 0$ against $H_1 : B \neq 0$.
2 The significance level $\alpha = .05$ and the sample size $n = 5$ have already been selected.
3 The statistic we use is T_{n-2}, whose observed value is given by (8-25). If the hypothesis is true, then $T_{n-2} = T_3$ has a t distribution with 3 degrees of freedom provided the five conditions discussed in step 3 of Example 8-2 are satisfied.
4 The critical region is $t_3 < t_{3;.025} = -3.182$ and $t_3 > t_{3;.975} = 3.182$.
5 From Example 8-1 we have $b = .72$, $s_{Y|x} = .35$, $\sum_{i=1}^{n} x_i^2 - n\bar{x}^2 = 2.50$. Thus the observed value of the statistic is

$$t_3 = \frac{.72 - 0}{.35} \sqrt{2.50} = (2.06)(1.58) = 3.25$$

6 We reject the hypothesis that the slope of the regression line is 0 and conclude that the x variable is worth using in further analysis.

EXAMPLE 8-6

Find the power of the test used in Example 8-5 if the slope is actually 2σ.

Solution

We have $B_0 = 0$, $B_1 = 2\sigma$, $f = 3$,

$$\delta = \frac{|0 - 2\sigma| \sqrt{2.50}}{\sigma} = 2\sqrt{2.50} = 2(1.58) = 3.16$$

Since the alternative is two-sided, we enter Appendix B10 with $\alpha/2 = .025$. We find that the power is approximately .56.

EXAMPLE 8-7

Using the observed results given for the grade-point problem of Example 8-1, find a confidence interval for B with confidence coefficient .95.

Solution

Since $b = .72$, $1 - \alpha/2 = .975$, $t_{3;.975} = 3.182$, $s_{Y|x} = .35$, $\sum_{i=1}^{5} x_i^2 - 5\bar{x}^2 = 2.50$, the interval is

$$\left[.72 - \frac{(3.182)(.35)}{\sqrt{2.50}}, .72 + \frac{(3.182)(.35)}{\sqrt{2.50}} \right]$$

or $(.72 - .70, .72 + .70)$, which reduces to $(.02, 1.42)$. Thus we infer that $\mu_{Y|x}$ increases somewhere between .02 and 1.42 units when x increases 1 unit.

Sometimes we like to have an interval which has a good chance of capturing one future observation of Y for a given x, say $x = x_0$. The interval used when either assumptions (8-10) or (8-11) are satisfied is

$$\left(a + bx_0 - t_{n-2;1-\alpha/2}\, s_{Y|x}\sqrt{1 + \frac{1}{n} + \frac{(x_0 - \bar{x})^2}{\sum_{i=1}^{n} x_i^2 - n\bar{x}^2}}, \right.$$

$$\left. a + bx_0 + t_{n-2;1-\alpha/2}\, s_{Y|x}\sqrt{1 + \frac{1}{n} + \frac{(x_0 - \bar{x})^2}{\sum_{i=1}^{n} x_i^2 - n\bar{x}^2}}\right) \tag{8-28}$$

The probability is $1 - \alpha$ that the random interval associated with (8-28) contains one future value of Y. An observed interval like (8-28), designed to capture one or more future observations of a random variable with a high degree of confidence, is usually called a *prediction interval* (as contrasted with a confidence interval designed to capture a parameter of a distribution). The probability $1 - \alpha$ could be called the *prediction coefficient*. The interval (8-28) may or may not contain the next observed value of Y. However, if many such intervals are computed, all based on samples followed by observing another Y for $x = x_0$, then about $1 - \alpha$ of the intervals will contain the next observation.

EXAMPLE 8-8

Using the observed results given for the grade-point problem of Example 8-1, find a prediction interval with prediction coefficient .95 for one future observation of college grade point, given the high school grade point is 2.40.

Solution

Since $\alpha = .05$, $1 - \alpha/2 = .975$, we need $t_{n-2;1-\alpha/2} = t_{3;.975} = 3.182$. Also $a + bx_0 = .49 + .72(2.40) = 2.22$, $s_{Y|x} = .35$, and

$$t_{n-2;1-\alpha/2}\, s_{Y|x}\sqrt{1 + \frac{1}{n} + \frac{(x_0 - \bar{x})^2}{\sum_{i=1}^{n} x_i^2 - n\bar{x}^2}}$$

$$= 3.182(.35)\sqrt{1 + \frac{1}{5} + \frac{(2.40 - 3.00)^2}{2.50}}$$

$$= 1.11\sqrt{1.344} = (1.11)(1.16) = 1.29$$

Hence the interval is $(2.22 - 1.29, 2.22 + 1.29)$ or $(.93, 3.51)$. The interval is wide because the prediction coefficient is large and n is small.

Inferences concerning A, which is the value of $\mu_{Y|x}$ if $x = 0$, can be handled as a special case of inferences about $\mu_{Y|x_0} = \mu_{Y|0}$. Inferences about

$\sigma_{Y|x}^2 = \sigma^2$ are handled according to the procedures of Sec. 5-7 using the statistic

$$Y_{n-2} = \frac{(n-2)S_{Y|x}^2}{\sigma^2} \tag{8-29}$$

instead of

$$Y_{n-1} = \frac{(n-1)S^2}{\sigma^2}$$

EXERCISES

◆ 8-6   Rework Example 8-2 using the observed results of Exercise 8-1 (instead of the observed results of Example 8-1).

◆ 8-7   Using the observed results of Exercise 8-1, find a confidence interval with confidence coefficient .95 for $\mu_{Y|x_0}$ if $x_0 = 3.5$.

◆ 8-8   Find the power of the test used in Exercise 8-6 if the true average college grade point is σ units below 2.00 when $x_0 = 2.00$.

◆ 8-9   Using the observed results of Exercise 8-1 and $\alpha = .05$, test the hypothesis $B = 0$ against the alternative $B \neq 0$. Suppose we wish to "prove" that the slope of the regression line is greater than .75. Formulate the hypothesis and the alternative and draw the appropriate conclusion using $\alpha = .05$.

8-10   Find the power of the first test conducted in Exercise 8-9 if $B = \sigma$. Find the power of the second test conducted in that exercise if the slope is σ units larger than .75.

8-11   Using the observed results of Exercise 8-1, find a confidence interval with confidence coefficient .95 for B.

◆ 8-12   Use the observed results of Exercise 8-1 to find a prediction interval with prediction coefficient .95 for the college grade point of a student randomly selected from the distribution which has high school grade point equal to 2.40.

◆ 8-13   With $\alpha = .05$ and the observed results of Exercise 8-1, test

$$H_0 : \sigma_{Y|x}^2 = \sigma^2 = .25 \qquad \text{against} \qquad H_1 : \sigma_{Y|x}^2 = \sigma^2 < .25.$$

8-14   Using the observed results given in Exercise 8-2, find a prediction interval with prediction coefficient .95 for the number of man-hours required to fill the orders if the mail on hand at 7 A.M. weighs 673 pounds.

8-15   Using the observed results of Exercise 8-2, find a confidence interval for B with confidence coefficient .95.

◆ 8-16   Using the observed results of Exercise 8-2, find a confidence region for the regression line with confidence coefficient .95. Then find the specific intervals that this region yields for $x = 6.00, 7.00, 8.00$. Give a verbal interpretation of the results.

◈ 8-17 For the snow-water runoff situation described in Exercise 8-3, it is hypothesized that a maximum snow depth of 20 feet produces on the average 40,000 acre feet of runoff. Using $\alpha = .01$ and the observed results of that exercise, test this hypothesis.

◈ 8-18 For the test used in Exercise 8-17 find the power if the average runoff differs from 40,000 by 1 standard deviation. (This power is correct only for the observed set of x's.)

◈ 8-19 Using the observed results given in Exercise 8-3, find a prediction interval with prediction coefficient .95 for next summer's water runoff in acre feet if this winter's maximum snow depth is 20.2 feet. Also find a confidence interval with confidence coefficient .95 for the $\mu_{Y|x_0}$ when $x_0 = 20.2$.

8-20 Consider again Exercise 8-4 concerning the heights of boys on their fourteenth and eighteenth birthdays. Using prediction coefficient .95 and the observed results given in that problem, find a prediction interval for a boy's height on his eighteenth birthday if he is 60 inches tall on his fourteenth birthday. Based upon this interval, does a 14-year-old who is 60 inches tall have much hope of being 6 feet 4 inches tall on his eighteenth birthday?

8-21 Using the observed results of Exercise 8-4 and a confidence coefficient of .95, find a confidence interval for the mean height of 18-year-old boys who were 60 inches tall on their fourteenth birthday.

8-22 Using $\alpha = .05$ and the observed results of Exercise 8-4, test the hypothesis that boys who are 60 inches tall on their fourteenth birthday will on the average be 70 inches tall on their eighteenth birthday.

8-23 Find the power of test used in Exercise 8-22 if the actual mean height differs from 70 inches by 1 standard deviation. (This power is correct only for the observed set of x's.)

◈ 8-24 Consider again the transistor problem of Exercise 8-5. Using $\alpha = .01$, we would like to be able to prove that the average life of a transistor will exceed 475 hours when operated at a temperature of 280° centigrade. Formulate the appropriate hypothesis testing situation, conduct the test, and draw the conclusions. For observed results, use those given in Exercise 8-5.

8-25 Find the power of the test used in Exercise 8-24 if the actual mean life at 280° centigrade is 1 standard deviation larger than 475 hours.

8-26 Using the observed results of Exercise 8-5, find a confidence interval with confidence coefficient .90 for the mean life of transistors operated at 400° centigrade.

◈ 8-27 Using the observed results of Exercise 8-5 and a prediction coefficient of .90, find a prediction interval for the length of life of a transistor to be operated at 400° centigrade.

8-5

INFERENCES IN CORRELATION ANALYSIS

In Sec. 8-3 we observed that the density function of a bivariate normal distribution contains a parameter ρ, called the correlation coefficient, which can be considered as a measure of relationship between the two random variables X and Y. It can be shown

1 $-1 \leqq \rho \leqq 1$, that is, ρ must be between -1 and $+1$ inclusive.

2 When X and Y have a bivariate normal distribution, X and Y are independent if, and only if, $\rho = 0$. [Recall our discussion of independent events in Chap. 1. Two events are independent when the occurrence or nonoccurrence of one event in no way influences the occurrence or nonoccurrence of the other. For X and Y to be independent, the experiment which generates a value x is in no way related to or dependent upon the experiment which generates a value y. For independent events A_1 and A_2 we had $\Pr(A_1 A_2) = \Pr(A_1)\Pr(A_2)$. For independent random variables X and Y we can write

$$\Pr(X \leqq x, Y \leqq y) = \Pr(X \leqq x)\Pr(Y \leqq y)$$

and

$$\Pr(X = x, Y = y) = \Pr(X = x)\Pr(Y = y)$$

for every x, y.]

3 If $\rho = -1$ or $\rho = +1$, then all points (x, y) lie on the regression line. In other words, for each x the distribution of Y is concentrated at a single point, so that actually Y does not have a distribution in the usual sense.

4 The distribution of Y for any given x is closely concentrated about the regression line (that is, Y has small variance) if ρ is close to -1 or $+1$. The variance $\sigma^2_{Y|x} = \sigma^2$ is largest if $\rho = 0$. The further ρ is from 0, the smaller is σ^2. When $\rho = -1$ or $\rho = +1$, then $\sigma^2 = 0$.

5 If ρ is positive, then the regression line upon which the $\mu_{Y|x}$ lie has a positive slope. Thus, as x is increased, so is $\mu_{Y|x}$; and the larger the value of x, the larger an observed value y tends to be. If ρ is negative, then the regression line has negative slope and $\mu_{Y|x}$ decreases as x increases. (Similar statements could be made concerning the line containing $\mu_{X|y}$.)

Thus in the bivariate normal situation, comments 2, 3, and 4 above imply that ρ can be regarded as a measure of relationship in the sense that it is an indicator of the closeness of concentration about the regression line for the distribution of Y given x. When $\rho = 0$, we regard X and Y as being completely unrelated. At the other extreme, when $\rho = -1$ or $\rho = +1$, we regard X and Y as being perfectly related. How one looks upon other values of ρ depends to a great extent upon the field of application and the accumulated experience of many research workers in that field, but a ρ close to zero indicates little relationship between X and Y while a ρ close to -1 or $+1$ indicates a strong relationship.

The parameter ρ, like other parameters we have considered, will generally be unknown. If $(X_1, Y_1), (X_2, Y_2), \ldots, (X_n, Y_n)$ is a random sample of pairs drawn from a bivariate normal distribution, then

$$R = \frac{1}{n-1} \sum_{i=1}^{n} \left(\frac{X_i - \overline{X}}{S_X} \right) \left(\frac{Y_i - \overline{Y}}{S_Y} \right) \tag{8-30}$$

called the *sample correlation coefficient*, is an unbiased estimator of ρ. Here S_X^2 and S_Y^2 are the sample variances of the X's and Y's, respectively. After performing the experiment we observe $(x_1, y_1), (x_2, y_2), \ldots, (x_n, y_n)$ from which we can compute the *observed sample correlation coefficient*

$$r = \frac{1}{n-1} \sum_{i=1}^{n} \left(\frac{x_i - \bar{x}}{s_X} \right) \left(\frac{y_i - \bar{y}}{s_Y} \right) \tag{8-31}$$

From (8-31) we can obtain the computational formula

$$r = \frac{\sum_{i=1}^{n} x_i y_i - n\bar{x}\bar{y}}{\sqrt{(\sum_{i=1}^{n} x_i^2 - n\bar{x}^2)(\sum_{i=1}^{n} y_i^2 - n\bar{y}^2)}} \tag{8-32}$$

(where again it is frequently convenient to replace $n\bar{x}^2$, $n\bar{y}^2$, $n\bar{x}\bar{y}$ by $(\sum_{i=1}^{n} x_i)^2/n$, $(\sum_{i=1}^{n} y_i)^2/n$ and $(\sum_{i=1}^{n} x_i)(\sum_{i=1}^{n} y_i)/n$, respectively). Like ρ, it can be shown that r must always be between -1 and $+1$ inclusive.

One of the most useful hypothesis testing situations is the test of independence of X and Y. In terms of ρ we test

$$H_0 : \rho = 0 \qquad \text{against} \qquad H_1 : \rho \neq 0 \tag{8-33}$$

The statistic we use for (8-33) is

$$T_{n-2} = \frac{R\sqrt{n-2}}{\sqrt{1-R^2}} \tag{8-34}$$

which has a t distribution with $n - 2$ degrees of freedom when the hypothesis $\rho = 0$ is true and (a) $(x_1, y_1), (x_2, y_2), \ldots, (x_n, y_n)$ is an observed random sample of n pairs and (b) the pairs are drawn from a bivariate normal distribution. A test with significance level α is obtained by rejecting when

$$t_{n-2} = \frac{r\sqrt{n-2}}{\sqrt{1-r^2}} < t_{n-2;\alpha/2} \qquad \text{or} \qquad t_{n-2} > t_{n-2;1-\alpha/2} \tag{8-35}$$

If conditions (a) and (b) are satisfied but $\rho \neq 0$, then T_{n-2} no longer has a t distribution, and special tables (mentioned later in the section) or graphs are needed to compute the power of the test. The actual numerical calculations in terms of the x_i's and the y_i's using (8-35) to test (8-33) are exactly the same as the ones performed when testing $H_0 : B = 0$ against $H_1 : B \neq 0$. We would, of course, expect that there would be no relationship between X and Y if the fact that one variable is given or known is of no assistance in drawing inferences about the other.

The random variable (8-34) can also be used for testing

$$H_0 : \rho = 0 \qquad \text{against} \qquad H_1 : \rho < 0 \tag{8-36}$$

and

$$H_0 : \rho = 0 \qquad \text{against} \qquad H_1 : \rho > 0 \tag{8-37}$$

rejecting when t_{n-2} is small in the first case, large in the second. Hence, a test with significance level α is obtained by rejecting when

$$t_{n-2} < t_{n-2;\alpha} \tag{8-38}$$

and

$$t_{n-2} > t_{n-2;1-\alpha} \tag{8-39}$$

respectively.

EXAMPLE 8-9

Eleven students are selected at random from those who have completed both the first course in English and the first course in physical education. Their final grades (in percentages) are (67, 92), (81, 65), (65, 81), (42, 75), (53, 85), (40, 78), (71, 77), (64, 79), (60, 81), (68, 82), (49, 85), where the first grade in each pair is the English grade and the second the physical education grade, both grades in a pair belonging to the same student. Assuming that both English and physical education grades are normally distributed, so that the bivariate normal is a reasonable model, test the hypothesis that a student's English grade is independent of his physical education grade. Use $\alpha = .05$.

Solution

We shall follow the six-step outline.

1 To test independence, or no relationship, when the pairs (X, Y) are governed by a bivariate normal distribution, we choose between $H_0 : \rho = 0$ and $H_1 : \rho \neq 0$.

2 The significance level $\alpha = .05$ and the sample size $n = 11$ have already been selected.

3 The statistic we use is T_{n-2} given by (8-34). This random variable has a t distribution with $n - 2 = 9$ degrees of freedom when H_0 is true provided (a) the pairs (students) are selected randomly and (b) the pairs (X, Y) have a bivariate normal distribution. We are given that (a) is satisfied and (b) is reasonable.

4 The critical region is $t_9 < t_{9;.025} = -2.262$ and $t_9 > t_{9;.975} = 2.262$.

5 We now perform the experiment and observe the above results. To evaluate the observed value of statistic (8-34) we need r and r^2. Letting the English grade be x_i and the physical education grade be y_i, we have

$$\bar{x} = \frac{67 + 81 + \cdots + 49}{11} = 60$$

$$\bar{y} = \frac{92 + 65 + \cdots + 85}{11} = 80$$

$$\sum_{i=1}^{11} x_i^2 = 67^2 + 81^2 + \cdots + 49^2 = 41{,}210$$

$$\sum_{i=1}^{11} y_i^2 = 92^2 + 65^2 + \cdots + 85^2 = 70{,}864$$

$$\sum_{i=1}^{11} x_i y_i = (67)(92) + (81)(65) + \cdots + (49)(85) = 52{,}593$$

$$r = \frac{52{,}593 - 11(60)(80)}{\sqrt{[41{,}210 - 11(60)^2][70{,}864 - 11(80)^2]}}$$

$$= \frac{-207}{\sqrt{(1{,}610)(464)}} = \frac{-207}{\sqrt{747{,}040}} = \frac{-207}{864.3} = -.240$$

$$r^2 = \frac{42{,}849}{747{,}040} = .0574 \qquad \sqrt{1 - r^2} = \sqrt{.9426} = .971$$

The observed value of T_9 is

$$t_9 = \frac{(-.240)\sqrt{9}}{.971} = \frac{-.720}{.971} = -.74$$

6 We accept the hypothesis of independence and conclude that there is no relationship between English and physical education grades.

Sometimes an experimenter may wish to test hypotheses about ρ other than those specifying that $\rho = 0$. The three standard hypothesis testing situations, which include (8-36), (8-37), and (8-33) as special cases, are respectively

$$H_0 : \rho = \rho_0 \qquad \text{against} \qquad H_1 : \rho < \rho_0 \qquad \qquad \text{(8-40)}$$

$$H_0 : \rho = \rho_0 \qquad \text{against} \qquad H_1 : \rho > \rho_0 \qquad \qquad \text{(8-41)}$$

and

$$H_0 : \rho = \rho_0 \qquad \text{against} \qquad H_1 : \rho \neq \rho_0 \qquad \qquad \text{(8-42)}$$

When $3 \leq n \leq 25$, special tables prepared by David [1] can be used to obtain critical regions and evaluate power. If n is at least 25, the approximate procedure which we are about to describe will be adequate in most practical situations.

For the testing situations (8-40) to (8-42), we can use the statistic

$$Z' = [z(R) - z(\rho_0)]\sqrt{n - 3} \qquad (8\text{-}43)$$

where $z(r)$ and $z(\rho_0)$ are read from Appendix B18. If $\rho = \rho_0$ and (a) the n pairs are selected at random and (b) the pairs are drawn from a bivariate normal distribution, then Z' has approximately a standard normal distribution. When r (or ρ_0) is negative, we use $z(-r) = -z(r)$. Thus, for example, $z(-.50) = -z(.50) = -.5493$. The alternative of (8-40) is supported by small values of r, and a critical region with approximate significance level α is obtained by rejecting when

$$z' = [z(r) - z(\rho_0)]\sqrt{n - 3} < z_\alpha \qquad (8\text{-}44)$$

Similarly, the alternative of (8-41) is supported by large values of r, and to obtain a critical region with significance level approximately α we reject when

$$z' > z_{1-\alpha} \qquad (8\text{-}45)$$

Finally, the two-sided alternative of (8-42) is supported by both small and large r, and the critical region usually selected is

$$z' < z_{\alpha/2} \quad \text{and} \quad z' > z_{1-\alpha/2} \qquad (8\text{-}46)$$

This two-side test then has significance level approximately α.

EXAMPLE 8-10

College entrance examinations include a section which supposedly tests mathematical ability. After students enroll in freshman mathematics, the mathematics department gives another examination to determine which of several levels is the proper one for each individual student. Suppose that testing experts have decided that if it can be proved that $\rho > .5$, then it is a waste of time to give the second examination. That is, results from the entrance examination might just as well be used for the determination of levels. From past records of students who have taken both examinations, a random sample of 103 pairs is observed and it is found that $r = .63$. With $\alpha = .05$ what conclusion can be drawn?

Solution

We shall follow the six-step outline.

1 The problem suggests that we test $H_0 : \rho = .50$ against $H_1 : \rho > .50$. If H_0 is rejected, this will be regarded as "proof" that $\rho > .50$.

2 The significance level $\alpha = .05$ and sample size $n = 103$ have already been determined.

3 The statistic we use is Z' given by (8-43). If H_0 is true, Z' has approximately a standard normal distribution provided (a) the students are selected randomly (which is given) and (b) the pairs of scores are governed by a bivariate normal distribution. Since it is not unreasonable to expect both sets of scores to be normally distributed, the latter assumption appears to be satisfied.

4 The critical region is $z' > z_{.95} = 1.645$.

5 After observing the sample we first calculate $r = .63$. We have $\rho_0 = .50$. Then, from Appendix B18 we find $z(r) = z(.63) = .7414$, $z(\rho_0) = z(.50) = .5493$. The observed value of the statistic is

$$z' = (.7414 - .5493)\sqrt{103 - 3} = 1.921.$$

6 Since the observed value of the statistic falls in the critical region, we reject H_0 and conclude that the extra test is unnecessary.

Approximate power can be calculated in the usual way for tests based upon the standard normal. Perhaps the easiest method is to use the $f = \infty$ curves of Appendix B10 with

$$\delta = |z(\rho_0) - z(\rho_1)|\sqrt{n - 3} \tag{8-47}$$

where ρ_1 is the value in the alternative for which we seek power. If it is desired to make the power at least $1 - \beta$ when $\rho = \rho_1$, find δ corresponding to power $1 - \beta$ and use the minimum n satisfying

$$n \geqq \left[\frac{\delta}{z(\rho_0) - z(\rho_1)}\right]^2 + 3 \tag{8-48}$$

[This is counterpart of inequality (5-14).]

EXAMPLE 8-11

Find the power of the test used in Example 8-10 if the actual value of ρ is .60. How large should n be to raise this to .90?

Solution

With $\rho_0 = .50$, $\rho_1 = .60$, $n = 103$, we find

$$\delta = |z(.50) - z(.60)|\sqrt{100} = |.5493 - .6931|(10) = 1.438$$

Turning to the page of Appendix B10 headed by $\alpha = .05$, the $f = \infty$ curve yields power $= .42$ for $\delta = 1.44$.

In order to have power $= .90$ with $\alpha = .05$ requires $\delta = 2.9$. We get from inequality (8-48)

$$n \geqq \left(\frac{2.9}{-.1438}\right)^2 + 3 = (20.17)^2 + 3 = 406.8 + 3 = 409.8$$

Thus, we need $n = 410$, approximately, to meet the specified power conditions.

Appendix B19 contains graphs from which a confidence interval for ρ of the bivariate normal can be obtained. Specifically, we can find two numbers w_1 and w_2 such that

$$\Pr(W_1 < \rho < W_2) = 1 - \alpha$$

for $\alpha = .05, .10$. (The David [1] publication from which these graphs were taken also includes curves for $\alpha = .02, .01$.) Then (w_1, w_2) is a confidence interval for ρ with confidence coefficient $1 - \alpha$. To use the graphs we need n, $1 - \alpha$, and the point estimate r. We turn to the page headed by $1 - \alpha$ and find the observed r on the horizontal axis. Draw (or imagine) a vertical line through r which intersects the two curves labeled with the appropriate n. These two intersections when connected to the vertical axis (labeled ρ) by horizontal lines yield the desired w_1 and w_2. Let us illustrate with an example.

EXAMPLE 8-12

Find a confidence interval with confidence coefficient .95 for the correlation coefficient in the bivariate normal situation encountered in Example 8-9.

Solution

We had $n = 11$, $r = .24$ (rounded to two significant figures). Turn to the set of graphs in Appendix B19 headed by confidence coefficient .95. Locate $r = .24$ on the horizontal scale, and imagine that a vertical line has been drawn through this point intersecting the curves above. Since curves are available for $n = 10$ and $n = 12$ but not $n = 11$, we estimate that a curve for $n = 11$ would be about halfway between the other two. The estimated curves for $n = 11$ appear to intersect the vertical line at $w_1 = -.71$ and $w_2 = .40$ (approximately), which are read from the vertical scale. Hence $(-.71, .40)$ is the desired interval.

We note that the confidence interval contains $\rho = 0$, the hypothesized value of Example 8-9. This is exactly what we should expect, since H_0 was accepted. Thus, again we have the usual relationship between confidence intervals and two-sided tests.

EXAMPLE 8-13

Find a confidence interval with confidence coefficient .95 for the correlation coefficient in the bivariate normal situation encountered in Example 8-10.

Solution

We had $n = 103$, $r = .63$.   Again we turn to the set of graphs in Appendix B19 headed by confidence coefficient .95.   Since there is no curve for $n = 103$, we shall use the curve for $n = 100$, the difference being so slight that interpolation is unnecessary.   The vertical line through $r = .63$ crosses the two curves labeled $n = 100$ at $w_1 = .49$, $w_2 = .73$ (approximately).   Hence the desired interval is (.49, .73).

The graphs of Appendix B19 can be used to obtain critical regions (but not power) for two-sided tests with significance levels $\alpha = .05$, .10, and for one-sided tests with significance levels $\alpha = .025$, .05.   For two-sided tests we use the page headed by $\alpha =$ significance level, while for one-sided tests we use the page headed by $\alpha =$ twice the significance level.   We enter the vertical axis scale with the value of $\rho = \rho_0$ specified by the hypothesis, draw a horizontal line through ρ_0 intersecting the curves with the appropriate n, and project these intersections onto the r scale.   In the two-sided case, the two r's so obtained form the boundary of the critical region. For the one-sided case (8-40) use only the small set of r's, while for (8-41) use only the large set of r's.   Let us illustrate by examples.

EXAMPLE 8-14

Use Appendix B19 to find the critical region in terms of r for the test of Example 8-9.

Solution

We had $\rho_0 = 0$, $\alpha = .05$ and previously found that one way to express the critical region was $t_9 < -2.262$ and $t_9 > 2.262$.   Turning to the graphs of Appendix B19 headed by $\alpha = .05$, we find that the horizontal line through $\rho = 0$ crosses the curves for $n = 11$ (halfway between those for $n = 10$ and $n = 12$) at approximately $r = -.60$ and $r = .60$.   Thus the critical region in terms of r is $r < -.60$ and $r > .60$.   If $\rho = 0, n = 11$, then $\Pr(R < -.60)$ $+ \Pr(R > .60) = .025 + .025 = .05$.   Since $r = -.240$, H_0 is accepted.

EXAMPLE 8-15

Use Appendix B19 to find the critical region for the problem of Example 8-10 in terms of r.

Solution

Turning to the graphs of Appendix B19 headed by $\alpha = .10$ (twice the significance level for a one-sided test), we find that the curves for $n = 100$ (used in place of 103) cross the horizontal line through $\rho = .5$ above $r = .37$ and $r = .62$. The critical region for testing $H_0 : \rho = .5$ against $H_1 : \rho > .5$ is $r > .62$, and $\Pr(R > .62) = .05$ if $\rho = .5$, $n = 100$. (The critical region $r < .37$ is used for testing $H_0 : \rho = .5$ against $H_1 : \rho < .5$, and $\Pr(R < .37) = .05$ if $\rho = .5$, $n = 100$.) Since the observed $r = .63$ falls in the critical region, the decision is again rejection.

Since $n = 103$ is fairly large, the approximate procedure is probably better because power can be readily calculated.

EXERCISES

◆ 8-28   Eighteen students are selected at random from those who have completed elementary statistics. The first number in each of the following pairs is a midterm grade, the second is the final-examination grade for the same student: $(54, 48)$, $(44, 64)$, $(31, 38)$, $(32, 33)$, $(36, 29)$, $(34, 41)$, $(53, 59)$, $(34, 45)$, $(12, 17)$, $(48, 28)$, $(36, 40)$, $(51, 44)$, $(50, 63)$, $(44, 38)$, $(39, 49)$, $(44, 57)$, $(52, 49)$, $(39, 34)$. Assuming that the pairs of grades have a bivariate normal distribution, test the hypothesis that midterm and final-examination grades are independent. Use the t test with $\alpha = .05$ and follow the six-step outline.

◆ 8-29   Use the graphs of Appendix B19 to find the critical region in Exercise 8-28 in terms of r.

◆ 8-30   Find a confidence interval with confidence coefficient .95 for the correlation coefficient in the bivariate normal situation encountered in Exercise 8-28. What is the relationship between the confidence interval and the test of Exercise 8-28?

   8-31   An instructor claims that the correlation coefficient between his midterm and final-examination grades is at least .6. A random sample of 403 past records yields $r = .54$. With $\alpha = .05$ test the appropriate hypothesis. Follow the six-step outline.

   8-32   For the test used in Exercise 8-31 find the power if $\rho = .5$. If it is desired that this power be at least .75, how large a sample should have been used?

   8-33   Using the observed sample results of Exercise 8-31, find a confidence interval with confidence coefficient .95 for the correlation coefficient.

◆ 8-34   Using the observed sample results given in Exercise 8-4, find a confidence interval for ρ with confidence coefficient .95.

◆ 8-35   Two different 1-hour examinations are used alternatively to cover certain topics in an elementary statistics course. Suppose that we would be willing to regard the two examinations as equivalent if it can be "proved" that the correlation coefficient between scores obtained on the two exceeds .80. To make a decision the questions from the two examinations are arranged in a random order, yielding one examination twice as long. Then, a class of 84 students takes the combined examinations, being allowed 2 hours total time. Each student's paper yields two scores, one on each of two 1-hour examinations. From these observed pairs calculations yield $r = .88$. Formulate the hypothesis and alternative which seems to be most appealing, and conduct the appropriate test with $\alpha = .01$. Follow the six-step outline.

◊ 8-36 Find the power of the test used in Exercise 8-35 if $\rho = .85$. If it is desired that this power be at least .80, how many students should have been used?

◊ 8-37 Using the observed sample results given in Exercise 8-35, find a confidence interval with confidence coefficient .90 for the correlation coefficient. *Hint:* Since the graphs are difficult to reproduce and to read for r's close to -1 or $+1$, it is better to use our standard method of obtaining a confidence interval here. Thus we convert

$$\Pr\{-z_{1-\alpha/2} < [z(R) - z(\rho)]\sqrt{n-3} < z_{1-\alpha/2}\} \cong 1 - \alpha$$

to

$$\Pr\left[z(R) - \frac{z_{1-\alpha/2}}{\sqrt{n-3}} < z(\rho) < z(R) + \frac{z_{1-\alpha/2}}{\sqrt{n-3}}\right] \cong 1 - \alpha$$

and the confidence interval for $z(\rho)$ is

$$\left[z(r) - \frac{z_{1-\alpha/2}}{\sqrt{n-3}}, \quad z(r) + \frac{z_{1-\alpha/2}}{\sqrt{n-3}}\right] = [z(w_1), z(w_2)]$$

Then w_1, w_2 can be found to the nearest .01 by entering Appendix B18 with $z(w_1)$, $z(w_2)$.

REFERENCES

1 David, F. N.: "Tables of the Ordinates and Probability Integral of the Distribution of the Correlation Coefficient in Small Samples," The Biometrika Office, London, 1938.

Summary of Results

DRAWING INFERENCES IN
REGRESSION AND CORRELATION
ANALYSIS

I. Point estimation of the regression line

A. Assuming that the means $\mu_{Y|x}$ lie on a straight line, so that $\mu_{Y|x} = A + Bx$, estimates for A and B obtained by the principle of least squares are

$$b = \frac{\sum_{i=1}^n x_i y_i - n\bar{x}\bar{y}}{\sum_{i=1}^n x_i^2 - n\bar{x}^2} \qquad a = \bar{y} - b\bar{x}$$

B. If we also assume that every Y distribution has the same variance, then an unbiased estimate of $\sigma_{Y|x}^2 = \sigma^2$ is

$$s_{Y|x}^2 = \frac{1}{n-2}\left[\sum_{i=1}^n y_i^2 - n\bar{y}^2 - \frac{(\sum_{i=1}^n x_i y_i - n\bar{x}\bar{y})^2}{\sum_{i=1}^n x_i^2 - n\bar{x}^2}\right]$$

II. Inferences concerning the mean of Y given $x = x_0$ (denoted by $\mu_{Y|x_0}$)

A. Standard hypothesis, alternatives, statistics, and critical regions (significance level α)

Hypothesis	Alternative	Statistic† (observed value)	Critical region
$H_0: \mu_{Y|x_0} = \mu_0$	$H_1: \mu_{Y|x_0} < \mu_0$	t_{n-2}	$t_{n-2} < t_{n-2;\alpha}$
$H_0: \mu_{Y|x_0} \geqq \mu_0$	$H_1: \mu_{Y|x_0} < \mu_0$	$= \dfrac{a + bx_0 - \mu_0}{s_{Y|x}\sqrt{\dfrac{1}{n} + \dfrac{(x_0 - \bar{x})^2}{\sum_{i=1}^n x_i^2 - n\bar{x}^2}}}$	$t_{n-2} < t_{n-2;\alpha}$
$H_0: \mu_{Y|x_0} = \mu_0$	$H_1: \mu_{Y|x_0} > \mu_0$		$t_{n-2} > t_{n-2;1-\alpha}$
$H_0: \mu_{Y|x_0} \leqq \mu_0$	$H_1: \mu_{Y|x_0} > \mu_0$		$t_{n-2} > t_{n-2;1-\alpha}$
$H_0: \mu_{Y|x_0} = \mu_0$	$H_1: \mu_{Y|x_0} \neq \mu_0$		$t_{n-2} < t_{n-2;\alpha/2}$, $t_{n-2} > t_{n-2;1-\alpha/2}$

† Note that whenever a statistic is given in this summary, certain assumptions are made. These assumptions, giving the statistic the required distribution, may be found in the text material.

B. Power calculations

1. Use Appendix B10 in the usual way with entries α ($\alpha/2$ for two-sided alternatives), $f = n - 2$,

$$\delta = \frac{|\mu_0 - \mu_1|}{\sigma\sqrt{\dfrac{1}{n} + \dfrac{(x_0 - \bar{x})^2}{\sum_{i=1}^n x_i^2 - n\bar{x}^2}}}$$

2. μ_1 must be in the range of the alternative.

C. Two-sided confidence interval with confidence coefficient $1 - \alpha$ for $\mu_{Y|x_0}$

$$\left(a + bx_0 - t_{n-2;1-\alpha/2}s_{Y|x}\sqrt{\frac{1}{n} + \frac{(x_0 - \bar{x})^2}{\sum_{i=1}^{n} x_i^2 - n\bar{x}^2}}\,,\right.$$

$$\left.a + bx_0 + t_{n-2;1-\alpha/2}s_{Y|x}\sqrt{\frac{1}{n} + \frac{(x_0 - \bar{x})^2}{\sum_{i=1}^{n} x_i^2 - n\bar{x}^2}}\right)$$

III. Confidence region for the regression line
   A. Confidence region with confidence coefficient $1 - \alpha$

$$\left(a + bx - s_{Y|x}\sqrt{2f_{2,n-2;1-\alpha}\left[\frac{1}{n} + \frac{(x - \bar{x})^2}{\sum_{i=1}^{n} x_i^2 - n\bar{x}^2}\right]}\,,\right.$$

$$\left.a + bx + s_{Y|x}\sqrt{2f_{2,n-2;1-\alpha}\left[\frac{1}{n} + \frac{(x - \bar{x})^2}{\sum_{i=1}^{n} x_i^2 - n\bar{x}^2}\right]}\right)$$

IV. Inferences concerning the slope B of the regression line
   A. Standard hypotheses, alternatives, statistics, and critical regions (significance level α)

Hypothesis	Alternative	Statistic (observed value)	Critical region
$H_0: B = B_0$	$H_1: B < B_0$	t_{n-2}	$t_{n-2} < t_{n-2;\alpha}$
$H_0: B \geq B_0$	$H_1: B < B_0$		$t_{n-2} < t_{n-2;\alpha}$
$H_0: B = B_0$	$H_1: B > B_0$	$= \dfrac{b - B_0}{s_{Y|x}}\sqrt{\sum_{i=1}^{n} x_i^2 - n\bar{x}^2}$	$t_{n-2} > t_{n-2;1-\alpha}$
$H_0: B \leq B_0$	$H_1: B > B_0$		$t_{n-2} > t_{n-2;1-\alpha}$
$H_0: B = B_0$	$H_1: B \neq B_0$		$t_{n-2} < t_{n-2;\alpha/2}$, $t_{n-2} > t_{n-2;1-\alpha/2}$

B. Power calculations
   1. Use Appendix B10 in the usual way with entries α ($\alpha/2$ for two-sided alternatives), $f = n - 2$,

$$\delta = \frac{|B_0 - B_1|}{\sigma}\sqrt{\sum_{i=1}^{n} x_i^2 - n\bar{x}^2}$$

   2. B_1 must be in the range of the alternative.
C. Two-sided confidence interval with confidence coefficient $1 - \alpha$

$$\left(b - \frac{t_{n-2;1-\alpha/2}\,s_{Y|x}}{\sqrt{\sum_{i=1}^{n} x_i^2 - n\bar{x}^2}}\,,\; b + \frac{t_{n-2;1-\alpha/2}\,s_{Y|x}}{\sqrt{\sum_{i=1}^{n} x_i^2 - n\bar{x}^2}}\right)$$

V. Prediction interval for one future observation of Y, given $x = x_0$

A. Two-sided interval with prediction coefficient $1 - \alpha$

$$\left(a + bx_0 - t_{n-2;1-\alpha/2}\, s_{Y|x}\sqrt{1 + \frac{1}{n} + \frac{(x_0 - \bar{x})^2}{\sum_{i=1}^{n} x_i^2 - n\bar{x}^2}}\,,\right.$$

$$\left. a + bx_0 + t_{n-2;1-\alpha/2}\, s_{Y|x}\sqrt{1 + \frac{1}{n} + \frac{(x_0 - \bar{x})^2}{\sum_{i=1}^{n} x_i^2 - n\bar{x}^2}}\right)$$

VI. Inferences concerning the variance of Y given x, $\sigma_{Y|x}^2$. Part IV of the Summary of Results of Chap. 5 applies if $n - 1$ is replaced by $n - 2$, s^2 by $s_{Y|x}^2$, and σ^2 by $\sigma_{Y|x}^2$. The assumptions for the use of the statistic become (8-10) or (8-11).

VII. Inferences concerning the correlation coefficient ρ

A. Standard hypotheses, alternatives, statistics, and critical regions (significance level α)

Hypothesis	Alternative	Statistic	Critical region
$H_0: \rho = \rho_0$	$H_1: \rho < \rho_0$	$T_{n-2} = \dfrac{R\sqrt{n-2}}{\sqrt{1-R^2}}$ if $\rho_0 = 0$	Reject if the appropriate statistic is small (for example, $z' < z_\alpha$)
$H_0: \rho \geqq \rho_0$	$H_1: \rho < \rho_0$	$Z' = [z(R) - z(\rho)]\sqrt{n-3}$ $n > 25$ or	Reject if the appropriate statistic is small (for example, $z' < z_\alpha$)
$H_0: \rho = \rho_0$	$H_1: \rho > \rho_0$	R [defined by (8-30)] $n \leqq 25$ (and the special tables of David [1] or the graphs of Appendix B19)	Reject if the appropriate statistic is large (for example, $z' > z_{1-\alpha}$)
$H_0: \rho \leqq \rho_0$	$H_1: \rho > \rho_0$	if $\rho_0 \neq 0$	Reject if the appropriate statistic is large (for example, $z' > z_{1-\alpha}$)
$H_0: \rho = \rho_0$	$H_1: \rho \neq \rho_0$		Reject if the appropriate statistic is either small or large (for example, $z' < z_{\alpha/2}$, $z' > z_{1-\alpha/2}$)

B. Power calculations

1. Omitted for first statistic listed in VII-A

2. For the second statistic listed in VII-A use Appendix B10 with entries α ($\alpha/2$ for two-sided alternatives), $f = \infty$,

$$\delta = |z(\rho_0) - z(\rho_1)|\sqrt{n - 3}$$

where ρ_1 is in the range of ρ specified by the alternative.

3. Omitted for the third statistic listed in VII-A but can be found in the David [1] tables.

C. Two-sided confidence interval with confidence coefficient $1 - \alpha$

1. Find the (w_1, w_2) from Appendix B19. Entries required are $1 - \alpha, n, r$.

9

DISTRIBUTION-FREE INFERENCES

9-1
INTRODUCTION

In Chaps. 5, 7, and 8 we considered a number of inference problems concerning distributions of continuous random variables. Probability statements were derived under distributional assumptions, specifically the assumption that the random variables had normal distributions. When inferences concerned means, probability statements were still approximately correct even though normality assumptions were not satisfied. On the other hand, tests and confidence intervals for the variance appeared to depend heavily on the fact that the normal model was appropriate. In this chapter we will consider some inferences about the distributions of *continuous* random variables that do not depend on the particular form of that distribution. Thus it will not be necessary to assume that the normal, the triangular, or any other continuous density is the one which governs the behavior of a random variable X. Inferences that are valid irrespective of the form of the distribution are called *distribution-free inferences.*

Since it may be difficult to choose the correct probability model for an experimental situation, it is appealing to have procedures which do not depend on that choice. However, as is frequently the case, when we gain in one place we are apt to lose in another. Distribution-free procedures have the advantage of requiring less in the way of assumptions and are often easier to apply. Generally they are less powerful, and power calculations may be very difficult (and, consequently, will usually be omitted).

In some of the following sections we will use the term *quantile of a distribution.* To define this concept let $f(x)$ be the density function of a continuous random variable X. Then Q_p is called the pth quantile, or the quantile of order p, if $\Pr(X < Q_p) = p$. The geometrical relationship between p and Q_p is shown in Fig. 9-1. Actually we have used quan-

406

FIGURE 9-1

The pth quantile Q_p of a continuous distribution.

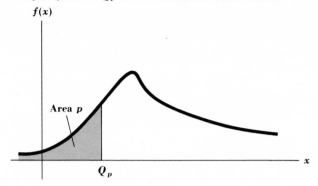

tiles many times before for the standard distributions.   Figure 2-1 shows that z_p is the pth quantile of the standard normal, while Figs. 2-10, 2-13, and 2-15 show that $t_{v;\,p}$, $\chi^2_{v;\,p}$, and $f_{v_1,v_2;\,p}$ are the pth quantiles for the t, chi-square, and F distributions respectively.   When $p = 1/2$, so that $Q_p = Q_{.50}$, the quantile is called the *median*.

In this chapter we will discuss only a few distribution-free techniques.   Recently a number of texts devoted almost entirely to distribution-free problems have been published.   For those interested in further information the books by Bradley [1], Conover [2], Gibbons [3], and Hájek [4] may be helpful.   Of the four, the first two require less mathematical sophistication of the reader.

9-2
THE BINOMIAL TEST (AND THE SIGN TEST), INFERENCES ABOUT THE QUANTILE OF ORDER p

It is frequently of interest, particularly in a field known as statistical quality control, to test hypotheses about the pth quantile of a distribution. Suppose that numbers L and p_0 are given or determined in advance and we would like to test the hypothesis that $Q_{p_0} = L$ against various alternatives.   The standard hypothesis testing situations could be written as

$$H_0 : Q_{p_0} = L \quad \text{against} \quad H_1 : Q_{p_0} < L \tag{9-1}$$

$$H_0 : Q_{p_0} = L \quad \text{against} \quad H_1 : Q_{p_0} > L \tag{9-2}$$

and

$$H_0 : Q_{p_0} = L \quad \text{against} \quad H_1 : Q_{p_0} \neq L \tag{9-3}$$

If we let $\Pr(X < L) = p$, then (9-1) to (9-3) are equivalent, respectively, to testing

$$H_0 : p = p_0 \qquad \text{against} \qquad H_1 : p > p_0 \tag{9-4}$$

$$H_0 : p = p_0 \qquad \text{against} \qquad H_1 : p < p_0 \tag{9-5}$$

and

$$H_0 : p = p_0 \qquad \text{against} \qquad H_1 : p \neq p_0 \tag{9-6}$$

In Sec. 6-2 we have already devoted considerable space to the latter three hypothesis testing situations. To relate our current problem to the former discussion, (a) let $X_1, X_2, \ldots, X_n$ denote a random sample from the X distribution and (b) assume that distribution is continuous [with density function $f(x)$]. If an observed value of X, say x, is such that $x < L$, we will say that we have observed a success, while if $x > L$ we have observed a failure. (In quality control successes and failures are called defectives and nondefectives, respectively.) After the observed random sample $x_1, x_2, \ldots, x_n$ is available, then we can determine y, the observed value of $Y =$ number of successes. It is easy to verify that Y has a binomial distribution with parameters n and p. (If H_0 is true, then $p = p_0$.) To check the binomial conditions (2-1) we see that

(a) On the basis of x the item we are observing can be classified as a success or a failure (ignoring the case $x = L$ which occurs with probability 0).
(b) The probability $p = \Pr(X < L)$ remains constant since each X has the same distribution.
(c) Since the sample is selected randomly, items are successes or failures independently of the classification of other items.
(d) The series of experiments in which success or failure is determined consists of n trials.

Hence, critical regions in terms of y can be determined as in Sec. 6-2 (where we used X instead of Y). As before, when considering (9-4) H_1 is supported if y is too large. For (9-5) H_1 is supported if y is too small. Finally, when testing (9-6) H_1 is supported by both small and large y. Again we will denote the "aimed-at" significance level by α_0.

Power calculations and sample size problems can be treated as in Chap. 6. Now, an alternative $p = p_1$ implies $L = Q_p = Q_{p_1}$. That is, the specified number L is the quantile of order p_1.

When $p_0 = .50$, then the hypothesis of (9-1) to (9-6) is that the given number L is the median of the distribution. For this case the test which we discussed above is usually called the *sign test*. The name arises from the fact that Y is the number of negative signs associated with the differences $X_1 - L, X_2 - L, \ldots, X_n - L$. If we are willing to assume that the density function of X is symmetric about the median, then the median $Q_{.50}$ and the mean μ (provided μ exists) are equal and the sign test can be regarded as a competitor of the t test of Sec. 5-5 for testing hypotheses about the mean.

EXAMPLE 9-1

A type of transistor is considered satisfactory if the maximum temperature at which it will operate is 200° centigrade or more. If the maximum operating temperature is less than 200, the transistor is regarded as a defective. A randomly selected sample of 20 transistors is tested, and the following maximum operating temperatures are recorded: 221, 210, 213, 190, 187, 245, 205, 183, 206, 231, 227, 179, 209, 217, 216, 235, 208, 195, 202, 220. If no more than 10 percent of the transistors are defective, we are satisfied with the quality of the product. Using $\alpha_0 = .05$, test the appropriate hypothesis using the distribution-free procedures of this section.

Solution

We shall follow the six-step outline.

1 The hypothesis and alternative suggested by the problem are H_0: $p = .10$ against H_1: $p > .10$ or, equivalently, H_1: $Q_{.10} = 200$ against H_1: $Q_{.10} < 200$. If H_0 is rejected, we will conclude that too many transistors are defective.

2 The significance level we hope to achieve is $\alpha_0 = .05$. According to the convention adopted in Chap. 6, the actual α will be less than .05 but as close to .05 as possible. The sample size $n = 20$ has already been selected.

3 The statistic we use is $Y =$ the number of defectives. The distribution of Y is binomial with parameters $n = 20$ and p if (a) $X_1, X_2, \ldots, X_{20}$ is a random sample from a distribution and (b) that distribution is continuous. Further, if H_0 is true, then $p = .10$.

4 With $n = 20$, $p = .10$, binomial tables yield

$$\Pr(Y \geq 4) = 1 - \Pr(Y \leq 3) = 1 - .86705 = .13295 > .05$$

$$\Pr(Y \geq 5) = 1 - \Pr(Y \leq 4) = 1 - .95683 = .04317 < .05$$

Hence, the critical region is $y \geq 5$.

5 We now observe the 20 temperatures and see that 5 numbers are less than 200. Thus, $y = 5$.

6 Since the observed value of the statistic falls in the critical region, we reject H_0. We conclude that the fraction of defective transistors is greater than .10, or alternatively, the .10th quantile is less than 200.

EXAMPLE 9-2

Consider again the light bulb problem of Example 5-1. Suppose now that the old brand has a median life of 1,000 hours. The company will switch to the new brand unless it is proved with level of significance no greater than $\alpha_0 = .05$ that the new brand has a smaller median life than the old brand. A random sample of 100 new-brand bulbs is tested, the quantities $x_1 - 1,000$, $x_2 - 1,000$, ..., $x_{100} - 1,000$ are calculated, and 60 negative signs are observed. If the sign test is used, what decision should be made?

Solution

We shall follow the six-step outline.

1 The problem suggests that we test $H_0: Q_{.50} = 1,000$ against $H_1: Q_{.50} < 1,000$. If H_0 is rejected, then the observed sample "proves" that the new-brand median is lower than the old-brand median.

2 The sample size $n = 100$ has already been selected, and it has been specified that the significance level should be no greater than $\alpha_0 = .05$.

3 The statistic we will use is $Y =$ the number of negative signs. The distribution of Y is binomial with parameters $n = 100$ and p if (a) the sample of light bulbs was randomly selected and (b) the distribution of life lengths is continuous. If H_0 is true, then $p = .50$.

4 We reject if too many negative signs are observed. With $n = 100$, $p = .50$, binomial tables yield

$$\Pr(Y \geq 58) = 1 - \Pr(Y \leq 57) = 1 - .93339 = .06661$$

$$\Pr(Y \geq 59) = 1 - \Pr(Y \leq 58) = 1 - .97156 = .02844$$

Hence, with $\alpha_0 = .05$, we reject when $y \geq 59$.

5 The sample is observed and 60 negative signs are counted. Thus, $y = 60$.

6 The hypothesis is rejected, and the company will continue to operate with the old brand.

EXAMPLE 9-3

Find the power of the test used in Example 9-1 if $200 = Q_{.25}$ so that one-fourth of the transistors are defective.

Solution

Now Y has a binomial distribution with $n = 20$, $p = .25$. The power is $\Pr(Y \geq 5) = 1 - \Pr(Y \leq 4) = 1 - \sum_{y=0}^{4} b(y; 20, p)$. With $p = .25$ we get power $= 1 - .41484 = .58516$.

EXAMPLE 9-4

Find the power of the test used in Example 9-2 if the new brand has $Q_{.60} = 1,000$. In other words, 60 percent will have a life length less than 1,000 hours.

Solution

Now Y has a binomial distribution with $n = 100$, $p = .60$. The power is $\Pr(Y \geq 59)$. Since $p > .50$, let $Z =$ the number of positive signs and $1 - p$ be the probability of a success. Then the power is $\Pr(Y \geq 59) = \Pr(Z \leq 41) = \sum_{z=0}^{41} b(z; 100, 1 - p)$. With $p = .60$, $1 - p = .40$, we find power $= .62253$.

If we had asked for the minimum n such that $\alpha_0 = .05$ when $p = .50$ and power $\geq .90$ when $p = .60$, this would be our usual sample size problem. The solution was discussed in Secs. 6-2 and 6-4.

Had we been willing to assume that the X's are normally distributed in either Example 9-1 or Example 9-2, a more powerful test is available (in the latter case the ordinary one-sample t test). What this means is that a larger n is required to solve our usual sample size problem if we use the distribution-free test based on the binomial. This is not surprising since we discard considerable information when we record only that an item is a success or failure and pay no attention to the magnitude of x.

The sign test can also be used with paired observations (which were discussed in Sec. 5-10). If we let the random sample of n pairs be $(X_{11}, X_{12}), (X_{21}, X_{22}), \ldots, (X_{n1}, X_{n2})$, then again we can form the differences $D_1, D_2, \ldots, D_n$ where $D_i = X_{i1} - X_{i2}$, $i = 1, 2, \ldots, n$. Letting $Q_{.50}$ be the median of the distribution of the random variable D and Y be the number of negative signs associated with $D_1, D_2, \ldots, D_n$, the sign test can be used as previously discussed to test

$H_0 : Q_{.50} = 0$	against	$H_1 : Q_{.50} < 0$	(9-7)
$H_0 : Q_{.50} = 0$	against	$H_1 : Q_{.50} > 0$	(9-8)

or

$H_0 : Q_{.50} = 0$	against	$H_1 : Q_{.50} \neq 0$	(9-9)

Once more we reject with large values of y in the first case, small values of y in the second, and with either small or large values in the third. This form of the sign test can be regarded as a competitor of the paired t test (used to test hypothesis about $\mu_D = \mu_1 - \mu_2$) of Sec. 5-10.

EXAMPLE 9-5

Use the sign test for the problem of Example 5-38 involving Army recruits and a physical training problem. Now the question we propose is "Does the program produce a distribution of weight changes with median equal to 0?"

Solution

We shall use the six-step outline.

1 Now we test $H_0 : Q_{.50} = 0$ against $H_1 : Q_{.50} \neq 0$ for the distribution of D.
2 The "aimed-at" significance level $\alpha_0 = .05$ and sample size $n = 10$ have already been chosen.
3 The statistic we use is $Y =$ the number of negative signs counted from $D_1, D_2, \ldots, D_{10}$. The distribution of Y is binomial with parameters $n = 10$ and p if (a) the sample of pairs (recruits) was randomly selected and (b) the distribution of differences is continuous. If H_0 is true, then $p = .50$.
4 With $n = 10$, $p = .50$, binomial tables yield

$$\Pr(Y \leq 1) = .01074 < .025 \qquad \Pr(Y \leq 2) = .05469 > .025$$

$$\Pr(Y \geq 9) = .01074 < .025 \qquad \Pr(Y \geq 8) = .05469 > .025$$

Following our usual convention, the critical region is $y \leq 1$ and $y \geq 9$.
5 We observe the sample and count the number of negative signs. The differences were $-8, -5, 2, -12, -4, 5, -4, -11, 3, -5$, so that $y = 7$.
6 Since the statistic does not fall in the critical region, H_0 is accepted. We conclude that the distribution of differences has median equal to zero. Alternatively, the probability of gaining weight is $1/2$ and the probability of losing weight is $1/2$ (which follows from the definition of median).

EXAMPLE 9-6

Find the power of the test used in Example 9-5 if the probability of gaining weight is .30 instead of .50.

Solution

We seek the power when $Q_{.30} = 0$. Now Y has a binomial distribution with $n = 10$, $p = .30$, and

$$\text{Power} = \Pr(Y \leq 1) + \Pr(Y \geq 9)$$

$$= \sum_{y=0}^{1} b(y; 10, .30) + \sum_{y=9}^{10} b(y; 10, .30)$$

$$= .14931 + .00014 = .14935$$

The power is small as we might expect with $n = 10$. The t test of Example 5-18 is, of course, more powerful than the sign test.

Although it is theoretically impossible to have $x = L$ (since $\Pr(X = L) = 0$), this situation may occur in practice due to the fact that measurements may not have been made as accurately as desired. If it should happen that $x = L$, so that $d = x - L = 0$ and d has no plus or minus sign, then we will not count this x as part of the sample.

Next, we consider the problem of obtaining a confidence interval for Q_p. Let $X_1, X_2, \ldots, X_n$ be a random sample from a continuous distribution. After the observed sample is available, we can arrange $x_1, x_2, \ldots, x_n$ in order of increasing magnitude. Denote the smallest value by $x_{(1)}$, the second smallest by $x_{(2)}, \ldots$, the rth smallest by $x_{(r)}, \ldots$, the largest by $x_{(n)}$. Then $x_{(1)} < x_{(2)} < \cdots < x_{(r)} < \cdots < x_{(n)}$ and for the corresponding random variables (called order statistics) we have

$$X_{(1)} < X_{(2)} < \cdots < X_{(r)} < \cdots < X_{(n)}.$$

If we could evaluate $\Pr[X_{(i)} < Q_p < X_{(j)}]$ for $i < j$, getting

$$\Pr[X_{(i)} < Q_p < X_{(j)}] = \gamma \tag{9-10}$$

then, to be consistent with our previous usage of the terms, we could say that $(x_{(i)}, x_{(j)})$ is a confidence interval for Q_p with confidence coefficient γ. To evaluate the left side of formula (9-10) we observe that $X_{(i)}$ will be less than Q_p if Q_p is either (1) between $X_{(i)}$ and $X_{(j)}$ or (2) greater than $X_{(j)}$ [and hence greater than $X_{(i)}$]. (See Fig. 9-2.) If we define

A_1 is the event $X_{(i)} < Q_p$

A_2 is the event $X_{(i)} < Q_p < X_{(j)}$

A_3 is the event $X_{(i)} < X_{(j)} < Q_p$

Then $A_1 = A_2 \cup A_3$ and A_2 and A_3 are mutually exclusive. Consequently, the addition theorem yields $\Pr(A_1) = \Pr(A_2) + \Pr(A_3)$ so that $\Pr(A_2) = \Pr(A_1) - \Pr(A_3)$. That is,

$$\Pr[X_{(i)} < Q_p < X_{(j)}] = \Pr[X_{(i)} < Q_p] - \Pr[X_{(j)} < Q_p] \tag{9-11}$$

FIGURE 9-2

Two mutually exclusive ways which result in $x_{(i)} < Q_p$.

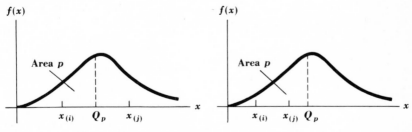

The two probabilities on the right-hand side of (9-11) are easy to evaluate. The event $X_{(i)} < Q_p$ occurs if i or more of the sample values are less than Q_p. The probability that each X_i is less than Q_p is p. It is easy to verify that all the binomial conditions (2-1) are satisfied, and hence

$\Pr[X_{(i)} < Q_p]$ = probability of i or more successes in a series of experiments satisfying the binomial conditions

$$= \sum_{y=i}^{n} b(y; n, p)$$

Since a similar result holds for $X_{(j)}$, we have

$$\Pr[X_{(i)} < Q_p < X_{(j)}] = \sum_{y=i}^{n} b(y; n, p) - \sum_{y=j}^{n} b(y; n, p) \qquad (9\text{-}12)$$

$$= \sum_{y=0}^{j-1} b(y; n, p) - \sum_{y=0}^{i-1} b(y; n, p) \qquad (9\text{-}13)$$

To determine a confidence interval with confidence coefficient at least $1 - \alpha$, we will choose j to be the smallest integer such that

$$\sum_{y=0}^{j-1} b(y; n, p) \geqq 1 - \frac{\alpha}{2} \qquad (9\text{-}14)$$

and i to be the largest integer such that

$$\sum_{y=0}^{i-1} b(y; n, p) \leqq \frac{\alpha}{2} \qquad (9\text{-}15)$$

Such a choice obviously makes the difference given by (9-13) greater than or equal to $1 - \alpha$. The confidence interval so obtained has the usual relationship to the two-sided hypothesis testing situation given by (9-3). That is, if the α_0 of the test and the α of the confidence interval are the same, then the confidence interval $(x_{(i)}, x_{(j)})$ obtained with $p = p_0$ will include $Q_{p_0} = L$ if, and only if, $H_0 : Q_{p_0} = L$ is accepted.

EXAMPLE 9-7

Using the observed results of Example 9-1, find a confidence interval for the .25th quantile having a confidence coefficient of at least .95.

Solution

We identify $n = 20$, $p = .25$, $1 - \alpha = .95$, $\alpha = .05$, $\alpha/2 = .025$. According to inequality (9-14) we seek the smallest value of j which satisfies

$$\sum_{y=0}^{j-1} b(y; 20, .25) \geqq 1 - .025 = .975$$

From Appendix B10 we find

$$\sum_{y=0}^{8} b(y; 20, .25) = .95907 \qquad \sum_{y=0}^{9} b(y; 20, .25) = .98614$$

so that $j - 1 = 9, j = 10$. According to inequality (9-15) we need the largest value of i which satisfies

$$\sum_{y=0}^{i-1} b(y; 20, .25) \leq .025$$

Appendix B10 yields

$$\sum_{y=0}^{1} b(y; 20, .25) = .02431 \qquad \sum_{y=0}^{2} b(y; 20, .25) = .09126$$

Hence, $i - 1 = 1$, $i = 2$. The desired interval is $(x_{(2)}, x_{(10)})$. With the observed sample of Example 9-1 it is easy to verify that $x_{(2)} = 183$, $x_{(10)} = 209$, so that the interval is (183, 209).

EXERCISES

◈ 9-1   A company cans and sells tomato juice in 46-ounce containers. It is undesirable to have the average content per can be either less than or greater than 46 ounces. An observed random sample of 25 containers yields the following results: 45.63, 45.82, 45.77, 45.80, 46.21, 46.07, 46.16, 45.91, 45.87, 46.03, 46.16, 45.92, 45.76, 45.89, 45.85, 46.16, 45.80, 45.79, 45.86, 46.10, 45.88, 45.91, 45.87, 45.96, 45.99. Use the sign test with $\alpha_0 = .05$ to draw the appropriate conclusion. Follow the six-step outline.

◈ 9-2   Find the power of the test used in Exercise 9-1 if the .60th quantile is equal to 46. How large should n be to make this probability at least .80? *Hint:* Use inequality (6-25) with α replaced by $\alpha/2, p_0 = .50$, $p_1 = .60$.

◈ 9-3   Suppose we had decided that tomato juice weights are normally distributed and had used the t test in Exercise 9-1. Now how large would n have to be to satisfy the power requirements of Exercise 9-2? *Hint:* Now $46 = Q_{.60}$. Letting μ_1 be the mean of the distribution, we have $(46 - \mu_1)/\sigma = z_{.60}$, so that formula (5-14) yields $n = (\delta/z_{.60})^2 = (\delta/.253)^2$. Now proceed as in Example 5-12.

◈ 9-4   Use the observed results of Exercise 9-1 and inequalities (9-14) and (9-15) to find a confidence interval for the median of the distribution of tomato juice weights having confidence coefficient at least .90.

◈ 9-5   Consider again the psychologist of Exercise 5-5 who raises rats. We recall that he was considering a variety which matures faster than the standard variety. Suppose that he decides that a change is worthwhile if the .75th quantile is no larger than 240 days. For an observed sample of 50 rats, 36 mature in less than 240 days while 14 require more than 240 days. Use $\alpha_0 = .10$ to test the appropriate hypothesis, and then recommend a course of action for the psychologist. Follow the six-step outline.

◈ 9-6   Suppose that for the situation described in Exercise 9-5 the psychologist desires a confidence interval with confidence coefficient at least .90 for the .75th quantile of the distribution of age at maturity. Which order statistics should he use? *Hint*: Rewrite (9-12) as $\sum_{z=0}^{50-i} b(z;50,.25) - \sum_{z=0}^{50-j} b(z;50,.25)$. Then, as with inequalities (9-14) and (9-15) make the first sum at least .95 and the second sum no greater than .05.

9-7   Consider again the statistics department of Exercise 5-8 which is contemplating the addition of extra problem sessions meeting twice a week. It feels that the addition is worthwhile if it can be proved that the median can be raised above 60. In a class of 100 students using the 5-meeting system, 38 had course grades below 60 while 62 had course grades above 60. Using the sign test, what conclusion is reached with $\alpha_0 = .05$? Follow the six-step outline.

9-8   Find the power of the test used in Exercise 9-7 if the .40th quantile of the distribution of grades for the 5-meeting system is equal to 60.

9-9   Consider again the problem of the merchandising chain discussed in Exercise 5-3. Suppose that it is unrealistic to assume that the amount of electricity used by a 100-watt bulb in a 10-minute period is normally distributed. The chain will buy a large number of bulbs if it can be convinced that the median of the distribution is less than 50 units. In order to make a decision, 50 bulbs are tested. Of these 35 use less than 50 units and 15 use more than 50 units. The appropriate hypothesis is to be tested with $\alpha_0 = .01$ using the sign test. Follow the six-step outline and recommend the appropriate action.

9-10   Find the power of the test used in Exercise 9-9 if the .70th quantile is equal to 50. How large should n be to make this probability at least .90? *Hint:* Use inequality (6-25) with $p_0 = .50$, $p_1 = .70$.

9-11   Suppose that the merchandising chain of Exercise 9-9 desires a confidence interval for the median with confidence coefficient at least .90. Which order statistics should be used?

9-12   Rework Exercise 5-99 using the sign test. (Since $n = 8$, the necessary probabilities can be quickly found without a table.)

◈ 9-13   Rework Exercise 5-108 using the sign test.

◈ 9-14   Find the power of test used in Exercise 9-13 if only one-fourth of the differences are negative so that $Q_{.25} = 0$.

9-15   Suppose that the ointments of Exercise 5-108 are tested with 50 individuals. Differences $d_1, d_2, \ldots, d_{50}$ are obtained by subtracting the new ointment recovery time from the old. If 14 negative signs are counted, what decision should be made using $\alpha_0 = .01$? Follow the six-step outline.

9-16   Find the power of test used in Exercise 9-15 if the .30th quantile of the distribution of differences is equal to zero.

9-3

THE WILCOXON SIGNED RANK TEST

The sign test of the previous section, used to test hypotheses about the median $Q_{.50}$, was based upon a statistic $Y =$ the number of negative signs found in $X_1 - L$, $X_2 - L$, ..., $X_n - L$, where L was the hypothesized value $Q_{.50}$. No use was made of the magnitude of X's as with the t test which is derived under the normality assumption. If we use more information provided by the sample, it seems reasonable to expect that a more powerful test could be constructed. One such test that does not depend upon any distributional assumptions other than that X has a continuous distribution which is symmetric about the median (which is then equal to the mean) is the Wilcoxon signed rank test.

Since $Q_{.50} = \mu = E(X)$ when the density of X is symmetric about the median, we can rewrite the standard hypothesis testing situations (9-1) to (9-3) as

$$H_0 : \mu = \mu_0 \quad \text{against} \quad H_1 : \mu < \mu_0 \tag{9-16}$$

$$H_0 : \mu = \mu_0 \quad \text{against} \quad H_1 : \mu > \mu_0 \tag{9-17}$$

and

$$H_0 : \mu = \mu_0 \quad \text{against} \quad H_1 : \mu \neq \mu_0 \tag{9-18}$$

(Although a continuous distribution always has a median, the mean may fail to exist. For such cases μ represents only the median.) From X_1, X_2, ..., X_n the differences D_1, D_2, ..., D_n can be obtained where $D_i = X_i - \mu_0$. The absolute values of these differences (obtained by disregarding the signs), say $|D_1|$, $|D_2|$, ..., $|D_n|$, can be ranked in order from smallest to largest. Denote the rank of the smallest such number by 1, the second smallest by 2, ..., the largest by n, and let R_1, R_2, ..., R_n be the ranks of $|D_1|$, $|D_2|$, ..., $|D_n|$, respectively. (When football teams or students are ranked, usually the highest score is rank 1, the second highest is rank 2, etc. Hence, for the Wilcoxon test ranking is done in reverse order of that ordinarily encountered.) Now attach the sign of the original difference to each of the ranks, obtaining R_1', R_2', ..., R_n'. Hence the sign of R_i' is the same as the sign of D_i. Let

$$T = \text{sum of the positive ranks} \tag{9-19}$$

Then T is a discrete random variable called the Wilcoxon statistic. If we let

$$U_i = 1 \quad \text{if } D_i > 0$$
$$= 0 \quad \text{if } D_i < 0$$

then a formula for (9-19) is

$$T = \sum_{i=1}^{n} U_i R_i \tag{9-20}$$

The distribution of T, derived under the assumptions (a) $X_1, X_2, \ldots, X_n$ is a random sample from a distribution, (b) the distribution is continuous, (c) the distribution is symmetric about the median, and (d) the median is μ_0, has been constructed and tabulated. A sample table giving $\Pr(T \leq t)$ for some t and $n = 5(1)20$ appears in Appendix B20. [The table published by the Institute of Mathematical Statistics [5] gives the distribution function for nearly all values of t and $n = 5(1)50$.]

We can use the statistic T for the hypothesis testing situations (9-16) to (9-18). If the differences are mostly negative, then t will be small indicating that $\mu < \mu_0$. Hence, a reasonable critical region for (9-16) consists of small values of t. With a significance level no greater than α_0, we will reject H_0 when $t \leq t_0$ where t_0 is the largest value of t such that $\Pr(T \leq t) \leq \alpha_0$. This is the same convention followed in Chap. 6 for tests based upon other statistics having discrete distributions. Similarly, when testing (9-17) large values of t support H_1, and with "aimed-at" significance level α_0 we reject when $t \geq t_0$ where t_0 is the smallest value of t such that $\Pr(T \geq t) \leq \alpha_0$. Finally, when testing (9-18) both small and large values of t support H_1. The critical region we will use to obtain a significance level no greater than α_0 is $t \leq t_1$ and $t \geq t_2$ where t_1 is the largest value of t such that $\Pr(T \leq t) \leq \alpha_0/2$ and t_2 is the smallest value of t such that $\Pr(T \geq t) \leq \alpha_0/2$.

Since the table of Appendix B20 gives $\Pr(T \leq t)$, we cannot solve $\Pr(T \geq t) \leq \alpha_0$ for the smallest value of t. However, it can be shown that the solution is accomplished by solving $\Pr(T \leq t) \leq \alpha_0$ for the largest t and replacing t by $[n(n + 1)/2] - t$. This is due to the fact that

$$\Pr(T \geq t) = \Pr\left[T \leq \frac{n(n + 1)}{2} - t\right] \tag{9-21}$$

For the two-sided case this implies that

$$t_2 = \frac{n(n + 1)}{2} - t_1 \tag{9-22}$$

and the actual level of significance is

$$\alpha = 2\Pr(T \leq t_1) \tag{9-23}$$

EXAMPLE 9-8

With the first 10 observations of Exercise 9-1 use the Wilcoxon signed rank test to test the hypothesis suggested by that problem. Take $\alpha_0 = .05$.

Solution

We shall follow the six-step outline.

1 The problem suggests that we test H_0: $\mu = 46$ against H_1: $\mu \neq 46$.

2 The aimed at significance level $\alpha_0 = .05$ and the sample size $n = 10$ have already been chosen.

3 The statistic we use is T, the sum of the positive ranks. Probabilities for the distribution of T may be obtained from Appendix B20 provided (a) the sample is selected randomly, (b) the distribution of weights is continuous, (c) the distribution is symmetric about the median, and (d) the median is $\mu_0 = 46$. Certainly the distribution of weights is continuous, and we are given that the sample is selected at random. It is reasonable to assume that the distribution of weights is symmetric about the median (equal to the mean).

4 To obtain the critical region we find $t_1 = $ the largest value of t such that $\Pr(T \leq t) \leq \alpha_0/2 = .05/2 = .025$ and $t_2 = $ the smallest value of t such that $\Pr(T \geq t) \leq .025$. From the table of Appendix B20 we find with $n = 10$

$$\Pr(T \leq 8) = .0244 \qquad \Pr(T \leq 9) = .0322$$

so that $t_1 = 8$. From the remarks immediately preceding this example and formula (9-22) we have that

$$t_2 = \frac{n(n + 1)}{2} - t_1 = \frac{10(11)}{2} - 8 = 47$$

Hence, the critical region is $t \leq 8$ and $t \geq 47$. The actual significance level of the test is

$$\Pr(T \leq 8) + \Pr(T \geq 47) = .0244 + .0244 = .0488$$

5 The observed sample values are

$x_1 = 45.63$ $\qquad x_2 = 45.82$ $\qquad x_3 = 45.77$ $\qquad x_4 = 45.80$

$x_5 = 46.21$ $\qquad x_6 = 46.07$ $\qquad x_7 = 46.16$ $\qquad x_8 = 45.91$

$x_9 = 45.87$ $\qquad x_{10} = 46.03$

Subtracting 46 from each observed value yields

$d_1 = -.37$ $\quad d_2 = -.18$ $\quad d_3 = -.23$ $\quad d_4 = -.20$ $\quad d_5 = .21$

$d_6 = .07$ $\quad\;\; d_7 = .16$ $\quad\;\; d_8 = -.09$ $\quad d_9 = -.13$ $\quad d_{10} = .03$

The absolute values of the d_i's are .37, .18, .23, .20, .21, .07, .16, .09, .13, .03, respectively. The observed ranks are

$r_1 = 10$ $\quad r_2 = 6$ $\quad r_3 = 9$ $\quad r_4 = 7$ $\quad r_5 = 8$

$r_6 = 2$ $\quad\;\; r_7 = 5$ $\quad r_8 = 3$ $\quad r_9 = 4$ $\quad r_{10} = 1$

and the signed ranks are

$r'_1 = -10$ $\quad r'_2 = -6$ $\quad r'_3 = -9$ $\quad r'_4 = -7$ $\quad r'_5 = 8$

$r'_6 = 2$ $\qquad\;\; r'_7 = 5$ $\qquad r'_8 = -3$ $\quad r'_9 = -4$ $\quad r'_{10} = 1$

Hence the observed value of T is $t = 8 + 2 + 5 + 1 = 16$.

A systematic method for determining the signed ranks consists of the following steps:
(a) Order the d_i by their absolute values from smallest to largest.
(b) Write the numbers $1, 2, \ldots, n$ below the n values of the d_i so ordered. (c) Transfer the sign of the d_i to the rank order. Hence, here we would write

d_i	.03	.07	−.09	−.13	.16	−.18	−.20	.21	−.23	−.37
r_i	1	2	3	4	5	6	7	8	9	10
r_i'	1	2	−3	−4	5	−6	−7	8	−9	−10

Thus, $t = 1 + 2 + 5 + 8 = 16$.
6 Since $t = 16$ does not fall in the critical region, we accept the hypothesis.

Like the sign test, the signed rank test can also be used with paired observations. As in Sec. 9-2, let the random sample of n pairs be $(X_{11}, X_{12}), (X_{21}, X_{22}), \ldots, (X_{n1}, X_{n2})$ and form differences $D_1, D_2, \ldots, D_n$ where $D_i = X_{i1} - X_{i2}$, $i = 1, 2, \ldots, n$. Now μ is the mean of the differences, and we test hypotheses (9-16) to (9-18) with $\mu_0 = 0$ using the D_i to obtain signed ranks as before. The assumptions needed for the use of the Wilcoxon statistic are the same as given previously [following formula (9-20) except that now the D_i replace the X_i and the median μ_0 is zero].

Although it is theoretically impossible for two or more of the $|d_i|$ to have the same value, this situation may occur in practice due to the fact that measurements have not been made as accurately as desired. When this happens, we say that ties have occurred. One easy method (but not necessarily the best) of handling ties is to disregard them. This will be done in the following example.

EXAMPLE 9-9

Consider again the problem of Example 5-38 involving army recruits and a physical training program. Use the signed rank test to draw the appropriate conclusion. (In Example 9-5 we used the sign test.)

Solution

We shall follow the six-step outline.

1 For the distribution of differences we test $H_0 : \mu = 0$ against $H_1 : \mu \neq 0$.
2 The aimed-at significance level has been chosen as $\alpha_0 = .05$. Although 10 observations were made, two pairs of ties were recorded (two -5's and two -4's) and will be disregarded. Hence we have $n = 6$. (The ties could have been avoided by recording the weights more accurately.)

3   The statistic we use is T, the sum of the positive ranks. Probabilities for the distribution of T may be obtained from Appendix B20 provided (a) the sample of differences is randomly selected, (b) the differences have a continuous distribution, (c) the distribution is symmetric about the median, (d) the median is 0. There is no question about assumptions (a) and (b), and (c) is probably reasonable.

4   To obtain the critical region we find $t_1 =$ the largest value of t such that $\Pr(T \leq t) \leq \alpha_0/2 = .025$. With $n = 6$ and Appendix B20 we find

$$\Pr(T \leq 0) = .0156 \qquad \Pr(T \leq 1) = .0313$$

Hence $t_1 = 0$, $t_2 = [n(n + 1)/2] - 0 = 6(7)/2 = 21$ and the critical region is $t \leq 0$ and $t \geq 21$ (or $t = 0$, $t = 21$).

5   From Example 5-38 the observed differences are $d_1 = -8$, $d_2 = 2$, $d_3 = -12$, $d_4 = 5$, $d_5 = -11$, $d_6 = 3$. The absolute values are 8, 2, 12, 5, 11, 3, respectively, and the observed ranks are $r_1 = 4$, $r_2 = 1$, $r_3 = 6$, $r_4 = 3$, $r_5 = 5$, $r_6 = 2$. The signed ranks are $r_1' = -4$, $r_2' = 1$, $r_3' = -6$, $r_4' = 3$, $r_5' = -5$, $r_6' = 2$. The observed value of the statistic is $t = 1 + 3 + 2 = 6$. As in Example 9-9 we can write

d_i	2	3	5	-8	-11	-12
r_i	1	2	3	4	5	6
r_i'	1	2	3	-4	-5	-6

and $t = 1 + 2 + 3 = 6$.

6   Since the observed value of the statistic does not fall in the critical region, we accept H_0. The experiment does not provide sufficient evidence to conclude that the program affects average weight.

When n exceeds the range of the existing tables, large-sample results can be used to conduct tests. It is known that

$$Z' = \frac{T - \mu_T}{\sigma_T} \tag{9-24}$$

where

$$\mu_T = \frac{n(n + 1)}{4} \qquad \sigma_T^2 = \frac{n(n + 1)(2n + 1)}{24}$$

has approximately a standard normal distribution. Using Z' instead of T, tests are performed as in Sec. 6-4 with the binomial and Poisson cases. However, for n this large and a symmetric continuous distribution, we can always use the ordinary one-sample t test (unless, of course, the mean does not exist) and be certain that probability statements are reasonably accurate.

The topics of power and confidence intervals for the signed rank test will not be discussed here.

EXERCISES

◆ 9-17   Rework Exercise 9-1 with the signed rank test using the last 14 weights as the observed random sample.

◆ 9-18   Consider again the problem of the merchandising chain discussed in Exercise 5-3.   Suppose that we are willing to assume a continuous distribution symmetric about the median is appropriate, but we do not want to assume normality.   The chain will buy a large number of bulbs if it can be convinced that the median of the distribution is less than 50 units.   Twelve bulbs are tested and the following amounts of electricity are consumed: 48.7, 49.6, 49.4, 49.0, 51.5, 49.7, 48.8, 49.1, 49.2, 51.8, 49.3, 48.9.   Use the signed rank test to test the appropriate hypothesis with $\alpha_0 = .01$.   Follow the six-step outline and recommend the appropriate action.

9-19   Each of two hybrid seed corn companies claims that its product is superior to the other's.   Three scientists are hired to settle the dispute.   They plant 1 acre of each kind in 10 different localities which represent 10 different soil and climate conditions.   When the corn is harvested, they record the yields in bushels per acre, obtaining 10 pairs of observed values.   These are (114.2, 107.3), (86.4, 94.5), (93.7, 86.2), (75.3, 70.0), (102.8, 90.0), (89.5, 82.2), (64.1, 73.7), (95.2, 81.2), (106.6, 105.5), (90.2, 100.4), where the first number in each pair is the yield for variety Long Ear and the second is the yield for variety Fat Kernel.   Use the signed rank test with $\alpha_0 = .05$ to test the appropriate hypothesis.   Follow the six-step outline and indicate the conclusion which should be expressed in the scientists' report.

9-20   Rework Exercise 5-108 concerning the old and new ointments using the signed rank test.   However, replace the observed results of that exercise by the following results obtained from 13 individuals: (14.63, 14.18), (9.81, 9.40), (15.04, 14.51), (17.79, 17.03), (11.44, 10.98), (8.69, 8.32), (12.60, 12.80), (13.22, 12.70), (10.38, 9.84), (8.93, 8.70), (11.84, 12.24), (12.16, 12.18), (15.12, 14.60). The first number in each pair is recovery time for the old ointment; the second is recovery time for the new ointment.

9-21   A company manufactures tires for automobiles.   Under a standard set of driving conditions its brand has a median life of 20,000 miles. A new process for making tires is being considered but the company cannot afford to convert to the new process unless there is outstanding evidence to indicate that a larger median will be achieved. It is decided that the significance level will be no larger than .01 and that the sample size should be 16.   A random sample of tires

produced by the new process yields the following numbers of service-
able miles:

20,600	19,900	23,000	21,800
20,800	22,200	18,900	21,300
21,200	21,700	20,500	18,400
19,800	20,700	21,500	20,900

If the signed rank test is used, what action should be taken? Follow
the six-step outline.

◈ 9-22 A psychologist claims that most students experience an increase in
pulse rate just before taking an examination. Specifically, he
hopes to prove that the median of the distribution of differences
(pulse rate before minus pulse rate 24 hours later at the same class)
is greater than zero. A random sample of 20 students taking
examinations during a month-long period yields the following pairs
of scores:

(82.4, 73.2)	(80.6, 73.2)	(80.3, 77.3)	(75.3, 70.0)
(79.1, 72.0)	(76.1, 72.0)	(75.2, 72.1)	(82.4, 72.4)
(74.3, 72.1)	(74.0, 72.1)	(72.1, 72.3)	(81.6, 72.1)
(72.5, 72.0)	(70.5, 71.9)	(72.0, 72.4)	(82.0, 76.0)
(71.4, 72.3)	(71.9, 72.0)	(76.1, 72.1)	(70.8, 72.0)

The first number in each pair is the pulse rate taken just before
an examination. Use the signed rank test with $\alpha_0 = .01$ to investigate the claim. Follow the six-step outline.

9-4
THE RANK SUM TEST

In Sec. 5-9 we considered $H_0 : \mu_1 = \mu_2$ by assuming that the two distributions under consideration were both normal and both had the same
variance. In other words, if H_0 is true, then the two distributions are
exactly the same. Now we wish to consider a distribution-free test of
the hypothesis that two independent continuous random variables, say X
and Y, have identical distributions.

One way to state the hypothesis testing situation is

H_0 : the random variables X and Y have the same distribution

against (9-25)

H_1 : the two distributions are not the same

Formulation (9-25) is the counterpart of our usual two-sided testing situation. Suppose we are willing to assume that the density functions of X and Y are exactly the same shape with the only possible difference being that the second is obtained from the first by displacing the first by a fixed amount. In other words, if we "push" the first density to the right or to the left we get the second density. (See Fig. 9-3). Then,

FIGURE 9-3

Two densities with the same shape but with different locations.

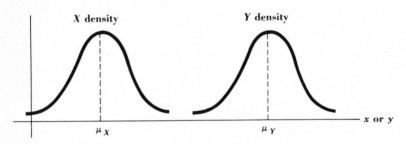

letting μ_X and μ_Y be the two mean values (median values if the means do not exist), the hypotheses and alternatives can be given in the usual way. That is, standard hypothesis testing situations are

$$H_0 : \mu_X = \mu_Y \quad \text{against} \quad H_1 : \mu_X < \mu_Y \qquad (9\text{-}26)$$

$$H_0 : \mu_X = \mu_Y \quad \text{against} \quad H_1 : \mu_X > \mu_Y \qquad (9\text{-}27)$$

and [equivalent to (9-25)]

$$H_0 : \mu_X = \mu_Y \quad \text{against} \quad H_1 : \mu_X \neq \mu_Y \qquad (9\text{-}28)$$

A distribution-free test frequently used for the above situations is known as the Wilcoxon two-sample test. Let $X_1, X_2, \ldots, X_m$ and $Y_1, Y_2, \ldots, Y_n$ be random samples from the X and Y distributions. If $m \neq n$, we will assume that $m < n$ so that the X distribution is the one from which the smaller sample is selected. After the samples are observed, obtaining $x_1, x_2, \ldots, x_m$ and $y_1, y_2, \ldots, y_n$, these numbers can be arranged in order of size from smallest to largest and assigned ranks from 1 to $m + n$. Then

$$T = \text{sum of the ranks assigned to the } X\text{'s} \qquad (9\text{-}29)$$

is the statistic used for the Wilcoxon two-sample test (usually called the rank sum test). For example, if the sample values are $x_1 = 3.2$, $x_2 = 5.7$,

$y_1 = 4.1$, $y_2 = 5.0$, $y_3 = 7.4$, then the five numbers arranged from smallest to largest, together with their ranks, are

3.2	4.1	5.0	5.7	7.4
<u>1</u>	2	3	<u>4</u>	5

where the underlined ranks arise from the X distribution. Here the observed value of T is $t = 1 + 4 = 5$. (Again, if ties occur, we will disregard the tied observations.) Obviously, T is a discrete random variable which can assume only integer values [the smallest of which is $1 + 2 + 3 + \cdots + m = m(m + 1)/2$ and the largest of which is $(n + 1) + (n + 2) + \cdots + (n + m) = m(m + 2n + 1)/2$]. The distribution of T, derived under the assumptions (a) both samples are selected at random, (b) both distributions are continuous, and (c) both distributions are the same, has been constructed and tabulated. A table giving $\Pr(T \leq t_1)$ and $\Pr(T \geq t_2)$ some t_1, t_2 needed for one-sided significance level no greater than $\alpha_0 = .05$, .01 (and two-sided significance levels no greater than $\alpha_0 = .10$, .02), and $m = 6(1)10$, $n = 6(1)19$ appears in Appendix B21. The table is entered with m, n, and $\alpha_0 = P$ and contains t_1, t_2, and p such that $\Pr(T \leq t_1) = \Pr(T \geq t_2) = p$. Of the two sets given for each entry combination, the first set contains the boundary (or boundaries) of the critical region and the significance level α ($\alpha/2$ in the two-sided case). Thus, for example, with $m = 8$, $n = 15$, $\alpha_0 = P = .05$ (one-side), we read 69, 123, .0437. Hence $\Pr(T \leq 69) = \Pr(T \geq 123) = .0437$. The critical region for (9-26) is $t \leq 69$, for (9-27) is $t \geq 123$, and for (9-28) is $t \leq 69$ and $t \geq 123$. The significance levels are respectively .0437, .0437, and $2(.0437) = .0874$. [The table published by the Institute of Mathematical Statistics [5] gives corresponding probabilities for $\alpha_0 = .05$, .025, .01, .005 for the one-sided cases (double these for the two-sided cases) and $m \leq n = 3(1)50$.]

We can use the statistic T given by (9-29) for the hypothesis testing situations (9-26) to (9-28). If the ranks associated with the x's are mostly small numbers, then t will be small. Such a result supports the alternative H_1: $\mu_X < \mu_Y$. Hence a reasonable critical region for (9-26) consists of small values of t. If the significance level is to be no greater than α_0, we will reject H_0 when $t \leq t_0$ where t_0 is the largest value of t such that $\Pr(T \leq t) \leq \alpha_0$. Similarly, when testing (9-27) large ranks for the x's resulting in a large value of t support H_1 : $\mu_X > \mu_Y$ and lead to rejection. If the significance level is to be no greater than α_0, we will reject H_0 when $t \geq t_0$ where t_0 is the smallest value of t such that $\Pr(T \geq t) \leq \alpha_0$. Finally, when testing (9-28), or (9-25), both small and large values of t support H_1. The critical region we will use to obtain significance level no greater than α_0 is $t \leq t_1$ and $t \geq t_2$ where t_1 is the largest value of t such that $\Pr(T \leq t) \leq \alpha_0/2$ and t_2 is the smallest value of t such that $\Pr(T \geq t) \leq \alpha_0/2$.

EXAMPLE 9-10

The large corporation of Examples 5-1 and 5-32 wants to choose between two brands of light bulbs. Brand 1 is slightly less expensive than brand 2. The company would like to buy brand 1 unless there is evidence to indicate that brand 2 tends to have a longer life. Suppose that it is reasonable to assume that both distributions have density functions with the same shape but unreasonable to believe that normal models with equal variances are appropriate. It is decided to use the rank sum test with $\alpha_0 = .05$. Seven brand 1 bulbs are tested and yield observed results (in hours of life) 981, 952, 1,342, 1,051, 1,005, 974, 1,216, while nine brand 2 bulbs yield 1,380, 1,004, 1,032, 1,263, 1,040, 990, 1,102, 1,170, 1,205. What decision should be made?

Solution

We shall follow the six-step outline.

1 The results for brand 1 constitute observations from the X distribution, while those for brand 2 are observations for the Y distribution (since there are fewer observations for brand 1). The problem suggests that we test $H_0: \mu_X = \mu_Y$ against $H_1: \mu_X < \mu_Y$ and buy brand 1 if H_0 is accepted.

2 The significance level will not exceed $\alpha_0 = .05$, and the sample sizes are $m = 7$, $n = 9$, numbers already chosen.

3 The statistic we use is T given by (9-29). Probabilities for the distribution of T may be obtained from Appendix B21 provided (a) both samples are selected at random, (b) both distributions are continuous, and (c) both distributions are the same (or $\mu_X = \mu_Y$ if both densities have the same shape). Sampling should be performed to make (a) realistic, and condition (b) is satisfied. The most difficult assumption to justify is that the densities have the same shape.

4 To obtain the critical region we identify $m = 7$, $n = 9$, $\alpha_0 = .05 = P$. We want t_0 where t_0 is the largest t such that $\Pr(T \le t) \le .05$. The table gives $t_0 = 43$, $\alpha = .0454$. Hence, we reject if $t \le 43$.

5 We observe the samples and order the observed results. This yields

952	974	981	990	1,004	1,005	1,032	1,040	1,051
1	2	3	4	5	6	7	8	9

1,102	1,170	1,205	1,216	1,263	1,342	1,380
10	11	12	13	14	15	16

The observed value of T is $t = 1 + 2 + 3 + 6 + 9 + 13 + 15 = 49$.

6 Since the observed value of the statistic does not fall in the critical region, we accept H_0. The cheaper brand will be purchased.

Had a two-sided critical region with $\alpha_0 = .10$ been required in Example 9-10, then we would reject when $t \leq 43$ or $t \geq 76$, and the actual significance level would be $\alpha = 2(.0454) = .0908$.

When n and (or) m exceed the range of existing tables, large sample results can be used to conduct tests. It is known that

$$Z' = \frac{T - \mu_T}{\sigma_T} \tag{9-30}$$

where Z' has approximately a standard normal distribution and

$$\mu_T = \frac{m(m + n + 1)}{2} \qquad \sigma_T^2 = \frac{mn(m + n + 1)}{12}$$

Using Z' instead of T, tests are performed as in Sec. 6-4 with the binomial and Poisson cases.

The calculation of power for the rank sum test involves mathematical details which will not be discussed here. If the conditions for the use of the test given in Sec. 5-9 are satisfied, then the rank sum test is not as powerful as the test given in that earlier section. However, even in this case for which the basis of comparison is most unfavorable to the rank sum test, this distribution-free test has a power curve which is nearly as good as the normal model test. For some X and Y distributions (whose densities are the same shape), the rank sum test is more powerful than the normal test of Sec. 5-9. For these reasons the rank sum test is often considered one of the best distribution-free tests.

A confidence interval problem can be solved using the statistic of the rank sum test, but that topic is left for more advanced courses.

EXERCISES

◊ 9-23 Two different types of wheat are being considered and it is desired to know which (if either) produces the greater average yield. Ten acres of each kind are planted with fairly uniform growing conditions. The results in bushels per acre are as follows: variety A—38.1, 36.2, 41.3, 31.4, 35.8, 32.3, 37.6, 37.8, 39.0, 34.0; variety B—34.3, 29.8, 31.6, 36.0, 34.2, 30.0, 33.7, 33.9, 37.1, 37.4. What conclusion should be drawn using $\alpha_0 = .10$ and the rank sum test? Follow the six-step outline.

9-24 A psychology professor believes that IQ can be increased by studying courses specifically designed to give experience in types of skills which questions on the IQ test seem to emphasize. To test his theory two randomly selected groups of tenth graders are used for one semester. Nine in one group take the standard high school schedule of courses, while eight in the second group study courses designed by the professor. At the end of the term all 17 students are given the same IQ test. The scores are the following numbers: ordinary schedule—103, 98, 90, 102, 105, 100, 101, 96, 82; special schedule—115, 111, 97, 94, 110, 114, 118, 107. If $\alpha_0 = .05$, has the professor proved his theory? Follow the six-step outline using the rank sum test.

◈ 9-25 Suppose that the merchandising chain mentioned in Exercise 5-3 is going to buy a large order of 100-watt bulbs. It has a choice between brand 1 and brand 2. The chain would rather buy brand 2, since it knows that the product is dependable, but will buy brand 1 if it can be proved that the average amount of electricity consumed over a 10 minute period is less for that brand. Fifteen brand 1 and ten brand 2 bulbs are tested. The following consumption results are recorded: brand 1—43.3, 48.0, 51.2, 50.0, 42.1, 46.3, 40.8, 49.7, 46.4, 43.9, 44.4, 46.5, 45.0, 48.6, 51.4; brand 2—52.1, 48.3, 50.6, 50.9, 53.7, 47.6, 47.8, 49.9, 52.0, 52.6. If the significance level is to be no larger than $\alpha_0 = .01$, which brand should be purchased? Follow the six-step outline using the rank sum test.

9-26 A cattle feeder would like to know whether or not to give his animals a diet supplement which is supposed to help them gain weight. He has 20 animals about the same size available for the experiment which are divided at random into two groups of 10 each. For a period of time group 1 is fed the standard diet, while group 2 is fed the standard diet plus the supplement. The gain in weight for each animal is recorded in pounds, yielding the following results: group 1—272, 243, 227, 254, 238, 249, 261, 251, 218, 270; group 2—263, 294, 277, 286, 265, 273, 246, 268, 259, 252. What conclusion should be drawn with $\alpha_0 = .01$? Use the rank sum test and follow the six-step outline.

◈ 9-27 A teacher has two sections of the same course. There are 8 students in the first section and 12 students in the second section. According to available records, the classes appear to be fairly evenly matched on the basis of ability. The first section is taught using lectures followed by a short quiz at the end of the period. The second section is taught by using lectures only. All 20 students take the same final examination at the end of the course. The teacher hopes to be able to prove that students subjected to the quizzes will have the higher average. Final examination scores are: first section—61, 52, 84, 57, 45, 66, 71, 75; second section—43, 51, 60, 38, 46, 39, 21, 42, 49, 37, 31, 50. Use the rank sum test with $\alpha_0 = .01$ to draw the appropriate conclusion. Follow the six-step outline.

9-28 Suppose that the experiment described in Exercise 9-27 is performed with 80 students in the first section and 120 students in the second. If the rank sum statistic has observed value 8,710, what decision is made with $\alpha_0 = .01$? With $\alpha_0 = .05$? *Hint:* Use statistic (9-30).

9-5
THE RANK CORRELATION TEST OF INDEPENDENCE

In Sec. 8-5 we considered two random variables X and Y whose joint density function was assumed to be bivariate normal. The first hypothesis testing situations which we discussed in that section concerned the independence of the two random variables. We learned that we could test the hypothesis of independence against various alternatives by using the statistic T_{n-2}, defined by (8-34). The distribution of the test statistic was derived under the assumption that the bivariate normal model was appropriate. Now we wish to discuss a test of the hypothesis that X and Y are independent which depends upon the assumption that the two random variables are continuous but does not require normality.

Perhaps the most interesting hypothesis testing situation (corresponding to $H_0 : \rho = 0$ against $H_1 : \rho \neq 0$ of Sec. 8-5) is

H_0 : random variables X and Y are independent

against   (9-31)

H_1 : X and Y are not independent

However, we might be interested in testing

H_0 : X and Y are independent

against   (9-32)

H_1 : small X and large Y tend to occur together as do large X and small Y

or

H_0 : X and Y are independent

against   (9-33)

H_1 : small X and small Y tend to occur together as do large X and large Y

[We see that (9-32) and (9-33) are the counterparts of (8-36) and 8-37), respectively.]

One statistic proposed for the above hypothesis testing situations is based upon the sample correlation coefficient defined by formula (8-30). Suppose (X_1, Y_1), (X_2, Y_2), ..., (X_n, Y_n) is a random sample of pairs drawn from a continuous bivariate distribution. If we rank both the X's and the Y's from 1 to n and compute the sample correlation coefficient

$$R = \frac{1}{n-1} \sum_{i=1}^{n} \left(\frac{X_i - \overline{X}}{S_X} \right) \left(\frac{Y_i - \overline{Y}}{S_Y} \right)$$

after each X_i is replaced by rank X_i and each Y_i is replaced by rank Y_i, then the resulting R is called the rank correlation coefficient (frequently referred to as Spearman's coefficient of rank correlation). With a little algebra the formula for R can be considerably simplified. If we let $D_i = $ rank $X_i - $ rank Y_i, then it can be shown that

$$R = 1 - \frac{6 \sum_{i=1}^{n} D_i^2}{n(n^2 - 1)} \tag{9-34}$$

Like other statistics based upon ranks, the rank correlation coefficient has a discrete distribution. The distribution of R, derived under the assumptions (a) (X_1, Y_1), (X_2, Y_2), ..., (X_n, Y_n) is a random sample of pairs, (b) the joint distribution of X and Y is continuous, (c) the variables X and Y are independent, has been constructed and tabulated in various forms. Appendix B22 gives r_0, which is the largest value of r which satisfies $\Pr(R < r) \leqq \alpha_0$. Hence, $\Pr(R < r_0) \leqq \alpha_0$ while $\Pr(R \leqq r_0) > \alpha_0$. The distribution of R is symmetric about 0 so that $\Pr(R < r) = \Pr(R > -r)$, a fact useful in determining two-sided critical regions.

When testing (9-32), small observed values of the statistic support the alternative. If the significance level is to be no greater than α_0, then we reject H_0 when

$$r < r_0 \tag{9-35}$$

where r_0 is read from Appendix B22 with the appropriate n and α_0. Similarly, when testing (9-33), large values of r support the alternative and a reasonable critical region with significance level no larger than α_0 is

$$r > -r_0 \tag{9-36}$$

where r_0 is read from the table with n and α_0. Finally, when testing (9-31), both small and large values of r support the alternative, and a critical region with significance level not exceeding α_0 is

$$r < r_0 \quad \text{and} \quad r > -r_0 \tag{9-37}$$

where r_0 is read from the table with $\alpha_0/2$ (instead of α_0) and n.

EXAMPLE 9-11

Ten students are selected at random from those who have completed both the first course in English and the first course in physical education. Their final grades (in percentages) are (67, 92), (81, 65), (65, 81), (42, 75), (53, 85), (40, 78), (71, 77), (64, 79), (60, 80), (68, 82) where the first grade in each pair is the English grade and the second the physical education grade, both grades in a pair belonging to the same student. Using the rank correlation statistic with $\alpha_0 = .05$, test the hypothesis that a student's English grade is independent of his physical education grade.

Solution

We shall follow the six-step outline.

1 We test

H_0 : English grades are independent of physical education grades

against

H_1: English grades are not independent of physical education grades

2 We are given that the significance level is to be no greater than $\alpha_0 = .05$ and the sample size is $n = 10$.

3 The statistic we use is R, the rank correlation coefficient. Critical regions for the test may be obtained from Appendix B22 provided (a) the students are selected at random, (b) the distribution of grades is continuous, and (c) H_0 is true so that English grades and physical education grades are independent. We are given that the selection was made at random. Actually, the distribution of grades is probably not continuous since they are rounded off to the nearest percentage. However, the grades could be recorded on a finer scale, so that it is not unreasonable to assume that a continuous distribution is appropriate. In other words, a continuous model is a reasonable approximation to the actual situation.

4 With $\alpha_0 = .05$, $\alpha_0/2 = .025$, $n = 10$, we read from Appendix B22 that $r_0 = -.6364$. Hence the critical region is $r < -.6364$ and $r > .6364$.

5 The grades are observed and recorded below. Under each x and under each y we record the ranks. We can write

Student i	1	2	3	4	5	6	7	8	9	10
English grade x_i	67	81	65	42	53	40	71	64	60	68
Rank x_i	7	10	6	2	3	1	9	5	4	8
Physical education grade y_i	92	65	81	75	85	78	77	79	80	82
Rank y_i	10	1	7	2	9	4	3	5	6	8
d_i	−3	9	−1	0	−6	−3	6	0	−2	0

We next obtain

$$\sum_{i=1}^{10} d_i^2 = (-3)^2 + 9^2 + (-1)^2 + 0^2 + (-6)^2 + (-3)^2 + 6^2 + 0^2$$
$$+ (-2)^2 + 0^2 = 176$$

Then

$$r = 1 - \frac{6(176)}{10(100 - 1)} = 1 - 1.067 = -.067$$

6 Since the observed value of the statistic does not fall in the critical region, we accept H_0 and conclude that English grades and physical education grades are independent.

When n exceeds the range of the table given in Appendix B22, large-sample results can be used. It is known that

$$Z' = R\sqrt{n-1} \tag{9-38}$$

has approximately a standard normal distribution. Using Z' instead of the rank correlation coefficient R, tests are performed as in Sec. 6-4 with the binomial and Poisson cases.

Again we omit sample size problems and power calculations.

EXERCISES

◈ 9-29 Consider again Exercise 9-19 and the 10 pairs of observed corn yields. Use the rank correlation coefficient to test the hypothesis of independence against the appropriate alternative. Take $\alpha_0 = .05$ and follow the six-step outline.

9-30 Consider again Exercise 9-20 and the 13 pairs of observed recovery times. Suppose that we would like to prove that large values of X tend to occur with large values of Y. Use the rank correlation coefficient to conduct the appropriate test. Take $\alpha_0 = .01$ and follow the six-step outline.

9-31 Twenty students are selected at random from those who have completed elementary statistics. The first number in each of the following pairs is a grade determined prior to the final examination and the second is the final-examination grade for the same student: (52, 32), (55, 27), (45, 44), (39, 15), (57, 37), (38, 5),(73, 42), (58, 69), (60, 70), (48, 43), (74, 60), (40, 30), (67, 51), (37, 16), (49, 25), (53, 52), (61, 53), (47, 31), (41, 29), (79, 67). Using the rank correlation coefficient, test the hypothesis that grades determined prior to the final examination are independent of final-examination grades. Take $\alpha_0 = .01$ and follow the six-step outline.

◈ 9-32 A psychologist has constructed a test which measures physical dexterity. Like an IQ test, the dexterity test measures a kind of ability, If IQ and physical dexterity scores are not independent, perhaps one of the examination scores would give some indication as to an individual's ability measured by the other. Consequently, it is of interest to investigate the hypothesis of independence. Fifteen individuals take both the IQ test and the dexterity test and their scores are (80, 34), (118, 49), (86, 39), (104, 37), (98, 36), (102, 48), (103, 54), (106, 56), (97, 52), (79, 62), (101, 46), (83, 63), (95, 55), (130, 33), (113, 57), where the first number in each pair is the IQ score. Use the rank correlation coefficient to test the appropriate hypothesis. Take $\alpha_0 = .05$ and follow the six-step outline.

◊ 9-33 An investigator in an educational study wishes to know if college grade-point average is independent of the student's reading rate. He is particularly interested in those students who have just completed the freshman year (successfully or unsuccessfully). A random sample of 12 such students are given a reading test and their scores are (in words per minute) 173, 205, 187, 230, 202, 141, 275, 180, 197, 133, 225, 161. The grade-point averages of these 12 students were, respectively, 1.67, 2.24, 2.54, 3.05, 3.40, 0.67, 2.80, 3.25, 1.31, 1.42, 2.06, 0.90. Use the rank correlation coefficient to test the hypothesis of independence against the alternative that higher reading rates tend to produce higher grade points. Take $\alpha_0 = .05$ and follow the six-step outline.

9-34 Suppose that the experiment described in Exercise 9-33 is performed with 82 students and calculations yield $\sum_{i=1}^{82} d_i^2 = 72,000$. What decision is made with $\alpha_0 = .05$? With $\alpha_0 = .01$? *Hint:* Use statistic (9-38).

REFERENCES

1 Bradley, James V.: "Distribution-free Statistical Tests," Prentice-Hall, Inc., Englewood Cliffs, N.J., 1968.

2 Conover, W. J.: "Practical Nonparametric Statistics," John Wiley & Sons, Inc., New York, 1971.

3 Gibbons, Jean Dickinson: "Nonparametric Statistical Inference," McGraw-Hill Book Company, New York, 1971.

4 Hájek, Jaroslav: "Nonparametric Statistics," Holden-Day, San Francisco, 1969.

5 Harter, H. L., and D. B. Owen (eds.): "Selected Tables in Mathematical Statistics," Markham Publishing Company, Chicago, 1970.

Summary of Results

DISTRIBUTION-FREE INFERENCES

I. Inferences about quantiles (Sec. 9-2)

A. Standard hypotheses, alternatives, statistics, and critical regions [significance level α less than or equal to α_0 (disregard difference between α, α_0 with large-sample procedures)]

Hypothesis	Alternative	Equivalent hypothesis	Equivalent alternative
$H_0: Q_{p_0} = L$	$H_1: Q_{p_0} < L$	$H_0: p = p_0$	$H_1: p > p_0$
$H_0: Q_{p_0} \geq L$	$H_1: Q_{p_0} < L$	$H_0: p \leq p_0$	$H_1: p > p_0$
$H_0: Q_{p_0} = L$	$H_1: Q_{p_0} > L$	$H_0: p = p_0$	$H_1: p < p_0$
$H_0: Q_{p_0} \leq L$	$H_1: Q_{p_0} > L$	$H_0: p \geq p_0$	$H_1: p < p_0$
$H_0: Q_{p_0} = L$	$H_1: Q_{p_0} \neq L$	$H_0: p = p_0$	$H_1: p \neq p_0$

The outline for the equivalent hypotheses and equivalent alternatives was given in the Summary of Results, I-A, Chap. 6. The only difference is that now instead of X we use $Y =$ number of values less than L. When $p_0 = .50$, then $Q_{p_0} = Q_{.50}$ is the median and the above hypotheses and alternatives are concerned with the median. Now Y is also the number of negative signs associated with $X_1 - L, X_2 - L, \ldots, X_n - L$.

B. Power calculations

If $Q_{p_1} = L$, then Y has a binomial distribution with parameters n and p_1. Hence, with X replaced by Y, the comments in Summary of Results, I-B, Chap. 6 are appropriate.

C. Paired observations

Letting $D_i = X_{i1} - X_{i2}$, the hypotheses given in A can be tested for the distribution of differences in exactly the same way. Now Y is the number of differences less than 0. Power calculations are performed as in B.

D. Confidence interval for Q_p with confidence coefficient at least $1 - \alpha$

The interval is $(x_{(i)}, x_{(j)})$ where $x_{(i)}$ is the ith smallest and $x_{(j)}$ is the jth smallest observed value. Choose i and j such that

$$\sum_{y=0}^{i-1} b(y; n, p) \leq \frac{\alpha}{2} \quad \text{and} \quad \sum_{y=0}^{j-1} b(y; n, p) \geq 1 - \frac{\alpha}{2}$$

and the sums are as close to $\alpha/2$, $1 - \alpha/2$ as possible.

II. The Wilcoxon signed rank test (Sec. 9-3)

A. Standard hypotheses, alternatives, statistics, and critical regions [significance level α less than or equal to α_0 (disregard differences between α, α_0 with large-sample procedures)]

Hypothesis	Alternative	Statistic†	Critical region
$H_0: \mu = \mu_0$ $H_0: \mu \geq \mu_0$	$H_1: \mu < \mu_0$ $H_1: \mu < \mu_0$	Small sample: $T =$ sum of positive ranks of $\|D_1\|, \|D_2\|, \ldots,$ $\|D_n\|$ with sign of $X_i - \mu_0$ attached to rank of each $\|D_i\|$ Large sample: $$Z' = \frac{T - \mu_T}{\sigma_T}$$ where $$\mu_T = \frac{n(n+1)}{4}$$ $$\sigma_T^2 = \frac{n(n+1)(2n+1)}{24}$$	Small sample: reject if $t \leq t_0$ where t_0 is the largest value of t such that $\Pr(T \leq t) \leq \alpha_0$. Large sample: reject if $z' < z_\alpha$.
$H_0: \mu = \mu_0$ $H_0: \mu \leq \mu_0$	$H_1: \mu > \mu_0$ $H_1: \mu > \mu_0$		Small sample: reject if $t \geq t_0$ where t_0 is the smallest value of t such that $\Pr(T \geq t) \leq \alpha_0$. Large sample: reject if $z' > z_{1-\alpha}$.
$H_0: \mu = \mu_0$	$H_1: \mu \neq \mu_0$		Small sample: reject if $t \leq t_1$, $t \geq t_2$ where t_1 is the largest value of t such that $\Pr(T \leq t) \leq \alpha_0/2$, $$t_2 = \frac{n(n+1)}{2} - t_1.$$ Large sample: reject if $z' < z_{\alpha/2}, z' > z_{1-\alpha/2}$.

† Assumptions for the use of the statistic may be found in Sec. 9-3.

III. The rank sum test (Sec. 9-4)

A. Standard hypotheses, alternatives, statistics, and critical regions [significance level α less than or equal to α_0 (disregard differences between α, α_0 with large-sample procedures)]

Hypothesis	Alternative	Statistic†	Critical region
$H_0: \mu_X = \mu_Y$ $H_0: \mu_X \geqq \mu_Y$	$H_1: \mu_X < \mu_Y$ $H_1: \mu_X < \mu_Y$		Small sample: reject if $t \leqq t_0$ where t_0 is the largest value of t such that $Pr(T \leqq t) \leqq \alpha_0$. Large sample: reject if $z' < z_\alpha$.
		Small sample: $T =$ sum of the ranks assigned to the X's	
$H_0: \mu_X = \mu_Y$ $H_0: \mu_X \leqq \mu_Y$	$H_1: \mu_X > \mu_Y$ $H_1: \mu_X > \mu_Y$	Large sample $$Z' = \frac{T - \mu_T}{\sigma_T}$$ where	Small sample: reject if $t \geqq t_0$ where t_0 is the smallest value of t such that $Pr(T \geqq t) \leqq \alpha_0$. Large sample: reject if $z' > z_{1-\alpha}$.
$H_0: \mu_X = \mu_Y$	$H_1: \mu_X \neq \mu_Y$	$$\mu_T = \frac{m(m + n + 1)}{2}$$ $$\sigma_T^2 = \frac{mn(m + n + 1)}{12}$$	Small sample: reject if $t \leqq t_1, t \geqq t_2$ where t_1 is the largest value of t such that $Pr(T \leqq t) \leqq \alpha_0/2$, t_2 is the smallest value of t such that $Pr(T \geqq t) \leqq \alpha_0/2$. Large sample: reject if $z' < z_{\alpha/2}$, $z' > z_{1-\alpha/2}$.

† Assumptions for the use of the statistic may be found in Sec. 9-4.

IV. The rank correlation test of independence (Sec. 9-5)
  A. Standard hypotheses, alternatives, statistics, and critical regions
     [significance level α less than or equal to α_0 (disregard difference
     between α, α_0 with large-sample procedures)]

Hypothesis	Alternative	Statistic†	Critical region
H_0: X and Y are independent	H_1: X and Y are not independent	Small sample: $$R = 1 - \frac{6 \sum_{i=1}^{n} D_i^2}{n(n^2 - 1)}$$ where $D_i = $ rank $X_i - $ rank Y_i	Small sample: reject if $r < r_0, r > -r_0$, where r_0 is the largest value of r such that $\Pr(R < r) \leqq \alpha_0/2$. Large sample: reject if $z' < z_{\alpha/2}, z' > z_{1-\alpha/2}$.
H_0: X and Y are independent	H_1: small X and large Y, large X and small Y tend to occur together	Large sample $$Z' = R\sqrt{n-1}$$	Small sample: reject if $r < r_0$ where r_0 is the largest value of r such that $\Pr(R < r) \leqq \alpha_0$. Large sample: reject if $z' < z_\alpha$.
H_0: X and Y are independent	H_1: small X and small Y, large X and large Y tend to occur together		Small sample: reject if $r > -r_0$ where r_0 is the largest value of r such that $\Pr(R < r_0) \leqq \alpha_0$. Large sample: reject if $z' > z_{1-\alpha}$.

† Assumptions for the use of the statistic may be found in Sec. 9-5.

APPENDIX A

SOME ADDITIONAL RESULTS

APPENDIX A1

SUMMATION NOTATION

A1-1

SINGLE SUMMATION

Summation signs are used to abbreviate lengthy mathematical expressions. By definition we have

$$\sum_{i=1}^{n} x_i = x_1 + x_2 + \cdots + x_n \tag{A1-1}$$

which is read as "the sum of the x_i, i going from 1 to n." Each x_i represents a number which may or may not be known later in the discussion. It is obviously more convenient to write the left-hand side of (A1-1) than to write out the right-hand side of that equation. As a slight generalization of (A1-1) we have

$$\sum_{i=1}^{n} u(x_i) = u(x_1) + u(x_2) + \cdots + u(x_n) \tag{A1-2}$$

If the upper limit on the sum is infinity, then we imply that the summing goes on forever. Alternatively, we can write

$$\sum_{i=1}^{\infty} x_i = x_1 + x_2 + x_3 + \cdots \tag{A1-3}$$

The expression

$$\sum_{i} u(x_i) \tag{A1-4}$$

without an upper limit on the summation implies that the various values $u(x_i)$ are summed over all values of i under consideration.

EXAMPLE A1-1

Write out $\sum_{i=1}^{6} y_i^2$.

440

Solution

By (A1-2) we get a sum of six terms with i taking on the values 1, 2, 3, 4, 5, 6. Thus

$$\sum_{i=1}^{6} y_i^2 = y_1^2 + y_2^2 + y_3^2 + y_4^2 + y_5^2 + y_6^2$$

EXAMPLE A1-2

If $x_1 = 3$, $x_2 = 5$, $x_3 = 11$, find $\sum_{i=1}^{3} x_i$, $(\sum_{i=1}^{3} x_i)^2$, $\sum_{i=1}^{3} x_i^2$.

Solution

$$\sum_{i=1}^{3} x_i = 3 + 5 + 11 = 19 \qquad \left(\sum_{i=1}^{3} x_i\right)^2 = 19^2 = 361$$

$$\sum_{i=1}^{3} x_i^2 = 3^2 + 5^2 + 11^2 = 9 + 25 + 121 = 155$$

Thus, the x_i being known numbers, the summations can be evaluated and yield specific numbers.

EXAMPLE A1-3

Write out $\sum_{i=1}^{5} (x_i + y_i)$ and show that it is equivalent to

$$\sum_{i=1}^{5} x_i + \sum_{i=1}^{5} y_i$$

Solution

The definition (A1-2) yields

$$\sum_{i=1}^{5} (x_i + y_i) = (x_1 + y_1) + (x_2 + y_2) + (x_3 + y_3) + (x_4 + y_4) + (x_5 + y_5)$$

But the right-hand side can be rewritten as

$$(x_1 + x_2 + x_3 + x_4 + x_5) + (y_1 + y_2 + y_3 + y_4 + y_5) = \sum_{i=1}^{5} x_i + \sum_{i=1}^{5} y_i$$

EXAMPLE A1-4

Write out $\sum_{i=1}^{4} C$.

Solution

By definition (A1-1), we get four terms in the sum. Since there is no subscript to change from term to term, each term is C. Thus

$$\sum_{i=1}^{4} C = C + C + C + C = 4C$$

EXAMPLE A1-5

Write out $\sum_{i=1}^{5} x_i y_i$.

Solution

By definition (A1-1) we get

$$\sum_{i=1}^{5} x_i y_i = x_1 y_1 + x_2 y_2 + x_3 y_3 + x_4 y_4 + x_5 y_5$$

EXAMPLE A1-6

Write out $\sum_{i=1}^{3} (x_i + y_i)^2$.

Solution

Definition (A1-1) yields

$$\sum_{i=1}^{3} (x_i + y_i)^2 = (x_1 + y_1)^2 + (x_2 + y_2)^2 + (x_3 + y_3)^2$$

If we so desire, each of three binomial terms can be expanded, giving

$$x_1^2 + 2x_1 y_1 + y_1^2 + x_2^2 + 2x_2 y_2 + y_2^2 + x_3^2 + 2x_3 y_3 + y_3^2$$

$$= (x_1^2 + x_2^2 + x_3^2) + 2(x_1 y_1 + x_2 y_2 + x_3 y_3) + (y_1^2 + y_2^2 + y_3^2)$$

$$= \sum_{i=1}^{3} x_i^2 + 2 \sum_{i=1}^{3} x_i y_i + \sum_{i=1}^{3} y_i^2$$

EXAMPLE A1-7

Show that $\sum_{i=1}^{4} 3x_i = 3 \sum_{i=1}^{4} x_i$.

Solution

Definition (A1-1) yields

$$\sum_{i=1}^{4} 3x_i = 3x_1 + 3x_2 + 3x_3 + 3x_4$$

$$= 3(x_1 + x_2 + x_3 + x_4)$$

$$= 3 \sum_{i=1}^{4} x_i$$

EXAMPLE A1-8

Show that $E[(X - \mu)^2] = E(X^2) - \mu^2$

Solution

From formula (1-29) we have

$$E[(X - \mu)^2] = \sum_i (x_i - \mu)^2 f(x_i)$$

$$= (x_1 - \mu)^2 f(x_1) + (x_2 - \mu)^2 f(x_2) + \cdots$$

Expanding each of the binomial quantities yields

$$E[(X - \mu)^2] = (x_1^2 - 2x_1\mu + \mu^2)f(x_1) + (x_2^2 - 2x_2\mu + \mu^2)f(x_2) + \cdots$$

$$= [x_1^2 f(x_1) + x_2^2 f(x_2) + \cdots]$$

$$- 2\mu[x_1 f(x_1) + x_2 f(x_2) + \cdots]$$

$$+ \mu^2[f(x_1) + f(x_2) + \cdots]$$

$$= E(X^2) - 2\mu[\mu] + \mu^2[1]$$

$$= E(X^2) - \mu^2$$

The preceding examples illustrate three useful theorems. All are proved by using definition (A1-1). They are

THEOREM I

$\sum_{i=1}^n C = nC$, where C is any quantity that does not have a summation subscript.

Proof

By definition

$$\sum_{i=1}^n C = \underbrace{C + C + \cdots + C}_{n \text{ terms}} = nC$$

THEOREM II

$$\sum_{i=1}^n Cx_i = C \sum_{i=1}^n x_i$$

Proof

By definition

$$\sum_{i=1}^n Cx_i = Cx_1 + Cx_2 + Cx_3 + \cdots + Cx_n$$

$$= C(x_1 + x_2 + \cdots + x_n)$$

$$= C \sum_{i=1}^n x_i$$

THEOREM III

$$\sum_{i=1}^{n} (x_i + y_i - z_i) = \sum_{i=1}^{n} x_i + \sum_{i=1}^{n} y_i - \sum_{i=1}^{n} z_i$$

Proof

By definition

$$\sum_{i=1}^{n} (x_i + y_i - z_i) = (x_1 + y_1 - z_1)$$

$$+ (x_2 + y_2 - z_2) + \cdots + (x_n + y_n - z_n)$$

$$= (x_1 + x_2 + \cdots + x_n) + (y_1 + y_2 + \cdots + y_n)$$

$$- (z_1 + z_2 + \cdots + z_n)$$

$$= \sum_{i=1}^{n} x_i + \sum_{i=1}^{n} y_i - \sum_{i=1}^{n} z_i$$

Theorem III generalizes in the obvious way.

EXAMPLE A1-9

Show that $E(cX) = cE(X)$.

Solution

From formula (1-28) we have

$$E(cX) = \sum_i cx_i f(x_i)$$

Then, by Theorem II we get

$$E(cX) = c \sum_i x_i f(x_i)$$

$$= cE(X) \qquad \text{by definition of } E(X)$$

In particular, if $c = 1/n$ we have $E(X/n) = (1/n)E(X)$.

A1-2
DOUBLE SUMMATION

Double summation does not require any new concepts. It can be regarded as a single summation on one subscript followed by a single summation on a second subscript. Double, rather than single, summation is used for convenience when working with a two-way table (such as Table 6-2 or Table 7-1).

We can evaluate $\sum_{j=1}^{r} \sum_{i=1}^{n_j} x_{ij}$ by applying definition (A1-1) twice. We get

$$\sum_{j=1}^{r} \sum_{i=1}^{n_j} x_{ij} = \sum_{j=1}^{r} (x_{1j} + x_{2j} + x_{3j} + \cdots + x_{n_j j})$$

$$= (x_{11} + x_{21} + \cdots + x_{n_1 1})$$

$$+ (x_{12} + x_{22} + \cdots + x_{n_2 2}) + \cdots$$

$$+ (x_{1r} + x_{2r} + \cdots + x_{n_r r}) \quad \text{(A1-5)}$$

Thus to evaluate the double sum, we can first sum each column and then sum the r column totals.

EXAMPLE A1-10

Evaluate $\sum_{j=1}^{2} \sum_{i=1}^{3} x_{ij}$ and $\sum_{i=1}^{3} \sum_{j=1}^{2} x_{ij}$.

Solution

Using definition (A1-1) twice, we get

$$\sum_{j=1}^{2} \sum_{i=1}^{3} x_{ij} = \sum_{j=1}^{2} (x_{1j} + x_{2j} + x_{3j})$$

$$= (x_{11} + x_{21} + x_{31}) + (x_{12} + x_{22} + x_{32})$$

$$\sum_{i=1}^{3} \sum_{j=1}^{2} x_{ij} = \sum_{i=1}^{3} (x_{i1} + x_{i2})$$

$$= (x_{11} + x_{12}) + (x_{21} + x_{22}) + (x_{31} + x_{32})$$

We note

$$\sum_{j=1}^{2} \sum_{i=1}^{3} x_{ij} = \sum_{i=1}^{3} \sum_{j=1}^{2} x_{ij}$$

Example A1-10 illustrates the following theorem:

THEOREM IV

$$\sum_{j=1}^{b} \sum_{i=1}^{a} x_{ij} = \sum_{i=1}^{a} \sum_{j=1}^{b} x_{ij}$$

That is, if both limits on the summation signs are constants, then the order of summation may be interchanged.

Proof

The theorem can be proved by following the scheme outlined in Example A1-10.

Corresponding to Theorem I, we have

THEOREM V

$$\sum_{j=1}^{r} \sum_{i=1}^{n_j} C = C \sum_{j=1}^{r} n_j = CN$$

Proof

By Theorem I, $\sum_{i=1}^{n_j} C = n_j C$. Hence

$$\sum_{j=1}^{r} \sum_{i=1}^{n_j} C = \sum_{j=1}^{r} Cn_j$$

$$= C \sum_{j=1}^{r} n_j \qquad \text{by Theorem II}$$

$$= CN \qquad \text{by definition of } N$$

The counterpart of Theorem II is

THEOREM VI

$$\sum_{j=1}^{r} \sum_{i=1}^{n_j} Cx_{ij} = C \sum_{j=1}^{r} \sum_{i=1}^{n_j} x_{ij}$$

a result which follows easily, since C will appear in every term of the sum and can be factored out.

Similar to Theorem II is

THEOREM VII

$$\sum_{j=1}^{r} \sum_{i=1}^{n_j} y_j x_{ij} = \sum_{j=1}^{r} y_j \sum_{i=1}^{n_j} x_{ij}$$

Thus if part of a product which is being summed involves only the outside index of summation, this part can be factored out of the inside summation sign.

Proof

By Theorem II

$$\sum_{i=1}^{n_j} y_j x_{ij} = y_j \sum_{i=1}^{n_j} x_{ij}$$

Hence the result.

The counterpart of Theorem III is

THEOREM VIII

$$\sum_{j=1}^{r} \sum_{i=1}^{n_j} (x_{ij} + y_{ij} - z_{ij}) = \sum_{j=1}^{r} \sum_{i=1}^{n_j} x_{ij} + \sum_{j=1}^{r} \sum_{i=1}^{n_j} y_{ij} - \sum_{j=1}^{r} \sum_{i=1}^{n_j} z_{ij}$$

Proof

The proof follows easily by expanding the sum as in formula (A1-5), followed by regrouping terms.

EXAMPLE A1-11

Prove $\sum_{j=1}^{r} \sum_{i=1}^{n_j} (x_{ij} - \bar{x}_{.j}) = 0$, where $\bar{x}_{.j} = \sum_{i=1}^{n_j} x_{ij}/n_j$ as defined in Sec. 7-2.

Solution

$$\sum_{j=1}^{r} \sum_{i=1}^{n_j} (x_{ij} - \bar{x}_{.j}) = \sum_{j=1}^{r} \sum_{i=1}^{n_j} x_{ij} - \sum_{j=1}^{r} \sum_{i=1}^{n_j} \bar{x}_{.j} \qquad \text{by Theorem VIII}$$

and

$$\sum_{j=1}^{r} \sum_{i=1}^{n_j} \bar{x}_{.j} = \sum_{j=1}^{r} n_j \bar{x}_{.j} \qquad \text{by Theorem I}$$

$$= \sum_{j=1}^{r} \sum_{i=1}^{n_j} x_{ij} \qquad \text{by the definition of } \bar{x}_{.j}$$

Consequently

$$\sum_{j=1}^{r} \sum_{i=1}^{n_j} (x_{ij} - \bar{x}_{.j}) = \sum_{j=1}^{r} \sum_{i=1}^{n_j} x_{ij} - \sum_{j=1}^{r} \sum_{i=1}^{n_j} x_{ij} = 0$$

EXAMPLE A1-12

Prove

$$\sum_{j=1}^{r} \sum_{i=1}^{n_j} (x_{ij} - \bar{x}_{..})^2 = \sum_{j=1}^{r} \sum_{i=1}^{n_j} x_{ij}^2 - \frac{t_{..}^2}{N}$$

where all symbols are as defined in Sec. 7-2.

Solution

$$\sum_{j=1}^{r} \sum_{i=1}^{n_j} (x_{ij} - \bar{x}_{..})^2 = \sum_{j=1}^{r} \sum_{i=1}^{n_j} (x_{ij}^2 - 2\bar{x}_{..} x_{ij} + \bar{x}_{..}^2)$$

$$= \sum_{j=1}^{r} \sum_{i=1}^{n_j} x_{ij}^2 - \sum_{j=1}^{r} \sum_{i=1}^{n_j} 2\bar{x}_{..} x_{ij} + \sum_{j=1}^{r} \sum_{i=1}^{n_j} \bar{x}_{..}^2$$

$$\text{by Theorem VIII}$$

But

$$-\sum_{j=1}^{r}\sum_{i=1}^{n_j} 2\bar{x}..x_{ij} = -2\bar{x}..\sum_{j=1}^{r}\sum_{i=1}^{n_j} x_{ij} \qquad \text{by Theorem VI}$$

$$= -2\frac{t..}{N}t.. = -2\frac{t^2..}{N} \qquad \text{by definition of } t.. \text{ and } \bar{x}..$$

and

$$\sum_{j=1}^{r}\sum_{i=1}^{n_j} \bar{x}^2.. = N\bar{x}^2.. \qquad \text{by Theorem V}$$

$$= N\left(\frac{t..}{N}\right)^2 \qquad \text{by definition of } \bar{x}..$$

Hence we have

$$\sum_{j=1}^{r}\sum_{i=1}^{n_j} (x_{ij} - \bar{x}..)^2 = \sum_{j=1}^{r}\sum_{i=1}^{n_j} x_{ij} - 2\frac{t^2..}{N} + \frac{t^2..}{N}$$

$$= \sum_{j=1}^{r}\sum_{i=1}^{n_j} x_{ij}^2 - \frac{t^2..}{N}$$

APPENDIX A2

INTERPOLATION

A2-1
LINEAR INTERPOLATION

Frequently we try to enter a table or a graph with a number x_1 and hope to read a corresponding value y_1. Instead of finding x_1, however, we find x_0 and x_2 (with corresponding y_0 and y_2), the two closest entries, where $x_0 < x_1$ and $x_2 > x_1$. The most frequently used method for approximating y_1 (by y_1') is called linear interpolation. The name arises from the assumption that y can be approximately represented by a straight line in a small interval as indicated in Fig. A2-1. We take for our estimate $y_1' = y_0 + BE$, where BE can be found by using properties of similar triangles. That is,

$$\frac{BE}{AB} = \frac{CD}{AC}$$

FIGURE A2-1

Linear interpolation.

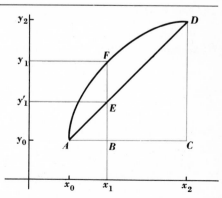

so that

$$\frac{y_1' - y_0}{x_1 - x_0} = \frac{y_2 - y_0}{x_2 - x_0} \qquad \text{(A2-1)}$$

Solving for y_1' yields

$$y_1' = y_0 + \frac{x_1 - x_0}{x_2 - x_0}(y_2 - y_0) \qquad \text{(A2-2)}$$

EXAMPLE A2-1

Approximate $t_{33;.975}$.

Solution

Appendix B6 yields

$$x_0 = 30 \qquad y_0 = t_{30;.975} = 2.042$$
$$x_1 = 33 \qquad y_1 = t_{33;.975} \text{ not given}$$
$$x_2 = 40 \qquad y_2 = t_{40;.975} = 2.021$$

Thus

$$y_1' = 2.042 + \frac{33 - 30}{40 - 30}(2.021 - 2.042)$$

$$= 2.042 + \frac{3}{10}(-.021)$$

$$= 2.042 - .006$$

$$= 2.036$$

That is, since x_1 is three-tenths of the way between x_0 and x_2, we take y_1' to be three-tenths of the way between y_0 and y_2.

EXAMPLE A2-2

Approximate $\Pr(Y_{20} < 22.8)$ where Y_{20} has a chi-square distribution with 20 degrees of freedom by using Appendix B7 and linear interpolation.

Solution

Appendix B7 yields

$$x_0 = 12.44 \qquad y_0 = \Pr(Y_{20} < 12.44) = .10$$

$$x_1 = 22.8 \qquad y_1 = \Pr(Y_{20} < 22.8) \text{ not given}$$

$$x_2 = 28.41 \qquad y_2 = \Pr(Y_{20} < 28.41) = .90$$

Thus

$$y_1' = .10 + \frac{22.8 - 12.44}{28.41 - 12.44}(.90 - .10)$$

$$= .10 + \frac{10.36}{15.97}(.80) = .10 + (.65)(.80)$$

$$= .10 + .52 = .62$$

Since we are interpolating over a relatively long interval, we might expect the approximation to be poor. In fact

$$Pr(Y_{20} < 22.8) = Pr(Y_{20}/20 < 1.14)$$
$$= .70 \quad \text{from Appendix B8}$$

EXAMPLE A2-3

Approximate $f_{4,18;.95}$ by using linear interpolation.

Solution

Appendix B9 yields

$$x_0 = 15 \quad y_0 = f_{4,15;.95} = 3.06$$

$$x_1 = 18 \quad y_1 = f_{4,18;.95} \text{ not given}$$

$$x_2 = 20 \quad y_2 = f_{4,20;.95} = 2.87$$

Thus

$$y_1' = 3.06 + \frac{18 - 15}{20 - 15}(2.87 - 3.06)$$

$$= 3.06 + (3/5)(-.19)$$

$$= 3.06 - .11 = 2.95$$

EXAMPLE A2-4

Read the power (value on vertical scale) from the graphs of Appendix B10 headed by $\alpha = .01$ if $\delta = 4, f = 15$.

Solution

From the set of graphs we read

$$x_0 = 12 \quad y_0 = \text{power} = .875$$

$$x_1 = 15 \quad y_1 = \text{power not given}$$

$$x_2 = 24 \quad y_2 = \text{power} = .925$$

Thus

$$y_1' = .875 + \frac{15 - 12}{24 - 12}(.925 - .875)$$

$$= .875 + (1/4)(.050) = .888 \text{ (or .89)}$$

If points E and F in Fig. A2-1 are close together, then the approximation will be very good. Of course, the closer the y curve (as represented by arc AFD) is to a straight line, the better the approximation. In most situations which we encounter, the approximation is quite good from a practical point of view, as demonstrated by Examples A2-1, A2-2, A2-4.

A2-2
INVERSE LINEAR INTERPOLATION

If $x_2 = \infty$, then Fig. A2-1 does not make sense and formulas (A2-1) and (A2-2) are of no assistance. When this happens, we can invert the x's and proceed with linear interpolation as before, except for using the reciprocals.

EXAMPLE A2-5

Approximate the power in Example A2-4 if $f = 50$.

Solution

The graphs of Appendix B10 yield

$x_0 = 24 \quad y_0 = \text{power} = .925$

$x_1 = 50 \quad y_1 = \text{power not given}$

$x_2 = \infty \quad y_2 = \text{power} = .95$

In a practical situation we would probably be satisfied to know that the power is between .925 and .95. However, we shall use inverse linear interpolation to demonstrate the procedure. We have

$$\frac{1}{x_0} = \frac{1}{24} = .042 \qquad y_0 = \text{power} = .925$$

$$\frac{1}{x_1} = \frac{1}{50} = .02 \qquad y_1 = \text{power not given}$$

$$\frac{1}{x_2} = 0 \qquad y_2 = \text{power} = .95$$

We use as our estimate

$$y_1' = .925 + \frac{.02 - .042}{0 - .042}(.95 - .925)$$

$$= .925 + \frac{.022}{.042}(.025) = .925 + .52(.025)$$

$$= .925 + .013 \cong .94$$

A2-3
DOUBLE LINEAR INTERPOLATION

Linear interpolation can be used on both entries of a double-entry table. We shall demonstrate by an example.

EXAMPLE A2-6

Approximate $f_{35, 52; .95}$ by using linear interpolation.

Solution

Appendix B9 yields

$$f_{30, 40; .95} = 1.74 \qquad f_{30, 60; .95} = 1.65$$

$$f_{40, 40; .95} = 1.69 \qquad f_{40, 60; .95} = 1.59$$

In a practical situation we would probably be satisfied to know that $f_{35, 52; .95}$ is between 1.59 and 1.74. However, we shall demonstrate double linear interpolation. First we shall approximate $f_{35, 40; .95}$ and $f_{35, 60; .95}$. For the former we write

$$x_0 = 30 \qquad y_0 = f_{30, 40; .95} = 1.74$$

$$x_1 = 35 \qquad y_1 = f_{35, 40; .95} \text{ not given}$$

$$x_2 = 40 \qquad y_2 = f_{40, 40; .95} = 1.69$$

and

$$y_1' = 1.74 + \frac{35 - 30}{40 - 30}(1.69 - 1.74)$$

$$= 1.74 + (1/2)(-.05) = 1.72$$

For the latter we get

$$x_0 = 30 \qquad y_0 = f_{30, 60; .95} = 1.65$$

$$x_1 = 35 \qquad y_1 = f_{35, 60; .95} \text{ not given}$$

$$x_2 = 40 \qquad y_2 = f_{40, 60; .95} = 1.59$$

and

$$y_1' = 1.65 + \frac{35 - 30}{40 - 30}(1.59 - 1.65)$$

$$= 1.65 + (1/2)(-.06) = 1.62$$

Finally, using the above approximations, we shall approximate $f_{35, 52; .95}$. We have

$$x_0 = 40 \qquad y_1 = f_{35, 40; .95} = 1.72 \text{ approximately}$$

$$x_1 = 52 \qquad y_1 = f_{35, 52; .95} \text{ unknown}$$

$$x_2 = 60 \qquad y_2 = f_{35, 60; .95} = 1.62 \text{ approximately}$$

and

$$y_1' = 1.72 + \frac{52 - 40}{60 - 40}(1.62 - 1.72)$$

$$= 1.72 + (.6)(-.10) = 1.66$$

Hence our estimate of $f_{35,52; .95}$ is 1.66. If we interpolate first between 40 and 60 and then between 35 and 40, we still get 1.66 as our estimate. However, in some situations slightly different estimates will be obtained from the two orders of interpolation. If this happens, use the average of the two estimates.

APPENDIX A3
SQUARE ROOT

A3-1
TABLES

One simple way to determine a square root is to read it out of a table. (Of course, the simplest way to determine square root is to punch the correct button on an electronic desk calculator if one is available.)   Since the exercises in Chaps. 1 to 8 require very few square roots (which can be found rather quickly by the procedure described in Sec. A3-2) and good square root tables are quite lengthy, none is given in the table appendix.

Square root tables are available in a number of mathematical and statistical books.   A good 40-page table is found in a reasonably priced paper bound edition of Herbert Arkin and Raymond Colton, "Tables for Statisticians" (Barnes and Noble, Inc., New York, 1950).   The table gives $\sqrt{N}$ and $\sqrt{10N}$ to four decimal places for

$$N = 1,000(1)10,000$$

Square roots of other numbers are obtained from the table by locating the decimal point properly.

EXAMPLE A3-1

From tables find $\sqrt{338.7}$.

Solution

The Arkin and Colton table yields 58.1979 and 184.0380 when entered with 3387.   We have to select the correct one and locate the decimal point. Since $18^2 = 324$ and $19^2 = 361$, the answer is 18.40380.   Of course, this is far greater accuracy than is needed for most statistical problems.

EXAMPLE A3-2

From tables find $\sqrt{.12}$.

Solution

The Arkin and Colton table yields 34.6410 and 109.5445 when entered with 1200. Since $(.3)^2 = .09$ and $(.4)^2 = .16$, the correct answer is .346410.

EXAMPLE A3-3

From tables find $\sqrt{2}$.

Solution

The Arkin and Colton table yields 44.7214 and 141.4214 when entered with 2000. Since $1^2 = 1$, $2^2 = 4$, the answer is 1.414214.

EXAMPLE A3-4

From tables find $\sqrt{42.678}$.

Solution

The Arkin and Colton table yields 65.3223 and 206.5672 when entered with 4267 and 65.3299 and 206.5914 when entered with 4268. Since $6^2 = 36$,

$7^2 = 49$, we have

$x_0 = 42.67 \qquad y_0 = 6.53223$

$x_1 = 42.678 \qquad y_1$ not given

$x_2 = 42.68 \qquad y_2 = 6.53299$

and by linear interpolation

$$y_1' = 6.53223 + \frac{42.678 - 42.67}{42.68 \ - 42.67}(6.53299 - 6.53223)$$

$$= 6.53223 + (8/10)(.00076)$$

$$= 6.53223 + .00061$$

$$= 6.53284$$

In a practical application we would probably be content to round off 42.678 to 42.68 and use 6.53 for the square root.

A3-2

HAND CALCULATION

If only one square root is needed, the result can be calculated in less time than it takes to locate a table. The numerical process, which is similar to long division, is best illustrated by examples.

EXAMPLE A3-5

Find $\sqrt{2}$.

Solution

The calculation is

```
          1. 4  1  4      answer
    ┌──────────────
1   │ 2.00′00′00
    │ 1
    ├──────
24  │ 100
    │  96
    ├──────
281 │   400
    │   281
    ├──────
2824│   11900
    │   11296
    ├──────
    │     604  etc.
```

We start at the decimal point and mark off digits two at a time with the primes moving in each direction. Then we examine the pair farthest to the left (in this case 02) and inquire which is the largest number, when squared, that is smaller than 2. The answer is 1, and a 1 is written above 2, below 2, and to the left of 2. Then we subtract as in long division except that two numbers are brought down (instead of one). Next the number on top of the horizontal line is doubled, a zero is added temporarily (giving 20), and the result is written to the left of 100. We inquire the number of times 20 goes into 100, which is, of course, 5. The temporary zero is replaced by a 5, another 5 is written behind the 1 above the horizontal line, and the product $5(25) = 125$ is written below 100. Since 125 is larger than 100, we try the calculation again with 4, writing $4(24) = 96$ below 100. Now we can subtract with a positive result, getting 4. The next two digits are brought down, as in the preceding step, and the process is repeated. Thus 14 is doubled and a zero is added temporarily, giving 280, which is written to the left of 400. Since 280 goes into 400 one time, the 0 is replaced by a 1, a 1 is written behind the 4 above the horizontal line, and 281 is written to the left of 400 and below 400. The process can be repeated indefinitely depending upon the number of places we seek in our answer (which is the number being formed above the horizontal line). The decimal point above the horizontal line is located above the decimal point below that line.

EXAMPLE A3-6

Find $\sqrt{338.7}$.

Solution

The calculation is

```
          1  8. 4  0  3        answer
    1  | 3'38.70'00'00'
       |  1
   28  | 238
       | 224
  364  |  1470
       |  1456
 3680  |    1400
36803  |   140000
       |   110409
       |        etc.
```

EXAMPLE A3-7

Find $\sqrt{.12}$.

Solution

The calculation is

```
        . 3  4  6  4        Answer
    3 | .12' 00' 00' 00'
      |  9
   64 |  300
      |  256
  686 |   4400
      |   4116
 6924 |    28400
      |    27696
      |       etc.
```

EXAMPLE A3-8

Find $\sqrt{42.678}$.

Solution

The calculation is

```
            6.  5   3   2       answer
     6  │ '42'. 67' 80' 00'
        │  36
   125  │   667
        │   625
  1303  │    4280
        │    3909
 13062  │     37100
        │     26124

                  etc.
```

APPENDIX A4
BINOMIAL EXPANSIONS

A4-1
BINOMIAL EXPANSION WITH
POSITIVE INTEGRAL EXPONENT

By actual multiplication we can verify that

$(a + b)^2 = (a + b)(a + b) = a^2 + 2ab + b^2$
$(a + b)^3 = (a + b)^2(a + b) = a^3 + 3a^2b + 3ab^2 + b^3$
$(a + b)^4 = (a + b)^3(a + b) = a^4 + 4a^3b + 6a^2b^2 + 4ab^3 + b^4$

Letting n represent the exponent for the three cases, we observe that

(a) The first term is a^n (and the last is b^n).

(b) As we proceed from term to term moving left to right, the exponent on a decreases by 1, the exponent on b increases by 1, and the sum of the two exponents is n.

(c) The coefficient of any term is obtained by multiplying the exponent in the preceding term by the coefficient in the preceding term followed by dividing the product by the number of the preceding term (that is, the first term is number 1, the second number 2, etc.).   (A4-1)

It can be shown that (A4-1) can be used to expand $(a + b)^n$ where n is any positive integer.

By counting permutations we are also able to obtain the coefficient for any term. Consider the coefficient of a^2b in $(a + b)^3 = (a + b)(a + b)$ $(a + b)$. Terms of this type result by selecting

1 a from the first term of the product, a from the second, and b from the third, or aab.

2 a from the first term, b from the second, and a from the third, or aba.

3 b from the first term, a from the second, and a from the third, or baa.

Hence a^2b terms can be formed in the number of ways three things taken three at a time can be permuted if two are alike and one is different.

That is, $3!/(2!)(1!) = 3$. In general, from

$$(a + b)^n = (a + b)(a + b) \cdots (a + b)$$

terms of the type $b^x a^{n-x}$ can be formed in the number of ways n things taken n at a time can be permuted if x are alike and $n - x$ are alike. This number is $n!/x!(n - x)! = \dbinom{n}{x}$. Hence

$$(a + b)^n = \binom{n}{0}a^n + \binom{n}{1}ba^{n-1} + \binom{n}{2}b^2 a^{n-2} + \cdots + \binom{n}{n}b^n \qquad \text{(A4-2)}$$

or, in summation form,

$$(a + b)^n = \sum_{x=0}^{n} \binom{n}{x} b^x a^{n-x} \qquad \text{(A4-3)}$$

In particular, if $p > 0$, $q > 0$, and $p + q = 1$, we have

$$\sum_{x=0}^{n} \binom{n}{x} p^x q^{n-x} = (q + p)^n = 1 \qquad \text{(A4-4)}$$

A4-2
BINOMIAL EXPANSION WITH NEGATIVE INTEGRAL EXPONENT

Let n be a positive integer, and consider $1/(1 - a)^n = (1 - a)^{-n}$. It seems natural to attempt to expand $(1 - a)^{-n}$ according to the procedure given by (A4-1). If we do this we get

$$(1 - a)^{-n} = 1^{-n} + (-n)1^{-n-1}(-a) + \frac{(-n)(-n - 1)}{2!} 1^{-n-2}(-a)^2$$

$$+ \frac{(-n)(-n - 1)(-n - 2)}{3!} 1^{-n-3}(-a)^3 + \cdots$$

$$= 1 + na + \frac{n(n + 1)}{2!} a^2 + \frac{n(n + 1)(n + 2)}{3!} a^3 + \cdots$$

If we multiply every coefficient in the last expression by $(n - 1)!/(n - 1)!$, then each coefficient represents the value of a combination symbol, and we can write

$$(1 - a)^{-n} = \binom{n - 1}{0} + \binom{n}{1} a + \binom{n + 1}{2} a^2 + \binom{n + 2}{3} a^3 + \cdots \qquad \text{(A4-5)}$$

$$= \sum_{x=0}^{\infty} \binom{n - 1 + x}{x} a^x \qquad \text{(A4-6)}$$

$$= \sum_{x=0}^{\infty} \binom{x - 1 + n}{n - 1} a^x \qquad \text{(A4-7)}$$

Since procedure (A4-1) was derived for positive integral exponent, it is natural to inquire as to whether or not (A4-5) is a valid formula. An additional complication arises from the fact that the expansion now contains an infinite number of terms. It can be shown that (A4-5) is correct provided $-1 < a < 1$.

It is easy to verify the following special cases:

$$(1 - a)^{-1} = 1 + a + a^2 + a^3 + \cdots \tag{A4-8}$$

$$(1 - a)^{-2} = 1 + 2a + 3a^2 + 4a^3 + \cdots \tag{A4-9}$$

$$(1 - a)^{-3} = 1 + 3a + 6a^2 + 10a^3 + \cdots \tag{A4-10}$$

By writing out both sums it is easy to verify that

$$\sum_{x=0}^{\infty} \binom{x - 1 + n}{n - 1} a^x = \sum_{y=n}^{\infty} \binom{y - 1}{n - 1} a^{y-n} \tag{A4-11}$$

If in the right side of (A4-11) we replace n by c and a by q, we have

$$(1 - q)^{-c} = \sum_{y=c}^{\infty} \binom{y - 1}{c - 1} q^{y-c} \tag{A4-12}$$

Multiplying both sides of (A4-12) by p^c gives

$$p^c(1 - q)^{-c} = \sum_{y=c}^{\infty} \binom{y - 1}{c - 1} p^c q^{y-c} \tag{A4-13}$$

Now, if $p > 0$, $q > 0$, and $p = 1 - q$, then (A4-13) yields

$$\sum_{y=c}^{\infty} \binom{y - 1}{c - 1} p^c q^{y-c} = p^c p^{-c} = 1$$

which proves, as we already know, that the negative binomial probability function summed over all values is 1.

APPENDIX B

TABLES AND GRAPHS

Appendix B1   *The cumulative binomial distribution*

n	r	p = .10	p = .20	p = .25	p = .30	p = .40	p = .50
5	0	.59049	.32768	.23730	.16807	.07776	.03125
	1	.91854	.73728	.63281	.52822	.33696	.18750
	2	.99144	.94208	.89648	.83692	.68256	.50000
	3	.99954	.99328	.98437	.96922	.91296	.81250
	4	.99999	.99968	.99902	.99757	.98976	.96875
	5	1.00000	1.00000	1.00000	1.00000	1.00000	1.00000
10	0	.34868	.10737	.05631	.02825	.00605	.00098
	1	.73610	.37581	.24403	.14931	.04636	.01074
	2	.92981	.67780	.52559	.38278	.16729	.05469
	3	.98720	.87913	.77588	.64961	.38228	.17187
	4	.99837	.96721	.92187	.84973	.63310	.37695
	5	.99985	.99363	.98027	.95265	.83376	.62305
	6	.99999	.99914	.99649	.98941	.94524	.82812
	7	1.00000	.99992	.99958	.99841	.98771	.94531
	8		1.00000	.99997	.99986	.99832	.98926
	9			1.00000	.99999	.99990	.99902
	10				1.00000	1.00000	1.00000
15	0	.20589	.03518	.01336	.00475	.00047	.00003
	1	.54904	.16713	.08018	.03527	.00517	.00049
	2	.81594	.39802	.23609	.12683	.02711	.00369
	3	.94444	.64816	.46129	.29687	.09050	.01758
	4	.98728	.83577	.68649	.51549	.21728	.05923
	5	.99775	.93895	.85163	.72162	.40322	.15088
	6	.99969	.98194	.94338	.86886	.60981	.30362
	7	.99997	.99576	.98270	.94999	.78690	.50000
	8	1.00000	.99921	.99581	.98476	.90495	.69638
	9		.99989	.99921	.99635	.96617	.84912
	10		.99999	.99988	.99933	.99065	.94077
	11		1.00000	.99999	.99991	.99807	.98242
	12			1.00000	.99999	.99972	.99631
	13				1.00000	.99997	.99951
	14					1.00000	.99997
	15						1.00000
20	0	.12158	.01153	.00317	.00080	.00004	.00000
	1	.39175	.06918	.02431	.00764	.00052	.00002
	2	.67693	.20608	.09126	.03548	.00361	.00020
	3	.86705	.41145	.22516	.10709	.01596	.00129
	4	.95683	.62965	.41484	.23751	.05095	.00591
	5	.98875	.80421	.61717	.41637	.12560	.02069
	6	.99761	.91331	.78578	.60801	.25001	.05766

n	r	p = .10	p = .20	p = .25	p = .30	p = .40	p = .50
20	7	.99958	.96786	.89819	.77227	.41589	.13159
	8	.99994	.99002	.95907	.88667	.59560	.25172
	9	.99999	.99741	.98614	.95204	.75534	.41190
	10	1.00000	.99944	.99606	.98286	.87248	.58810
	11		.99990	.99906	.99486	.94347	.74828
	12		.99998	.99982	.99872	.97897	.86841
	13		1.00000	.99997	.99974	.99353	.94234
	14			1.00000	.99996	.99839	.97931
	15				.99999	.99968	.99409
	16				1.00000	.99995	.99871
	17					.99999	.99980
	18					1.00000	.99998
	19						1.00000
25	0	.07179	.00378	.00075	.00013	.00000	.00000
	1	.27121	.02739	.00702	.00157	.00005	.00000
	2	.53709	.09823	.03211	.00896	.00043	.00001
	3	.76359	.23399	.09621	.03324	.00237	.00008
	4	.90201	.42067	.21374	.09047	.00947	.00046
	5	.96660	.61669	.37828	.19349	.02936	.00204
	6	.99052	.78004	.56110	.34065	.07357	.00732
	7	.99774	.89088	.72651	.51185	.15355	.02164
	8	.99954	.95323	.85056	.67693	.27353	.05388
	9	.99992	.98267	.92867	.81056	.42462	.11476
	10	.99999	.99445	.97033	.90220	.58577	.21218
	11	1.00000	.99846	.98027	.95575	.73228	.34502
	12		.99963	.99663	.98253	.84623	.50000
	13		.99992	.99908	.99401	.92220	.65498
	14		.99999	.99979	.99822	.96561	.78782
	15		1.00000	.99996	.99955	.98683	.88524
	16			.99999	.99990	.99567	.94612
	17			1.00000	.99998	.99879	.97836
	18				1.00000	.99972	.99268
	19					.99995	.99796
	20					.99999	.99954
	21					1.00000	.99992
	22						.99999
	23						1.00000
50	0	.00515	.00001	.00000	.00000		
	1	.03379	.00019	.00001	.00000		
	2	.11173	.00129	.00009	.00000		
	3	.25029	.00566	.00050	.00003		

n	r	p = .10	p = .20	p = .25	p = .30	p = .40	p = .50
50	4	.43120	.01850	.00211	.00017		
	5	.61612	.04803	.00705	.00072	.00000	
	6	.77023	.10340	.01939	.00249	.00001	
	7	.87785	.19041	.04526	.00726	.00006	
	8	.94213	.30733	.09160	.01825	.00023	
	9	.97546	.44374	.16368	.04023	.00076	.00000
	10	.99065	.58356	.26220	.07885	.00220	.00001
	11	.99678	.71067	.38162	.13904	.00569	.00005
	12	.99900	.81394	.51099	.22287	.01325	.00015
	13	.99971	.88941	.63704	.32788	.02799	.00047
	14	.99993	.93928	.74808	.44683	.05396	.00130
	15	.99998	.96920	.83692	.56918	.09550	.00330
	16	1.00000	.98556	.90169	.68388	.15609	.00767
	17		.99374	.94488	.78219	.23688	.01642
	18		.99749	.97127	.85944	.33561	.03245
	19		.99907	.98608	.91520	.44648	.05946
	20		.99968	.99374	.95224	.56103	.10132
	21		.99990	.99738	.97491	.67014	.16112
	22		.99997	.99898	.98772	.76602	.23994
	23		.99999	.99963	.99441	.84383	.33591
	24		1.00000	.99988	.99763	.90219	.44386
	25			.99996	.99907	.94266	.55614
	26			.99999	.99966	.96859	.66409
	27			1.00000	.99988	.98397	.76006
	28				.99996	.99238	.83888
	29				.99999	.99664	.89868
	30				1.00000	.99863	.94054
	31					.99948	.96755
	32					.99982	.98358
	33					.99994	.99233
	34					.99998	.99670
	35					1.00000	.99870
	36						.99953
	37						.99985
	38						.99995
	39						.99999
	40						1.00000
100	0	.00003					
	1	.00032					
	2	.00194					
	3	.00784					
	4	.02371	.00000				

n	r	p = .10	p = .20	p = .25	p = .30	p = .40	p = .50
100	5	.05758	.00002				
	6	.11716	.00008				
	7	.20605	.00028	.00000			
	8	.32087	.00086	.00001			
	9	.45129	.00233	.00004			
	10	.58316	.00570	.00014	.00000		
	11	.70303	.01257	.00039	.00001		
	12	.80182	.02533	.00103	.00002		
	13	.87612	.04691	.00246	.00006		
	14	.92743	.08044	.00542	.00016		
	15	.96011	.12851	.01108	.00040		
	16	.97940	.19234	.02111	.00097		
	17	.98999	.27119	.03763	.00216		
	18	.99542	.36209	.06301	.00452	.00000	
	19	.99802	.46016	.09953	.00889	.00001	
	20	.99919	.55946	.14883	.01646	.00002	
	21	.99969	.65403	.21144	.02883	.00004	
	22	.99989	.73893	.28637	.04787	.00011	
	23	.99996	.81091	.37018	.07553	.00025	
	24	.99999	.86865	.46167	.11357	.00056	
	25	1.00000	.91252	.55347	.16313	.00119	
	26		.94417	.64174	.22440	.00240	
	27		.96585	.72238	.29637	.00460	.00000
	28		.97998	.79246	.37678	.00843	.00001
	29		.98875	.85046	.46234	.01478	.00002
	30		.99394	.89621	.54912	.02478	.00004
	31		.99687	.93065	.63311	.03985	.00009
	32		.99845	.95540	.71072	.06150	.00020
	33		.99926	.97241	.77926	.09125	.00044
	34		.99966	.98357	.83714	.13034	.00089
	35		.99985	.99059	.88392	.17947	.00176
	36		.99994	.99482	.92012	.23861	.00332
	37		.99998	.99725	.94695	.30681	.00602
	38		.99999	.99860	.96602	.38219	.01049
	39		1.00000	.99931	.97901	.46208	.01760
	40			.99968	.98750	.54329	.02844
	41			.99985	.99283	.62253	.04431
	42			.99994	.99603	.69674	.06661
	43			.99997	.99789	.76347	.09667
	44			.99999	.99891	.82110	.13563
	45			1.00000	.99946	.86891	.18410
	46				.99974	.90702	.24206
	47				.99988	.93621	.30865

Appendix B1 (*Continued*)

n	r	p = .10	p = .20	p = .25	p = .30	p = .40	p = .50
100	48				.99995	.95770	.38218
	49				.99998	.97290	.46021
	50				.99999	.98324	.53979
	51				1.00000	.98999	.61782
	52					.99424	.69135
	53					.99680	.75794
	54					.99829	.81590
	55					.99912	.86437
	56					.99956	.90333
	57					.99979	.93339
	58					.99990	.95569
	59					.99996	.97156
	60					.99998	.98240
	61					.99999	.98951
	62					1.00000	.99398
	63						.99668
	64						.99824
	65						.99911
	66						.99956
	67						.99980
	68						.99991
	69						.99996
	70						.99998
	71						.99999
	72						1.00000

**Row
number**

00000	10097 32533	76520 13586	34673 54876	80959 09117	39292 74945
00001	37542 04805	64894 74296	24805 24037	20636 10402	00822 91665
00002	08422 68953	19645 09303	23209 02560	15953 34764	35080 33606
00003	99019 02529	09376 70715	38311 31165	88676 74397	04436 27659
00004	12807 99970	80157 36147	64032 36653	98951 16877	12171 76833
00005	66065 74717	34072 76850	36697 36170	65813 39885	11199 29170
00006	31060 10805	45571 82406	35303 42614	86799 07439	23403 09732
00007	85269 77602	02051 65692	68665 74818	73053 85247	18623 88579
00008	63573 32135	05325 47048	90553 57548	28468 28709	83491 25624
00009	73796 45753	03529 64778	35808 34282	60935 20344	35273 88435
00010	98520 17767	14905 68607	22109 40558	60970 93433	50500 73998
00011	11805 05431	39808 27732	50725 68248	29405 24201	52775 67851
00012	83452 99634	06288 98033	13746 70078	18475 40610	68711 77817
00013	88685 40200	86507 58401	36766 67951	90364 76493	29609 11062
00014	99594 67348	87517 64969	91826 08928	93785 61368	23478 34113
00015	65481 17674	17468 50950	58047 76974	73039 57186	40218 16544
00016	80124 35635	17727 08015	45318 22374	21115 78253	14385 53763
00017	74350 99817	77402 77214	43236 00210	45521 64237	96286 02655
00018	69916 26803	66252 29148	36936 87203	76621 13990	94400 56418
00019	09893 20505	14225 68514	46427 56788	96297 78822	54382 14598
00020	91499 14523	68479 27686	46162 83554	94750 89923	37089 20048
00021	80336 94598	26940 36858	70297 34135	53140 33340	42050 82341
00022	44104 81949	85157 47954	32979 26575	57600 40881	22222 06413
00023	12550 73742	11100 02040	12860 74697	96644 89439	28707 25815
00024	63606 49329	16505 34484	40219 52563	43651 77082	07207 31790
00025	61196 90446	26457 47774	51924 33729	65394 59593	42582 60527
00026	15474 45266	95270 79953	59367 83848	82396 10118	33211 59466
00027	94557 28573	67897 54387	54622 44431	91190 42592	92927 45973
00028	42481 16213	97344 08721	16868 48767	03071 12059	25701 46670
00029	23523 78317	73208 89837	68935 91416	26252 29663	05522 82562
00030	04493 52494	75246 33824	45862 51025	61962 79335	65337 12472
00031	00549 97654	64051 88159	96119 63896	54692 82391	23287 29529
00032	35963 15307	26898 09354	33351 35462	77974 50024	90103 39333
00033	59808 08391	45427 26842	83609 49700	13021 24892	78565 20106
00034	46058 85236	01390 92286	77281 44077	93910 83647	70617 42941

Row
number

00035	32179 00597	87379 25241	05567 07007	86743 17157	85394 11838
00036	69234 61406	20117 45204	15956 60000	18743 92423	97118 96338
00037	19565 41430	01758 75379	40419 21585	66674 36806	84962 85207
00038	45155 14938	19476 07246	43667 94543	59047 90033	20826 69541
00039	94864 31994	36168 10851	34888 81553	01540 35456	05014 51176
00040	98086 24826	45240 28404	44999 08896	39094 73407	35441 31880
00041	33185 16232	41941 50949	89435 48581	88695 41994	37548 73043
00042	80951 00406	96382 70774	20151 23387	25016 25298	94624 61171
00043	79752 49140	71961 28296	69861 02591	74852 20539	00387 59579
00044	18633 32537	98145 06571	31010 24674	05455 61427	77938 91936
00045	74029 43902	77557 32270	97790 17119	52527 58021	80814 51748
00046	54178 45611	80993 37143	05335 12969	56127 19255	36040 90324
00047	11664 49883	52079 84827	59381 71539	09973 33440	88461 23356
00048	48324 77928	31249 64710	02295 36870	32307 57546	15020 09994
00049	69074 94138	87637 91976	35584 04401	10518 21615	01848 76938
00050	09188 20097	32825 39527	04220 86304	83389 87374	64278 58044
00051	90045 85497	51981 50654	94938 81997	91870 76150	68476 64659
00052	73189 50207	47677 26269	62290 64464	27124 67018	41361 82760
00053	75768 76490	20971 87749	90429 12272	95375 05871	93823 43178
00054	54016 44056	66281 31003	00682 27398	20714 53295	07706 17813
00055	08358 69910	78542 42785	13661 58873	04618 97553	31223 08420
00056	28306 03264	81333 10591	40510 07893	32604 60475	94119 01840
00057	53840 86233	81594 13628	51215 90290	28466 68795	77762 20791
00058	91757 53741	61613 62269	50263 90212	55781 76514	83483 47055
00059	89415 92694	00397 58391	12607 17646	48949 72306	94541 37408
00060	77513 03820	86864 29901	68414 82774	51908 13980	72893 55507
00061	19502 37174	69979 20288	55210 29773	74287 75251	65344 67415
00062	21818 59313	93278 81757	05686 73156	07082 85046	31853 38452
00063	51474 66499	68107 23621	94049 91345	42836 09191	08007 45449
00064	99559 68331	62535 24170	69777 12830	74819 78142	43860 72834
00065	33713 48007	93584 72869	51926 64721	58303 29822	93174 93972
00066	85274 86893	11303 22970	28834 34137	73515 90400	71148 43643
00067	84133 89640	44035 52166	73852 70091	61222 60561	62327 18423
00068	56732 16234	17395 96131	10123 91622	85496 57560	81604 18880
00069	65138 56806	87648 85261	34313 65861	45875 21069	85644 47277

Appendix B2 (*Continued*)

Row
number

00070	38001 02176	81719 11711	71602 92937	74219 64049	65584 49698
00071	37402 96397	01304 77586	56271 10086	47324 62605	40030 37438
00072	97125 40348	87083 31417	21815 39250	75237 62047	15501 29578
00073	21826 41134	47143 34072	64638 85902	49139 06441	03856 54552
00074	73135 42742	95719 09035	85794 74296	08789 88156	64691 19202
00075	07638 77929	03061 18072	96207 44156	23821 99538	04713 66994
00076	60528 83441	07954 19814	59175 20695	05533 52139	61212 06455
00077	83596 35655	06958 92983	05128 09719	77433 53783	92301 50498
00078	10850 62746	99599 10507	13499 06319	53075 71839	06410 19362
00079	39820 98952	43622 63147	64421 80814	43800 09351	31024 73167
00080	59580 06478	75569 78800	88835 54486	23768 06156	04111 08408
00081	38508 07341	23793 48763	90822 97022	17719 04207	95954 49953
00082	30692 70668	94688 16127	56196 80091	82067 63400	05462 69200
00083	65443 95659	18238 27437	49632 24041	08337 65676	96299 90836
00084	27267 50264	13192 72294	07477 44606	17985 48911	97341 30358
00085	91307 06991	19072 24210	36699 53728	28825 35793	28976 66252
00086	68434 94688	84473 13622	62126 98408	12843 82590	09815 93146
00087	48908 15877	54745 24591	35700 04754	83824 52692	54130 55160
00088	06913 45197	42672 78601	11883 09528	63011 98901	14974 40344
00089	10455 16019	14210 33712	91342 37821	88325 80851	43667 70883
00090	12883 97343	65027 61184	04285 01392	17974 15077	90712 26769
00091	21778 30976	38807 36961	31649 42096	63281 02023	08816 47449
00092	19523 59515	65122 59659	86283 68258	69572 13798	16435 91529
00093	67245 52670	35583 16563	79246 86686	76463 34222	26655 90802
00094	60584 47377	07500 37992	45134 26529	26760 83637	41326 44344
00095	53853 41377	36066 94850	58838 73859	49364 73331	96240 43642
00096	24637 38736	74384 89342	52623 07992	12369 18601	03742 83873
00097	83080 12451	38992 22815	07759 51777	97377 27585	51972 37867
00098	16444 24334	36151 99073	27493 70939	85130 32552	54846 54759
00099	60790 18157	57178 65762	11161 78576	45819 52979	65130 04860
00100	03991 10461	93716 16894	66083 24653	84609 58232	88618 19161
00101	38555 95554	32886 59780	08355 60860	29735 47762	71299 23853
00102	17546 73704	92052 46215	55121 29281	59076 07936	27954 58909
00103	32643 52861	95819 06831	00911 98936	76355 93779	80863 00514
00104	69572 68777	39510 35905	14060 40619	29549 69616	33564 60780

Appendix B2 (*Continued*)

**Row
number**

00105	24122	66591	27699	06494	14845	46672	61958	77100	90899	75754
00106	61196	30231	92962	61773	41839	55382	17267	70943	78038	70267
00107	30532	21704	10274	12202	39685	23309	10061	68829	55986	66485
00108	03788	97599	75867	20717	74416	53166	35208	33374	87539	08823
00109	48228	63379	85783	47619	53152	67433	35663	52972	16818	60311
00110	60365	94653	35075	33949	42614	29297	01918	28316	98953	73231
00111	83799	42402	56623	34442	34994	41374	70071	14736	09958	18065
00112	32960	07405	36409	83232	99385	41600	11133	07586	15917	06253
00113	19322	53845	57620	52606	66497	68646	78138	66559	19640	99413
00114	11220	94747	07399	37408	48509	23929	27482	45476	85244	35159
00115	31751	57260	68980	05339	15470	48355	88651	22596	03152	19121
00116	88492	99382	14454	04504	20094	98977	74843	93413	22109	78508
00117	30934	47744	07481	83828	73788	06533	28597	20405	94205	20380
00118	22888	48893	27499	98748	60530	45128	74022	84617	82037	10268
00119	78212	16993	35902	91386	44372	15486	65741	14014	87481	37220
00120	41849	84547	46850	52326	34677	58300	74910	64345	19325	81549
00121	46352	33049	69248	93460	45305	07521	61318	31855	14413	70951
00122	11087	96294	14013	31792	59747	67277	76503	34513	39663	77544
00123	52701	08337	56303	87315	16520	69676	11654	99893	02181	68161
00124	57275	36898	81304	48585	68652	27376	92852	55866	88448	03584
00125	20857	73156	70284	24326	79375	95220	01159	63267	10622	48391
00126	15633	84924	90415	93614	33521	26665	55823	47641	86225	31704
00127	92694	48297	39904	02115	59589	49067	66821	41575	49767	04037
00128	77613	19019	88152	00080	20554	91409	96277	48257	50816	97616
00129	38688	32486	45134	63545	59404	72059	43947	51680	43852	59693
00130	25163	01889	70014	15021	41290	67312	71857	15957	68971	11403
00131	65251	07629	37239	33295	05870	01119	92784	26340	18477	65622
00132	36815	43625	18637	37509	82444	99005	04921	73701	14707	93997
00133	64397	11692	05327	82162	20247	81759	45197	25332	83745	22567
00134	04515	25624	95096	67946	48460	85558	15191	18782	16930	33361
00135	83761	60873	43253	84145	60833	25983	01291	41349	20368	07126
00136	14387	06345	80854	09279	43529	06318	38384	74761	41196	37480
00137	51321	92246	80088	77074	88722	56736	66164	49431	66919	31678
00138	72472	00008	80890	18002	94813	31900	54155	83436	35352	54131
00139	05466	55306	93128	18464	74457	90561	72848	11834	79982	68416

472

Row
number

00140	39528 72484	82474 25593	48545 35247	18619 13674	18611 19241
00141	81616 18711	53342 44276	75122 11724	74627 73707	58319 15997
00142	07586 16120	82641 22820	92904 13141	32392 19763	61199 67940
00143	90767 04235	13574 17200	69902 63742	78464 22501	18627 90872
00144	40188 28193	29593 88627	94972 11598	62095 36787	00441 58997
00145	34414 82157	86887 55087	19152 00023	12302 80783	32624 68691
00146	63439 75363	44989 16822	36024 00867	76378 41605	65961 73488
00147	67049 09070	93399 45547	94458 74284	05041 49807	20288 34060
00148	79495 04146	52162 90286	54158 34243	46978 35482	59362 95938
00149	91704 30552	04737 21031	75051 93029	47665 64382	99782 93478
00150	94015 46874	32444 48277	59820 96163	64654 25843	41145 42820
00151	74108 88222	88570 74015	25704 91035	01755 14750	48968 38603
00152	62880 87873	95160 59221	22304 90314	72877 17334	39283 04149
00153	11748 12102	80580 41867	17710 59621	06554 07850	73950 79552
00154	17944 05600	60478 03343	25852 58905	57216 39618	49856 99326
00155	66067 42792	95043 52680	46780 56487	09971 59481	37006 22186
00156	54244 91030	45547 70818	59849 96169	61459 21647	87417 17198
00157	30945 57589	31732 57260	47670 07654	46376 25366	94746 49580
00158	69170 37403	86995 90307	94304 71803	26825 05511	12459 91314
00159	08345 88975	35841 85771	08105 59987	87112 21476	14713 71181
00160	27767 43584	85301 88977	29490 69714	73035 41207	74699 09310
00161	13025 14338	54066 15243	47724 66733	47431 43905	31048 56699
00162	80217 36292	98525 24335	24432 24896	43277 58874	11466 16082
00163	10875 62004	90391 61105	57411 06368	53856 30743	08670 84741
00164	54127 57326	26629 19087	24472 88779	30540 27886	61732 75454
00165	60311 42824	37301 42678	45990 43242	17374 52003	70707 70214
00166	49739 71484	92003 98086	76668 73209	59202 11973	02902 33250
00167	78626 51594	16453 94614	39014 97066	83012 09832	25571 77628
00168	66692 13986	99837 00582	81232 44987	09504 96412	90193 79568
00169	44071 28091	07362 97703	76447 42537	98524 97831	65704 09514
00170	41468 85149	49554 17994	14924 39650	95294 00556	70481 06905
00171	94559 37559	49678 53119	70312 05682	66986 34099	74474 20740
00172	41615 70360	64114 58660	90850 64618	80620 51790	11436 38072
00173	50273 93113	41794 86861	24781 89683	55411 85667	77535 99892
00174	41396 80504	90670 08289	40902 05069	95083 06783	28102 57816

**Row
number**

00175	25807 24260	71529 78920	72682 07385	90726 57166	98884 08583
00176	06170 97965	88302 98041	21443 41808	68984 83620	89747 98882
00177	60808 54444	74412 81105	01176 28838	36421 16489	18059 51061
00178	80940 44893	10408 36222	80582 71944	92638 40333	67054 16067
00179	19516 90120	46759 71643	13177 55292	21036 82808	77501 97427
00180	49386 54480	23604 23554	21785 41101	91178 10174	29420 90438
00181	06312 88940	15995 69321	47458 64809	98189 81851	29651 84215
00182	60942 00307	11897 92674	40405 68032	96717 54244	10701 41393
00183	92329 98932	78284 46347	71209 92061	39448 93136	25722 08564
00184	77936 63574	31384 51924	85561 29671	58137 17820	22751 36518
00185	38101 77756	11657 13897	95889 57067	47648 13885	70669 93406
00186	39641 69457	91339 22502	92613 89719	11947 56203	19324 20504
00187	84054 40455	99396 63680	67667 60631	69181 96845	38525 11600
00188	47468 03577	57649 63266	24700 71594	14004 23153	69249 05747
00189	43321 31370	28977 23896	76479 68562	62342 07589	08899 05985
00190	64281 61826	18555 64937	13173 33365	78851 16499	87064 13075
00191	66847 70495	32350 02985	86716 38746	26313 77463	55387 72681
00192	72461 33230	21529 53424	92581 02262	78438 66276	18396 73538
00193	21032 91050	13058 16218	12470 56500	15292 76139	59526 52113
00194	95362 67011	06651 16136	01016 00857	55018 56374	35824 71708
00195	49712 97380	10404 55452	34030 60726	75211 10271	36633 68424
00196	58275 61764	97586 54716	50259 46345	87195 46092	26787 60939
00197	89514 11788	68224 23417	73959 76145	30342 40277	11049 72049
00198	15472 50669	48139 36732	46874 37088	73465 09819	58869 35220
00199	12120 86124	51247 44302	60883 52109	21437 36786	49226 77837

† Extracted with permission from the Rand Corporation publication "A Million Random Digits," The Free Press of Glencoe, New York, 1955.

Appendix B3 The hypergeometric distribution†

N	n	k	r or x	P(r)	p(x)	N	n	k	r or x	P(r)	p(x)
10	1	1	0	0.900000	0.900000	10	5	3	0	0.083333	0.083333
10	1	1	1	1.000000	0.100000	10	5	3	1	0.500000	0.416667
10	2	1	0	0.800000	0.800000	10	5	3	2	0.916667	0.416667
10	2	1	1	1.000000	0.200000	10	5	3	3	1.000000	0.083333
10	2	2	0	0.622222	0.622222	10	5	4	0	0.023810	0.023810
10	2	2	1	0.977778	0.355556	10	5	4	1	0.261905	0.238095
10	2	2	2	1.000000	0.022222	10	5	4	2	0.738095	0.476190
10	3	1	0	0.700000	0.700000	10	5	4	3	0.976190	0.238095
10	3	2	1	1.000000	0.300000	10	5	4	4	1.000000	0.023810
10	3	2	0	0.466667	0.466667	10	5	5	0	0.003968	0.003968
10	3	2	1	0.933333	0.466667	10	5	5	1	0.103175	0.099206
10	3	2	2	1.000000	0.066667	10	5	5	2	0.500000	0.396825
10	3	3	0	0.291667	0.291667	10	5	5	3	0.896825	0.396825
10	3	3	1	0.816667	0.525000	10	5	5	4	0.996032	0.099206
10	3	3	2	0.991667	0.175000	10	5	5	5	1.000000	0.003968
10	3	3	3	1.000000	0.008333	10	6	1	0	0.400000	0.400000
10	4	1	0	0.600000	0.600000	10	6	1	1	1.000000	0.600000
10	4	1	1	1.000000	0.400000	10	6	2	0	0.133333	0.133333
10	4	2	0	0.333333	0.333333	10	6	2	1	0.666667	0.533333
10	4	2	1	0.866667	0.533333	10	6	2	2	1.000000	0.333333
10	4	2	2	1.000000	0.133333	10	6	3	0	0.033333	0.033333
10	4	3	0	0.166667	0.166667	10	6	3	1	0.333333	0.300000
10	4	3	1	0.666667	0.500000	10	6	3	2	0.833333	0.500000
10	4	3	2	0.966667	0.300000	10	6	3	3	1.000000	0.166667
10	4	3	3	1.000000	0.033333	10	6	4	0	0.004762	0.004762
10	4	4	0	0.071429	0.071429	10	6	4	1	0.119048	0.114286
10	4	4	1	0.452381	0.380952	10	6	4	2	0.547619	0.428571
10	4	4	2	0.880952	0.428571	10	6	4	3	0.928571	0.380952
10	4	4	3	0.995238	0.114286	10	6	4	4	1.000000	0.071429
10	4	4	4	1.000000	0.004762	10	6	5	1	0.023810	0.023810
10	5	1	0	0.500000	0.500000	10	6	5	2	0.261905	0.238095
10	5	1	1	1.000000	0.500000	10	6	5	3	0.738095	0.476190
10	5	2	0	0.222222	0.222222	10	6	5	4	0.976190	0.238095
10	5	2	1	0.777778	0.555556	10	6	5	5	1.000000	0.023810
10	5	2	2	1.000000	0.222222	10	6	6	2	0.071429	0.071429

N	n	k	r or x	P(r)	p(x)	N	n	k	r or x	P(r)	p(x)
10	6	6	3	0.452381	0.380952	10	8	3	2	0.533333	0.466667
10	6	6	4	0.880952	0.428571	10	8	3	3	1.000000	0.466667
10	6	6	5	0.995238	0.114286	10	8	4	2	0.133333	0.133333
10	6	6	6	1.000000	0.004762	10	8	4	3	0.666667	0.533333
10	7	1	0	0.300000	0.300000	10	8	4	4	1.000000	0.333333
10	7	1	1	1.000000	0.700000	10	8	5	3	0.222222	0.222222
10	7	2	0	0.066667	0.066667	10	8	5	4	0.777778	0.555556
10	7	2	1	0.533333	0.466667	10	8	5	5	1.000000	0.222222
10	7	2	2	1.000000	0.466667	10	8	6	4	0.333333	0.333333
10	7	3	0	0.008333	0.008333	10	8	6	5	0.866667	0.533333
10	7	3	1	0.183333	0.175000	10	8	6	6	1.000000	0.133333
10	7	3	2	0.708333	0.525000	10	8	7	5	0.466667	0.466667
10	7	3	3	1.000000	0.291667	10	8	7	6	0.933333	0.466667
10	7	4	1	0.033333	0.033333	10	8	7	7	1.000000	0.066667
10	7	4	2	0.333333	0.300000	10	8	8	6	0.622222	0.622222
10	7	4	3	0.833333	0.500000	10	8	8	7	0.977778	0.355556
10	7	4	4	1.000000	0.166667	10	8	8	8	1.000000	0.022222
10	7	5	2	0.083333	0.083333	10	9	1	0	0.100000	0.100000
10	7	5	3	0.500000	0.416667	10	9	1	1	1.000000	0.900000
10	7	5	4	0.916667	0.416667	10	9	2	1	0.200000	0.200000
10	7	5	5	1.000000	0.083333	10	9	2	2	1.000000	0.800000
10	7	6	3	0.166667	0.166667	10	9	3	2	0.300000	0.300000
10	7	6	4	0.666667	0.500000	10	9	3	3	1.000000	0.700000
10	7	6	5	0.966667	0.300000	10	9	4	3	0.400000	0.400000
10	7	6	6	1.000000	0.033333	10	9	4	4	1.000000	0.600000
10	7	7	4	0.291667	0.291667	10	9	5	4	0.500000	0.500000
10	7	7	5	0.816667	0.525000	10	9	5	5	1.000000	0.500000
10	7	7	6	0.991667	0.175000	10	9	6	5	0.600000	0.600000
10	7	7	7	1.000000	0.008333	10	9	6	6	1.000000	0.400000
10	8	1	0	0.200000	0.200000	10	9	7	6	0.700000	0.700000
10	8	1	1	1.000000	0.800000	10	9	7	7	1.000000	0.300000
10	8	2	0	0.022222	0.022222	10	9	8	7	0.800000	0.800000
10	8	2	1	0.377778	0.355556	10	9	8	8	1.000000	0.200000
10	8	2	2	1.000000	0.622222	10	9	9	8	0.900000	0.900000
10	8	3	1	0.066667	0.066667	10	9	9	9	1.000000	0.100000

† Extracted with permission from Gerald J. Lieberman and Donald B. Owen, "Tables of the Hypergeometric Probability Distribution," Stanford University Press, Stanford, Calif., 1961.

Appendix B4   The cumulative Poisson distribution†

r	$\mu = .1$	$\mu = .2$	$\mu = .3$	$\mu = .4$	$\mu = .5$
0	.90484	.81873	.74082	.67302	.60653
1	.99532	.98248	.96306	.93845	.90980
2	.99985	.99885	.99640	.99207	.98561
3	1.00000	.99994	.99973	.99922	.99825
4		1.00000	.99998	.99994	.99983
5			1.00000	1.00000	.99999
6					1.00000

r	$\mu = .6$	$\mu = .7$	$\mu = .8$	$\mu = .9$	$\mu = 1.0$
0	.54881	.49658	.44933	.40657	.36788
1	.87810	.84419	.80879	.77248	.73576
2	.97688	.96586	.95258	.93714	.91970
3	.99664	.99425	.99092	.98654	.98101
4	.99961	.99921	.99859	.99766	.99634
5	.99996	.99991	.99982	.99966	.99941
6	1.00000	.99999	.99998	.99996	.99992
7		1.00000	1.00000	1.00000	.99999
8					1000000

r	$\mu = 2$	$\mu = 3$	$\mu = 4$	$\mu = 5$	$\mu = 6$
0	.13534	.04979	.01832	.00674	.00248
1	.40601	.19915	.09158	.04043	.01735
2	.67668	.42319	.23810	.12465	.06197
3	.85712	.64723	.43347	.26503	.15120
4	.94735	.81526	.62884	.44049	.28506
5	.98344	.91608	.78513	.61596	.44568
6	.99547	.96649	.88933	.76218	.60630
7	.99890	.98810	.94887	.86663	.74398
8	.99976	.99620	.97864	.93191	.84724
9	.99995	.99890	.99187	.96817	.91608
10	.99999	.99971	.99716	.98630	.95738
11	1.00000	.99993	.99908	.99455	.97991
12		.99998	.99973	.99798	.99117
13		1.00000	.99992	.99930	.99637
14			.99998	.99977	.99860
15			1.00000	.99993	.99949
16				.99998	.99982
17				1.00000	.99994
18					.99998
19					1.00000

r	$\mu = 7$	$\mu = 8$	$\mu = 9$	$\mu = 10$
0	.00091	.00033	.00012	.00004
1	.00730	.00302	.00123	.00050
2	.02964	.01375	.00623	.00277
3	.08176	.04238	.02123	.01034
4	.17299	.09963	.05496	.02925
5	.30071	.19124	.11569	.06709
6	.44971	.31337	.20678	.13014
7	.59871	.45296	.32390	.22022
8	.72909	.59255	.45565	.33282
9	.83050	.71662	.58741	.45793
10	.90148	.81589	.70599	.58304
11	.94665	.88808	.80301	.69678
12	.97300	.93620	.87577	.79156
13	.98719	.96582	.92615	.86446
14	.99428	.98274	.95853	.91654
15	.99759	.99177	.97796	.95126
16	.99904	.99628	.98889	.97296
17	.99964	.99841	.99468	.98572
18	.99987	.99935	.99757	.99281
19	.99996	.99975	.99894	.99655
20	.99999	.99991	.99956	.99841
21	1.00000	.99997	.99982	.99930
22		.99999	.99993	.99970
23		1.00000	.99998	.99988
24			.99999	.99995
25			1.00000	.99998
26				.99999
27				1.00000

† Extracted with permission from E. C. Molina, "Poisson's Binomial Exponential Limit," D. Van Nostrand Company, Inc., Princeton, N.J., 1949.

Appendix B5 The cumulative standardized normal distribution function†   [Tabulated value = $\Pr(Z < z_p) = p$]

z_p	.00	.01	.02	.03	.04	.05	.06	.07	.08	.09
−.0	.5000	.4960	.4920	.4880	.4840	.4801	.4761	.4721	.4681	.4641
−.1	.4602	.4562	.4522	.4483	.4443	.4404	.4364	.4325	.4286	.4247
−.2	.4207	.4168	.4129	.4090	.4052	.4013	.3974	.3936	.3897	.3859
−.3	.3821	.3783	.3745	.3707	.3669	.3632	.3594	.3557	.3520	.3483
−.4	.3446	.3409	.3372	.3336	.3300	.3264	.3228	.3192	.3156	.3121
−.5	.3085	.3050	.3015	.2981	.2946	.2912	.2877	.2843	.2810	.2776
−.6	.2743	.2709	.2676	.2643	.2611	.2578	.2546	.2514	.2483	.2451
−.7	.2420	.2389	.2358	.2327	.2297	.2266	.2236	.2206	.2177	.2148
−.8	.2119	.2090	.2061	.2033	.2005	.1977	.1949	.1922	.1894	.1867
−.9	.1841	.1814	.1788	.1762	.1736	.1711	.1685	.1660	.1635	.1611
−1.0	.1587	.1562	.1539	.1515	.1492	.1469	.1446	.1423	.1401	.1379
−1.1	.1357	.1335	.1314	.1292	.1271	.1251	.1230	.1210	.1190	.1170
−1.2	.1151	.1131	.1112	.1093	.1075	.1056	.1038	.1020	.1003	.09853
−1.3	.09680	.09510	.09342	.09176	.09012	.08851	.08691	.08534	.08379	.08226
−1.4	.08076	.07927	.07780	.07636	.07493	.07353	.07215	.07078	.06944	.06811
−1.5	.06681	.06552	.06426	.06301	.06178	.06057	.05938	.05821	.05705	.05592
−1.6	.05480	.05370	.05262	.05155	.05050	.04947	.04846	.04746	.04648	.04551
−1.7	.04457	.04363	.04272	.04182	.04093	.04006	.03920	.03836	.03754	.03673
−1.8	.03593	.03515	.03438	.03362	.03288	.03216	.03144	.03074	.03005	.02938
−1.9	.02872	.02807	.02743	.02680	.02619	.02559	.02500	.02442	.02385	.02330
−2.0	.02275	.02222	.02169	.02118	.02068	.02018	.01970	.01923	.01876	.01831
−2.1	.01786	.01743	.01700	.01659	.01616	.01578	.01539	.01500	.01463	.01426
−2.2	.01390	.01355	.01321	.01287	.01255	.01222	.01191	.01160	.01130	.01101
−2.3	.01072	.01044	.01017	$.0^2 9903$	$.0^2 9642$	$.0^2 9387$	$.0^2 9137$	$.0^2 8894$	$.0^2 8656$	$.0^2 8424$
−2.4	$.0^2 8198$	$.0^2 7976$	$.0^2 7760$	$.0^2 7549$	$.0^2 7344$	$.0^2 7143$	$.0^2 6947$	$.0^2 6756$	$.0^2 6569$	$.0^2 6387$
−2.5	$.0^2 6210$	$.0^2 6037$	$.0^2 5868$	$.0^2 5703$	$.0^2 5543$	$.0^2 5386$	$.0^2 5234$	$.0^2 5085$	$.0^2 4940$	$.0^2 4799$
−2.6	$.0^2 4661$	$.0^2 4527$	$.0^2 4396$	$.0^2 4269$	$.0^2 4145$	$.0^2 4025$	$.0^2 3907$	$.0^2 3793$	$.0^2 3681$	$.0^2 3573$
−2.7	$.0^2 3467$	$.0^2 3364$	$.0^2 3264$	$.0^2 3167$	$.0^2 3072$	$.0^2 2980$	$.0^2 2890$	$.0^2 2803$	$.0^2 2718$	$.0^2 2635$

.2	.5793	.5832	.5871	.5910	.5948	.5987	.6026	.6064	.6103	.6141
.3	.6179	.6217	.6255	.6293	.6331	.6368	.6406	.6443	.6480	.6517
.4	.6554	.6591	.6628	.6664	.6700	.6736	.6772	.6808	.6844	.6879
.5	.6915	.6950	.6985	.7019	.7054	.7088	.7123	.7157	.7190	.7224
.6	.7257	.7291	.7324	.7357	.7389	.7422	.7454	.7486	.7517	.7549
.7	.7580	.7611	.7642	.7673	.7703	.7734	.7764	.7794	.7823	.7852
.8	.7881	.7910	.7939	.7967	.7995	.8023	.8051	.8078	.8106	.8133
.9	.8159	.8186	.8212	.8238	.8264	.8289	.8315	.8340	.8365	.8389
1.0	.8413	.8438	.8461	.8485	.8508	.8531	.8554	.8577	.8599	.8621
1.1	.8643	.8665	.8686	.8708	.8729	.8749	.8770	.8790	.8810	.8830
1.2	.8849	.8869	.8888	.8907	.8925	.8944	.8962	.8980	.8997	.90147
1.3	.90320	.90490	.90658	.90824	.90988	.91149	.91309	.91466	.91621	.91774
1.4	.91924	.92073	.92220	.92364	.92507	.92647	.92785	.92922	.93056	.93189
1.5	.93319	.93448	.93574	.93669	.93822	.93943	.94062	.94179	.94295	.94408
1.6	.94520	.94630	.94738	.94845	.94950	.95053	.95154	.95254	.95352	.95449
1.7	.95543	.95637	.95728	.95818	.95907	.95994	.96080	.96164	.96246	.96327
1.8	.96407	.96485	.96562	.96638	.96712	.96784	.96856	.96926	.96995	.97062
1.9	.97128	.97193	.97257	.97320	.97381	.97441	.97500	.97558	.97615	.97670
2.0	.97725	.97778	.97831	.97882	.97932	.97982	.98030	.98077	.98124	.98169
2.1	.98214	.98257	.98300	.98341	.98382	.98422	.98461	.98500	.98537	.98574
2.2	.98610	.98645	.98679	.98713	.98745	.98778	.98809	.98840	.98870	.98899
2.3	.98928	.98956	.98983	$.9^{2}0097$	$.9^{2}0358$	$.9^{2}0613$	$.9^{2}0863$	$.9^{2}1106$	$.9^{2}1344$	$.9^{2}1576$
2.4	$.9^{2}1802$	$.9^{2}2024$	$.9^{2}2240$	$.9^{2}2451$	$.9^{2}2656$	$.9^{2}2857$	$.9^{2}3053$	$.9^{2}3244$	$.9^{2}3431$	$.9^{2}3613$
2.5	$.9^{2}3790$	$.9^{2}3963$	$.9^{2}4132$	$.9^{2}4297$	$.9^{2}4457$	$.9^{2}4614$	$.9^{2}4766$	$.9^{2}4915$	$.9^{2}5060$	$.9^{2}5201$
2.6	$.9^{2}5339$	$.9^{2}5473$	$.9^{2}5604$	$.9^{2}5731$	$.9^{2}5855$	$.9^{2}5975$	$.9^{2}6093$	$.9^{2}6207$	$.9^{2}6319$	$.9^{2}6427$
2.7	$.9^{2}6533$	$.9^{2}6636$	$.9^{2}6736$	$.9^{2}6833$	$.9^{2}6928$	$.9^{2}7020$	$.9^{2}7110$	$.9^{2}7197$	$.9^{2}7282$	$.9^{2}7365$
2.8	$.9^{2}7445$	$.9^{2}7523$	$.9^{2}7599$	$.9^{2}7673$	$.9^{2}7744$	$.9^{2}7814$	$.9^{2}7882$	$.9^{2}7948$	$.9^{2}8012$	$.9^{2}8074$
2.9	$.9^{2}8134$	$.9^{2}8193$	$.9^{2}8250$	$.9^{2}8305$	$.9^{2}8359$	$.9^{2}8411$	$.9^{2}8462$	$.9^{2}8511$	$.9^{2}8559$	$.9^{2}8605$
3.0	$.9^{2}8650$	$.9^{2}8694$	$.9^{2}8736$	$.9^{2}8777$	$.9^{2}8817$	$.9^{2}8856$	$.9^{2}8893$	$.9^{2}8930$	$.9^{2}8965$	$.9^{2}8999$

† Reprinted with permission from A. Hald, "Statistical Tables and Formulas," John Wiley & Sons, Inc., New York, 1952.

Note: $.0^{2}1350 = .001350$ $.9^{2}8650 = .998650$

Tabulated value $= t_{v;p}$ where $\Pr(T_v < t_{v;p}) = p$ $t_{v;1-p} = -t_{v;p}$

$\diagdown$ p v $\diagdown$	.60	.70	.80	.90	.95	.975	.990	.995	.999	.9995
1	.325	.727	1.376	3.078	6.314	12.71	31.82	63.66	318.3	636.6
2	.289	.617	1.061	1.886	2.920	4.303	6.965	9.925	22.33	31.60
3	.277	.584	.978	1.638	2.353	3.182	4.541	5.841	10.22	12.94
4	.271	.569	.941	1.533	2.132	2.776	3.747	4.604	7.173	8.610
5	.267	.559	.920	1.476	2.015	2.571	3.365	4.032	5.893	6.859
6	.265	.553	.906	1.440	1.943	2.447	3.143	3.707	5.208	5.959
7	.263	.549	.896	1.415	1.895	2.365	2.998	3.499	4.785	5.405
8	.262	.546	.889	1.397	1.860	2.306	2.896	3.355	4.501	5.041
9	.261	.543	.883	1.383	1.833	2.262	2.821	3.250	4.297	4.781
10	.260	.542	.879	1.372	1.812	2.228	2.764	3.169	4.144	4.587
11	.260	.540	.876	1.363	1.796	2.201	2.718	3.106	4.025	4.437
12	.259	.539	.873	1.356	1.782	2.179	2.681	3.055	3.930	4.318
13	.259	.538	.870	1.350	1.771	2.160	2.650	3.012	3.852	4.221
14	.258	.537	.868	1.345	1.761	2.145	2.624	2.977	3.787	4.140
15	.258	.536	.866	1.341	1.753	2.131	2.602	2.947	3.733	4.073
16	.258	.535	.865	1.337	1.746	2.120	2.583	2.921	3.686	4.015
17	.257	.534	.863	1.333	1.740	2.110	2.567	2.898	3.646	3.965
18	.257	.534	.862	1.330	1.734	2.101	2.552	2.878	3.611	3.922
19	.257	.533	.861	1.328	1.729	2.093	2.539	2.861	3.579	3.883
20	.257	.533	.860	1.325	1.725	2.086	2.528	2.845	3.552	3.850
21	.257	.532	.859	1.323	1.721	2.080	2.518	2.831	3.527	3.819
22	.256	.532	.858	1.321	1.717	2.074	2.508	2.819	3.505	3.792
23	.256	.532	.858	1.319	1.714	2.069	2.500	2.807	3.485	3.767
24	.256	.531	.857	1.318	1.711	2.064	2.492	2.797	3.467	3.745
25	.256	.531	.856	1.316	1.708	2.060	2.485	2.787	3.450	3.725
26	.256	.531	.856	1.315	1.706	2.056	2.479	2.779	3.435	3.707
27	.256	.531	.855	1.314	1.703	2.052	2.473	2.771	3.421	3.690
28	.256	.530	.855	1.313	1.701	2.048	2.467	2.763	3.408	3.674
29	.256	.530	.854	1.311	1.699	2.045	2.462	2.756	3.396	3.659
30	.256	.530	.854	1.310	1.697	2.042	2.457	2.750	3.385	3.646
40	.255	.529	.851	1.303	1.684	2.021	2.423	2.704	3.307	3.551
50	.255	.528	.849	1.298	1.676	2.009	2.403	2.678	3.262	3.495
60	.254	.527	.848	1.296	1.671	2.000	2.390	2.660	3.232	3.460
80	.254	.527	.846	1.292	1.664	1.990	2.374	2.639	3.195	3.415
100	.254	.526	.845	1.290	1.660	1.984	2.365	2.626	3.174	3.389
200	.254	.525	.843	1.286	1.653	1.972	2.345	2.601	3.131	3.339
500	.253	.525	.842	1.283	1.648	1.965	2.334	2.586	3.106	3.310
∞	.253	.524	.842	1.282	1.645	1.960	2.326	2.576	3.090	3.291

† Reprinted by permission from A. Hald, "Statistical Tables and Formulas," John Wiley & Sons, Inc., New York, 1952.

Appendix B7 The chi-square distribution†

Tabulated value $= \chi^2_{v;p}$ where $\Pr(Y^2_v < \chi^2_{v;p}) = p$

v \ p	.005	.010	.025	.050	.100	.900	.950	.975	.990	.995
1	0.0⁴393	0.0³157	0.0³982	0.0²393	0.0158	2.71	3.84	5.02	6.63	7.88
2	0.0100	0.0201	0.0506	0.103	0.211	4.61	5.99	7.38	9.21	10.60
3	0.072	0.115	0.216	0.352	0.584	6.25	7.81	9.35	11.34	12.84
4	0.207	0.297	0.484	0.711	1.064	7.78	9.49	11.14	13.28	14.86
5	0.412	0.554	0.831	1.145	1.61	9.24	11.07	12.83	15.09	16.75
6	0.676	0.872	1.24	1.64	2.20	10.64	12.59	14.45	16.81	18.55
7	0.989	1.24	1.69	2.17	2.83	12.02	14.07	16.01	18.48	20.28
8	1.34	1.65	2.18	2.73	3.49	13.36	15.51	17.53	20.09	21.96
9	1.73	2.09	2.70	3.33	4.17	14.68	16.92	19.02	21.67	23.59
10	2.16	2.56	3.25	3.94	4.87	15.99	18.31	20.48	23.21	25.19
11	2.60	3.05	3.82	4.57	5.58	17.28	19.68	21.92	24.72	26.76
12	3.07	3.57	4.40	5.23	6.30	18.55	21.03	23.34	26.22	28.30
13	3.57	4.11	5.01	5.89	7.04	19.81	22.36	24.74	27.69	29.82
14	4.07	4.66	5.63	6.57	7.79	21.06	23.68	26.12	29.14	31.32
15	4.60	5.23	6.26	7.26	8.55	22.31	25.00	27.49	30.58	32.80
16	5.14	5.81	6.91	7.96	9.31	23.54	26.30	28.85	32.00	34.27
17	5.70	6.41	7.56	8.67	10.09	24.77	27.59	30.19	33.41	35.72
18	6.26	7.01	8.23	9.39	10.86	25.99	28.87	31.53	34.81	37.16
19	6.84	7.63	8.91	10.12	11.65	27.20	30.14	32.85	36.19	38.58
20	7.43	8.26	8.59	10.85	12.44	28.41	31.41	34.17	37.57	40.00
21	8.03	8.90	10.28	11.59	13.24	29.62	32.67	35.48	38.93	41.40
22	8.64	9.54	10.98	12.34	14.04	30.81	33.92	36.78	40.29	42.80
23	9.26	10.20	11.69	13.09	14.85	32.01	35.17	38.08	41.64	44.18
24	9.89	10.86	12.40	13.85	15.66	33.20	36.42	39.36	42.98	45.56
25	10.52	11.52	13.12	14.61	16.47	34.38	37.65	40.65	44.31	46.93
26	11.16	12.20	13.84	15.38	17.29	35.56	38.89	41.92	45.64	48.29
27	11.81	12.88	14.57	16.15	18.11	36.74	40.11	43.19	46.96	49.64
28	12.46	13.56	15.31	16.93	18.94	37.92	41.34	44.46	48.28	50.99
29	13.21	14.26	16.05	17.71	19.77	39.09	42.56	45.72	49.59	52.34
30	13.79	14.95	16.79	18.49	20.60	40.26	43.77	46.98	50.89	53.67
40	20.71	22.16	24.43	26.51	29.05	51.80	55.76	59.34	63.69	66.77
50	27.99	29.71	32.36	34.76	37.69	63.17	67.50	71.42	76.15	79.49
60	35.53	37.48	40.48	43.19	46.46	74.40	79.08	83.30	88.38	91.95
70	43.28	45.44	48.76	51.74	55.33	85.53	90.53	95.02	100.4	104.2
80	51.17	53.54	57.15	60.39	64.28	96.58	101.9	106.6	112.3	116.3
90	59.20	61.75	65.65	69.13	73.29	107.6	113.1	118.1	124.1	128.3
100	67.33	70.06	74.22	77.93	82.36	118.5	124.3	129.6	135.8	140.2

† Extracted with permission from H. L. Harter, A New Table of Percentage Points of the Chi-square Distribution, *Biometrika*, June, 1964.

Appendix B8 Chi-square divided by degrees of freedom†

Tabulated value $= \chi^2_{v;p}/v$ where $\Pr(Y^2_v/v < \chi^2_{v;p}/v) = p$

p / v	.0005	.001	.005	.01	.025	.05	.10	.20	.30	.40	.50	.60	.70	.80	.90	.95	.975	.99	.995	.999	.9995
1	$.0^639$	$.0^5157$	$.0^439$	$.0^416$	$.0^398$	$.0^339$	.016	.064	.148	.275	.455	.708	1.07	1.64	2.71	3.84	5.02	6.64	7.88	10.83	12.12
2	.001	.001	.005	.010	.025	.052	.106	.223	.356	.511	.693	.916	1.20	1.61	2.30	3.00	3.69	4.61	5.30	6.91	7.60
3	.005	.008	.024	.038	.072	.117	.195	.335	.475	.623	.789	.982	1.22	1.55	2.08	2.60	3.12	3.78	4.28	5.42	5.91
4	.016	.023	.052	.074	.121	.178	.266	.412	.549	.688	.839	1.011	1.22	1.50	1.94	2.37	2.79	3.32	3.72	4.62	5.00
5	.032	.042	.082	.111	.166	.229	.322	.469	.600	.731	.870	1.03	1.21	1.46	1.85	2.21	2.57	3.02	3.35	4.10	4.42
6	.050	.064	.113	.145	.206	.272	.367	.512	.638	.762	.891	1.04	1.21	1.43	1.77	2.10	2.41	2.80	3.09	3.74	4.02
7	.069	.085	.141	.177	.241	.310	.405	.546	.667	.785	.907	1.04	1.20	1.40	1.72	2.01	2.29	2.64	2.90	3.47	3.72
8	.089	.107	.168	.206	.272	.342	.436	.574	.691	.803	.918	1.04	1.19	1.38	1.67	1.94	2.19	2.51	2.74	3.27	3.48
9	.108	.128	.193	.232	.300	.369	.463	.598	.710	.817	.927	1.05	1.18	1.36	1.63	1.88	2.11	2.41	2.62	3.10	3.30
10	.126	.148	.216	.256	.325	.394	.487	.618	.727	.830	.934	1.05	1.18	1.34	1.60	1.83	2.05	2.32	2.52	2.96	3.14
11	.144	.167	.237	.278	.347	.416	.507	.635	.741	.840	.940	1.05	1.17	1.33	1.57	1.79	1.99	2.25	2.43	2.84	3.01
12	.161	.184	.256	.298	.367	.436	.525	.651	.753	.848	.945	1.05	1.17	1.32	1.55	1.75	1.94	2.18	2.36	2.74	2.90
13	.177	.201	.274	.316	.385	.453	.542	.664	.764	.856	.949	1.05	1.16	1.31	1.52	1.72	1.90	2.13	2.29	2.66	2.81
14	.193	.217	.291	.333	.402	.469	.556	.676	.773	.863	.953	1.05	1.16	1.30	1.50	1.69	1.87	2.08	2.24	2.58	2.72
15	.207	.232	.307	.349	.418	.484	.570	.687	.781	.869	.956	1.05	1.15	1.29	1.49	1.67	1.83	2.04	2.19	2.51	2.65
16	.221	.246	.321	.363	.432	.498	.582	.697	.789	.874	.959	1.05	1.15	1.28	1.47	1.64	1.80	2.00	2.14	2.45	2.58
17	.234	.260	.335	.377	.445	.510	.593	.706	.796	.879	.961	1.05	1.15	1.27	1.46	1.62	1.78	1.97	2.10	2.40	2.52
18	.247	.272	.348	.390	.457	.522	.604	.714	.802	.883	.963	1.05	1.14	1.26	1.44	1.60	1.75	1.93	2.06	2.35	2.47
19	.258	.285	.360	.402	.469	.532	.613	.722	.808	.887	.965	1.05	1.14	1.26	1.43	1.59	1.73	1.90	2.03	2.31	2.42
20	.270	.296	.372	.413	.480	.543	.622	.729	.813	.890	.967	1.05	1.14	1.25	1.42	1.57	1.71	1.88	2.00	2.27	2.37
22	.291	.317	.393	.434	.499	.561	.638	.742	.823	.897	.970	1.05	1.13	1.24	1.40	1.54	1.67	1.83	1.95	2.19	2.30
24	.310	.337	.412	.452	.517	.577	.652	.753	.831	.902	.972	1.05	1.13	1.23	1.38	1.52	1.64	1.79	1.90	2.13	2.23
26	.328	.355	.429	.469	.532	.592	.665	.762	.838	.907	.974	1.05	1.12	1.22	1.37	1.50	1.61	1.76	1.86	2.08	2.17
28	.345	.371	.445	.484	.547	.605	.676	.771	.845	.911	.976	1.04	1.12	1.22	1.35	1.48	1.59	1.72	1.82	2.03	2.12
30	.360	.386	.460	.498	.560	.616	.687	.779	.850	.915	.978	1.04	1.12	1.21	1.34	1.46	1.57	1.70	1.79	1.99	2.07

Appendix B8 (Continued)

p / v	.0005	.001	.005	.01	.025	.05	.10	.20	.30	.40	.50	.60	.70	.80	.90	.95	.975	.99	.995	.999	.9995
35	.394	.420	.491	.529	.588	.642	.708	.795	.862	.922	.981	1.04	1.11	1.19	1.32	1.42	1.52	1.64	1.72	1.90	1.98
40	.423	.448	.518	.554	.611	.663	.726	.809	.872	.928	.983	1.04	1.10	1.18	1.30	1.39	1.48	1.59	1.67	1.84	1.90
45	.448	.472	.540	.576	.630	.680	.741	.820	.880	.933	.985	1.04	1.10	1.17	1.28	1.37	1.45	1.55	1.63	1.78	1.84
50	.469	.494	.560	.594	.647	.695	.754	.829	.886	.937	.987	1.04	1.09	1.16	1.26	1.35	1.43	1.52	1.59	1.73	1.79
55	.488	.512	.577	.610	.662	.708	.765	.837	.892	.941	.988	1.04	1.09	1.16	1.25	1.33	1.41	1.50	1.56	1.69	1.75
60	.506	.529	.592	.625	.675	.720	.774	.844	.897	.944	.989	1.03	1.09	1.15	1.24	1.32	1.39	1.47	1.53	1.66	1.71
70	.535	.558	.618	.649	.697	.739	.790	.856	.905	.949	.990	1.03	1.08	1.14	1.22	1.29	1.36	1.43	1.49	1.60	1.65
80	.560	.582	.640	.669	.714	.755	.803	.865	.911	.952	.992	1.03	1.08	1.13	1.21	1.27	1.33	1.40	1.45	1.56	1.60
90	.581	.602	.658	.686	.729	.768	.814	.873	.917	.955	.993	1.03	1.07	1.12	1.20	1.26	1.31	1.38	1.43	1.52	1.56
100	.619	.619	.673	.701	.742	.779	.824	.879	.921	.958	.993	1.03	1.07	1.12	1.18	1.24	1.30	1.36	1.40	1.49	1.53
120	.629	.648	.699	.724	.763	.798	.839	.890	.929	.962	.994	1.03	1.06	1.11	1.17	1.22	1.27	1.32	1.36	1.45	1.48
140	.653	.671	.719	.743	.780	.812	.850	.898	.934	.965	.995	1.03	1.06	1.10	1.16	1.20	1.25	1.30	1.33	1.41	1.44
160	.673	.690	.736	.758	.793	.824	.860	.905	.939	.968	.996	1.02	1.06	1.09	1.15	1.19	1.23	1.28	1.31	1.38	1.41
180	.689	.706	.749	.771	.804	.833	.868	.910	.942	.970	.996	1.02	1.05	1.09	1.14	1.18	1.22	1.26	1.29	1.36	1.38
200	.703	.719	.761	.782	.814	.841	.874	.915	.945	.972	.997	1.02	1.05	1.08	1.13	1.17	1.21	1.25	1.28	1.34	1.36
250	.732	.746	.785	.804	.832	.858	.887	.924	.951	.975	.997	1.02	1.04	1.07	1.12	1.15	1.18	1.22	1.25	1.30	1.32
300	.753	.767	.802	.820	.846	.870	.897	.931	.956	.977	.998	1.02	1.04	1.07	1.11	1.14	1.17	1.20	1.22	1.27	1.29
350	.770	.783	.816	.833	.857	.879	.904	.936	.959	.979	.998	1.02	1.04	1.06	1.10	1.13	1.15	1.18	1.21	1.25	1.27
400	.784	.796	.827	.843	.866	.887	.911	.940	.962	.981	.998	1.02	1.04	1.06	1.09	1.12	1.14	1.17	1.19	1.24	1.25
450	.795	.807	.837	.852	.874	.893	.916	.944	.964	.982	.999	1.02	1.03	1.06	1.09	1.11	1.13	1.16	1.18	1.22	1.23
500	.805	.816	.845	.859	.880	.898	.920	.946	.966	.983	.999	1.01	1.03	1.05	1.08	1.11	1.13	1.15	1.17	1.21	1.22
750	.839	.848	.872	.884	.901	.917	.934	.956	.972	.986	.999	1.01	1.03	1.04	1.07	1.09	1.10	1.12	1.14	1.17	1.18
1000	.859	.868	.889	.899	.914	.928	.943	.962	.976	.988	.999	1.01	1.02	1.04	1.06	1.07	1.09	1.11	1.12	1.14	1.15
5000	.936	.939	.949	.954	.961	.967	.974	.983	.989	.995	1.00	1.00	1.01	1.02	1.02	1.03	1.04	1.05	1.05	1.06	1.07
∞	1	1	1	1	1	1	1	1	1	1	1	1	1	1	1	1	1	1	1	1	1

† Reprinted with permission from W. J. Dixon and F. J. Massey, Jr., "Introduction to Statistical Analysis," 3d ed., McGraw-Hill Book Company, New York, 1969.

Appendix B9 The F distribution†

Tabulated value $= F_{\nu_1,\nu_2;p}$ where $\Pr(F_{\nu_1,\nu_2} < F_{\nu_1,\nu_2;p}) = p$

ν_2	p	1	2	3	4	5	6	7	8	9	10	11	12	p
1	.0005	$.0^6 62$	$.0^3 50$	$.0^2 38$	$.0^2 94$	.016	.022	.027	.032	.036	.039	.042	.045	.0005
	.001	$.0^5 25$	$.0^2 10$	$.0^2 60$	.013	.021	.028	.034	.039	.044	.048	.051	.054	.001
	.005	$.0^4 62$	$.0^2 51$	.018	.032	.041	.054	.062	.068	.073	.078	.082	.085	.005
	.010	$.0^3 25$	.010	.029	.047	.062	.073	.082	.089	.095	.100	.104	.107	.010
	.025	$.0^3 15$	.026	.057	.082	.100	.113	.124	.132	.139	.144	.149	.153	.025
	.05	$.0^2 62$	.054	.099	.130	.151	.167	.179	.188	.195	.201	.207	.211	.05
	.10	.025	.117	.181	.220	.246	.265	.279	.289	.298	.304	.310	.315	.10
	.25	.172	.389	.494	.553	.591	.617	.637	.650	.661	.670	.680	.684	.25
	.50	1.00	1.50	1.71	1.82	1.89	1.94	1.98	2.00	2.03	2.04	2.05	2.07	.50
	.75	5.83	7.50	8.20	8.58	8.82	8.98	9.10	9.19	9.26	9.32	9.36	9.41	.75
	.90	39.9	49.5	53.6	55.8	57.2	58.2	58.9	59.4	59.9	60.2	60.5	60.7	.90
	.95	161	200	216	225	230	234	237	239	241	242	243	244	.95
	.975	648	800	864	900	922	937	948	957	963	969	973	977	.975
	.99	405^1	500^1	540^1	562^1	576^1	586^1	593^1	598^1	602^1	606^1	608^1	611^1	.99
	.995	162^2	200^2	216^2	225^2	231^2	234^2	237^2	239^2	241^2	242^2	243^2	244^2	.995
	.999	406^3	500^3	540^3	562^3	576^3	586^3	593^3	598^3	602^3	606^3	609^3	611^3	.999
	.9995	162^4	200^4	216^4	225^4	231^4	234^4	237^4	239^4	241^4	242^4	243^4	244^4	.9995
2	.0005	$.0^6 50$	$.0^3 50$	$.0^2 42$	.011	.020	.029	.037	.044	.050	.056	.061	.065	.0005
	.001	$.0^5 20$	$.0^2 10$	$.0^2 68$	.016	.027	.037	.046	.054	.061	.067	.072	.077	.001
	.005	$.0^4 50$	$.0^2 50$	.020	.038	.055	.069	.081	.091	.099	.106	.112	.118	.005
	.01	$.0^3 20$	.010	.032	.056	.075	.092	.105	.116	.125	.132	.139	.144	.01
	.025	$.0^3 13$	.026	.062	.094	.119	.138	.153	.165	.175	.183	.190	.196	.025
	.05	$.0^2 50$	.053	.105	.144	.173	.194	.211	.224	.235	.244	.251	.257	.05
	.10	.020	.111	.183	.231	.265	.289	.307	.321	.333	.342	.350	.356	.10
	.25	.133	.333	.439	.500	.540	.568	.588	.604	.616	.626	.633	.641	.25
	.50	.667	1.00	1.13	1.21	1.25	1.28	1.30	1.32	1.33	1.34	1.35	1.36	.50
	.75	2.57	3.00	3.15	3.23	3.28	3.31	3.34	3.35	3.37	3.38	3.39	3.39	.75
	.90	8.53	9.00	9.16	9.24	9.29	9.33	9.35	9.37	9.38	9.39	9.40	9.41	.90
	.95	18.5	19.0	19.2	19.2	19.3	19.3	19.4	19.4	19.4	19.4	19.4	19.4	.95
	.975	38.5	39.0	39.2	39.2	39.3	39.3	39.4	39.4	39.4	39.4	39.4	39.4	.975
	.99	98.5	99.0	99.2	99.2	99.3	99.3	99.4	99.4	99.4	99.4	99.4	99.4	.99
	.995	198	199	199	199	199	199	199	199	199	199	199	199	.995
	.999	998	999	999	999	999	999	999	999	999	999	999	999	.999
	.9995	200^1	200^1	200^1	200^1	200^1	200^1	200^1	200^1	200^1	200^1	200^1	200^1	.9995
3	.0005	$.0^6 46$	$.0^3 50$	$.0^2 44$	.012	.023	.033	.043	.052	.060	.067	.074	.079	.0005
	.0001	$.0^5 19$	$.0^2 10$	$.0^2 71$	.018	.030	.042	.053	.063	.072	.079	.086	.093	.001
	.005	$.0^4 46$	$.0^2 50$	.021	.041	.060	.077	.092	.104	.115	.124	.132	.138	.005
	.01	$.0^3 19$	.010	.034	.060	.083	.102	.118	.132	.143	.153	.161	.168	.01
	.025	$.0^3 12$	.026	.065	.100	.129	.152	.170	.185	.197	.207	.216	.224	.025
	.05	$.0^2 46$	.052	.108	.152	.185	.210	.230	.246	.259	.270	.279	.287	.05
	.10	.019	.109	.185	.239	.276	.304	.325	.342	.356	.367	.376	.384	.10
	.25	.122	.317	.424	.489	.531	.561	.582	.600	.613	.624	.633	.641	.25
	.50	.585	.881	1.00	1.06	1.10	1.13	1.15	1.16	1.17	1.18	1.19	1.20	.50
	.75	2.02	2.28	2.36	2.39	2.41	2.42	2.43	2.44	2.44	2.44	2.45	2.45	.75
	.90	5.54	5.46	5.39	5.34	5.31	5.28	5.27	5.25	5.24	5.23	5.22	5.22	.90
	.95	10.1	9.55	9.28	9.12	9.01	8.94	8.89	8.85	8.81	8.79	8.76	8.74	.95
	.975	17.4	16.0	15.4	15.1	14.9	14.7	14.6	14.5	14.5	14.4	14.4	14.3	.975
	.99	34.1	30.8	29.5	28.7	28.2	27.9	27.7	27.5	27.3	27.2	27.1	27.1	.99
	.995	55.6	49.8	47.5	46.2	45.4	44.8	44.4	44.1	43.9	43.7	43.5	43.4	.995
	.999	167	149	141	137	135	133	132	131	130	129	129	128	.999
	.9995	266	237	225	218	214	211	209	208	207	206	204	204	.9995

Read $.0^5 6$ as .00056, 200^1 as 2000, 162^4 as 1620000, etc.

ν_1 / p	15	20	24	30	40	50	60	100	120	200	500	∞	p	ν_2
.0005	.051	.058	.062	.066	.069	.072	.074	.077	.078	.080	.081	.083	.0005	1
.001	.060	.067	.071	.075	.079	.082	.084	.087	.088	.089	.091	.092	.001	
.005	.093	.101	.105	.109	.113	.116	.118	.121	.122	.124	.126	.127	.005	
.01	.115	.124	.128	.132	.137	.139	.141	.145	.146	.148	.150	.151	.01	
.025	.161	.170	.175	.180	.184	.187	.189	.193	.194	.196	.198	.199	.025	
.05	.220	.230	.235	.240	.245	.248	.250	.254	.255	.257	.259	.261	.05	
.10	.325	.336	.342	.347	.353	.356	.358	.362	.364	.366	.368	.370	.10	
.25	.698	.712	.719	.727	.734	.738	.741	.747	.749	.752	.754	.756	.25	
.50	2.09	2.12	2.13	2.15	2.16	2.17	2.17	2.18	2.18	2.19	2.19	2.20	.50	
.75	9.49	9.58	9.63	9.67	9.71	9.74	9.76	9.78	9.80	9.82	9.84	9.85	.75	
.90	61.2	61.7	62.0	62.3	62.5	62.7	62.8	63.0	63.1	63.2	63.3	63.3	.90	
.95	246	248	249	250	251	252	252	253	253	254	254	254	.95	
.975	985	993	997	100^1	101^1	101^1	101^1	101^1	101^1	102^1	102^1	102^1	.975	
.99	616^1	621^1	623^1	626^1	629^1	630^1	631^1	633^1	634^1	635^1	636^1	637^1	.99	
.995	246^2	248^2	249^2	250^2	251^2	252^2	253^2	253^2	254^2	254^2	254^2	255^2	.995	
.999	616^3	621^3	623^3	626^3	629^3	630^3	631^3	633^3	634^3	635^3	636^3	637^3	.999	
.9995	246^4	248^4	249^4	250^4	251^4	252^4	252^4	253^4	253^4	253^4	254^4	254^4	.9995	
.0005	.076	.088	.094	.101	.108	.113	.116	.122	.124	.127	.130	.132	.0005	2
.001	.088	.100	.107	.114	.121	.126	.129	.135	.137	.140	.143	.145	.001	
.005	.130	.143	.150	.157	.165	.169	.173	.179	.181	.184	.187	.189	.005	
.01	.157	.171	.178	.186	.193	.198	.201	.207	.209	.212	.215	.217	.01	
.025	.210	.224	.232	.239	.247	.251	.255	.261	.263	.266	.269	.271	.025	
.05	.272	.286	.294	.302	.309	.314	.317	.324	.326	.329	.332	.334	.05	
.10	.371	.386	.394	.402	.410	.415	.418	.424	.426	.429	.433	.434	.10	
.25	.657	.672	.680	.689	.697	.702	.705	.711	.713	.716	.719	.721	.25	
.50	1.38	1.39	1.40	1.41	1.42	1.42	1.43	1.43	1.43	1.44	1.44	1.44	.50	
.75	3.41	3.43	3.43	3.44	3.45	3.45	3.46	3.47	3.47	3.48	3.48	3.48	.75	
.90	9.42	9.44	9.45	9.46	9.47	9.47	9.47	9.48	9.48	9.49	9.49	9.49	.90	
.95	19.4	19.4	19.5	19.5	19.5	19.5	19.5	19.5	19.5	19.5	19.5	19.5	.95	
.975	39.4	39.4	39.5	39.5	39.5	39.5	39.5	39.5	39.5	39.5	39.5	39.5	.975	
.99	99.4	99.4	99.5	99.5	99.5	99.5	99.5	99.5	99.5	99.5	99.5	99.5	.99	
.995	199	199	199	199	199	199	199	199	199	199	199	200	.995	
.999	999	999	999	999	999	999	999	999	999	999	999	999	.999	
.9995	200^1	200^1	200^1	200^1	200^1	200^1	200^1	200^1	200^1	200^1	200^1	200^1	.9995	
.0005	.093	.109	.117	.127	.136	.143	.147	.156	.158	.162	.166	.169	.0005	3
.001	.107	.123	.132	.142	.152	.158	.162	.171	.173	.177	.181	.184	.001	
.005	.154	.172	.181	.191	.201	.207	.211	.220	.222	.227	.231	.234	.005	
.01	.185	.203	.212	.222	.232	.238	.242	.251	.253	.258	.262	.264	.01	
.025	.241	.259	.269	.279	.280	.295	.299	.308	.310	.314	.318	.321	.025	
.05	.304	.323	.332	.342	.352	.358	.363	.370	.373	.377	.382	.384	.05	
.10	.402	.420	.430	.439	.449	.455	.459	.467	.469	.474	.476	.480	.10	
.25	.658	.675	.684	.693	.702	.708	.711	.719	.721	.724	.728	.730	.25	
.50	1.21	1.23	1.23	1.24	1.25	1.25	1.25	1.26	1.26	1.26	1.27	1.27	.50	
.75	2.46	2.46	2.46	2.47	2.47	2.47	2.47	2.47	2.47	2.47	2.47	2.47	.75	
.90	5.20	5.18	5.18	5.17	5.16	5.15	5.15	5.14	5.14	5.14	5.14	5.13	.90	
.95	8.70	8.66	8.63	8.62	8.59	8.58	8.57	8.55	8.55	8.54	8.53	8.53	.95	
.975	14.3	14.2	14.1	14.1	14.0	14.0	14.0	14.0	13.9	13.9	13.9	13.9	.975	
.99	26.9	26.7	26.6	26.5	26.4	26.4	26.3	26.2	26.2	26.2	26.1	26.1	.99	
.995	43.1	42.8	42.6	42.5	42.3	42.2	42.1	42.0	42.0	41.9	41.9	41.8	.995	
.999	127	126	126	125	125	125	124	124	124	124	124	123	.999	
.9995	203	201	200	199	199	198	198	197	197	197	196	196	.9995	

ν_2	p	1	2	3	4	5	6	7	8	9	10	11	12	p
4	.0005	$.0^{6}44$	$.0^{5}50$	$.0^{2}46$	.013	.024	.036	.047	.057	.066	.075	.082	.089	.0005
	.001	$.0^{5}18$	$.0^{2}10$	$.0^{2}73$	.019	.032	.046	.058	.069	.079	.089	.097	.104	.001
	.005	$.0^{4}44$	$.0^{2}50$	.022	.043	.064	.083	.100	.114	.126	.137	.145	.153	.005
	.01	$.0^{3}18$	.010	.035	.063	.088	.109	.127	.143	.156	.167	.176	.185	.01
	.025	$.0^{2}11$	.026	.066	.104	.135	.161	.181	.198	.212	.224	.234	.243	.025
	.05	$.0^{2}44$	.052	.110	.157	.193	.221	.243	.261	.275	.288	.298	.307	.05
	.10	.018	.108	.187	.243	.284	.314	.338	.356	.371	.384	.394	.403	.10
	.25	.117	.309	.418	.484	.528	.560	.583	.601	.615	.627	.637	.645	.25
	.50	.549	.828	.941	1.00	1.04	1.06	1.08	1.09	1.10	1.11	1.12	1.13	.50
	.75	1.81	2.00	2.05	2.06	2.07	2.08	2.08	2.08	2.08	2.08	2.08	2.08	.75
	.90	4.54	4.32	4.19	4.11	4.05	4.01	3.98	3.95	3.94	3.92	3.91	3.90	.90
	.95	7.71	6.94	6.59	6.39	6.26	6.16	6.09	6.04	6.00	5.96	5.94	5.91	.95
	.975	12.2	10.6	9.98	9.60	9.36	9.20	9.07	8.98	8.90	8.84	8.79	8.75	.975
	.99	21.2	18.0	16.7	16.0	15.5	15.2	15.0	14.8	14.7	14.5	14.4	14.4	.99
	.995	31.3	26.3	24.3	23.2	22.5	22.0	21.6	21.4	21.1	21.0	20.8	20.7	.995
	.999	74.1	61.2	56.2	53.4	51.7	50.5	49.7	49.0	48.5	48.0	47.7	47.4	.999
	.9995	106	87.4	80.1	76.1	73.6	71.9	70.6	69.7	68.9	68.3	67.8	67.4	.9995
5	.0005	$.0^{6}43$	$.0^{5}50$	$.0^{2}47$	.014	.025	.038	.050	.061	.070	.081	.089	.096	.0005
	.001	$.0^{5}17$	$.0^{2}10$	$.0^{2}75$	.019	.034	.048	.062	.074	.085	.095	.104	.112	.001
	.005	$.0^{4}43$	$.0^{2}50$	.022	.045	.067	.087	.105	.120	.134	.146	.156	.165	.005
	.01	$.0^{3}17$	.010	.035	.064	.091	.114	.134	.151	.165	.177	.188	.197	.01
	.025	$.0^{2}11$	.025	.067	.107	.140	.167	.189	.208	.223	.236	.248	.257	.025
	.05	$.0^{2}43$	.052	.111	.160	.198	.228	.252	.271	.287	.301	.313	.322	.05
	.10	.017	.108	.188	.247	.290	.322	.347	.367	.383	.397	.408	.418	.10
	.25	.113	.305	.415	.483	.528	.560	.584	.604	.618	.631	.641	.650	.25
	.50	.528	.799	.907	.965	1.00	1.02	1.04	1.05	1.06	1.07	1.08	1.09	.50
	.75	1.69	1.85	1.88	1.89	1.89	1.89	1.89	1.89	1.89	1.89	1.89	1.89	.75
	.90	4.06	3.78	3.62	3.52	3.45	3.40	3.37	3.34	3.32	3.30	3.28	3.27	.90
	.95	6.61	5.79	5.41	5.19	5.05	4.95	4.88	4.82	4.77	4.74	4.71	4.68	.95
	.975	10.0	8.43	7.76	7.39	7.15	6.98	6.85	6.76	6.68	6.62	6.57	6.52	.975
	.99	16.3	13.3	12.1	11.4	11.0	10.7	10.5	10.3	10.2	10.1	9.96	9.89	.99
	.995	22.8	18.3	16.5	15.6	14.9	14.5	14.2	14.0	13.8	13.6	13.5	13.4	.995
	.999	47.2	37.1	33.2	31.1	29.7	28.8	28.2	27.6	27.2	26.9	26.6	26.4	.999
	.9995	63.6	49.8	44.4	41.5	39.7	38.5	37.6	36.9	36.4	35.9	35.6	35.2	.9995
6	.0005	$.0^{6}43$	$.0^{5}50$	$.0^{2}47$	.014	.026	.039	.052	.064	.075	.085	.094	.103	.0005
	.001	$.0^{5}17$	$.0^{2}10$	$.0^{2}75$	.020	.035	.050	.064	.078	.090	.101	.111	.119	.001
	.005	$.0^{4}43$	$.0^{2}50$	.022	.045	.069	.090	.109	.126	.140	.153	.164	.174	.005
	.01	$.0^{3}17$	.010	.036	.066	.094	.118	.139	.157	.172	.186	.197	.207	.01
	.025	$.0^{2}11$	.025	.068	.109	.143	.172	.195	.215	.231	.246	.258	.268	.025
	.05	$.0^{2}43$	.052	.112	.162	.202	.233	.259	.279	.296	.311	.324	.334	.05
	.10	.017	.107	.189	.249	.294	.327	.354	.375	.392	.406	.418	.429	.10
	.25	.111	.302	.413	.481	.524	.561	.586	.606	.622	.635	.645	.654	.25
	.50	.515	.780	.886	.942	.977	1.00	1.02	1.03	1.04	1.05	1.05	1.06	.50
	.75	1.62	1.76	1.78	1.79	1.79	1.78	1.78	1.78	1.77	1.77	1.77	1.77	.75
	.90	3.78	3.46	3.29	3.18	3.11	3.05	3.01	2.98	2.96	2.94	2.92	2.90	.90
	.95	5.99	5.14	4.76	4.53	4.39	4.28	4.21	4.15	4.10	4.06	4.03	4.00	.95
	.975	8.81	7.26	6.60	6.23	5.99	5.82	5.70	5.60	5.52	5.46	5.41	5.37	.975
	.99	13.7	10.9	9.78	9.15	8.75	8.47	8.26	8.10	7.98	7.87	7.79	7.72	.99
	.995	18.6	14.5	12.9	12.0	11.5	11.1	10.8	10.6	10.4	10.2	10.1	10.0	.995
	.999	35.5	27.0	23.7	21.9	20.8	20.0	19.5	19.0	18.7	18.4	18.2	18.0	.999
	.9995	46.1	34.8	30.4	28.1	26.6	25.6	24.9	24.3	23.9	23.5	23.2	23.0	.9995

p \\ ν_1	15	20	24	30	40	50	60	100	120	200	500	∞	p	ν_2
.0005	.105	.125	.135	.147	.159	.166	.172	.183	.186	.191	.196	.200	.0005	4
.001	.121	.141	.152	.163	.176	.183	.188	.200	.202	.208	.213	.217	.001	
.005	.172	.193	.204	.216	.229	.237	.242	.253	.255	.260	.266	.269	.005	
.01	.204	.226	.237	.249	.261	.269	.274	.285	.287	.293	.298	.301	.01	
.025	.263	.284	.296	.308	.320	.327	.332	.342	.346	.351	.356	.359	.025	
.05	.327	.349	.360	.372	.384	.391	.396	.407	.409	.413	.418	.422	.05	
.10	.424	.445	.456	.467	.478	.485	.490	.500	.502	.508	.510	.514	.10	
.25	.664	.683	.692	.702	.712	.718	.722	.731	.733	.737	.740	.743	.25	
.50	1.14	1.15	1.16	1.16	1.17	1.18	1.18	1.18	1.18	1.19	1.19	1.19	.50	
.75	2.08	2.08	2.08	2.08	2.08	2.08	2.08	2.08	2.08	2.08	2.08	2.08	.75	
.90	3.87	3.84	3.83	3.82	3.80	3.80	3.79	3.78	3.78	3.77	3.76	3.76	.90	
.95	5.86	5.80	5.77	5.75	5.72	5.70	5.69	5.66	5.66	5.65	5.64	5.63	.95	
.975	8.66	8.56	8.51	8.46	8.41	8.38	8.36	8.32	8.31	8.29	8.27	8.26	.975	
.99	14.2	14.0	13.9	13.8	13.7	13.7	13.7	13.6	13.6	13.5	13.5	13.5	.99	
.995	20.4	20.2	20.0	19.9	19.8	19.7	19.6	19.5	19.5	19.4	19.4	19.3	.995	
.999	46.8	46.1	45.8	45.4	45.1	44.9	44.7	44.5	44.4	44.3	44.1	44.0	.999	
.9995	66.5	65.5	65.1	64.6	64.1	63.8	63.6	63.2	63.1	62.9	62.7	62.6	.9995	
.0005	.115	.137	.150	.163	.177	.186	.192	.205	.209	.216	.222	.226	.0005	5
.001	.132	.155	.167	.181	.195	.204	.210	.223	.227	.233	.239	.244	.001	
.005	.186	.210	.223	.237	.251	.260	.266	.279	.282	.288	.294	.299	.005	
.01	.219	.244	.257	.270	.285	.293	.299	.312	.315	.322	.328	.331	.01	
.025	.280	.304	.317	.330	.344	.353	.359	.370	.374	.380	.386	.390	.025	
.05	.345	.369	.382	.395	.408	.417	.422	.432	.437	.442	.448	.452	.05	
.10	.440	.463	.476	.488	.501	.508	.514	.524	.527	.532	.538	.541	.10	
.25	.669	.690	.700	.711	.722	.728	.732	.741	.743	.748	.752	.755	.25	
.50	1.10	1.11	1.12	1.12	1.13	1.13	1.14	1.14	1.14	1.15	1.15	1.15	.50	
.75	1.89	1.88	1.88	1.88	1.88	1.88	1.87	1.87	1.87	1.87	1.87	1.87	.75	
.90	3.24	3.21	3.19	3.17	3.16	3.15	3.14	3.13	3.12	3.12	3.11	3.10	.90	
.95	4.62	4.56	4.53	4.50	4.46	4.44	4.43	4.41	4.40	4.39	4.37	4.36	.95	
.975	6.43	6.33	6.28	6.23	6.18	6.14	6.12	6.08	6.07	6.05	6.03	6.02	.975	
.99	9.72	9.55	9.47	9.38	9.29	9.24	9.20	9.13	9.11	9.08	9.04	9.02	.99	
.995	13.1	12.9	12.8	12.7	12.5	12.5	12.4	12.3	12.3	12.2	12.2	12.1	.995	
.999	25.9	25.4	25.1	24.9	24.6	24.4	24.3	24.1	24.1	23.9	23.8	23.8	.999	
.9995	34.6	33.9	33.5	33.1	32.7	32.5	32.3	32.1	32.0	31.8	31.7	31.6	.9995	
.0005	.123	.148	.162	.177	.193	.203	.210	.225	.229	.236	.244	.249	.0005	6
.001	.141	.166	.180	.195	.211	.222	.229	.243	.247	.255	.262	.267	.001	
.005	.197	.224	.238	.253	.269	.279	.286	.301	.304	.312	.318	.324	.005	
.01	.232	.258	.273	.288	.304	.313	.321	.334	.338	.346	.352	.357	.01	
.025	.293	.320	.334	.349	.364	.375	.381	.394	.398	.405	.412	.415	.025	
.05	.358	.385	.399	.413	.428	.437	.444	.457	.460	.467	.472	.476	.05	
.10	.453	.478	.491	.505	.519	.526	.533	.546	.548	.556	.559	.564	.10	
.25	.675	.696	.707	.718	.729	.736	.741	.751	.753	.758	.762	.765	.25	
.50	1.07	1.08	1.09	1.10	1.10	1.11	1.11	1.11	1.12	1.12	1.12	1.12	.50	
.75	1.76	1.76	1.75	1.75	1.75	1.75	1.74	1.74	1.74	1.74	1.74	1.74	.75	
.90	2.87	2.84	2.82	2.80	2.78	2.77	2.76	2.75	2.74	2.73	2.73	2.72	.90	
.95	3.94	3.87	3.84	3.81	3.77	3.75	3.74	3.71	3.70	3.69	3.68	3.67	.95	
.975	5.27	5.17	5.12	5.07	5.01	4.98	4.96	4.92	4.90	4.88	4.86	4.85	.975	
.99	7.56	7.40	7.31	7.23	7.14	7.09	7.06	6.99	6.97	6.93	6.90	6.88	.99	
.995	9.81	9.59	9.47	9.36	9.24	9.17	9.12	9.03	9.00	8.95	8.91	8.88	.995	
.999	17.6	17.1	16.9	16.7	16.4	16.3	16.2	16.0	16.0	15.9	15.8	15.7	.999	
.9995	22.4	21.9	21.7	21.4	21.1	20.9	20.7	20.5	20.4	20.3	20.2	20.1	.9995	

Appendix B9 (Continued)

ν_2	p \ ν_1	1	2	3	4	5	6	7	8	9	10	11	12	p
7	.0005	$.0^642$	$.0^550$	$.0^248$	.014	.027	.040	.053	.066	.078	.088	.099	.108	.0005
	.001	$.0^517$	$.0^210$	$.0^276$	.020	.035	.051	.067	.081	.093	.105	.115	.125	.001
	.005	$.0^442$	$.0^250$	.023	.046	.070	.093	.113	.130	.145	.159	.171	.181	.005
	.01	$.0^317$	.010	.036	.067	.096	.121	.143	.162	.178	.192	.205	.216	.01
	.025	$.0^210$	.025	.068	.110	.146	.176	.200	.221	.238	.253	.266	.277	.025
	.05	$.0^242$	.052	.113	.164	.205	.238	.264	.286	.304	.319	.332	.343	.05
	.10	.017	.107	.190	.251	.297	.332	.359	.381	.399	.414	.427	.438	.10
	.25	.110	.300	.412	.481	.528	.562	.588	.608	.624	.637	.649	.658	.25
	.50	.506	.767	.871	.926	.960	.983	1.00	1.01	1.02	1.03	1.04	1.04	.50
	.75	1.57	1.70	1.72	1.72	1.71	1.71	1.70	1.70	1.69	1.69	1.69	1.68	.75
	.90	3.59	3.26	3.07	2.96	2.88	2.83	2.78	2.75	2.72	2.70	2.68	2.67	.90
	.95	5.59	4.74	4.35	4.12	3.97	3.87	3.79	3.73	3.68	3.64	3.60	3.57	.95
	.975	8.07	6.54	5.89	5.52	5.29	5.12	4.99	4.90	4.82	4.76	4.71	4.67	.975
	.99	12.2	9.55	8.45	7.85	7.46	7.19	6.99	6.84	6.72	6.62	6.54	6.47	.99
	.995	16.2	12.4	10.9	10.0	9.52	9.16	8.89	8.68	8.51	8.38	8.27	8.18	.995
	.999	29.2	21.7	18.8	17.2	16.2	15.5	15.0	14.6	14.3	14.1	13.9	13.7	.999
	.9995	37.0	27.2	23.5	21.4	20.2	19.3	18.7	18.2	17.8	17.5	17.2	17.0	.9995
8	.0005	$.0^642$	$.0^550$	$.0^248$	.014	.027	.041	.055	.068	.081	.092	.102	.112	.0005
	.001	$.0^517$	$.0^210$	$.0^276$	.020	.036	.053	.068	.083	.096	.109	.120	.130	.001
	.005	$.0^442$	$.0^250$	.027	.047	.072	.095	.115	.133	.149	.164	.176	.187	.005
	.01	$.0^317$	.010	.036	.068	.097	.123	.146	.166	.183	.198	.211	.222	.01
	.025	$.0^210$	.025	.069	.111	.148	.179	.204	.226	.244	.259	.273	.285	.025
	.05	$.0^242$	.052	.113	.166	.208	.241	.268	.291	.310	.326	.339	.351	.05
	.10	.017	.107	.190	.253	.299	.335	.363	.386	.405	.421	.435	.445	.10
	.25	.109	.298	.411	.481	.529	.563	.589	.610	.627	.640	.654	.661	.25
	.50	.499	.757	.860	.915	.948	.971	.988	1.00	1.01	1.02	1.02	1.03	.50
	.75	1.54	1.66	1.67	1.66	1.66	1.65	1.64	1.64	1.64	1.63	1.63	1.62	.75
	.90	3.46	3.11	2.92	2.81	2.73	2.67	2.62	2.59	2.56	2.54	2.52	2.50	.90
	.95	5.32	4.46	4.07	3.84	3.69	3.58	3.50	3.44	3.39	3.35	3.31	3.28	.95
	.975	7.57	6.06	5.42	5.05	4.82	4.65	4.53	4.43	4.36	4.30	4.24	4.20	.975
	.99	11.3	8.65	7.59	7.01	6.63	6.37	6.18	6.03	5.91	5.81	5.73	5.67	.99
	.995	14.7	11.0	9.60	8.81	8.30	7.95	7.69	7.50	7.34	7.21	7.10	7.01	.995
	.999	25.4	18.5	15.8	14.4	13.5	12.9	12.4	12.0	11.8	11.5	11.4	11.2	.999
	.9995	31.6	22.8	19.4	17.6	16.4	15.7	15.1	14.6	14.3	14.0	13.8	13.6	.9995
9	.0005	$.0^641$	$.0^550$	$.0^248$	.015	.027	.042	.056	.070	.083	.094	.105	.115	.0005
	.001	$.0^517$	$.0^210$	$.0^277$	.021	.037	.054	.070	.085	.099	.112	.123	.134	.001
	.005	$.0^442$	$.0^250$	.023	.047	.073	.096	.117	.136	.153	.168	.181	.192	.005
	.01	$.0^317$	.010	.037	.068	.098	.125	.149	.169	.187	.202	.216	.228	.01
	.025	$.0^210$	.025	.069	.112	.150	.181	.207	.230	.248	.265	.279	.291	.025
	.05	$.0^240$	.052	.113	.167	.210	.244	.272	.296	.315	.331	.345	.358	.05
	.10	.017	.107	.191	.254	.302	.338	.367	.390	.410	.426	.441	.452	.10
	.25	.108	.297	.410	.480	.529	.564	.591	.612	.629	.643	.654	.664	.25
	.50	.494	.749	.852	.906	.939	.962	.978	.990	1.00	1.01	1.01	1.02	.50
	.75	1.51	1.62	1.63	1.63	1.62	1.61	1.60	1.60	1.59	1.59	1.58	1.58	.75
	.90	3.36	3.01	2.81	2.69	2.61	2.55	2.51	2.47	2.44	2.42	2.40	2.38	.90
	.95	5.12	4.26	3.86	3.63	3.48	3.37	3.29	3.23	3.18	3.14	3.10	3.07	.95
	.975	7.21	5.71	5.08	4.72	4.48	4.32	4.20	4.10	4.03	3.96	3.91	3.87	.975
	.99	10.6	8.02	6.99	6.42	6.06	5.80	5.61	5.47	5.35	5.26	5.18	5.11	.99
	.995	13.6	10.1	8.72	7.96	7.47	7.13	6.88	6.69	6.54	6.42	6.31	6.23	.995
	.999	22.9	16.4	13.9	12.6	11.7	11.1	10.7	10.4	10.1	9.89	9.71	9.57	.999
	.9995	28.0	19.9	16.8	15.1	14.1	13.3	12.8	12.4	12.1	11.8	11.6	11.4	.9995

p \ ν_1	15	20	24	30	40	50	60	100	120	200	500	∞	p	ν_2
.0005	.130	.157	.172	.188	.206	.217	.225	.242	.246	.255	.263	.268	.0005	7
.001	.148	.176	.191	.208	.225	.237	.245	.261	.266	.274	.282	.288	.001	
.005	.206	.235	.251	.267	.285	.296	.304	.319	.324	.332	.340	.345	.005	
.01	.241	.270	.286	.303	.320	.331	.339	.355	.358	.366	.373	.379	.01	
.025	.304	.333	.348	.364	.381	.392	.399	.413	.418	.426	.433	.437	.025	
.05	.369	.398	.413	.428	.445	.455	.461	.476	.479	.485	.493	.498	.05	
.10	.463	.491	.504	.519	.534	.543	.550	.562	.566	.571	.578	.582	.10	
.25	.679	.702	.713	.725	.737	.745	.749	.760	.762	.767	.772	.775	.25	
.50	1.05	1.07	1.07	1.08	1.08	1.09	1.09	1.10	1.10	1.10	1.10	1.10	.50	
.75	1.68	1.67	1.67	1.66	1.66	1.66	1.65	1.65	1.65	1.65	1.65	1.65	.75	
.90	2.63	2.59	2.58	2.56	2.54	2.52	2.51	2.50	2.49	2.48	2.48	2.47	.90	
.95	3.51	3.44	3.41	3.38	3.34	3.32	3.30	3.27	3.27	3.25	3.24	3.23	.95	
.975	4.57	4.47	4.42	4.36	4.31	4.28	4.25	4.21	4.20	4.18	4.16	4.14	.975	
.99	6.31	6.16	6.07	5.99	5.91	5.86	5.82	5.75	5.74	5.70	5.67	5.65	.99	
.995	7.97	7.75	7.65	7.53	7.42	7.35	7.31	7.22	7.19	7.15	7.10	7.08	.995	
.999	13.3	12.9	12.7	12.5	12.3	12.2	12.1	11.9	11.9	11.8	11.7	11.7	.999	
.9995	16.5	16.0	15.7	15.5	15.2	15.1	15.0	14.7	14.7	14.6	14.5	14.4	.9995	
.0005	.136	.164	.181	.198	.218	.230	.239	.257	.262	.271	.281	.287	.0005	8
.001	.155	.184	.200	.218	.238	.250	.259	.277	.282	.292	.300	.306	.001	
.005	.214	.244	.261	.279	.299	.311	.319	.337	.341	.351	.358	.364	.005	
.01	.250	.281	.297	.315	.334	.346	.354	.372	.376	.385	.392	.398	.01	
.025	.313	.343	.360	.377	.395	.407	.415	.431	.435	.442	.450	.456	.025	
.05	.379	.409	.425	.441	.459	.469	.477	.493	.496	.505	.510	.516	.05	
.10	.472	.500	.515	.531	.547	.556	.563	.578	.581	.588	.595	.599	.10	
.25	.684	.707	.718	.730	.743	.751	.756	.767	.769	.775	.780	.783	.25	
.50	1.04	1.05	1.06	1.07	1.07	1.07	1.08	1.08	1.08	1.09	1.09	1.09	.50	
.75	1.62	1.61	1.60	1.60	1.59	1.59	1.59	1.58	1.58	1.58	1.58	1.58	.75	
.90	2.46	2.42	2.40	2.38	2.36	2.35	2.34	2.32	2.32	2.31	2.30	2.29	.90	
.95	3.22	3.15	3.12	3.08	3.04	3.02	3.01	2.97	2.97	2.95	2.94	2.93	.95	
.975	4.10	4.00	3.95	3.89	3.84	3.81	3.78	3.74	3.73	3.70	3.68	3.67	.975	
.99	5.52	5.36	5.28	5.20	5.12	5.07	5.03	4.96	4.95	4.91	4.88	4.86	.99	
.995	6.81	6.61	6.50	6.40	6.29	6.22	6.18	6.09	6.06	6.02	5.98	5.95	.995	
.999	10.8	10.5	10.3	10.1	9.92	9.80	9.73	9.57	9.54	9.46	9.39	9.34	.999	
.9995	13.1	12.7	12.5	12.2	12.0	11.8	11.8	11.6	11.5	11.4	11.4	11.3	.9995	
.0005	.141	.171	.188	.207	.228	.242	.251	.270	.276	.287	.297	.303	.0005	9
.001	.160	.191	.208	.228	.249	.262	.271	.291	.296	.307	.316	.323	.001	
.005	.220	.253	.271	.290	.310	.324	.332	.351	.356	366	.376	.382	.005	
.01	.257	.289	.307	.326	.346	.358	.368	.386	.391	.400	.410	.415	.01	
.025	.320	.352	.370	.388	.408	.420	.428	.446	.450	.459	.467	.473	.025	
.05	.386	.418	.435	.452	.471	.483	.490	.508	.510	.518	.526	.532	.05	
.10	.479	.509	.525	.541	.558	.568	.575	.588	.594	.602	.610	.613	.10	
.25	.687	.711	.723	.736	.749	.757	.762	.773	.776	.782	.787	.791	.25	
.50	1.03	1.04	1.05	1.05	1.06	1.06	1.07	1.07	1.07	1.08	1.08	1.08	.50	
.75	1.57	1.56	1.56	1.55	1.55	1.54	1.54	1.53	1.53	1.53	1.53	1.53	.75	
.90	2.34	2.30	2.28	2.25	2.23	2.22	2.21	2.19	2.18	2.17	2.17	2.16	.90	
.95	3.01	2.94	2.90	2.86	2.83	2.80	2.79	2.76	2.75	2.73	2.72	2.71	.95	
.975	3.77	3.67	3.61	3.56	3.51	3.47	3.45	3.40	3.39	3.37	3.35	3.33	.975	
.99	4.96	4.81	4.73	4.65	4.57	4.52	4.48	4.42	4.40	4.36	4.33	4.31	.99	
.995	6.03	5.83	5.73	5.62	5.52	5.45	5.41	5.32	5.30	5.26	5.21	5.19	.995	
.999	9.24	8.90	8.72	8.55	8.37	8.26	8.19	8.04	8.00	7.93	7.86	7.81	.999	
.9995	11.0	10.6	10.4	10.2	9.94	9.80	9.71	9.53	9.49	9.40	9.32	9.26	.9995	

ν_2	ν_1 / p	1	2	3	4	5	6	7	8	9	10	11	12	p
10	.0005	$.0^641$	$.0^550$	$.0^249$	.015	.028	.043	.057	.071	.085	.097	.108	.119	.0005
	.001	$.0^617$	$.0^210$	$.0^277$	.021	.037	.054	.071	.087	.101	.114	.126	.137	.001
	.005	$.0^441$	$.0^250$	.023	.048	.073	.098	.119	.139	.156	.171	.185	.197	.005
	.01	$.0^417$	.010	.037	.069	.100	.127	.151	.172	.190	.206	.220	.233	.01
	.025	$.0^210$	.025	.069	.113	.151	.183	.210	.233	.252	.269	.283	.296	.025
	.05	$.0^241$	.052	.114	.168	.211	.246	.275	.299	.319	.336	.351	.363	.05
	.10	.017	.106	.191	.255	.303	.340	.370	.394	.414	.430	.444	.457	.10
	.25	.107	.296	.409	.480	.529	.565	.592	.613	.631	.645	.657	.667	.25
	.50	.490	.743	.845	.899	.932	.954	.971	.983	.992	1.00	1.01	1.01	.50
	.75	1.49	1.60	1.60	1.59	1.59	1.58	1.57	1.56	1.56	1.55	1.55	1.54	.75
	.90	3.28	2.92	2.73	2.61	2.52	2.46	2.41	2.38	2.35	2.32	2.30	2.28	.90
	.95	4.96	4.10	3.71	3.48	3.33	3.22	3.14	3.07	3.02	2.98	2.94	2.91	.95
	.975	6.94	5.46	4.83	4.47	4.24	4.07	3.95	3.85	3.78	3.72	3.66	3.62	.975
	.99	10.0	7.56	6.55	5.99	5.64	5.39	5.20	5.06	4.94	4.85	4.77	4.71	.99
	.995	12.8	9.43	8.08	7.34	6.87	6.54	6.30	6.12	5.97	5.85	5.75	5.66	.995
	.999	21.0	14.9	12.6	11.3	10.5	9.92	9.52	9.20	8.96	8.75	8.58	8.44	.999
	.9995	25.5	17.9	15.0	13.4	12.4	11.8	11.3	10.9	10.6	10.3	10.1	9.93	.9995
11	.0005	$.0^641$	$.0^550$	$.0^249$	.015	.028	.043	.058	.072	.086	.099	.111	.121	.0005
	.001	$.0^616$	$.0^210$	$.0^278$	.021	.038	.055	.072	.088	.103	.116	.129	.140	.001
	.005	$.0^440$	$.0^250$	.023	.048	.074	.099	.121	.141	.158	.174	.188	.200	.005
	.01	$.0^316$	.010	.037	.069	.100	.128	.153	.175	.193	.210	.224	.237	.01
	.025	$.0^210$	.025	.069	.114	.152	.185	.212	.236	.256	.273	.288	.301	.025
	.05	$.0^241$	.052	.114	.168	.212	.248	.278	.302	.323	.340	.355	.368	.05
	.10	.017	.106	.192	.256	.305	.342	.373	.397	.417	.435	.448	.461	.10
	.25	.107	.295	.408	.481	.529	.565	.592	.614	.633	.645	.658	.667	.25
	.50	.486	.739	.840	.893	.926	.948	.964	.977	.986	.994	1.00	1.01	.50
	.75	1.47	1.58	1.58	1.57	1.56	1.55	1.54	1.53	1.53	1.52	1.52	1.51	.75
	.90	3.23	2.86	2.66	2.54	2.45	2.39	2.34	2.30	2.27	2.25	2.23	2.21	.90
	.95	4.84	3.98	3.59	3.36	3.20	3.09	3.01	2.95	2.90	2.85	2.82	2.79	.95
	.975	6.72	5.26	4.63	4.28	4.04	3.88	3.76	3.66	3.59	3.53	3.47	3.43	.975
	.99	9.65	7.21	6.22	5.67	5.32	5.07	4.89	4.74	4.63	4.54	4.46	4.40	.99
	.995	12.2	8.91	7.60	6.88	6.42	6.10	5.86	5.68	5.54	5.42	5.32	5.24	.995
	.999	19.7	13.8	11.6	10.3	9.58	9.05	8.66	8.35	8.12	7.92	7.76	7.62	.999
	.9995	23.6	16.4	13.6	12.2	11.2	10.6	10.1	9.76	9.48	9.24	9.04	8.88	.9995
12	.0005	$.0^641$	$.0^550$	$.0^249$	.015	.028	.044	.058	.073	.087	.101	.113	.124	.0005
	.001	$.0^616$	$.0^210$	$.0^278$	.021	.038	.056	.073	.089	.104	.118	.131	.143	.001
	.005	$.0^439$	$.0^250$	.023	.048	.075	.100	.122	.143	.161	.177	.191	.204	.005
	.01	$.0^316$	.010	.037	.070	.101	.130	.155	.176	.196	.212	.227	.241	.01
	.025	$.0^210$	.025	.070	.114	.153	.186	.214	.238	.259	.276	.292	.305	.025
	.05	$.0^241$	.052	.114	.169	.214	.250	.280	.305	.325	.343	.358	.372	.05
	.10	.016	.106	.192	.257	.306	.344	.375	.400	.420	.438	.452	.466	.10
	.25	.106	.295	.408	.480	.530	.566	.594	.616	.633	.649	.662	.671	.25
	.50	.484	.735	.835	.888	.921	.943	.959	.972	.981	.989	.995	1.00	.50
	.75	1.46	1.56	1.56	1.55	1.54	1.53	1.52	1.51	1.51	1.50	1.50	1.49	.75
	.90	3.18	2.81	2.61	2.48	2.39	2.33	2.28	2.24	2.21	2.19	2.17	2.15	.90
	.95	4.75	3.89	3.49	3.26	3.11	3.00	2.91	2.85	2.80	2.75	2.72	2.69	.95
	.975	6.55	5.10	4.47	4.12	3.89	3.73	3.61	3.51	3.44	3.37	3.32	3.28	.975
	.99	9.33	6.93	5.95	5.41	5.06	4.82	4.64	4.50	4.39	4.30	4.22	4.16	.99
	.995	11.8	8.51	7.23	6.52	6.07	5.76	5.52	5.35	5.20	5.09	4.99	4.91	.995
	.999	18.6	13.0	10.8	9.63	8.89	8.38	8.00	7.71	7.48	7.29	7.14	7.01	.999
	.9995	22.2	15.3	12.7	11.2	10.4	9.74	9.28	8.94	8.66	8.43	8.24	8.08	.9995

p / ν_1	15	20	24	30	40	50	60	100	120	200	500	∞	p	ν_2
.0005	.145	.177	.195	.215	.238	.251	.262	.282	.288	.299	.311	.319	.0005	10
.001	.164	.197	.216	.236	.258	.272	.282	.303	.309	.321	.331	.338	.001	
.005	.226	.260	.279	.299	.321	.334	.344	.365	.370	.380	.391	.397	.005	
.01	.263	.297	.316	.336	.357	.370	.380	.400	.405	.415	.424	.431	.01	
.025	.327	.360	.379	.398	.419	.431	.441	.459	.464	.474	.483	.488	.025	
.05	.393	.426	.444	.462	.481	.493	.502	.518	.523	.532	.541	.546	.05	
.10	.486	.516	.532	.549	.567	.578	.586	.602	.605	.614	.621	.625	.10	
.25	.691	.714	.727	.740	.754	.762	.767	.779	.782	.788	.793	.797	.25	
.50	1.02	1.03	1.04	1.05	1.05	1.06	1.06	1.06	1.06	1.07	1.07	1.07	.50	
.75	1.53	1.52	1.52	1.51	1.51	1.50	1.50	1.49	1.49	1.49	1.48	1.48	.75	
.90	2.24	2.20	2.18	2.16	2.13	2.12	2.11	2.09	2.08	2.07	2.06	2.06	.90	
.95	2.85	2.77	2.74	2.70	2.66	2.64	2.62	2.59	2.58	2.56	2.55	2.54	.95	
.975	3.52	3.42	3.37	3.31	3.26	3.22	3.20	3.15	3.14	3.12	3.09	3.08	.975	
.99	4.56	4.41	4.33	4.25	4.17	4.12	4.08	4.01	4.00	3.96	3.93	3.91	.99	
.995	5.47	5.27	5.17	5.07	4.97	4.90	4.86	4.77	4.75	4.71	4.67	4.64	.995	
.999	8.13	7.80	7.64	7.47	7.30	7.19	7.12	6.98	6.94	6.87	6.81	6.76	.999	
.9995	9.56	9.16	8.96	8.75	8.54	8.42	8.33	8.16	8.12	8.04	7.96	7.90	.9995	
.0005	.148	.182	.201	.222	.246	.261	.271	.293	.299	.312	.324	.331	.0005	11
.001	.168	.202	.222	.243	.266	.282	.292	.313	.320	.332	.343	.353	.001	
.005	.231	.266	.286	.308	.330	.345	.355	.376	.382	.394	.403	.412	.005	
.01	.268	.304	.324	.344	.366	.380	.391	.412	.417	.427	.439	.444	.01	
.025	.332	.368	.386	.407	.429	.442	.450	.472	.476	.485	.495	.503	.025	
.05	.398	.433	.452	.469	.490	.503	.513	.529	.535	.543	.552	.559	.05	
.10	.490	.524	.541	.559	.578	.588	.595	.614	.617	.625	.633	.637	.10	
.25	.694	.719	.730	.744	.758	.767	.773	.780	.788	.794	.799	.803	.25	
.50	1.02	1.03	1.03	1.04	1.05	1.05	1.05	1.06	1.06	1.06	1.06	1.06	.50	
.75	1.50	1.49	1.49	1.48	1.47	1.47	1.47	1.46	1.46	1.46	1.45	1.45	.75	
.90	2.17	2.12	2.10	2.08	2.05	2.04	2.03	2.00	2.00	1.99	1.98	1.97	.90	
.95	2.72	2.65	2.61	2.57	2.53	2.51	2.49	2.46	2.45	2.43	2.42	2.40	.95	
.975	3.33	3.23	3.17	3.12	3.06	3.03	3.00	2.96	2.94	2.92	2.90	2.88	.975	
.99	4.25	4.10	4.02	3.94	3.86	3.81	3.78	3.71	3.69	3.66	3.62	3.60	.99	
.995	5.05	4.86	4.76	4.65	4.55	4.49	4.45	4.36	4.34	4.29	4.25	4.23	.995	
.999	7.32	7.01	6.85	6.68	6.52	6.41	6.35	6.21	6.17	6.10	6.04	6.00	.999	
.9995	8.52	8.14	7.94	7.75	7.55	7.43	7.35	7.18	7.14	7.06	6.98	6.93	.9995	
.0005	.152	.186	.206	.228	.253	.269	.280	.305	.311	.323	.337	.345	.0005	12
.001	.172	.207	.228	.250	.275	.291	.302	.326	.332	.344	.357	.365	.001	
.005	.235	.272	.292	.315	.339	.355	.365	.388	.393	.405	.417	.424	.005	
.01	.273	.310	.330	.352	.375	.391	.401	.422	.428	.441	.450	.458	.01	
.025	.337	.374	.394	.416	.437	.450	.461	.481	.487	.498	.508	.514	.025	
.05	.404	.439	.458	.478	.499	.513	.522	.541	.545	.556	.565	.571	.05	
.10	.496	.528	.546	.564	.583	.595	.604	.621	.625	.633	.641	.647	.10	
.25	.695	.721	.734	.748	.762	.771	.777	.789	.792	.799	.804	.808	.25	
.50	1.01	1.02	1.03	1.03	1.04	1.04	1.05	1.05	1.05	1.05	1.06	1.06	.50	
.75	1.48	1.47	1.46	1.45	1.45	1.44	1.44	1.43	1.43	1.43	1.42	1.42	.75	
.90	2.11	2.06	2.04	2.01	1.99	1.97	1.96	1.94	1.93	1.92	1.91	1.90	.90	
.95	2.62	2.54	2.51	2.47	2.43	2.40	2.38	2.35	2.34	2.32	2.31	2.30	.95	
.975	3.18	3.07	3.02	2.96	2.91	2.87	2.85	2.80	2.79	2.76	2.74	2.72	.975	
.99	4.01	3.86	3.78	3.70	3.62	3.57	3.54	3.47	3.45	3.41	3.38	3.36	.99	
.995	4.72	4.53	4.43	4.33	4.23	4.17	4.12	4.04	4.01	3.97	3.93	3.90	.995	
.999	6.71	6.40	6.25	6.09	5.93	5.83	5.76	5.63	5.59	5.52	5.46	5.42	.999	
.9995	7.74	7.37	7.18	7.00	6.80	6.68	6.61	6.45	6.41	6.33	6.25	6.20	.9995	

Appendix B9 *(Continued)*

ν_2	ν_1 p	1	2	3	4	5	6	7	8	9	10	11	12	p
15	.0005	$.0^641$	$.0^350$	$.0^249$	.015	.029	.045	.061	.076	.091	.105	.117	.129	.0005
	.001	$.0^516$	$.0^210$	$.0^279$	.021	.039	.057	.075	.092	.108	.123	.137	.149	.001
	.005	$.0^439$	$.0^250$	.023	.049	.076	.102	.125	.147	.166	.183	.198	.212	.005
	.01	$.0^316$	.010	.037	.070	.103	.132	.158	.181	.202	.219	.235	.249	.01
	.025	$.0^210$	.025	.070	.116	.156	.190	.219	.244	.265	.284	.300	.315	.025
	.05	$.0^241$	.051	.115	.170	.216	.254	.285	.311	.333	.351	.368	.382	.05
	.10	.016	.106	.192	.258	.309	.348	.380	.406	.427	.446	.461	.475	.10
	.25	.105	.293	.407	.480	.531	.568	.596	.618	.637	.652	.667	.676	.25
	.50	.478	.726	.826	.878	.911	.933	.948	.960	.970	.977	.984	.989	.50
	.75	1.43	1.52	1.52	1.51	1.49	1.48	1.47	1.46	1.46	1.45	1.44	1.44	.75
	.90	3.07	2.70	2.49	2.36	2.27	2.21	2.16	2.12	2.09	2.06	2.04	2.02	.90
	.95	4.54	3.68	3.29	3.06	2.90	2.79	2.71	2.64	2.59	2.54	2.51	2.48	.95
	.975	6.20	4.76	4.15	3.80	3.58	3.41	3.29	3.20	3.12	3.06	3.01	2.96	.975
	.99	8.68	6.36	5.42	4.89	4.56	4.32	4.14	4.00	3.89	3.80	3.73	3.67	.99
	.995	10.8	7.70	6.48	5.80	5.37	5.07	4.85	4.67	4.54	4.42	4.33	4.25	.995
	.999	16.6	11.3	9.34	8.25	7.57	7.09	6.74	6.47	6.26	6.08	5.93	5.81	.999
	.9995	19.5	13.2	10.8	9.48	8.66	8.10	7.68	7.36	7.11	6.91	6.75	6.60	.9995
20	.0005	$.0^640$	$.0^350$	$.0^250$	.015	.029	.046	.063	.079	.094	.109	.123	.136	.0005
	.001	$.0^516$	$.0^210$	$.0^279$	.022	.039	.058	.077	.095	.112	.128	.143	.156	.001
	.005	$.0^439$	$.0^250$	.023	.050	.077	.104	.129	.151	.171	.190	.206	.221	.005
	.01	$.0^316$	.010	.037	.071	.105	.135	.162	.187	.208	.227	.244	.259	.01
	.025	$.0^210$	.025	.071	.117	.158	.193	.224	.250	.273	.292	.310	.325	.025
	.05	$.0^240$	.051	.115	.172	.219	.258	.290	.318	.340	.360	.377	.393	.05
	.10	.016	.106	.193	.260	.312	.353	.385	.412	.435	.454	.472	.485	.10
	.25	.104	.292	.407	.480	.531	.569	.598	.622	.641	.656	.671	.681	.25
	.50	.472	.718	.816	.868	.900	.922	.938	.950	.959	.966	.972	.977	.50
	.75	1.40	1.49	1.48	1.47	1.45	1.44	1.43	1.42	1.41	1.40	1.39	1.39	.75
	.90	2.97	2.59	2.38	2.25	2.16	2.09	2.04	2.00	1.96	1.94	1.91	1.89	.90
	.95	4.35	3.49	3.10	2.87	2.71	2.60	2.51	2.45	2.39	2.35	2.31	2.28	.95
	.975	5.87	4.46	3.86	3.51	3.29	3.13	3.01	2.91	2.84	2.77	2.72	2.68	.975
	.99	8.10	5.85	4.94	4.43	4.10	3.87	3.70	3.56	3.46	3.37	3.29	3.23	.99
	.995	9.94	6.99	5.82	5.17	4.76	4.47	4.26	4.09	3.96	3.85	3.76	3.68	.995
	.999	14.8	9.95	8.10	7.10	6.46	6.02	5.69	5.44	5.24	5.08	4.94	4.82	.999
	.9995	17.2	11.4	9.20	8.02	7.28	6.76	6.38	6.08	5.85	5.66	5.51	5.38	.9995
24	.0005	$.0^640$	$.0^350$	$.0^250$	.015	.030	.046	.064	.080	.096	.112	.126	.139	.0005
	.001	$.0^516$	$.0^210$	$.0^279$	.022	.040	.059	.079	.097	.115	.131	.146	.160	.001
	.005	$.0^440$	$.0^250$	.023	.050	.078	.106	.131	.154	.175	.193	.210	.226	.005
	.01	$.0^316$	.010	.038	.072	.106	.137	.165	.189	.211	.231	.249	.264	.01
	.025	$.0^210$	.025	.071	.117	.159	.195	.227	.253	.277	.297	.315	.331	.025
	.05	$.0^240$	.051	.116	.173	.221	.260	.293	.321	.345	.365	.383	.399	.05
	.10	.016	.106	.193	.261	.313	.355	.388	.416	.439	.459	.476	.491	.10
	.25	.104	.291	.406	.480	.532	.570	.600	.623	.643	.659	.671	.684	.25
	.50	.469	.714	.812	.863	.895	.917	.932	.944	.953	.961	.967	.972	.50
	.75	1.39	1.47	1.46	1.44	1.43	1.41	.140	1.39	1.38	1.38	1.37	1.36	.75
	.90	2.93	2.54	2.33	2.19	2.10	2.04	1.98	1.94	1.91	1.88	1.85	1.83	.90
	.95	4.26	3.40	3.01	2.78	2.62	2.51	2.42	2.36	2.30	2.25	2.21	2.18	.95
	.975	5.72	4.32	3.72	3.38	3.15	2.99	2.87	2.78	2.70	2.64	2.59	2.54	.975
	.99	7.82	5.61	4.72	4.22	3.90	3.67	3.50	3.36	3.26	3.17	3.09	3.03	.99
	.995	9.55	6.66	5.52	4.89	4.49	4.20	3.99	3.83	3.69	3.59	3.50	3.42	.995
	.999	14.0	9.34	7.55	6.59	5.98	5.55	5.23	4.99	4.80	4.64	4.50	4.39	.999
	.9995	16.2	10.6	8.52	7.39	6.68	6.18	5.82	5.54	5.31	5.13	4.98	4.85	.9995

p \ ν_1	15	20	24	30	40	50	60	100	120	200	500	∞	p	ν_2
.0005	.159	.197	.220	.244	.272	.290	.303	.330	.339	.353	.368	.377	.0005	15
.001	.181	.219	.242	.266	.294	.313	.325	.352	.360	.375	.388	.398	.001	
.005	.246	.286	.308	.333	.360	.377	.389	.415	.422	.435	.448	.457	.005	
.01	.284	.324	.346	.370	.397	.413	.425	.450	.456	.469	.483	.490	.01	
.025	.349	.389	.410	.433	.458	.474	.485	.508	.514	.526	.538	.546	.025	
.05	.416	.454	.474	.496	.519	.535	.545	.565	.571	.581	.592	.600	.05	
.10	.507	.542	.561	.581	.602	.614	.624	.641	.647	.658	.667	.672	.10	
.25	.701	.728	.742	.757	.772	.782	.788	.802	.805	.812	.818	.822	.25	
.50	1.00	1.01	1.02	1.02	1.03	1.03	1.03	1.04	1.04	1.04	1.04	1.05	.50	
.75	1.43	1.41	1.41	1.40	1.39	1.39	1.38	1.38	1.37	1.37	1.36	1.36	.75	
.90	1.97	1.92	1.90	1.87	1.85	1.83	1.82	1.79	1.79	1.77	1.76	1.76	.90	
.95	2.40	2.33	2.39	2.25	2.20	2.18	2.16	2.12	2.11	2.10	2.08	2.07	.95	
.975	2.86	2.76	2.70	2.64	2.59	2.55	2.52	2.47	2.46	2.44	2.41	2.40	.975	
.99	3.52	3.37	3.29	3.21	3.13	3.08	3.05	2.98	2.96	2.92	2.89	2.87	.99	
.995	4.07	3.88	3.79	3.69	3.59	3.52	3.48	3.39	3.37	3.33	3.29	3.26	.995	
.999	5.54	5.25	5.10	4.95	4.80	4.70	4.64	4.51	4.47	4.41	4.35	4.31	.999	
.9995	6.27	5.93	5.75	5.58	5.40	5.29	5.21	5.06	5.02	4.94	4.87	4.83	.9995	
.0005	.169	.211	.235	.263	.295	.316	.331	.364	.375	.391	.408	.422	.0005	20
.001	.191	.233	.258	.286	.318	.339	.354	.386	.395	.413	.429	.441	.001	
.005	.258	.301	.327	.354	.385	.405	.419	.448	.457	.474	.490	.500	.005	
.01	.297	.340	.365	.392	.422	.441	.455	.483	.491	.508	.521	.532	.01	
.025	.363	.406	.430	.456	.484	.503	.514	.541	.548	.562	.575	.585	.025	
.05	.430	.471	.493	.518	.544	.562	.572	.595	.603	.617	.629	.637	.05	
.10	.520	.557	.578	.600	.623	.637	.648	.671	.675	.685	.694	.704	.10	
.25	.708	.736	.751	.767	.784	.794	.801	.816	.820	.827	.835	.840	.25	
.50	.989	1.00	1.01	1.01	1.02	1.02	1.02	1.03	1.03	1.03	1.03	1.03	.50	
.75	1.37	1.36	1.35	1.34	1.33	1.33	1.32	1.31	1.31	1.30	1.30	1.29	.75	
.90	1.84	1.79	1.77	1.74	1.71	1.69	1.68	1.65	1.64	1.63	1.62	1.61	.90	
.95	2.20	2.12	2.08	2.04	1.99	1.97	1.95	1.91	1.90	1.88	1.86	1.84	.95	
.975	2.57	2.46	2.41	2.35	2.29	2.25	2.22	2.17	2.16	2.13	2.10	2.09	.975	
.99	3.09	2.94	2.86	2.78	2.69	2.64	2.61	2.54	2.52	2.48	2.44	2.42	.99	
.995	3.50	3.32	3.22	3.12	3.02	2.96	2.92	2.83	2.81	2.76	2.72	2.69	.995	
.999	4.56	4.29	4.15	4.01	3.86	3.77	3.70	3.58	3.54	3.48	3.42	3.38	.999	
.9995	5.07	4.75	4.58	4.42	4.24	4.15	4.07	3.93	3.90	3.82	3.75	3.70	.9995	
.0005	.174	.218	.244	.274	.309	.331	.349	.384	.395	.416	.434	.449	.0005	24
.0001	.196	.241	.268	.298	.332	.354	.371	.405	.417	.437	.455	.469	.001	
.005	.264	.310	.337	.367	.400	.422	.437	.469	.479	.498	.515	.527	.005	
.01	.304	.350	.376	.405	.437	.459	.473	.505	.513	.529	.546	.558	.01	
.025	.370	.415	.441	.468	.498	.518	.531	.562	.568	.585	.599	.610	.025	
.05	.437	.480	.504	.530	.558	.575	.588	.613	.622	.637	.649	.659	.05	
.10	.527	.566	.588	.611	.635	.651	.662	.685	.691	.704	.715	.723	.10	
.25	.712	.741	.757	.773	.791	.802	.809	.825	.829	.837	.844	.850	.25	
.50	.983	.994	1.00	1.01	1.01	1.02	1.02	1.02	1.02	1.02	1.03	1.03	.50	
.75	1.35	1.33	1.32	1.31	1.30	1.29	1.29	1.28	1.28	1.27	1.27	1.26	.75	
.90	1.78	1.73	1.70	1.67	1.64	1.62	1.61	1.58	1.57	1.56	1.54	1.53	.90	
.95	2.11	2.03	1.98	1.94	1.89	1.86	1.84	1.80	1.79	1.77	1.75	1.73	.95	
.975	2.44	2.33	2.27	2.21	2.15	2.11	2.08	2.02	2.01	1.98	1.95	1.94	.975	
.99	2.89	2.74	2.66	2.58	2.49	2.44	2.40	2.33	2.31	2.27	2.24	2.21	.99	
.995	3.25	3.06	2.97	2.87	2.77	2.70	2.66	2.57	2.55	2.50	2.46	2.43	.995	
.999	4.14	3.87	3.74	3.59	3.45	3.35	3.29	3.16	3.14	3.07	3.01	2.97	.999	
.9995	4.55	4.25	4.09	3.93	3.76	3.66	3.59	3.44	3.41	3.33	3.27	3.22	.9995	

ν_2	ν_1 / p	1	2	3	4	5	6	7	8	9	10	11	12	p
30	.0005	$.0^6 40$	$.0^3 50$	$.0^2 50$	.015	.030	.047	.065	.082	.098	.114	.129	.143	.0005
	.001	$.0^5 16$	$.0^2 10$	$.0^2 80$	.022	.040	.060	.080	.099	.117	.134	.150	.164	.001
	.005	$.0^4 40$	$.0^2 50$	.024	.050	.079	.107	.133	.156	.178	.197	.215	.231	.005
	.01	$.0^4 16$	.010	.038	.072	.107	.138	.167	.192	.215	.235	.254	.270	.01
	.025	$.0^2 10$	.025	.071	.118	.161	.197	.229	.257	.281	.302	.321	.337	.025
	.05	$.0^2 40$	.051	.116	.174	.222	.263	.296	.325	.349	.370	.389	.406	.05
	.10	.016	.106	.193	.262	.315	.357	.391	.420	.443	.464	.481	.497	.10
	.25	.103	.290	.406	.480	.532	.571	.601	.625	.645	.661	.676	.688	.25
	.50	.466	.709	.807	.858	.890	.912	.927	.939	.948	.955	.961	.966	.50
	.75	1.38	1.45	1.44	1.42	1.41	1.39	1.38	1.37	1.36	1.35	1.35	1.34	.75
	.90	2.88	2.49	2.28	2.14	2.05	1.98	1.93	1.88	1.85	1.82	1.79	1.77	.90
	.95	4.17	3.32	2.92	2.69	2.53	2.42	2.33	2.27	2.21	2.16	2.13	2.09	.95
	.975	5.57	4.18	3.59	3.25	3.03	2.87	2.75	2.65	2.57	2.51	2.46	2.41	.975
	.99	7.56	5.39	4.51	4.02	3.70	3.47	3.30	3.17	3.07	2.98	2.91	2.84	.99
	.995	9.18	6.35	5.24	4.62	4.23	3.95	3.74	3.58	3.45	3.34	3.25	3.18	.995
	.999	13.3	8.77	7.05	6.12	5.53	5.12	4.82	4.58	4.39	4.24	4.11	4.00	.999
	.9995	15.2	9.90	7.90	6.82	6.14	5.66	5.31	5.04	4.82	4.65	4.51	4.38	.9995
40	.0005	$.0^6 40$	$.0^3 50$	$.0^2 50$	.016	.030	.048	.066	.084	.100	.117	.132	.147	.0005
	.001	$.0^5 16$	$.0^2 10$	$.0^2 80$	.022	.042	.061	.081	.101	.119	.137	.153	.169	.001
	.005	$.0^4 40$	$.0^2 50$	.024	.051	.080	.108	.135	.159	.181	.201	.220	.237	.005
	.01	$.0^4 16$	.010	.038	.073	.108	.140	.169	.195	.219	.240	.259	.276	.01
	.025	$.0^3 99$	.025	.071	.119	.162	.199	.232	.260	.285	.307	.327	.344	.025
	.05	$.0^2 40$	.051	.116	.175	.224	.265	.299	.329	.354	.376	.395	.412	.05
	.10	.016	.106	.194	.263	.317	.360	.394	.424	.448	.469	.488	.504	.10
	.25	.103	.290	.405	.480	.533	.572	.603	.627	.647	.664	.680	.691	.25
	.50	.463	.705	.802	.854	.885	.907	.922	.934	.943	.950	.956	.961	.50
	.75	1.36	1.44	1.42	1.40	1.39	1.37	1.36	1.35	1.34	1.33	1.32	1.31	.75
	.90	2.84	2.44	2.23	2.09	2.00	1.93	1.87	1.83	1.79	1.76	1.73	1.71	.90
	.95	4.08	3.23	2.84	2.61	2.45	2.34	2.25	2.18	2.12	2.08	2.04	2.00	.95
	.975	5.42	4.05	3.46	3.13	2.90	2.74	2.62	2.53	2.45	2.39	2.33	2.29	.975
	.99	7.31	5.18	4.31	3.83	3.51	3.29	3.12	2.99	2.89	2.80	2.73	2.66	.99
	.995	8.83	6.07	4.98	4.37	3.99	3.71	3.51	3.35	3.22	3.12	3.03	2.95	.995
	.999	12.6	8.25	6.60	5.70	5.13	4.73	4.44	4.21	4.02	3.87	3.75	3.64	.999
	.9995	14.4	9.25	7.33	6.30	5.64	5.19	4.85	4.59	4.38	4.21	4.07	3.95	.9995
60	.0005	$.0^6 40$	$.0^3 50$	$.0^2 51$	.016	.031	.048	.067	.085	.103	.120	.136	.152	.0005
	.001	$.0^5 16$	$.0^2 10$	$.0^2 80$	.022	.041	.062	.083	.103	.122	.140	.157	.174	.001
	.005	$.0^4 40$	$.0^2 50$	.024	.051	.081	.110	.137	.162	.185	.206	.225	.243	.005
	.01	$.0^4 16$	.010	.038	.073	.109	.142	.172	.199	.223	.245	.265	.283	.01
	.025	$.0^3 99$	.025	.071	.120	.163	.202	.235	.264	.290	.313	.333	.351	.025
	.05	$.0^2 40$	.051	.116	.176	.226	.267	.303	.333	.359	.382	.402	.419	.05
	.10	.016	.106	.194	.264	.318	.362	.398	.428	.453	.475	.493	.510	.10
	.25	.102	.289	.405	.480	.534	.573	.604	.629	.650	.667	.680	.695	.25
	.50	.461	.701	.798	.849	.880	.901	.917	.928	.937	.945	.951	.956	.50
	.75	1.35	1.42	1.41	1.38	1.37	1.35	1.33	1.32	1.31	1.30	1.29	1.29	.75
	.90	2.79	2.39	2.18	2.04	1.95	1.87	1.82	1.77	1.74	1.71	1.68	1.66	.90
	.95	4.00	3.15	2.76	2.53	2.37	2.25	2.17	2.10	2.04	1.99	1.95	1.92	.95
	.975	5.29	3.93	3.34	3.01	2.79	2.63	2.51	2.41	2.33	2.27	2.22	2.17	.975
	.99	7.08	4.98	4.13	3.65	3.34	3.12	2.95	2.82	2.72	2.63	2.56	2.50	.99
	.995	8.49	5.80	4.73	4.14	3.76	3.49	3.29	3.13	3.01	2.90	2.82	2.74	.995
	.999	12.0	7.76	6.17	5.31	4.76	4.37	4.09	3.87	3.69	3.54	3.43	3.31	.999
	.9995	13.6	8.65	6.81	5.82	5.20	4.76	4.44	4.18	3.98	3.82	3.69	3.57	.9995

p \ ν_1	15	20	24	30	40	50	60	100	120	200	500	∞	p	ν_2
.0005	.179	.226	.254	.287	.325	.350	.369	.410	.420	.444	.467	.483	.0005	30
.001	.202	.250	.278	.311	.348	.373	.391	.431	.442	.465	.488	.503	.001	
.005	.271	.320	.349	.381	.416	.441	.457	.495	.504	.524	.543	.559	.005	
.01	.311	.360	.388	.419	.454	.476	.493	.529	.538	.559	.575	.590	.01	
.025	.378	.426	.453	.482	.515	.535	.551	.585	.592	.610	.625	.639	.025	
.05	.445	.490	.516	.543	.573	.592	.606	.637	.644	.658	.676	.685	.05	
.10	.534	.575	.598	.623	.649	.667	.678	.704	.710	.725	.735	.746	.10	
.25	.716	.746	.763	.780	.798	.810	.818	.835	.839	.848	.856	.862	.25	
.50	.978	.989	.994	1.00	1.01	1.01	1.01	1.02	1.02	1.02	1.02	1.02	.50	
.75	1.32	1.30	1.29	1.28	1.27	1.26	1.26	1.25	1.24	1.24	1.23	1.23	.75	
.90	1.72	1.67	1.64	1.61	1.57	1.55	1.54	1.51	1.50	1.48	1.47	1.46	.90	
.95	2.01	1.93	1.89	1.84	1.79	1.76	1.74	1.70	1.68	1.66	1.64	1.62	.95	
.975	2.31	2.20	2.14	2.07	2.01	1.97	1.94	1.88	1.87	1.84	1.81	1.79	.975	
.99	2.70	2.55	2.47	2.39	2.30	2.25	2.21	2.13	2.11	2.07	2.03	2.01	.99	
.995	3.01	2.82	2.73	2.63	2.52	2.46	2.42	2.32	2.30	2.25	2.21	2.18	.995	
.999	3.75	3.49	3.36	3.22	3.07	2.98	2.92	2.79	2.76	2.69	2.63	2.59	.999	
.9995	4.10	3.80	3.65	3.48	3.32	3.22	3.15	3.00	2.97	2.89	2.82	2.78	.9995	
.0005	.185	.236	.266	.301	.343	.373	.393	.441	.453	.480	.504	.525	.0005	40
.001	.209	.259	.290	.326	.367	.396	.415	.461	.473	.500	.524	.545	.001	
.005	.279	.331	.362	.396	.436	.463	.481	.524	.534	.559	.581	.599	.005	
.01	.319	.371	.401	.435	.473	.498	.516	.556	.567	.592	.613	.628	.01	
.025	.387	.437	.466	.498	.533	.556	.573	.610	.620	.641	.662	.674	.025	
.05	.454	.502	.529	.558	.591	.613	.627	.658	.669	.685	.704	.717	.05	
.10	.542	.585	.609	.636	.664	.683	.696	.724	.731	.747	.762	.772	.10	
.25	.720	.752	.769	.787	.806	.819	.828	.846	.851	.861	.870	.877	.25	
.50	.972	.983	.989	.994	1.00	1.00	1.01	1.01	1.01	1.01	1.02	1.02	.50	
.75	1.30	1.28	1.26	1.25	1.24	1.23	1.22	1.21	1.21	1.20	1.19	1.19	.75	
.90	1.66	1.61	1.57	1.54	1.51	1.48	1.47	1.43	1.42	1.41	1.39	1.38	.90	
.95	1.92	1.84	1.79	1.74	1.69	1.66	1.64	1.59	1.58	1.55	1.53	1.51	.95	
.975	2.18	2.07	2.01	1.94	1.88	1.83	1.80	1.74	1.72	1.69	1.66	1.64	.975	
.99	2.52	2.37	2.29	2.20	2.11	2.06	2.02	1.94	1.92	1.87	1.83	1.80	.99	
.995	2.78	2.60	2.50	2.40	2.30	2.23	2.18	2.09	2.06	2.01	1.96	1.93	.995	
.999	3.40	3.15	3.01	2.87	2.73	2.64	2.57	2.44	2.41	2.34	2.28	2.23	.999	
.9995	3.68	3.39	3.24	3.08	2.92	2.82	2.74	2.60	2.57	2.49	2.41	2.37	.9995	
.0005	.192	.246	.278	.318	.365	.398	.421	.478	.493	.527	.561	.585	.0005	60
.001	.216	.270	.304	.343	.389	.421	.444	.497	.512	.545	.579	.602	.001	
.005	.287	.343	.376	.414	.458	.488	.510	.559	.572	.602	.633	.652	.005	
.01	.328	.383	.416	.453	.495	.524	.545	.592	.604	.633	.658	.679	.01	
.025	.396	.450	.481	.515	.555	.581	.600	.641	.654	.680	.704	.720	.025	
.05	.463	.514	.543	.575	.611	.633	.652	.690	.700	.719	.746	.759	.05	
.10	.550	.596	.622	.650	.682	.703	.717	.750	.758	.776	.793	.806	.10	
.25	.725	.758	.776	.796	.816	.830	.840	.860	.865	.877	.888	.896	.25	
.50	.967	.978	.983	.989	.994	.998	1.00	1.00	1.01	1.01	1.01	1.01	.50	
.75	1.27	1.25	1.24	1.22	1.21	1.20	1.19	1.17	1.17	1.16	1.15	1.15	.75	
.90	1.60	1.54	1.51	1.48	1.44	1.41	1.40	1.36	1.35	1.33	1.31	1.29	.90	
.95	1.84	1.75	1.70	1.65	1.59	1.56	1.53	1.48	1.47	1.44	1.41	1.39	.95	
.975	2.06	1.94	1.88	1.82	1.74	1.70	1.67	1.60	1.58	1.54	1.51	1.48	.975	
.99	2.35	2.20	2.12	2.03	1.94	1.88	1.84	1.75	1.73	1.68	1.63	1.60	.99	
.995	2.57	2.39	2.29	2.19	2.08	2.01	1.96	1.86	1.83	1.78	1.73	1.69	.995	
.999	3.08	2.83	2.69	2.56	2.41	2.31	2.25	2.11	2.09	2.01	1.93	1.89	.999	
.9995	3.30	3.02	2.87	2.71	2.55	2.45	2.38	2.23	2.19	2.11	2.03	1.98	.9995	

ν_2	ν_1 \ p	1	2	3	4	5	6	7	8	9	10	11	12	p
120	.0005	$.0^640$	$.0^550$	$.0^251$	.016	.031	.049	.067	.087	.105	.123	.140	.156	.0005
	.001	$.0^516$	$.0^210$	$.0^281$	.023	.042	.063	.084	.105	.125	.144	.162	.179	.001
	.005	$.0^439$	$.0^250$	.024	.051	.081	.111	.139	.165	.189	.211	.230	.249	.005
	.01	$.0^316$	.010	.038	.074	.110	.143	.174	.202	.227	.250	.271	.290	.01
	.025	$.0^399$	.025	.072	.120	.165	.204	.238	.268	.295	.318	.340	.359	.025
	.05	$.0^239$	.051	.117	.177	.227	.270	.306	.337	.364	.388	.408	.427	.05
	.10	.016	.105	.194	.265	.320	.365	.401	.432	.458	.480	.500	.518	.10
	.25	.102	.288	.405	.481	.534	.574	.606	.631	.652	.670	.685	.699	.25
	.50	.458	.697	.793	.844	.875	.896	.912	.923	.932	.939	.945	.950	.50
	.75	1.34	1.40	1.39	1.37	1.35	1.33	1.31	1.30	1.29	1.28	1.27	1.26	.75
	.90	2.75	2.35	2.13	1.99	1.90	1.82	1.77	1.72	1.68	1.65	1.62	1.60	.90
	.95	3.92	3.07	2.68	2.45	2.29	2.18	2.09	2.02	1.96	1.91	1.87	1.83	.95
	.975	5.15	3.80	3.23	2.89	2.67	2.52	2.39	2.30	2.22	2.16	2.10	2.05	.975
	.99	6.85	4.79	3.95	3.48	3.17	2.96	2.79	2.66	2.56	2.47	2.40	2.34	.99
	.995	8.18	5.54	4.50	3.92	3.55	3.28	3.09	2.93	2.81	2.71	2.62	2.54	.995
	.999	11.4	7.32	5.79	4.95	4.42	4.04	3.77	3.55	3.38	3.24	3.12	3.02	.999
	.9995	12.8	8.10	6.34	5.39	4.79	4.37	4.07	3.82	3.63	3.47	3.34	3.22	.9995
∞	.0005	$.0^639$	$.0^550$	$.0^251$	.016	.032	.050	.069	.088	.108	.127	.144	.161	.0005
	.001	$.0^516$	$.0^210$	$.0^281$	.023	.042	.063	.085	.107	.128	.148	.167	.185	.001
	.005	$.0^439$	$.0^250$	.024	.052	.082	.113	.141	.168	.193	.216	.236	.256	.005
	.01	$.0^316$	.010	.038	.074	.111	.145	.177	.206	.232	.256	.278	.298	.01
	.025	$.0^398$	.025	.072	.121	.166	.206	.241	.272	.300	.325	.347	.367	.025
	.05	$.0^239$	.051	.117	.178	.229	.273	.310	.342	.369	.394	.417	.436	.05
	.10	.016	.105	.195	.266	.322	.367	.405	.436	.463	.487	.508	.525	.10
	.25	.102	.288	.404	.481	.535	.576	.608	.634	.655	.674	.690	.703	.25
	.50	.455	.693	.789	.839	.870	.891	.907	.918	.927	.934	.939	.945	.50
	.75	1.32	1.39	1.37	1.35	1.33	1.31	1.29	1.28	1.27	1.25	1.24	1.24	.75
	.90	2.71	2.30	2.08	1.94	1.85	1.77	1.72	1.67	1.63	1.60	1.57	1.55	.90
	.95	3.84	3.00	2.60	2.37	2.21	2.10	2.01	1.94	1.88	1.83	1.79	1.75	.95
	.975	5.02	3.69	3.12	2.79	2.57	2.41	2.29	2.19	2.11	2.05	1.99	1.94	.975
	.99	6.63	4.61	3.78	3.32	3.02	2.80	2.64	2.51	2.41	2.32	2.25	2.18	.99
	.995	7.88	5.30	4.28	3.72	3.35	3.09	2.90	2.74	2.62	2.52	2.43	2.36	.995
	.999	10.8	6.91	5.42	4.62	4.10	3.74	3.47	3.27	3.10	2.96	2.84	2.74	.999
	.9995	12.1	7.60	5.91	5.00	4.42	4.02	3.72	3.48	3.30	3.14	3.02	2.90	.9995

p	15	20	24	30	40	50	60	100	120	200	500	∞	p	ν_2
.0005	.199	.256	.293	.338	.390	.429	.458	.524	.543	.578	.614	.676	.0005	120
.001	.223	.282	.319	.363	.415	.453	.480	.542	.568	.595	.631	.691	.001	
.005	.297	.356	.393	.434	.484	.520	.545	.605	.623	.661	.702	.733	.005	
.01	.338	.397	.433	.474	.522	.556	.579	.636	.652	.688	.725	.755	.01	
.025	.406	.464	.498	.536	.580	.611	.633	.684	.698	.729	.762	.789	.025	
.05	.473	.527	.559	.594	.634	.661	.682	.727	.740	.767	.785	.819	.05	
.10	.560	.609	.636	.667	.702	.726	.742	.781	.791	.815	.838	.855	.10	
.25	.730	.765	.784	.805	.828	.843	.853	.877	.884	.897	.911	.923	.25	
.50	.961	.972	.978	.983	.989	.992	.994	1.00	1.00	1.00	1.01	1.01	.50	
.75	1.24	1.22	1.21	1.19	1.18	1.17	1.16	1.14	1.13	1.12	1.11	1.10	.75	
.90	1.55	1.48	1.45	1.41	1.37	1.34	1.32	1.27	1.26	1.24	1.21	1.19	.90	
.95	1.75	1.66	1.61	1.55	1.50	1.46	1.43	1.37	1.35	1.32	1.28	1.25	.95	
.975	1.95	1.82	1.76	1.69	1.61	1.56	1.53	1.45	1.43	1.39	1.34	1.31	.975	
.99	2.19	2.03	1.95	1.86	1.76	1.70	1.66	1.56	1.53	1.48	1.42	1.38	.99	
.995	2.37	2.19	2.09	1.98	1.87	1.80	1.75	1.64	1.61	1.54	1.48	1.43	.995	
.999	2.78	2.53	2.40	2.26	2.11	2.02	1.95	1.82	1.76	1.70	1.62	1.54	.999	
.9995	2.96	2.67	2.53	2.38	2.21	2.11	2.01	1.88	1.84	1.75	1.67	1.60	.9995	
.0005	.207	.270	.311	.360	.422	.469	.505	.599	.624	.704	.804	1.00	.0005	∞
.001	.232	.296	.338	.386	.448	.493	.527	.617	.649	.719	.819	1.00	.001	
.005	.307	.372	.412	.460	.518	.559	.592	.671	.699	.762	.843	1.00	.005	
.01	.349	.413	.452	.499	.554	.595	.625	.699	.724	.782	.858	1.00	.01	
.025	.418	.480	.517	.560	.611	.645	.675	.741	.763	.813	.878	1.00	.025	
.05	.484	.543	.577	.617	.663	.694	.720	.781	.797	.840	.896	1.00	.05	
.10	.570	.622	.652	.687	.726	.752	.774	.826	.838	.877	.919	1.00	.10	
.25	.736	.773	.793	.816	.842	.860	.872	.901	.910	.932	.957	1.00	.25	
.50	.956	.967	.972	.978	.983	.987	.989	.993	.994	.997	.999	1.00	.50	
.75	1.22	1.19	1.18	1.16	1.14	1.13	1.12	1.09	1.08	1.07	1.04	1.00	.75	
.90	1.49	1.42	1.38	1.34	1.30	1.26	1.24	1.18	1.17	1.13	1.08	1.00	.90	
.95	1.67	1.57	1.52	1.46	1.39	1.35	1.32	1.24	1.22	1.17	1.11	1.00	.95	
.975	1.83	1.71	1.64	1.57	1.48	1.43	1.39	1.30	1.27	1.21	1.13	1.00	.975	
.99	2.04	1.88	1.79	1.70	1.59	1.52	1.47	1.36	1.32	1.25	1.15	1.00	.99	
.995	2.19	2.00	1.90	1.79	1.67	1.59	1.53	1.40	1.36	1.28	1.17	1.00	.995	
.999	2.51	2.27	2.13	1.99	1.84	1.73	1.66	1.49	1.45	1.34	1.21	1.00	.999	
.9995	2.65	2.37	2.22	2.07	1.91	1.79	1.71	1.53	1.48	1.36	1.22	1.00	.9995	

† Reprinted with permission from W. J. Dixon and F. J. Massey, Jr., "Introduction to Statistical Analysis," 3d ed., McGraw-Hill Book Company, New York, 1969. Part of this table was extracted from (1) A. Hald, "Statistical Tables and Formulas," John Wiley & Sons, Inc., New York, 1952; (2) M. Merrington and C. M. Thompson, "Tables of the Percentage Points of the Inverted Beta Distribution," *Biometrika*, vol. 33, p. 73, 1943; (3) C. C. Colcord and L. S. Deming, "The One-Tenth Percent Level of Z," *Sankhya*, vol. 2, p. 423, 1936. Permission to reprint the needed entries was granted in each case.

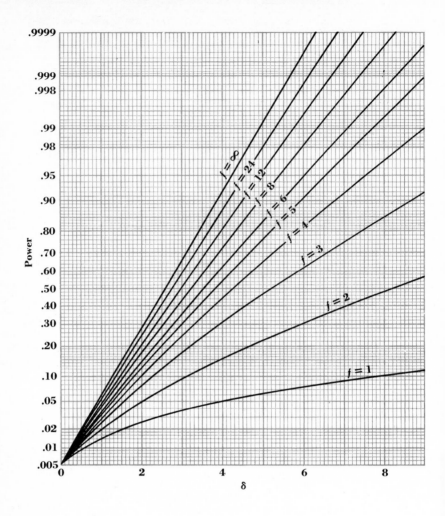

$$\alpha = 0.01 \qquad \delta = |\mu - \mu_0| \, \sqrt{n}/\sigma; f = \text{degrees of freedom}$$

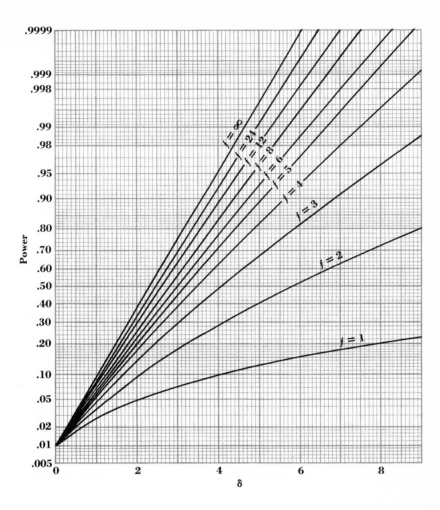

$$\alpha = 0.025 \qquad \delta = |\mu - \mu_0| \sqrt{n}/\sigma; f = \text{degrees of freedom}$$

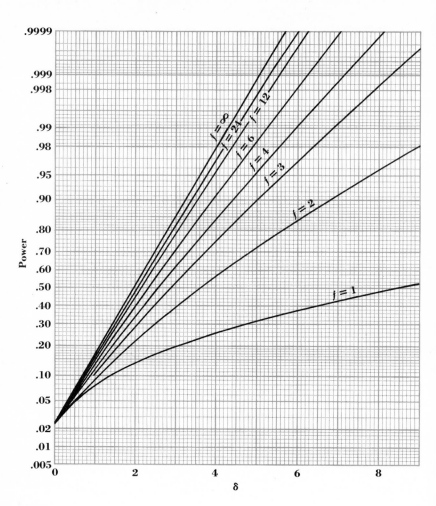

$$\alpha = 0.05 \qquad \delta = |\mu - \mu_0| \sqrt{n}/\sigma; f = \text{degrees of freedom}$$

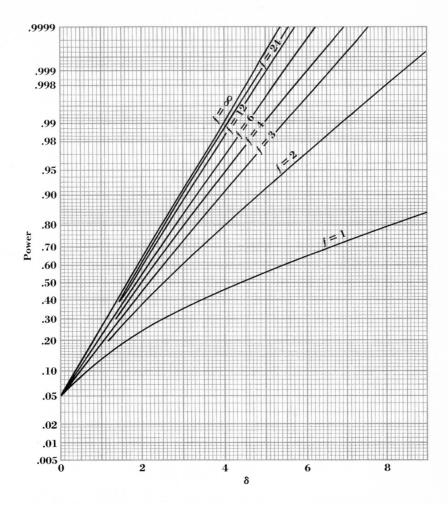

† Reproduced with permission from D. B. Owen, "Handbook of Statistical Tables," Addison-Wesley Publishing Company, Inc., Reading Mass., 1962.

Appendix B11 *Graphs of sample sizes required to ensure with a given*
probability that a confidence interval for the mean
with confidence coefficient $1 - \alpha$ *will be shorter than L*

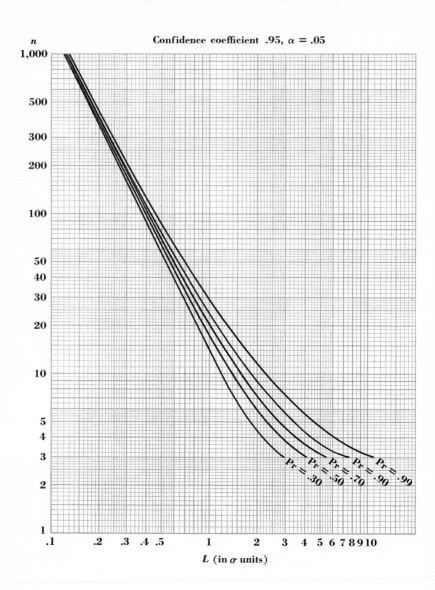

Confidence coefficient .95, $\alpha = .05$

n

L (in σ units)

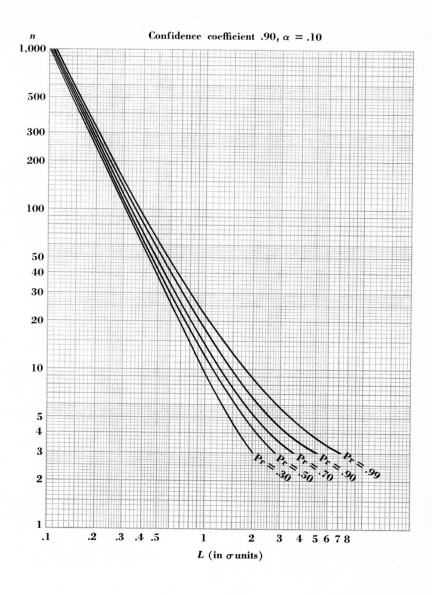

n

Confidence coefficient .90, $\alpha = .10$

L (in σ units)

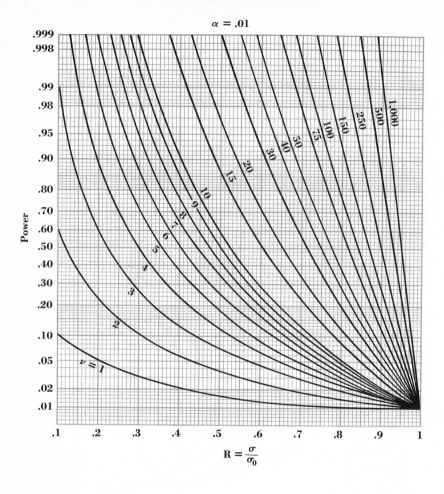

Appendix B12 *Power curves for testing hypotheses about σ^2 of the normal distribution $H_0 : \sigma^2 = \sigma_0^2$ against $H_1 : \sigma^2 < \sigma_0^2$*

$\alpha = .01$

Power

$R = \dfrac{\sigma}{\sigma_0}$

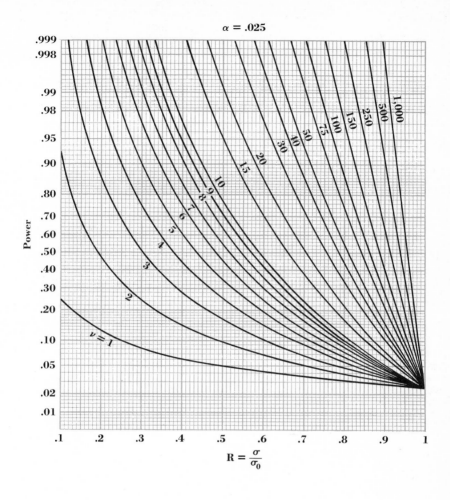

$\alpha = .025$

Power

$$R = \frac{\sigma}{\sigma_0}$$

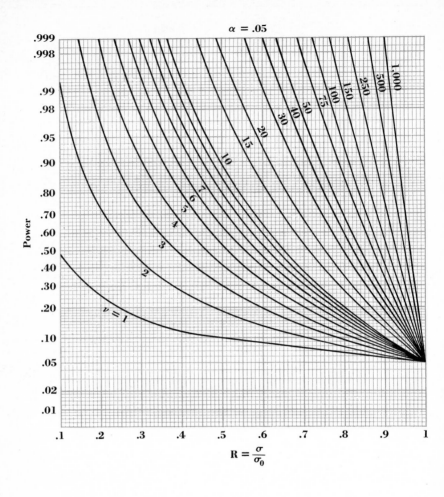

$\alpha = .05$

Power

$R = \dfrac{\sigma}{\sigma_0}$

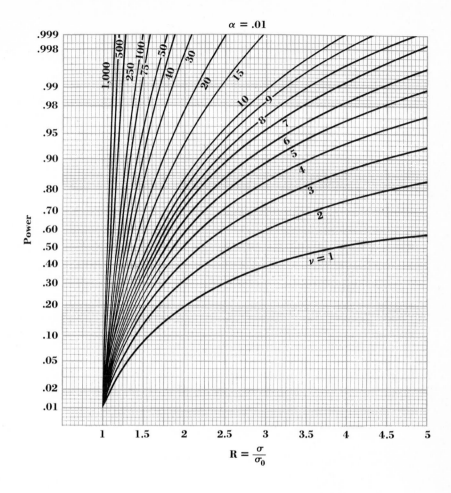

$$\alpha = .01$$

$$R = \frac{\sigma}{\sigma_0}$$

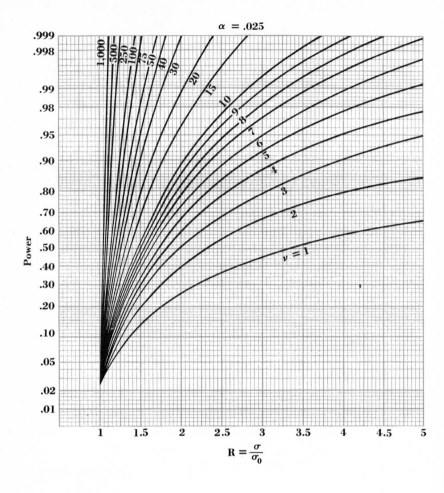

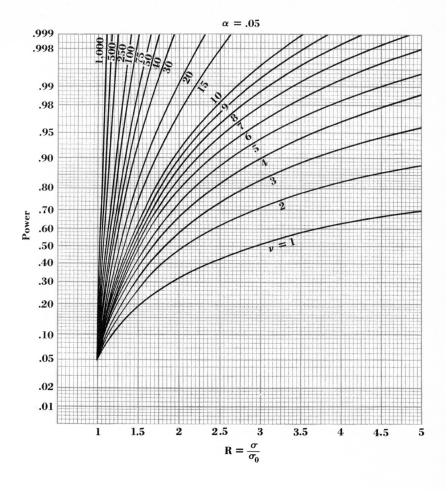

Appendix B13 *Graphs of sample sizes required to ensure with a given*
probability that a confidence interval for the
standard deviation with confidence coefficient $1 - \alpha$
will be shorter than L

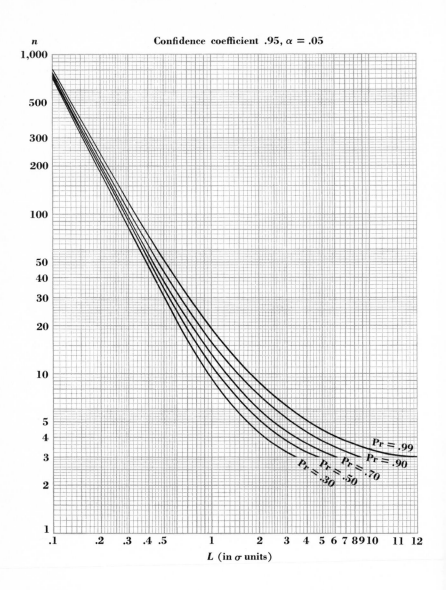

n — Confidence coefficient .95, $\alpha = .05$

L (in σ units)

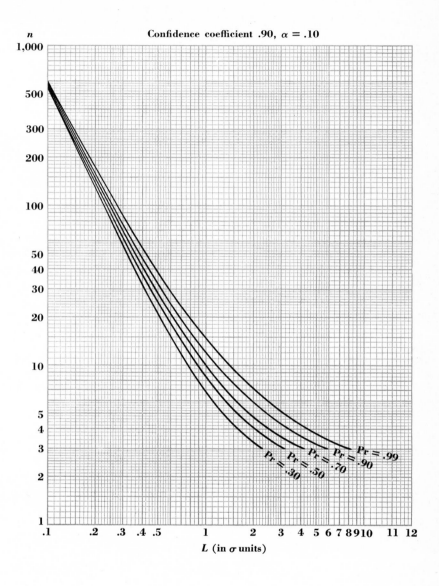

n   Confidence coefficient .90, $\alpha = .10$

L (in σ units)

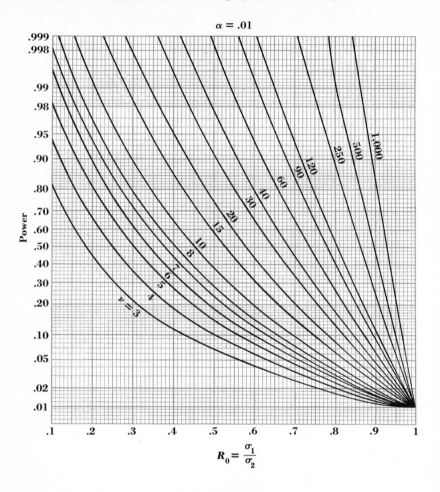

$$R_0 = \frac{\sigma_1}{\sigma_2}$$

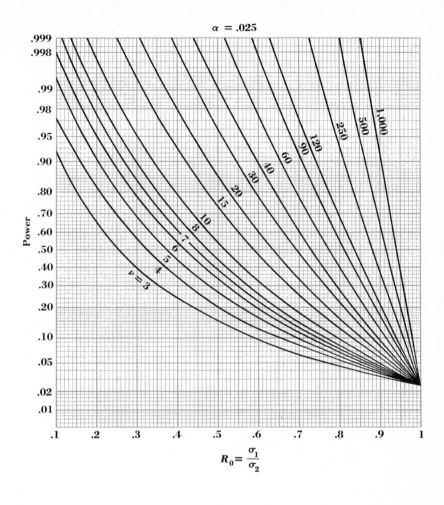

$$\alpha = .025$$

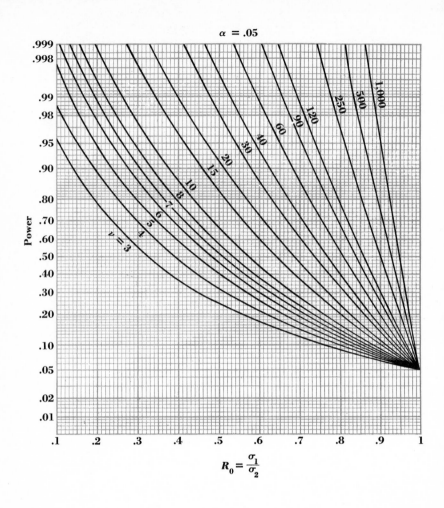

$$\alpha = .05$$

Power

$$R_0 = \frac{\sigma_1}{\sigma_2}$$

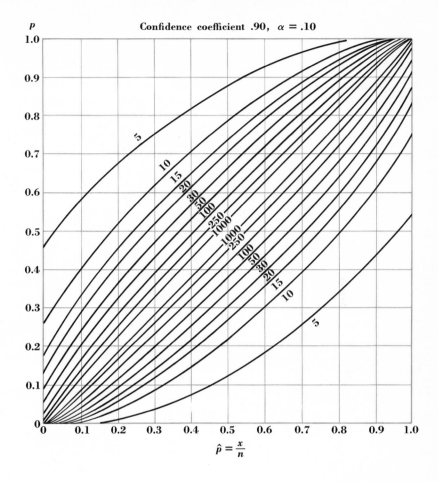

Confidence coefficient .90, $\alpha = .10$

$$\hat{p} = \frac{x}{n}$$

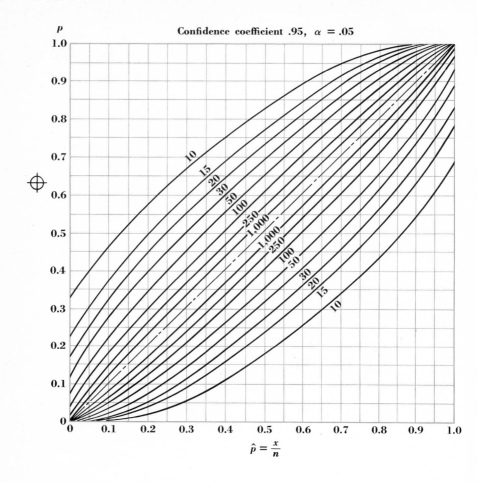

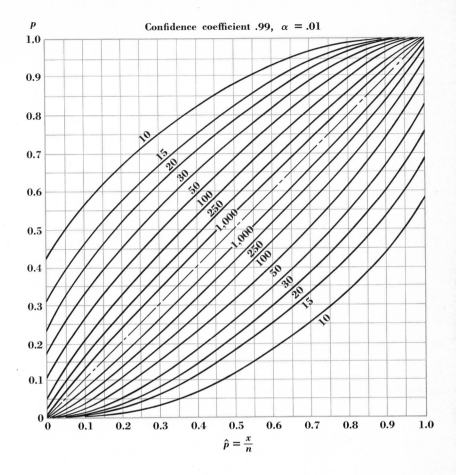

p

Confidence coefficient .99, $\alpha = .01$

$$\hat{p} = \frac{x}{n}$$

† The graphs for confidence coefficient .95 and .99 are reproduced with permission from E. S. Pearson and C. J. Clopper, The Use of Confidence Intervals or Fiducial Limits Illustrated in the Case of the Binomial, *Biometrika*, vol. 26, p. 404, 1934. The graph for confidence coefficient .90 is reproduced with permission from W. J. Dixon and F. J. Massey, Jr., " Introduction to Statistical Analysis," 3d ed., McGraw-Hill Book Company, New York, 1969.

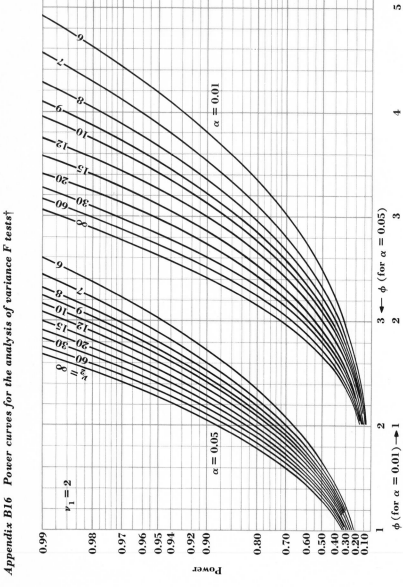

Appendix B16   Power curves for the analysis of variance F tests†

520

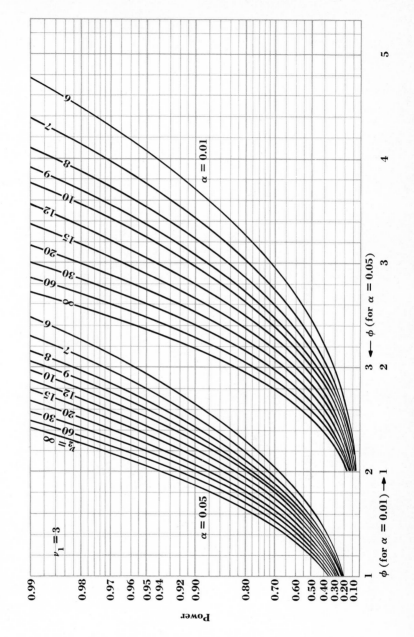

Appendix B16 (*Continued*)

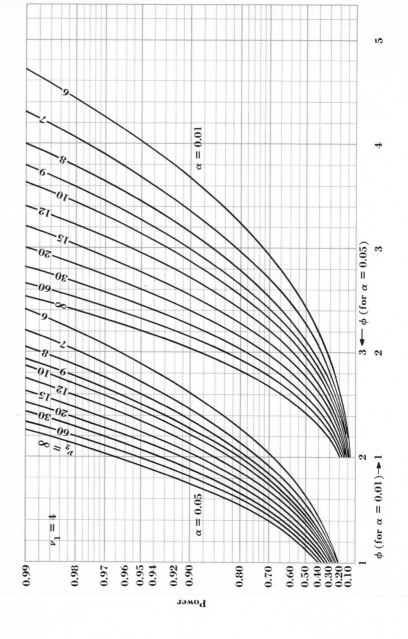

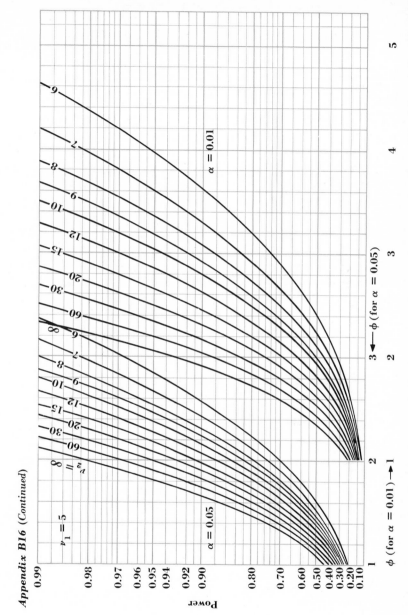

† Reproduced with permission from E. S. Pearson and H. O. Hartley, Charts of the Power Function for Analysis of Variance Tests, Derived from the Non-central F Distribution, *Biometrika*, vol. 38, p. 112, 1951.

Appendix B17 *Distribution of Cochran's statistic*†

Tabulated value $= R_{n,r;.95}$ where $\Pr(\text{statistic} < R_{n,r;.95}) = .95$

r \ n	2	3	4	5	6	7	8	9	10	11	17	37	145	∞
2	0.9985	0.9750	0.9392	0.9057	0.8772	0.8534	0.8332	0.8159	0.8010	0.7880	0.7341	0.6602	0.5813	0.5000
3	0.9669	0.8709	0.7977	0.7457	0.7071	0.6771	0.6530	0.6333	0.6167	0.6025	0.5466	0.4748	0.4031	0.3333
4	0.9065	0.7679	0.6841	0.6287	0.5895	0.5598	0.5365	0.5175	0.5017	0.4884	0.4366	0.3720	0.3093	0.2500
5	0.8412	0.6838	0.5981	0.5441	0.5065	0.4783	0.4564	0.4387	0.4241	0.4118	0.3645	0.3066	0.2513	0.2000
6	0.7808	0.6161	0.5321	0.4803	0.4447	0.4184	0.3980	0.3817	0.3682	0.3568	0.3135	0.2612	0.2119	0.1667
7	0.7271	0.5612	0.4800	0.4307	0.3974	0.3726	0.3535	0.3384	0.3259	0.3154	0.2756	0.2278	0.1833	0.1429
8	0.6798	0.5157	0.4377	0.3910	0.3595	0.3362	0.3185	0.3043	0.2926	0.2829	0.2462	0.2022	0.1616	0.1250
9	0.6385	0.4775	0.4027	0.3584	0.3286	0.3067	0.2901	0.2768	0.2659	0.2568	0.2226	0.1820	0.1446	0.1111
10	0.6020	0.4450	0.3733	0.3311	0.3029	0.2823	0.2666	0.2541	0.2439	0.2353	0.2032	0.1655	0.1308	0.1000
12	0.5410	0.3924	0.3264	0.2880	0.2624	0.2439	0.2299	0.2187	0.2098	0.2020	0.1737	0.1403	0.1100	0.0833
15	0.4709	0.3346	0.2758	0.2419	0.2195	0.2034	0.1911	0.1815	0.1736	0.1671	0.1429	0.1144	0.0889	0.0667
20	0.3894	0.2705	0.2205	0.1921	0.1735	0.1602	0.1501	0.1422	0.1357	0.1303	0.1108	0.0879	0.0675	0.0500
24	0.3434	0.2354	0.1907	0.1656	0.1493	0.1374	0.1286	0.1216	0.1160	0.1113	0.0942	0.0743	0.0567	0.0417
30	0.2929	0.1980	0.1593	0.1377	0.1237	0.1137	0.1061	0.1002	0.0958	0.0921	0.0771	0.0604	0.0457	0.0333
40	0.2370	0.1576	0.1259	0.1082	0.0968	0.0887	0.0827	0.0780	0.0745	0.0713	0.0595	0.0462	0.0347	0.0250
60	0.1737	0.1131	0.0895	0.0765	0.0682	0.0623	0.0583	0.0552	0.0520	0.0497	0.0411	0.0316	0.0234	0.0167
120	0.0998	0.0632	0.0495	0.0419	0.0371	0.0337	0.0312	0.0292	0.0279	0.0266	0.0218	0.0165	0.0120	0.0083
∞	0	0	0	0	0	0	0	0	0	0	0	0	0	0

Appendix B17 (Continued)

Tabulated value $= R_{n,r;.99}$ where $\Pr(\text{statistic} < R_{n,r;.99}) = .99$

r \ n	2	3	4	5	6	7	8	9	10	11	17	37	145	∞
2	0.9999	0.9950	0.9794	0.9586	0.9373	0.9172	0.8988	0.8823	0.8674	0.8539	0.7949	0.7067	0.6062	0.5000
3	0.9933	0.9423	0.8831	0.8335	0.7933	0.7606	0.7335	0.7107	0.6912	0.6743	0.6059	0.5153	0.4230	0.3333
4	0.9676	0.8643	0.7814	0.7212	0.6761	0.6410	0.6129	0.5897	0.5702	0.5536	0.4884	0.4057	0.3251	0.2500
5	0.9279	0.7885	0.6957	0.6329	0.5875	0.5531	0.5259	0.5037	0.4854	0.4697	0.4094	0.3351	0.2644	0.2000
6	0.8828	0.7218	0.6258	0.5635	0.5195	0.4866	0.4608	0.4401	0.4229	0.4084	0.3529	0.2858	0.2229	0.1667
7	0.8376	0.6644	0.5685	0.5080	0.4659	0.4347	0.4105	0.3911	0.3751	0.3616	0.3105	0.2494	0.1929	0.1429
8	0.7945	0.6152	0.5209	0.4627	0.4226	0.3932	0.3704	0.3522	0.3373	0.3248	0.2779	0.2214	0.1700	0.1250
9	0.7544	0.5727	0.4810	0.4251	0.3870	0.3592	0.3378	0.3207	0.3067	0.2950	0.2514	0.1992	0.1521	0.1111
10	0.7175	0.5358	0.4469	0.3934	0.3572	0.3308	0.3106	0.2945	0.2813	0.2704	0.2297	0.1811	0.1376	0.1000
12	0.6528	0.4751	0.3919	0.3428	0.3099	0.2861	0.2680	0.2535	0.2419	0.2320	0.1961	0.1535	0.1157	0.0833
15	0.5747	0.4069	0.3317	0.2882	0.2593	0.2386	0.2228	0.2104	0.2002	0.1918	0.1612	0.1251	0.0934	0.0667
20	0.4799	0.3297	0.2654	0.2288	0.2048	0.1877	0.1748	0.1646	0.1567	0.1501	0.1248	0.0960	0.0709	0.0500
24	0.4247	0.2871	0.2295	0.1970	0.1759	0.1608	0.1495	0.1406	0.1338	0.1283	0.1060	0.0810	0.0595	0.0417
30	0.3632	0.2412	0.1913	0.1635	0.1454	0.1327	0.1232	0.1157	0.1100	0.1054	0.0867	0.0658	0.0480	0.0333
40	0.2940	0.1915	0.1508	0.1281	0.1135	0.1033	0.0957	0.0898	0.0853	0.0816	0.0668	0.0503	0.0363	0.0250
60	0.2151	0.1371	0.1069	0.0902	0.0796	0.0722	0.0668	0.0625	0.0594	0.0567	0.0461	0.0344	0.0245	0.0167
120	0.1225	0.0759	0.0585	0.0489	0.0429	0.0387	0.0357	0.0334	0.0316	0.0302	0.0242	0.0178	0.0125	0.0083
∞	0	0	0	0	0	0	0	0	0	0	0	0	0	0

† Reprinted with permission from C. Eisenhart, M. W. Hastay, W. A. Wallis, "Techniques of Statistical Analysis," McGraw-Hill Book Company, New York, 1947.

Appendix B18 The z transformation for the correlation coefficient†

r or ρ	z(r) or z(ρ)	r or ρ	z(r) or z(ρ)	r or ρ	z(r) or z(ρ)
0.00	0.0000	0.35	0.3654	0.70	0.8673
0.01	0.0100	0.36	0.3769	0.71	0.8872
0.02	0.0200	0.37	0.3884	0.72	0.9076
0.03	0.0300	0.38	0.4001	0.73	0.9287
0.04	0.0400	0.39	0.4118	0.74	0.9505
0.05	0.0500	0.40	0.4236	0.75	0.9730
0.06	0.0601	0.41	0.4356	0.76	0.9962
0.07	0.0701	0.42	0.4477	0.77	1.0203
0.08	0.0802	0.43	0.4599	0.78	1.0454
0.09	0.0902	0.44	0.4722	0.79	1.0714
0.10	0.1003	0.45	0.4847	0.80	1.0986
0.11	0.1104	0.46	0.4973	0.81	1.1270
0.12	0.1206	0.47	0.5101	0.82	1.1568
0.13	0.1307	0.48	0.5230	0.83	1.1881
0.14	0.1409	0.49	0.5361	0.84	1.2212
0.15	0.1511	0.50	0.5493	0.85	1.2562
0.16	0.1614	0.51	0.5627	0.86	1.2933
0.17	0.1717	0.52	0.5763	0.87	1.3331
0.18	0.1820	0.53	0.5901	0.88	1.3758
0.19	0.1923	0.54	0.6042	0.89	1.4219
0.20	0.2027	0.55	0.6184	0.90	1.4722
0.21	0.2132	0.56	0.6328	0.91	1.5275
0.22	0.2237	0.57	0.6475	0.92	1.5890
0.23	0.2342	0.58	0.6625	0.93	1.6584
0.24	0.2448	0.59	0.6777	0.94	1.7380
0.25	0.2554	0.60	0.6931	0.95	1.8318
0.26	0.2661	0.61	0.7089	0.96	1.9459
0.27	0.2769	0.62	0.7250	0.961	1.9588
0.28	0.2877	0.63	0.7414	0.962	1.9721
0.29	0.2986	0.64	0.7582	0.963	1.9857
0.30	0.3095	0.65	0.7753	0.964	1.9996
0.31	0.3205	0.66	0.7928	0.965	2.0139
0.32	0.3316	0.67	0.8107	0.966	2.0287
0.33	0.3428	0.68	0.8291	0.967	2.0439
0.34	0.3541	0.69	0.8480	0.968	2.0595

r or ρ	$z(r)$ or $z(\rho)$	r or ρ	$z(r)$ or $z(\rho)$	r or ρ	$z(r)$ or $z(\rho)$
0.969	2.0756	0.979	2.2729	0.989	2.5987
0.970	2.0923	0.980	2.2976	0.990	2.6467
0.971	2.1095	0.981	2.3235	0.991	2.6996
0.972	2.1273	0.982	2.3507	0.992	2.7587
0.973	2.1457	0.983	2.3796	0.993	2.8257
0.974	2.1649	0.984	2.4101	0.994	2.9031
0.975	2.1847	0.985	2.4427	0.995	2.9945
0.976	2.2054	0.986	2.4774	0.996	3.1063
0.977	2.2269	0.987	2.5147	0.997	3.2504
0.978	2.2494	0.988	2.5550	0.998	3.4534

† Reproduced with permission from D. B. Owen, " Handbook of Statistical Tables," Addison-Wesley Publishing Company, Inc., Reading, Mass., 1962.

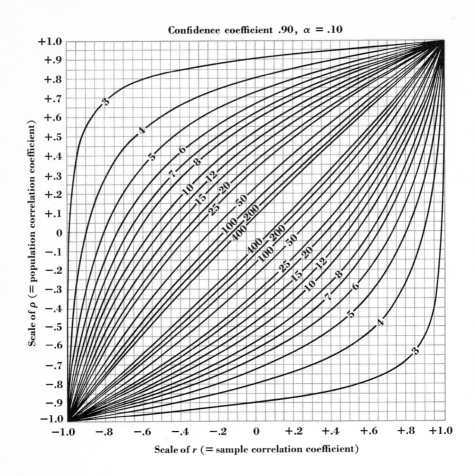

Confidence coefficient .90, $\alpha = .10$

Scale of ρ (= population correlation coefficient)

Scale of r (= sample correlation coefficient)

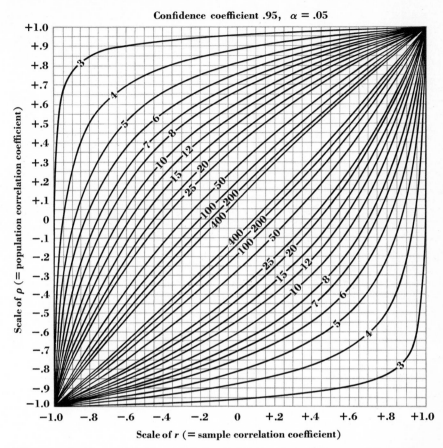

Confidence coefficient .95, $\alpha = .05$

Scale of ρ (= population correlation coefficient)

Scale of r (= sample correlation coefficient)

† Reproduced with permission from F. N. David, "Tables of the Ordinates and Probability Integral of the Distribution of the Correlation Coefficient in Small Samples," *Biometrika*, London, 1938.

Appendix B20 Cumulative probabilities for the Wilcoxon signed rank statistic

$n = 5$

t	$\Pr(T \le t)$
0	.0313
1	.0625
2	.0938
3	.1563

$n = 6$

t	$\Pr(T \le t)$
0	.0156
1	.0313
2	.0469
3	.0781
4	.1094

$n = 7$

t	$\Pr(T \le t)$
0	.0078
1	.0156
2	.0234
3	.0391
4	.0547
5	.0781
6	.1094

$n = 8$

t	$\Pr(T \le t)$
0	.0039
1	.0078
2	.0117
3	.0195
4	.0273
5	.0391
6	.0547
7	.0742
8	.0977
9	.1250

$n = 9$

t	$\Pr(T \le t)$
1	.0039
2	.0059
3	.0098
4	.0137
5	.0195
6	.0273
7	.0371
8	.0488
9	.0645
10	.0820
11	.1016

$n = 10$

t	$\Pr(T \le t)$
3	.0049
4	.0068
5	.0098
6	.0137
7	.0186
8	.0244
9	.0322
10	.0420
11	.0527
12	.0654
13	.0801
14	.0967
15	.1162

$n = 11$

t	$\Pr(T \le t)$
5	.0049
6	.0068
7	.0093
8	.0122
9	.0161
10	.0210
11	.0269
12	.0337
13	.0415
14	.0508
15	.0615
16	.0737
17	.0874
18	.1030

$n = 12$

t	$\Pr(T \le t)$
7	.0046
8	.0061
9	.0081
10	.0105
11	.0134
12	.0171
13	.0212
14	.0261
15	.0320
16	.0386
17	.0461
18	.0549
19	.0647
20	.0757
21	.0881
22	.1018

$n = 13$

t	$\Pr(T \le t)$
9	.0040
10	.0052
11	.0067
12	.0085
13	.0107
14	.0133
15	.0164
16	.0199
17	.0239
18	.0287
19	.0341
20	.0402
21	.0471
22	.0549
23	.0636
24	.0732
25	.0839
26	.0955
27	.1082

$n = 14$

t	$\Pr(T \le t)$
12	.0043
13	.0054
14	.0067
15	.0083
16	.0101
17	.0123
18	.0148
19	.0176
20	.0209
21	.0247
22	.0290
23	.0338
24	.0392
25	.0453
26	.0520
27	.0594
28	.0676
29	.0765
30	.0863
31	.0969
32	.1083

$n = 15$

t	$\Pr(T \le t)$
15	.0042
16	.0051
17	.0062
18	.0075
19	.0090
20	.0108
21	.0128
22	.0151
23	.0177
24	.0206
25	.0240
26	.0277
27	.0319
28	.0365
29	.0416
30	.0473
31	.0535
32	.0603
33	.0677
34	.0757
35	.0844
36	.0938
37	.1039

$n = 16$

t	$\Pr(T \le t)$
19	.0046
20	.0055
21	.0065
22	.0078
23	.0091
24	.0107
25	.0125
26	.0145
27	.0168
28	.0193
29	.0222
30	.0253
31	.0288
32	.0327
33	.0370
34	.0416
35	.0467
36	.0523
37	.0583
38	.0649
39	.0719
40	.0795
41	.0877
42	.0964
43	.1057

$n = 17$

t	$\Pr(T \le t)$
23	.0047
24	.0055
25	.0064
26	.0075
27	.0087
28	.0101
29	.0116
30	.0133
31	.0153
32	.0174
33	.0198
34	.0224
35	.0253
36	.0284
37	.0319
38	.0357
39	.0398
40	.0443
41	.0492

$n=17$		$n=18$		$n=19$		$n=20$	
t	$\Pr(T \leqq t)$	t	$\Pr(T \leqq t)$	t	$\Pr(T \leqq t)$	t	$\Pr(T \leqq t)$
42	.0544	44	.0368	44	.0201	43	.0096
43	.0601	45	.0407	45	.0223	44	.0107
44	.0662	46	.0449	46	.0247	45	.0120
45	.0727	47	.0494	47	.0273	46	.0133
46	.0797	48	.0542	48	.0301	47	.0148
47	.0871	49	.0594	49	.0331	48	.0164
48	.0950	50	.0649	50	.0364	49	.0181
49	.1034	51	.0708	51	.0399	50	.0200
		52	.0770	52	.0437	51	.0220
		53	.0837	53	.0478	52	.0242
$n=18$		54	.0907	54	.0521	53	.0266
27	.0045	55	.0982	55	.0567	54	.0291
28	.0052	56	.1061	56	.0616	55	.0319
29	.0060			57	.0668	56	.0348
30	.0069			58	.0723	57	.0379
31	.0080	$n=19$		59	.0782	58	.0413
32	.0091	32	.0047	60	.0844	59	.0448
33	.0104	33	.0054	61	.0909	60	.0487
34	.0118	34	.0062	62	.0978	61	.0527
35	.0134	35	.0070	63	.1051	62	.0570
36	.0152	36	.0080			63	.0615
37	.0171	37	.0090	$n=20$		64	.0664
38	.0192	38	.0102	37	.0047	65	.0715
39	.0216	39	.0115	38	.0053	66	.0768
40	.0241	40	.0129	39	.0060	67	.0825
41	.0269	41	.0145	40	.0068	68	.0884
42	.0300	42	.0162	41	.0077	69	.0947
43	.0333	43	.0180	42	.0086	70	.1012

Extracted with permission from "Selected Tables in Mathematical Statistics," vol. I (sponsored by the Institute of Mathematical Statistics and edited by H. L. Harter and D. B. Owen), Markham Publishing Company, Chicago, 1970.

Appendix B21 Critical values and probability levels for the Wilcoxon rank sum test

$$\alpha_0 = P = .05 \text{ one-sided}$$
$$\alpha_0 = P = .10 \text{ two-sided}$$

n	m = 6		m = 7		m = 8		m = 9		m = 1(	
6	28, 50	.0465								
	29, 49	.0660								
7	29, 55	.0367	39, 66	.0487						
	30, 54	.0507	40, 65	.0641						
8	31, 59	.0406	41, 71	.0469	51, 85	.0415				
	32, 58	.0539	42, 70	.0603	52, 84	.0524				
9	33, 63	.0440	43, 76	.0454	54, 90	.0464	66, 105	.0470		
	34, 62	.0567	44, 75	.0571	55, 89	.0570	67, 104	.0567		
10	35, 67	.0467	45, 81	.0439	56, 96	.0416	69, 111	.0474	82, 128	
	36, 66	.0589	46, 80	.0544	57, 95	.0506	70, 110	.0564	83, 127	
11	37, 71	.0491	47, 86	.0427	59, 101	.0454	72, 117	.0476	86, 134	
	38, 70	.0608	48, 85	.0521	60, 100	.0543	73, 116	.0560	87, 133	
12	38, 76	.0415	49, 91	.0416	62, 106	.0489	75, 123	.0477	89, 141	
	39, 75	.0512	50, 90	.0501	63, 105	.0576	76, 122	.0555	90, 140	
13	40, 80	.0437	52, 95	.0484	64, 112	.0445	78, 129	.0478	92, 148	
	41, 79	.0530	53, 94	.0573	65, 111	.0521	79, 128	.0551	93, 147	
14	42, 84	.0457	54, 100	.0469	67, 117	.0475	81, 135	.0478	96, 154	
	43, 83	.0547	55, 99	.0550	68, 116	.0550	82, 134	.0547	97, 153	
15	44, 88	.0474	56, 105	.0455	69, 123	.0437	84, 141	.0478	99, 161	
	45, 87	.0561	57, 104	.0531	70, 122	.0503	85, 140	.0542	100, 160	
16	46, 92	.0490	58, 110	.0443	72, 128	.0463	87, 147	.0477	103, 167	
	47, 91	.0574	59, 109	.0513	73, 127	.0528	88, 146	.0538	104, 166	
17	47, 97	.0433	61, 114	.0497	75, 133	.0487	90, 153	.0476	106, 174	
	48, 96	.0505	62, 113	.0569	76, 132	.0552	91, 152	.0534	107, 173	
18	49, 101	.0448	63, 119	.0484	77, 139	.0452	93, 159	.0475	110, 180	
	50, 100	.0518	64, 118	.0550	78, 138	.0510	94, 158	.0531	111, 179	
19	51, 105	.0462	65, 124	.0471	80, 144	.0475	96, 165	.0474	113, 187	
	52, 104	.0530	66, 123	.0533	81, 143	.0532	97, 164	.0527	114, 186	

$$\alpha_0 = P = .01 \text{ one-sided}$$
$$\alpha_0 = P = .02 \text{ two-sided}$$

m = 6		*m* = 7		*m* = 8		*m* = 9		*m* = 10	
24, 54	.0076								
25, 53	.0130								
25, 59	.0070	34, 71	.0087						
26, 58	.0111	35, 70	.0131						
27, 63	.0100	35, 77	.0070	45, 91	.0074				
28, 62	.0147	36, 76	.0103	46, 90	.0103				
28, 68	.0088	37, 82	.0082	47, 97	.0076	59, 112	.0094		
29, 67	.0128	38, 81	.0115	48, 96	.0103	60, 111	.0122		
29, 73	.0080	39, 87	.0093	49, 103	.0078	61, 119	.0086	74, 136	.0093
30, 72	.0112	40, 86	.0125	50, 102	.0103	62, 118	.0110	75, 135	.0116
30, 78	.0073	40, 93	.0077	51, 109	.0079	63, 126	.0079	77, 143	.0098
31, 77	.0101	41, 92	.0102	52, 108	.0102	64, 125	.0100	78, 142	.0121
32, 82	.0091	42, 98	.0085	53, 115	.0079	66, 132	.0092	79, 151	.0084
33, 81	.0122	43, 97	.0111	54, 114	.0101	67, 131	.0114	80, 150	.0103
33, 87	.0084	44, 103	.0093	56, 120	.0099	68, 139	.0085	82, 158	.0089
34, 86	.0110	45, 102	.0118	57, 119	.0123	69, 138	.0104	83, 157	.0107
34, 92	.0077	45, 109	.0079	58, 126	.0098	71, 145	.0096	85, 165	.0093
35, 91	.0100	46, 108	.0100	59, 125	.0120	72, 144	.0115	86, 164	.0110
36, 96	.0092	47, 114	.0086	60, 132	.0097	73, 152	.0089	88, 172	.0096
37, 95	.0117	48, 113	.0106	61, 131	.0117	74, 151	.0106	89, 171	.0113
37, 101	.0085	49, 119	.0092	62, 138	.0096	76, 158	.0098	91, 179	.0099
38, 100	.0107	50, 118	.0112	63, 137	.0115	77, 157	.0116	92, 178	.0115
39, 105	.0099	51, 124	.0097	64, 144	.0095	78, 165	.0091	93, 187	.0088
40, 104	.0122	52, 123	.0118	65, 143	.0113	79, 164	.0107	94, 186	.0102
40, 110	.0091	52, 130	.0085	66, 150	.0094	81, 171	.0100	96, 194	.0090
41, 109	.0112	53, 129	.0103	67, 149	.0110	82, 170	.0116	97, 193	.0104
41, 115	.0085	54, 135	.0090	68, 156	.0093	83, 178	.0093	99, 201	.0093
42, 114	.0104	55, 134	.0108	69, 155	.0108	84, 177	.0108	100, 200	.0106

acted with permission from "Selected Tables in Mathematical Statistics," vol. I (sponsored ie Institute of Mathematical Statistics and edited by H. L. Harter and D. B. Owen), Markham ishing Company, Chicago, 1970.

Appendix B22 Distribution of the rank correlation coefficient

Tabulated value $= r_0$, the largest value of r such that $\Pr(R < r) \leq \alpha_0$.

n	$\alpha_0 = .10$	$\alpha_0 = .05$	$\alpha_0 = .025$	$\alpha_0 = .01$	$\alpha_0 = .005$
4	$-.8000$	$-.8000$			
5	$-.7000$	$-.8000$	$-.9000$	$-.9000$	
6	$-.6000$	$-.7714$	$-.8286$	$-.8857$	$-.9429$
7	$-.5357$	$-.6786$	$-.7450$	$-.8571$	$-.8929$
8	$-.5000$	$-.6190$	$-.7143$	$-.8095$	$-.8571$
9	$-.4667$	$-.5833$	$-.6833$	$-.7667$	$-.8167$
10	$-.4424$	$-.5515$	$-.6364$	$-.7333$	$-.7818$
11	$-.4182$	$-.5273$	$-.6091$	$-.7000$	$-.7455$
12	$-.3986$	$-.4965$	$-.5804$	$-.6713$	$-.7273$
13	$-.3791$	$-.4780$	$-.5549$	$-.6429$	$-.6978$
14	$-.3626$	$-.4593$	$-.5341$	$-.6220$	$-.6747$
15	$-.3500$	$-.4429$	$-.5179$	$-.6000$	$-.6536$
16	$-.3382$	$-.4265$	$-.5000$	$-.5824$	$-.6324$
17	$-.3260$	$-.4118$	$-.4853$	$-.5637$	$-.6152$
18	$-.3148$	$-.3994$	$-.4716$	$-.5480$	$-.5975$
19	$-.3070$	$-.3895$	$-.4579$	$-.5333$	$-.5825$
20	$-.2977$	$-.3789$	$-.4451$	$-.5203$	$-.5684$
21	$-.2909$	$-.3688$	$-.4351$	$-.5078$	$-.5545$
22	$-.2829$	$-.3597$	$-.4241$	$-.4963$	$-.5426$
23	$-.2767$	$-.3518$	$-.4150$	$-.4852$	$-.5306$
24	$-.2704$	$-.3435$	$-.4061$	$-.4748$	$-.5200$
25	$-.2646$	$-.3362$	$-.3977$	$-.4654$	$-.5100$
26	$-.2588$	$-.3299$	$-.3894$	$-.4564$	$-.5002$
27	$-.2540$	$-.3236$	$-.3822$	$-.4481$	$-.4915$
28	$-.2490$	$-.3175$	$-.3749$	$-.4401$	$-.4828$
29	$-.2443$	$-.3113$	$-.3685$	$-.4320$	$-.4744$
30	$-.2400$	$-.3059$	$-.3620$	$-.4251$	$-.4665$

Extracted with permission from ''Practical Nonparametric Statistics'' by
W. J. Conover, John Wiley & Sons, Inc., New York, 1971.

ANSWERS

1-1 A reasonable sample space consists of six points which we can label 1, 2, 3, 4, 5, 6, where the numbers correspond to the number of spots on the various sides of the die. The weight $1/6$ for each point is reasonable. Since two points have designations greater than 4, the probability that the number is greater than 4 is $2/6$.

1-3 With five answers use a sample space consisting of five points each with weight $1/5$. This is the probability of a correct answer. Similarly, if three answers are eliminated, then the probability is $1/2$.

1-5 Using a seven-point sample space with equal weights of $1/7$ yields the probability $2/7$. In the second case the sample points are all labeled with women's names, and the probability is 1.

1-7 We can use a three-point sample space, designating the points by A, B, C. The best weights we can assign are .60, .30, .10. The probability that A loses is $.30 + .10 = .40$.

1-9 Denote the five categories by A, B, C, D, E, respectively. The best weights we can assign are .35, .25, .20, .15, .05. The probability that the individual is 35 or over is $.20 + .15 + .05 = .40$.

1-11 Twelve sample points are suggested by the problem. Denoting wind condition by w or c, rainfall by m or d, and temperature by a, n, or b, the sample points could be designated by wma, wmn, wmb, wda, wdn, wdb, cma, cmn, cmb, cda, cdn, cdb. To obtain weights we could examine the weather bureau records and assign as weights the fraction of July 4's which have fallen into each category.

1-13 Letting the weights be w, w, w, $2w$, the equation $w + w + w + 2w = 1$ yields $w = 1/5$. Hence the weights are $1/5$, $1/5$, $1/5$, $2/5$.

1-15 Use a sample space with points labeled 3, 2, 1, 0 and weights .133, .382, .367, .118, respectively. The probability of two or more boys is $.133 + .382 = .515$.

1-17 Formula (1-9) yields $9/20$. The occurrence of a tie can be regarded to have probability zero.

1-19 Formula (1-8) yields $5/12$.

1-21 $3/8$

1-23 $8/663$

1-25 $(5/6)^3(1/6) = 125/1,296$

1-27 $1 - (1/2)(1/3)(1/4) = 23/24$

1-29 $1/3$

1-31 Use the multiplication theorem for independent events.

1-33 $77/102$

1-35 $1,260$, $362,880$

1-37 $5/28$

1-39 $2^{10} = 1,024$

1-41 $21/575 = .0365$

1-43
$$\left[\binom{5}{3}\binom{21}{10} + \binom{5}{2}\binom{21}{11}\right] \Big/ \binom{26}{13} = 156/230 = .678$$

$$\left[\binom{5}{4}\binom{21}{9} + \binom{5}{1}\binom{21}{12}\right] \Big/ \binom{26}{13} = 65/230 = .283$$

$$\left[\binom{5}{5}\binom{21}{8} + \binom{5}{0}\binom{21}{13}\right] \Big/ \binom{26}{13} = 9/230 = .039$$

1-45 $5/21 = .238$

1-47 $9/35 = .257$

1-49

x	2	3	4	5	6	7	8	9	10	11	12
$f(x)$	1/36	2/36	3/36	4/36	5/36	6/36	5/36	4/36	3/36	2/36	1/36

r	2	3	4	5	6	7	8	9	10	11	12
$F(r)$	1/36	3/36	6/36	10/36	15/36	21/36	26/36	30/36	33/36	35/36	1

$E(X) = 7$

1-51

x	1	2	3	4	5	6
$F(x)$	1/21	3/21	6/21	10/21	15/21	1

$E(X) = 13/3$, variance of $X = 20/9$

1-53 (a) .84, .31, .57

(b)

x	0	1	2	3	4	5
$f(x)$	.13	.14	.26	.31	.08	.08

1-55 $-\$1/18 \cong -\$.056$

1-57 2

1-59 1

1-61 $-\$17/216 \cong -\$.08$

1-63 $f(x) = (3/4)(1/4)^{x-1}$ $x = 1, 2, 3, \ldots$
$E(X) = 4/3$

1-65 (a) .10   (b) .19

2-1 .16308, .13230. Since some players appear to hit in streaks, one might doubt that p remains constant and successive trials are independent.

2-3 .00163. One is apt to doubt the claim.

2-5 .07257. Perhaps one should improve with practice, so that p may change from trial to trial. In addition, successive trials may not be independent, since a thrower may tend to follow a poor result by another poor result. $\mu = 10$, $\sigma^2 = 9$, $\sigma = 3$.

2-7 .00004. It is reasonable to believe that the new cure is successful more than .30 of the time.

2-9 p is not the same for each experiment.

2-11 The (*b*) and (*c*) conditions are apt to be violated. *p* will be different for each server.

2-13 .81394

2-15 .16286

2-17 .36689, 320/3 or about 107, $\sigma_Y^2 = 2,240/9$

2-19 .00001. We are apt to conclude that the probability of a success is less than .90.

2-21 .92743

2-23 .50, .875

2-25 .77484, .89291

2-27 $[5!/(1!\,1!\,3!)](.5)(.3)(.2)^3 = .024$. If drawing is done without replacement, probabilities associated with the three categories will change slightly from selection to selection, and successive draws will be dependent. However, with a large number of students neither of these objections is serious. Both can be overcome by drawing with replacement. If X_1 is the number on the committee from fraternities and sororities, then the marginal distribution of X_1 is a binomial with $n = 5$, $p = .20$, and $\Pr(X_1 \geq 3) = 1 - \sum_{x_1=0}^{2} b(x_1; 5, .20) = 1 - .94208 = .05792$.

2-29 .12, $11/16 = .6875$

2-31 .0384, .05792

2-33 .1296

2-35 We start in row 121, column 20. The first number is 0, which we could let correspond to 10 and use one-digit numbers. Below 0 are 2 and 5. Hence, individuals with numbers 2, 5, and 10 prepare the report. If we use two-digit numbers, then the first number is 04 and we use individuals 4, 2, and 5.

2-37 Let numbers 1, 2, 3 be rewarded one step, 4 and 5 be rewarded two steps, 6 and 7 be rewarded three steps, 8 be rewarded 4 steps, 9 be rewarded 5 steps, and 10 be rewarded six steps. Then the probabilities of being rewarded the various number of steps are identical with the spinner. Since the first five numbers obtained from the table are 3, 3, 9, 2, 7, the corresponding number of steps rewarded are 1, 1, 5, 1, 3.

2-39 The first two numbers 395, 816 are too large. The next number 075 is in the desired range. Continuing, we next find 344 and 117 in the desired range. Hence day 75 (March 15), day 117 (April 26), and day 344 (December 9) are the first three birthdays.

2-41 28, 50, 70, 6, 32, 37, 64, 39, 26, 31

2-43 .476190, .738095

2-45 Letting X be the number of frozen oranges in the sample, we seek $\Pr(X \geq 7) = 1 - \Pr(X \leq 6) = 1 - P(100, 10, 20, 6)$. The approximation yields

$$\sum_{x=7}^{10} b(x; 10, .20) = 1 - \sum_{x=0}^{6} b(x; 10, .20) = 1 - .99914 = .00086$$

Since this probability is so small, one is apt to conclude that the box contains more than 20 frozen oranges.

2-47 $p(10, 4, 3, 3) = .033333$, $P(10, 4, 3, 3) = 1$ (obviously), $P(10, 4, 3, 2) = .966667$

2-49

x	1	2	3	4	5
$\Pr(X = x)$	.023810	.238095	.476190	.238095	.023810

2-51 We need $1 - P(500, 25, 50, 4)$. Using (2-35) yields .09799 as an approximate answer.

2-53 $p(10, 2, 1, 0) = .800000$, $p(10, 2, 5, 0) = p(10, 5, 2, 0) = .222222$

2-55 .38404

2-57 Using the Poisson approximation to the binomial yields .60653.

2-59 $1 - P(3; 4) = .18474$

2-61 $P(6; 2) = .06197$, $1 - P(6; 9) = .08392$

2-63 (a) 2.326 (b) 1.282 (c) -1.960 (d) -1.645

2-65 (a) .3250 (b) .6935 (c) .0403

2-67 .74 (.75 with the improved approximation)

2-69 About .748 without the 1/2, about .798 with the 1/2. The exact value is .80182.

2-71 .9545, .0027

2-73 .1587, .0250

2-75 .3897

2-77 .34304 of an ounce

2-79 (a) 2.528 (b) -1.753 (c) 0

2-81 (a) 29.71 (b) 76.15

2-83 (a) 21.6 (b) 27.1

2-85 (a) .100 (b) 5.64 (c) 1.00

3-1 $t = .28$. The probabilities are .99161, .96846, .96480. The minimum occurs at $p = .50$ and is .96480.

3-3 323

3-5 $t = .3$

3-7 For fraction defective $t = 2/25 = .08$. For the number of defectives in the box $t = 1000(.08) = 80$.

3-9 $t_1 = .20$, $t_2 = .16$, $t_3 = .04$, $t_4 = .14$, $t_5 = .12$, $t_6 = .06$, $t_7 = .28$

3-11 $t = 109.4$ (We would probably take $t = 109$.) We have to assume that at any given time each locomotive has an equal chance of being observed.

3-13 6,623

3-15 $\bar{x} = 575$

3-17 $\bar{x} = 48.4$, $\mu = 50.5$

3-19 $\Pr(Z < -5) =$ practically 0. We conclude that the company claims too high an average.

3-21 .1587

3-23 $n = 44$

3-25 $\bar{x} = 1.8$

3-27 $.26503$, $\Pr(Z > 1) = .159$

3-29 $\Pr(Z > 1.058) = .144$

3-31 $\Pr(Z < -1.5) = .067$

3-33 $\Pr(Z > 1.5) = .067$

3-35 $n = 97$

3-37 The sample space consists of six pairs $(1, 2), (1, 3), (2, 1), (2, 3), (3, 1),$ $(3, 2)$, each pair having weight $1/6$. The probability distribution of $\bar{X}$ is

$\bar{x}$	1.5	2.0	2.5
$f(\bar{x})$	2/6	2/6	2/6

We find $E(X) = E(\bar{X}) = 2.0$, $\sigma_X^2 = 2/3$, $\sigma_{\bar{X}}^2 = 1/6$, $N = 3$.

3-39 Again $s^2 = 14{,}361$.

3-41 With (3-25) we get $s^2 = 9.32/9 = 1.036$. To use (3-28) we need $\sum_{i=1}^{10} x_i^2 = 20{,}530.22$, $\sum_{i=1}^{10} x_i = 453.0$. Then, $s = 1.018$.

3-43 $\sum_{i=1}^{50} x_i^2 = 220$, $\sum_{i=1}^{50} x_i = 90$, $s^2 = 58/49 = 1.184$, $s = 1.088$

3-45 $\bar{x} = 2$, $s^2 = 140/99 = 1.414$, $s = 1.189$

3-47 The probability distribution of S^2 is

s^2	.5	2.0
$f(s^2)$	4/6	2/6

$E(2S^2/3) = 2/3 = \sigma_X^2$

3-49 $\Pr(Y_{10} > 18.31) = .05$

4-1 $.21996, .43890, .65935, .92643, .99268$. The power is decreased and β is increased for all $p > .10$.

4-3 $.982, .804, .346, .05$. The power is increased and β is decreased for all $\mu < 30$.

4-5 $.912, .639, .263, .10, .263, .639, .912$. The power is increased and β is decreased.

4-7 The critical region is $x \geq 7$ with $\alpha = .00948$. We reject $H_0 : p = .10$.

5-1 Test $H_0 : \mu = 12$ against $H_1 : \mu \neq 12$. Since $z = -1.7$ is not in the critical region $z < -1.960$, $z > 1.960$, H_0 is accepted.

5-3 Test $H_0 : \mu = 50$ against $H_1 : \mu < 50$. Since $z = -1.925$ falls in the critical region $z < -1.645$, H_0 is rejected and the bulbs are purchased.

5-5 Test $H_0 : \mu = 220$ against $H_1 : \mu > 220$. Since $z = 1.2$ does not fall in the critical region $z > 1.282$, H_0 is accepted and the psychologist will switch to the new variety.

5-6 Test $H_0 : \mu = 240$ against $H_1 : \mu < 240$. Since $z = -1.467$ falls in the critical region $z < -1.282$, H_0 is rejected and the psychologist will switch to the new variety.

5-7 Test $H_0 : \mu = 90$ against $H_1 : \mu > 90$. Since $z = 1.5$ does not fall in the critical region $z > 1.645$, H_0 is accepted and the corn grower will not switch to fat kernel.

5-9   Test $H_0 : \mu = 50$ against $H_1 : \mu \neq 50$. Since $z = -3.6$ falls in the critical region $z < -2.576$, $z > 2.576$, H_0 is rejected. The conclusion is that the mean on the test has been lowered by teaching new math.

5-11   Test $H_0 : \mu = 10$ against $H_1 : \mu > 10$. Since $z = 3.2$ falls in the critical region $z > 2.326$, H_0 is rejected and it is concluded that brand F meets the fisherman's requirements.

5-12   We find $\delta = 2$. Since $\alpha = .05$ and the test is two-sided, use the graphs labeled $\alpha = .025$ and read power $= .52$. For the power to be .99, when $\mu = 11.8$ the graphs yield $\delta = 4.3$ and we need $n \cong 116$. Using (5-15) yields $\delta = 4.286$, $n = 115$.

5-14   We find $\delta = 3.5$, power $= .97$. To have power $= .90$ if $\mu = 48$ we need $\delta = 2.9$, $n = 34$. Formula (5-15) yields $n = 35$.

5-16   Power $= \Pr(Z < .051) = .52$. To have power $= .90$ if $\mu = 230$ we need $n = 60$.

5-18   We find $\delta = 4$, power $= .925$. To have power $= .80$ if μ differs from 50 by 1 unit, we need $\delta = 3.4$, $n = 289$. Formula (5-15) yields $n = 293$.

5-20   (11.63, 12.03). The interval includes 12, and the H_0 of Exercise 5-1 was accepted. $n = 385$.

5-22   (47.96, 49.84), $n = 174$

5-24   (214.3, 243.7), $n = 62$

5-25   (89.71, 96.29), $n = 44$, (90.52, 95.48)

5-27   (46.91, 49.49). The interval does not contain 50, and the H_0 of Exercise 5-9 was rejected. $n = 664$. (47.20, 49.20).

5-29   Since $t_{24} = 2$ does not fall in the critical region $t_{24} < -2.064$, $t_{24} > 2.064$, accept H_0.

5-30   We find $\delta = 2.5$. The graphs labeled $\alpha = .025$ yields power $= .67$ when $f = 24$. Following the procedure of Example 5-12, we find that the minimum n required to achieve power $= .99$ is in the range $73.96 \leq n \leq 81$. (It can be shown that the exact minimum n is 76.)

5-31   (11.99, 12.49). The interval includes $\mu = 12$, as it should, since the hypothesis of Exercise 5-29 was accepted.

5-32   The graphs of Appendix B11 yield probability .30. To raise this to .99 requires $n = 45$.

5-37   We find $\delta = 2.4$. The graphs labeled $\alpha = .01$ yields power $= .54$ with $f = \infty$, power $= .47$ with $f = 24$. We would probably guess that power is about .50 with $f = 35$. (Inverse linear interpolation yields power $= .49$.) Following the procedure of Example 5-12, we find that the minimum n required to achieve power $= .90$ is in the range $81 \leq n \leq 90.82$. (It can be shown that the exact minimum n is 85.) In Exercise 5-13 we needed $n = 81$ or 82. Hence the curve for $f = \infty$ very nearly yields the δ which produces the minimum n.

5-38   Since $t_{48} = -2.2$ falls in the critical region $t_{48} < -1.68$, reject H_0.

5-39   We find $\delta = 1.75$. The graph labeled $\alpha = .05$ yields power $= .55$

when $f = \infty$, power $= .53$ when $f = 24$. Hence we would probably guess that power $\cong .54$ when $f = 48$. Following the procedure of Example 5-12, we find that the minimum n required to achieve power $= .80$ is in the range $99.20 \leq n \leq 103.22$. (It can be shown that the exact minimum n is 101.)

5-40   (48.06, 49.74)

5-41   The probability (read from Appendix B11) is about .70. To raise this to .99 requires about $n = 63$.

5-45   Since $t_{15} = 1.5$ falls in the critical region $t_{15} > 1.341$, H_0 is rejected and the new variety will not be adopted.

5-46   (221.0, 237.0)

5-51 $k = (.12228)^2 = .014952$, $n \geq (.014952)(90.2)^2 = 121.6$. Thus $n = 122$. The observed value of the statistic is $t_{19} = -3.67$, which falls in the critical region $t_{19} < -1.729$. Hence, H_0 is rejected and the company will not switch to the new brand.

5-52 $k = (.08372)^2 = .007009$, $n \geq (.007009)(90.2)^2 = 57.03$. Thus $n = 58$. The interval is (960, 1010).

5-55 $k = (2.000)^2 = 4.000$, $n \geq (4.000)(15.9) = 63.6$. Hence $n = 64$, so that 50 additional observations are required. The observed value of the statistic is $t_{13} = -1.00$, which does not fall in the critical region $t_{13} < -2.650$. Thus H_0 is accepted and the bulbs will not be purchased.

5-56 $k = (3.542)^2 = 12.55$, $n \geq 12.55(15.9) = 199.5$. Hence $n = 200$. The interval is (49.0, 50.0).

5-57   Test $H_0 : \sigma^2 \leq .25$ against $H_1 : \sigma^2 > .25$. Since $y_{24} = 34.56$ does not fall in the critical region $y_{24} > 36.42$, H_0 is accepted. The company continues to operate (at least for the time being) under the assumption that the standard is being maintained.

5-58   Use Appendix B12 with $R = .60/.50 = 1.2$. With $v = 20$ the power is about .38, with $v = 30$ the power is about .44. Hence, for $v = 24$ the power is about .40. If $\sigma = .75$, then $R = 1.5$. With $v = 20$, power $= .83$, and with $v = 30$ power $= .93$. Hence, for $v = 24$, the power is about .87.

5-59   With Appendix B12 we find power $= .93$ if $v = 30$, power $= .97$ if $v = 40$. Hence, to have power $= .95$ when $R = 1.5$ we need $v = 35$, $n = 36$. By using the inequalities in Example 5-22, it may be shown that the exact minimum n is 35.

5-60   For σ^2 the interval is (.2195, .6968). Taking square roots yields (.469, .835) as the interval for σ.

5-61   From Appendix B13, the probability is slightly less than .90. We might estimate the probability to be about .86. To raise the power to .90 requires $n = 26$.

5-66   Test $H_0 : \sigma^2 = 2,500$ against $H_1 : \sigma^2 < 2,500$. Since $y_{30} = 14.7$ falls in the critical region $y_{30} < 14.95$, H_0 is rejected and the new method is adopted.

5-67   With $\alpha = .01$, $R = .8$, $v = 30$, Appendix B12 yields power $= .20$.

To make power $= .60$ requires $v = 75$, $n = 76$. To raise the power to .95 we see that v must be larger than 150 yet less than 250. Since power $= .93$ if $v = 150$ and power $= .996$ if $v = 250$, linear interpolation yields $v = 180$, $n = 181$ as the required sample size. Since interpolation is over such a wide span, we might expect the result to not be very accurate. By using the inequalities in Example 5-22, it may be shown that the exact minimum n is 165.

5-68   For σ^2 the interval is $(.0782, .219)$. Taking square roots yields $(.280, .468)$ as the interval for σ.

5-69   The graphs of Appendix B13 show that when $n = 31$, $L = .5\sigma$, the probability is about .30. To raise this to .99 requires an n of about 52.

5-73   Test $H_0 : \sigma_1^2 = \sigma_2^2$ against $H_1 : \sigma_1^2 < \sigma_2^2$. Since $f_{40, 40} = .650$ does not fall in the critical region $f_{40, 40} < .591$, accept H_0 and use method 2.

5-74   With $\alpha = .05$, $R_0 = .5$, $v = 40$, the graphs of Appendix B14 yield power $= .996$. If $R_0 = .8$, then power $= .40$. To raise this to .75, v must be between 90 and 120. Since power $= .67$ if $v = 90$ and power $= .78$ if $v = 120$, linear interpolation yields $v = 112$, $n_1 = n_2 = 113$ as the minimum equal sample size.

5-75   The interval for σ_1^2/σ_2^2 is $(.346, 1.22)$, and the interval for σ_1/σ_2 is $(.588, 1.10)$.

5-79   Test $H_0 : \sigma_1^2 = \sigma_2^2$ against $H_1 : \sigma_1^2 \neq \sigma_2^2$. Since $f_{60, 120} = .444$, reject H_0 even with $\alpha = .001$.

5-80   The interval for σ_1^2/σ_2^2 is $(.310, .651)$, and the interval for σ_1/σ_2 is $(.557, .807)$.

5-83   Test $H_0 : \mu_1 = \mu_2$ against $H_1 : \mu_1 \neq \mu_2$. Since the observed value of the statistic $t_{62} = -3.6$ falls in the critical region $t_{62} < -2.66$, $t_{62} > 2.66$, H_0 is rejected. Undoubtedly we would conclude that variety B has a higher mean yield.

5-84   We find $\delta = 4$. Since the test is two-sided, we use the graphs of Appendix B10 labeled $\alpha = .005$. We read power $= .88$ if $f = 24$, power $= .92$ if $f = \infty$. Hence for $f = 62$ we have power $\cong .90$. (Inverse linear interpolation could be used.) Following the procedure of Example 5-36, we find that the minimum n required to achieve power $= .99$ is in the range $48.02 \leq n \leq 56.18$. We would probably take $n = 50$. (It can be shown that the exact minimum n is 50.)

5-85 $(-4.7, -.7)$. The interval does not include 0, and $H_0 : \mu_1 - \mu_2 = 0$ is rejected.

5-89   Test $H_0 : \mu_1 = \mu_2$ against $H_1 : \mu_1 \neq \mu_2$. We get $s_p = 9.33$, $t_{98} = 2.06$. Since the observed value of the statistic falls in the critical region $t_{98} < -1.986$, $t_{98} > 1.986$, H_0 is rejected and we conclude that the mean for method 1 is greater.

5-90   For $\sigma/2$ we find $\delta = 2.4$, power $\cong .65$. For $3\sigma/4$ we find $\delta = 3.6$, power $\cong .95$. To obtain these values enter the graphs with $\alpha/2 = .025$.

5-91   (.142, 7.86). The interval does not contain $\mu_1 - \mu_2 = 0$ and the test rejects $H_0 : \mu_1 - \mu_2 = 0$.

5-96   We would probably test $H_0 : \mu_1 = \mu_2$ against $H_1 : \mu_1 < \mu_2$ and require that the supplement "prove itself" before being adopted. We get $s_p = 31.82$, $t_{12} = -1.05$. Since the observed value of the statistic does not fall in the critical region $t_{18} < -2.552$, H_0 is accepted and the supplement is not adopted.

5-97   We find $\delta = 2.236$. With $f = 24$ the power is .41, while with $f = 12$ the power is .37. Hence with $f = 18$ the power is approximately .39. Formula (5-83) yields $n = 2\delta^2$. With the procedure of Example 5-36 we find that the minimum n is in the range $19.94 \leqq n \leqq 22.71$. (It can be shown that the exact minimum n is 22.)

5-98 $(-44.9, 14.9)$

5-99   Test $H_0 : \mu_1 = \mu_2$ against $H_1 : \mu_1 \neq \mu_2$ (or $H_0 : \mu_D = 0$ against $H_1 : \mu_D \neq 0$). We find $\bar{d} = 4.375$, $s_D = 8.48$, $t_7 = 1.46$. Since the observed value of the statistic does not fall in the critical region $t_7 < -2.365$, $t_7 > 2.365$, H_0 is accepted. No difference in mean yield has been demonstrated by the experiment.

5-100   We find $\delta = 2.83$. The graphs of Appendix B10 are entered with $\alpha/2 = .025$. With $f = 6$ the power is about .66, while with $f = 12$ the power is about .74. Hence with $f = 7$ the power is approximately .67. With $n = 25$, $\delta = 5$, and the power is slightly over .997. To find the minimum n required to raise the probability to .90, we use (5-14), which yields $n = \delta^2$. The chosen n is decreased until it is less than the computed n. With chosen $n = \infty$ ($f = \infty$) we find $\delta = 3.2$, computed $n = 10.24$. With $n = 25$ ($f = 24$) we find $\delta = 3.4$, computed $n = 11.56$. With $n = 13$ ($f = 12$) we find $\delta = 3.52$, computed $n = 12.39$. With $n = 7$ ($f = 6$) we find $\delta = 3.90$, computed $n = 15.21$. Hence the minimum n is in the range $12.39 \leqq n \leqq 15.21$ and we would probably take $n = 13$ or 14. (It can be shown that the actual minimum n is 13.)

5-101 $(-2.72, 11.47)$. The interval contains 0 and we accept $H_0 : \mu_D = 0$.

5-105   Test $H_0 : \mu_1 = \mu_2$ against $H_1 : \mu_1 \neq \mu_2$. We find $\bar{d} = -3.33$, $s_D = 3.75$, $t_{11} = -3.08$. The critical region is $t_{11} < -2.201$, $t_{11} > 2.201$. Hence H_0 is rejected, and we conclude that blood pressure measurements taken when the person is lying down have a higher mean than measurements taken when he is standing.

5-106   With $\delta = 1.732$, $\alpha/2 = .025$, $f = 11$, we find the power is about .35. To make power $= .80$ we use (5-14), which yields $n = 4\delta^2$. With $n = \infty$ ($f = \infty$), we find $\delta = 2.80$, computed $n = 31.36$. With $n = 25$ ($f = 24$) we find computed $n = 34.11$. Hence the minimum n is in the range $31.36 \leqq n \leqq 34.11$, and we would probably take $n = 32$, 33, or 34. (It may be shown that the actual minimum n is 34.)

5-107 $(-5.7, -1.0)$

6-1 Letting p be the fraction of seeds which will not germinate, test $H_0 : p = .10$ against $H_1 : p > .10$. The critical region is $x \geq 16$ with $\alpha = .03989$. Since the observed value of the statistic is $x = 19$, reject H_0 and do not believe the company's claim.

6-2 .87149, .98892

6-3 (.12, .29), obtained from Appendix B15.

6-7 $n = 25$ with critical region $x \geq 9$. When $p = .20$, power $= .04677$, and when $p = .50$, power $= .94612$. With $n = 20$ both conditions cannot be satisfied.

6-9 Letting p be the fraction which fail when the tutorial service is available, test $H_0 : p = .25$ against $H_1 : p < .25$. The critical region is $x \leq 5$ with $\alpha = .00705$. Since the observed value of the statistic is $x = 8$, H_0 is not rejected and the tutorial service will not be added.

6-12 Letting p be the fraction which fail the new examination, we test $H_0 : p = .10$ against $H_1 : p > .10$. The value of $\alpha_0 = .05$. The critical region is $x \geq 10$ with $\alpha = .02454$. When $p = .30$, $\Pr(X \geq 10) = .95977 > .95$. Hence $n = 50$ satisfies both requirements. Since the observed value of the statistic is $x = 9$, H_0 is accepted and the new examination is considered to be satisfactory. If it is decided that the examination is unsatisfactory if p is too small as well as when p is too large, then we should test $H_0 : p = .10$ against $H_1 : p \neq .10$. Now the critical region is $x \leq 0$ and $x \geq 10$ with significance level $.00515 + .02454 = .02969$. When $p = .30$ the power is $.95977 + .00000 = .95977$. Hence $n = 50$ meets both requirements, and again H_0 is accepted.

6-13 (.07, .37), obtained from Appendix B15.

6-16 Test $H_0: \mu = 1$ against $H_1 : \mu > 1$ where μ is the mean number of errors per page for the secretary. The critical region is $y \geq 10$ with $\alpha = .03183$ (computed with $n\mu_0 = 5$). Since the observed value of the statistic is $y = 13$, H_0 is rejected and the secretary is not hired.

6-17 When $\mu = 1.6$, $n\mu = 5(1.6) = 8$ and $\Pr(Y \geq 10) = .28338$. When $\mu = 2$, $n\mu = 5(2) = 10$ and $\Pr(Y \geq 10) = .54207$.

6-18 (1.38, 4.45)

6-19 We use formula (6-16). With $y_0 = 15$ we get

$$\frac{\chi^2_{30; .90}}{4} \leq n \leq \frac{\chi^2_{30; .05}}{2} \qquad \text{or} \qquad 10.06 \leq n \leq 9.24$$

so that y_0 is too small. With $y_0 = 20$ we get

$$\frac{\chi^2_{40; .90}}{4} \leq n \leq \frac{\chi^2_{40; .05}}{2} \qquad \text{or} \qquad 12.95 \leq n \leq 13.26$$

Hence we take $n = 13$ and critical region $y \geq 20$. The entries of Appendix B7 do not permit us to try $y_0 = 16, 17, 18, 19$ (at least, without interpolation).

6-23   Test $H_0 : \mu = 1$ against $H_1 : \mu < 1$. The critical region is $y \leq 3$.
With $n\mu_0 = 8(1) = 8$ we get $\alpha = \Pr(Y \leq 3) = .04238$. Since the
observed value of the statistic is $y = 3$, H_0 is rejected and the new
loom is purchased.

6-24   With $n\mu = 8(1/2) = 4$ we get power $= \Pr(Y \leq 3) = .43347$. With
$n\mu = 8(1/4) = 2$ we get power $= \Pr(Y \leq 3) = .85712$.

6-25   We use formula (6-15). With $y_0 = 8$ we get

$$\frac{\chi^2_{18; \, .95}}{2} \leq n \leq \frac{\chi^2_{18; \, .01}}{1/2} \qquad \text{or} \qquad 14.43 \leq n \leq 14.02$$

so that y_0 is too small. With $y_0 = 9$ we get

$$\frac{\chi^2_{20; \, .95}}{2} \leq n \leq \frac{\chi^2_{20; \, .01}}{1/2} \qquad \text{or} \qquad 15.70 \leq n \leq 16.52$$

Hence we take $n = 16$ and critical region $y \leq 9$. With $y = 8$,
H_0 is rejected and the new loom is purchased.

6-26 $(.102, .969)$

6-27   The critical region is $y \geq 3$. With $n\mu_0 = 60(.01) = .6$ we get
$\alpha = \Pr(Y \geq 3) = .02312$. With $n\mu = 60(.10) = 6$ we get power $=$
$\Pr(Y \geq 3) = .93803$. Hence, $n = 60$ is large enough to meet the
requirements. With (6-16) and $y_0 = 2$ we get

$$\frac{\chi^2_{4; \, .90}}{.20} \leq n \leq \frac{\chi^2_{4; \, .05}}{.02} \qquad \text{or} \qquad 38.90 \leq n \leq 35.55$$

so that y_0 is too small. With $y_0 = 3$ we get

$$\frac{\chi^2_{6; \, .90}}{.20} \leq n \leq \frac{\chi^2_{6; \, .05}}{.02} \qquad \text{or} \qquad 53.20 \leq n \leq 82.0$$

Hence the minimum $n = 54$ and the accompanying $y_0 = 3$.

6-31   With (6-16) and $y_0 = 20$ we get

$$\frac{\chi^2_{40; \, .95}}{8} \leq n \leq \frac{\chi^2_{40; \, .05}}{4} \qquad \text{or} \qquad 6.97 \leq n \leq 6.63$$

so that y_0 is too small. With $y_0 = 25$ we get

$$\frac{\chi^2_{50; \, .95}}{8} \leq n \leq \frac{\chi^2_{50; \, .05}}{4} \qquad \text{or} \qquad 8.44 \leq n \leq 8.63$$

Even though the upper bound on n is larger than the lower bound,
the interval does not contain an integer. Hence we try $y_0 = 30$
and get

$$\frac{\chi^2_{60; \, .95}}{8} \leq n \leq \frac{\chi^2_{60; \, .05}}{4} \qquad \text{or} \qquad 9.89 \leq n \leq 10.79$$

Hence, using the chi-square table of Appendix B7, we find $n = 10$

with critical region $y \geqq 30$. If the observed value of the statistic is $y = 24$, H_0 is accepted and a contract is signed with the publishing company. (With a more extensive chi-square table it may be shown that $n = 9$ with critical region $y \geqq 26$.)

6-32   The observed value of the statistic is $z' = 2.00$ and the critical region is $z' > 1.645$. Thus $H_0 : p = .10$ is rejected in favor of $H_1 : p > .10$.

6-33   With (6-24) we find power $= \Pr(Z < .94) = .83$. To raise this probability to .90, we find from (6-25) that n must be approximately 2,070.

6-34 $(.102, .128)$

6-38   The observed value of the statistic is $z' = 1.8$ and the critical region is $z' > 1.645$. Hence $H_0 : \mu = 1$ is rejected in favor of $H_1 : \mu > 1$, and the secretary is not hired.

6-39   When $\mu = 1.21$ formula (6-34) yields power $= \Pr(Z < 1.37) = .91$. To raise this probability to .95, we need $n = 271$, found from formula (6-35).

6-40 $(.98, 1.26)$

6-41   Letting p be the fraction of blacks selected for jury duty, test $H_0 : p = .01$ against $H_1 : p < .01$. The observed value of the statistic is $z' = -4.34$ and the critical region is $z' < -3.090$. Hence H_0 is rejected and we conclude $p < .01$. It is probably more accurate to conclude that non-property owners rather than blacks are discriminated against.

6-45   Test $H_0 : k = 3$ against $H_1 : k > 3$ where k is the number of defectives in the lot. The critical region is $x \geqq 3$ with $\alpha = .083333$, found with (6-40) and $N = 10$, $n = 5$, $k = 3$. Since the observed value of the statistic is $x = 2$, H_0 and the lot are accepted.

6-46   With $k = 5$ the power is $1 - P(10, 5, 5, 2) = .500000$.
       With $k = 7$ the power is $1 - P(10, 7, 5, 2) = .916667$.

6-47   Now (6-41) and (6-42) become $P(10, n, 1, c) \geqq .90$ and $P(10, n, 5, c) \leqq .20$. As in Example 6-22 we find that with $c = 0$ we need $n = 1$ and $n \geqq 3$, which is impossible. With $c = 1$ we need $n \leqq 10$ and $n \geqq 5$. Hence $n = 5$ is the minimum sample size, and the accompanying critical region is $x \geqq 2$.

6-50   Test $H_0 : k = 1$ against $H_1 : k > 1$. The critical region is $x \geqq 1$ with $\alpha = .300000$, found with (6-40) and $N = 10$, $n = 3$, $k = 1$. Since the observed value of the statistic is $x = 1$, reject H_0 and the lot.

6-51   With $k = 3$ the power is $1 - P(10, 3, 3, 0) = 1 - .291667 = .708333$.
       With $k = 5$ the power is $1 - P(10, 3, 5, 0) = 1 - .083333 = .916667$.

6-52   Now (6-41) and (6-42) become

$$P(100, n, 5, c) \geqq .95 \quad \text{and} \quad P(100, n, 20, c) \leqq .10.$$

As in Example 6-22 we find that with $c = 0$ we need $n \leqq 1$ and $n \geqq 10$ with $c = 1$ we need $n \leqq 8$ and $n \geqq 17$, with $c = 2$ we need $n \leqq 19$

and $n \geq 24$, with $c = 3$ we need $n \leq 35$ and $n \geq 29$. Hence the minimum n is 29, and the accompanying c is 3.

6-54 Test $H_0 : \mu_1 = \mu_2$ against $H_1 : \mu_1 < \mu_2$. Equivalently we test $H_0 : p = 1/2$ against $H_1 : p < 1/2$. The critical region is $y_1 \leq 5$. If H_0 is true, then Y_1 given $t = 20$ has a binomial distribution with $n = 20, p = .50$. Since the observed value of the statistic is $y_1 = 5$, H_0 is rejected, and the wholesaler will buy from firm 1.

6-55 Use formula (6-25) with $\alpha = .05$, $\beta = .10$, $p_0 = .50$, $p_1 = (\mu_2/4)/[(\mu_2/4) + \mu_2] = (1/4)/(5/4) = .20$. We get $t \geq 19.7$. (With the binomial table it can be observed that we should have $t \geq 23$.) We sample cloth 2 yards at a time, one from each brand, terminating when the total number of defects exceeds 19.7.

6-56 Entering the graphs of Appendix B15 with $\hat{p} = .25$, $t = 20$, we find $w_1 = .08$, $w_2 = .51$. Then (6-53) yields (.087, 1.04).

6-59 Letting μ_1, μ_2 be the means for heavy and light patrolling, respectively, we wish to test $H_0 : \mu_1 = \mu_2$ against $H_1 : \mu_1 < \mu_2$. Equivalently, we test $H_0 : p = .50$ against $H_1 : p < .50$ with Y_1 given $t = 50$. The critical region is $y_1 \leq 16$. Since the observed value of the statistic is $y_1 = 15$, H_0 is rejected, and heavy patrolling is adopted.

6-60 Entering the graphs of Appendix B15 with $\hat{p} = .30$, $t = 50$ yields $w_1 = .17$, $w_2 = .46$. Then (6-53) yields (.20, .85).

6-61 Use formula (6-25) with $\alpha = .01$, $\beta = .20$, $p_0 = .50$. $p_1 = [(2/3)\mu_2]/[(2/3)\mu_2 + \mu_2] = .4$. We get $t \geq 248.2$. Sampling would be continued until the number of traffic deaths exceeds 248.2.

6-65 Letting p_1 be the fraction of those who die from natural causes and were smokers and p_2 be the fraction of those who die from heart disease and were smokers, test $H_0 : p_1 = p_2$ against $H_1 : p_1 < p_2$. We identify $n_1 = 4$, $n_2 = 6$, $x = 0$, $y = 5$, $t = 5$. Using (6-59), we find that the critical region is $x = 0$ with $\alpha = .023810$. Since the observed value of X given t is $x = 0$, H_0 is rejected, and the claim is justified.

6-66 We identify $n_1 = 80$, $n_2 = 120$, $x = 3$, $y = 12$, $t = 15$. Now we need $P(200, 15, 80, x) \cong B(x; 15, .40) \leq .05$. The critical region is $x \leq 2$ with $\alpha = .027$. Since the observed value of the statistic is $x = 3$, H_0 is not rejected.

6-67 Letting p_1 be the fraction of property owners and p_2 be the fraction of non-property owners, test $H_0 : p_1 = p_2$ against $H_1 : p_1 < p_2$. The observed value of the statistic is $z' = -3$ and the critical region is $z' < -2.326$. Hence H_0 is rejected, and the anticipated result is supported.

6-71 Test $H_0 : p_1 = 1/8$, $p_2 = 3/8$, $p_3 = 3/8$, $p_4 = 1/8$ against $H_1 :$ the p's are not all as given under H_0. Since the observed value of the statistic is $y_3' = 4.44$ and the critical region is $y_3' > 7.81$, H_0 is accepted, and the theory is not contradicted.

6-73   Test $H_0 : p_1 = 9/16$, $p_2 = 3/16$, $p_3 = 3/16$, $p_4 = 1/16$ against H_1 : not all p's are as given.   Since the observed value of the statistic is $y_3' = .47$, H_0 is not rejected at any reasonable significance level.

6-75   Test $H_0 : p_1 = .4, p_2 = .4, p_3 = .2$ against H_1 : not all p's are as given. The observed value of the statistic is $y_2' = 17.5$ and the critical region is $y_2' > 9.21$.   Hence H_0 is rejected, and we conclude that the professor is not conforming with university percentages.

6-77   Test H_0 : the p's for Saturday and Sunday are 2/9, other p's are 1/9 against H_1 : the p's are not as given.   The observed value of the statistic is $y_6' = 8.2$ and the critical region is $y_6' > 12.59$.   Hence the observed results do not contradict the hypothesis.

6-79   We find $\bar{x} = 2$.   Then the estimated expected frequencies are 13.534, 27.067, 27.067, 18.044, 9.023, 5.265 (for five or more).   The observed value of the statistic is $y_4' = 6.62 + .95 + 4.53 + .00 + 3.96 + .10 = 16.16$.   The critical region is $y_4' > 9.49$, and H_0 is rejected.   We may have suspected this since the independence condition does not appear to be satisfied.

6-81   We test the hypothesis of homogeneity.   The estimated expected numbers are 60 for each cell in the first column, 100 for each cell in the second, 40 for each cell in the third.   The observed value of the statistic is $y_6' = 79.0$.   Since the probability of obtaining a value of the statistic this large or larger is practically zero, H_0 is rejected, and we conclude that the classes are not alike in their preferences.

6-83   We test the hypothesis of independence.   The estimated expected numbers are 80, 160, 120, 40 for the first column, 20, 40, 30, 10 for the second, 60, 120, 90, 30 for the third, and 40, 80, 60, 20 for the fourth.   These yield $y_9' = .31 + .31 + .53 + .90 + .45 + .40 + .53 + .90 + .27 + .53 + 1.34 + 7.50 + .40 + .31 + 3.75 + 28.80 = 47.23$.   We reject H_0 even with $\alpha = .005$.   We conclude that there is a relationship between smoking and drinking habits.

6-85   We test the hypothesis of independence.   The observed value of the statistic is $y_4' = 29.22$ and the critical region is $y_4' > 13.28$.   We reject H_0 and conclude that there is some dependence between grades in elementary statistics and grades in the prerequisite mathematics course.

6-87   We test the hypothesis of homogeneity.   The observed value of the statistic is $y_2' = 8.75$ and the critical region is $y_2' > 9.21$.   Hence H_0 is accepted, which implies that Ambitol is no more effective than water for making a person ambitious.

7-1 $$\sum_{j=1}^{r} \sum_{i=1}^{n_j} (x_{ij} - \bar{x}_{\cdot j})(\bar{x}_{\cdot j} - \bar{x}_{\cdot \cdot}) = \sum_{j=1}^{r} (\bar{x}_{\cdot j} - \bar{x}_{\cdot \cdot}) \sum_{i=1}^{n_j} (x_{ij} - \bar{x}_{\cdot j})$$

For every j we have $\sum_{i=1}^{n_j}(x_{ij} - \bar{x}_{\cdot j}) = \sum_{i=1}^{n_j} x_{ij} - n_j \bar{x}_{\cdot j}$.   Since $\bar{x}_{\cdot j} = \sum_{i=1}^{n_j} x_{ij}/n_j$ by definition, $\sum_{i=1}^{n_j} x_{ij} = n_j \bar{x}_{\cdot j}$.   Consequently, for every j we have $\sum_{i=1}^{n_j} x_{ij} - n_j \bar{x}_{.j} = n_j \bar{x}_{.j} - n_j \bar{x}_{.j} = 0$, and the original double sum is equal to $\sum_{j=1}^{r} (\bar{x}_{.j} - \bar{x}_{..})(0) = 0$.

7-2 We test the hypothesis that the three varieties have equal mean yields against the alternative that they do not. Calculations yield $SS_A = 1,877$, $SS_W = 4,753$, and the observed value of the statistic is $f_{2,30} = 5.94$. Since the critical region is $f_{2,30} > 3.32$, H_0 is rejected.

7-4 We test the hypothesis that the mean score associated with each method of giving instructions is the same against the alternative that the means are not equal. The observed value of the statistic is $f_{3,36} = 2.5$. Since $f_{3,30;\,.95} = 2.92$, $f_{3,40;\,.95} = 2.84$, so that $f_{3,33;\,.95} = 2.90$, the hypothesis of equal means is accepted.

7-6 We test the hypothesis that the means associated with the four locations are equal against the alternative that they are not equal. The observed value of the statistic is $f_{3,20} = 6.75$ and the critical region is $f_{3,20} > 4.94$. Hence H_0 is rejected.

7-8 We test the hypothesis that the mean score for each type of engine is the same against the alternative that all means are not the same. The observed value of the statistic is $f_{4,25} = 5.05$ and the critical region is $f_{4,25} > 4.19$ (obtained by linear interpolation). Hence H_0 is rejected.

7-9 Letting $\mu_1 = \mu_2$, $\mu_3 = \mu_1 + 1.2\sigma$, we get $\phi = \sqrt{(11/3)(.96)} = 1.88$. Then, with $v_1 = 2$, $v_2 = 30$, $\alpha = .05$, the graphs of Appendix B16 yield power approximately equal to .81.

7-10 Formula (7-22) yields $n = (1/.96)3\phi^2 = 3.125\phi^2$. With $v_1 = 2$, $\alpha = .05$, $v_2 = \infty$, the graphs of Appendix B16 yield power = .95 if $\phi = 2.25$ so that the computed $n = 15.82$. With $v_2 = 60$, chosen $n = (60/3) + 1 = 21$, power = .95 if $\phi = 2.33$, and computed $n = 16.96$. Since the computed n is still smaller than the chosen n, we continue the calculations. With $v_2 = 30$, chosen $n = (30/3) + 1 = 11$, power = .95 if $\phi = 2.40$ and computed $n = 18.00$. Since the computed n is larger than the chosen n, we terminate calculations and have that the minimum n is in the range $16.96 \leq n \leq 18.00$. Either $n = 17$ or 18 would be satisfactory choices.

7-11 With $a = 1.2$, $n = 11$, $r = 3$, formula (7-23) yields $\phi = 1.62$. Then with $v_1 = 2$, $v_2 = 30$, $\alpha = .05$, the graphs of Appendix B16 yield power $\cong .66$.

7-15 Letting $\mu_1 = \mu_2$, $\mu_3 = \mu_1 + 1.5\sigma$, $\mu_4 = \mu_1 + 1.5\sigma$, we get $\phi = \sqrt{(6/4)(36/16)} = 1.84$. With $\alpha = .01$, $v_1 = 3$, $v_2 = 20$, the graphs of Appendix B16 yield power $\cong .55$. With $n = 16$ we get $\phi = \sqrt{(16/4)(36/16)} = 3$, $v_2 = 60$, and power $> .99$. Formula (7-22) yields $n = (16/36)4\phi^2 = 1.78\phi^2$. With $v_1 = 3$, $v_2 = \infty$, $\alpha = .01$, power = .90 if $\phi = 2.18$, computed $n = 8.22$. With $v_2 = 60$, $n = 16$, power = .90 if $\phi = 2.28$, computed $n = 9.25$. With

$v_2 = 30$, $n = (30/4) + 1 = 8.5$, power $= .90$ if $\phi = 2.40$, computed $n = 10.25$. Hence the minimum n is in the range $9.25 \leqq n \leqq 10.25$, so that we take $n = 10$.

7-18   Preliminary calculations yield $\bar{x}_{.1} = 100.5$, $\bar{x}_{.2} = 111.4$, $\bar{x}_{.3} = 93.0$, $MS_W = 158.4$, $S^2 = 6.64$. The intervals are $(-24.7, 2.9)$ for $\mu_1 - \mu_2$, $(-6.3, 21.3)$ for $\mu_1 - \mu_3$, $(4.6, 32.2)$ for $\mu_2 - \mu_3$, $(-27.4, 20.6)$ for $2\mu_1 - \mu_2 - \mu_3$, $(-53.3, -5.3)$ for $\mu_1 - 2\mu_2 + \mu_3$, $(1.9, 49.9)$ for $\mu_1 + \mu_2 - 2\mu_3$. Consequently, we conclude (a) the mean of distribution 2 is larger than the mean of distribution 3, (b) the mean for distribution 2 is larger than the average of the means for distributions 1 and 3, (c) the mean for distribution 3 is smaller than the average of the means for distributions 1 and 2.

7-20   Preliminary calculations yield $MS_W = .00077$, $S^2 = 14.82$, $S\hat{\sigma}_L = .06$ for all six contrasts. The intervals are $(.01, .13)$ for $\mu_1 - \mu_2$, $(-.03, .09)$ for $\mu_1 - \mu_3$, $(-.04, .08)$ for $\mu_1 - \mu_4$, $(-.10, .02)$ for $\mu_2 - \mu_3$, $(-.11, .01)$ for $\mu_2 - \mu_4$, $(-.07, .05)$ for $\mu_3 - \mu_4$. We conclude that the mean for location 1 is larger than the mean for location 2, but other differences are not indicated.

7-21   Now $S\hat{\sigma}_L = .09$ and the interval is $(-.01, .17)$. Hence, the sample results do not indicate that the average of the means for the East Coast is different from the average of the means for the Gulf Coast.

7-24   The observed value of the statistic is $R_{8,4} = .527$ and the critical region is $R_{8,4} > .5365$. Accept H_0.

7-26   The observed value of the statistic is $R_{21,3} = .498$. To obtain $R_{21,3;.95}$ we find $R_{17,3;.95} = .5466$, $R_{37,3;.95} = .4748$. Then linear interpolation on $12/\sqrt{n-1}$ yields

$$R_{21,3;.95} = .4748 + \frac{2.68 - 2}{3 - 2}(.5466 - .4748) = .5236$$

Hence, the critical region is $R_{21,3} > .5236$ and H_0 is accepted.

7-27   We get $s_1^2 = .00056$, $s_2^2 = .00032$, $s_3^2 = .00104$, $s_4^2 = .00116$, $R_{6,4} = .377$. Since the critical region is $R_{6,4} > .6761$, H_0 is accepted.

8-1   (a) $\hat{\mu}_{Y|x} = -.5 + x$ (b) 2.5   (c) .16   (d) .40

8-3   Preliminary calculations yield

$$\sum_{i=1}^{10} x_i = 188.8 \qquad \sum_{i=1}^{10} x_i^2 = 3,795.20 \qquad \sum_{i=1}^{10} y_i = 365.3$$

$$\sum_{i=1}^{10} y_i^2 = 14,109.37 \qquad \sum_{i=1}^{10} x_i y_i = 7,312.47$$

Then $\hat{\mu}_{Y|x} = 2.55 + 1.80x$ and with $x = 20.2$ we get $\hat{\mu}_{Y|x} = 38.91$. The estimates of σ^2, σ are $s_{Y|x}^2 = 2.01$, $s_{Y|x} = 1.42$.

8-5 $\hat{\mu}_{Y|x} = 924 - 1.49x$ and with $x = 280$ we get $\hat{\mu}_{Y|x} = 507$. The estimates of σ^2, σ are $s_{Y|x}^2 = 941$, $s_{Y|x} = 36.8$.

8-6 The observed value of the statistic is $t_{40} = -4.20$, which is significant even with $\alpha = .0005$. Thus we conclude that the mean college grade point is less than 2.00 for those whose high school grade point is 2.00.

8-7 (2.84, 3.16)

8-8 With $\delta = 3.36$, $f = 40$, $\alpha = .05$, the graphs of Appendix B10 yield a power slightly over .95.

8-9 The observed value of the statistic is $t_{40} = 9.81$. Thus, H_0 is rejected even with $\alpha = .0005$. In the second case we test $H_0 : B = .75$ against $H_1 : B > .75$. Now the observed value of the statistic is $t_{40} = 2.45$ and the critical region is $t_{40} > 1.684$. Hence, H_0 is rejected, and we conclude that the slope is greater than .75.

8-12 (1.08, 2.72)

8-13 The observed value of the statistic is $y_{40} = 25.6$ and the critical region is $y_{40} < 26.51$. Hence H_0 is rejected.

8-16 The region is $2.72 + 1.91x \pm 1.15 \sqrt{8.92\left[\dfrac{1}{10} + \dfrac{(x - 6.84)^2}{8.24}\right]}$.

With $x = 6$, 7, 8 the intervals are (12.70, 15.66), (14.99, 17.19), (16.24, 19.76). If the experiment is repeated many times and the above region computed for each set of sample results, then about .95 of the time all $\mu_{Y|x}$ will be contained in the region.

8-17 Test $H_0 : \mu_{Y|x_0} = 40$ against $H_1 : \mu_{Y|x_0} \neq 40$ where $x_0 = 20$. The observed value of the statistic is $t_8 = -3.15$ and the critical region is $t_8 < -3.355$, $t_8 > 3.355$. Accept H_0.

8-18 With $\delta = 3.09$, $f = 8$, $\alpha/2 = .005$, Appendix B10 gives power $\cong .47$.

8-19 The prediction interval is (35.47, 42.35). The confidence interval is (37.85, 39.97). End points are in thousands of acre feet.

8-24 Test $H_0 : \mu_{Y|x_0} = 475$ against $H_1 : \mu_{Y|x_0} > 475$ where $x_0 = 280$. The observed value of the statistic is $t_{25} = 4.32$ and the critical region is $t_{25} > 2.485$. Hence H_0 is rejected, and we conclude that the average life exceeds 475.

8-27 (261.4, 394.6)

8-28 Test $H_0 : \rho = 0$ against $H_1 : \rho \neq 0$. Preliminary calculations yield

$$\sum_{i=1}^{18} x_i = 733 \qquad \sum_{i=1}^{18} x_i^2 = 31{,}717 \qquad \sum_{i=1}^{18} y_i = 776$$

$$\sum_{i=1}^{18} y_i^2 = 36{,}190 \qquad \sum_{i=1}^{18} x_i y_i = 33{,}084 \qquad r = .66$$

The observed value of the statistic is $t_{16} = 3.52$ and the critical region is $t_{16} < -2.120$, $t_{16} > 2.120$. We reject H_0 and conclude pairs of grades are dependent (or related).

8-29 The critical region is $r < -.47$ and $r > .47$. Since the observed value of r was $r = .66$, H_0 is rejected.

8-30   (.27, .85). This interval does not include 0 and the test rejects $H_0 : \rho = 0$.

8-34   We find $r = .78$. Then with $n = 12$, $1 - \alpha = .95$, the graphs of Appendix B19 yield (.39, .92).

8-35   Test $H_0 : \rho = .80$ against $H_1 : \rho > .80$. The observed value of the statistic is $z' = 2.49$ and the critical region is $z' > 2.326$. Hence, H_0 is rejected, and the examinations are regarded as equivalent.

8-36   With $\delta = 1.42$, $\alpha = .01$, the graphs of Appendix B10 yield power $\cong$ .19. To make power $= .80$ requires $\delta = 3.15$. Then formula (8-48) yields $n \geq 402.5$.

8-37   With $r = .88$, $1 - \alpha = .90$, $n = 84$, the graphs of Appendix B19 yield (.83, .91). Alternatively, $z(w_1) = 1.193$, $z(w_2) = 1.569$, and Appendix B18 yields the interval (.83, .92).

9-1   We test $H_0 : Q_{.50} = 46$ against $H_1 : Q_{.50} \neq 46$. The observed value of the statistic is $y = 18$ and the critical region is $y \leq 7$, $y \geq 18$. We reject H_0 and conclude that the median is below 46.

9-2   We need $\Pr(Y \leq 7) + \Pr(Y \geq 18)$ computed with $n = 25$, $p = .60$. This is equal to $\Pr(Z \leq 7) + \Pr(Z \geq 18)$ where Z has a binomial distribution with $n = 25, p = .40$. Using the latter we find power $= .15355 + .00121 = .15476$. To raise the power to .80 requires $n \geq 193.9$.

9-3   Using $\alpha/2 = .025$ to enter the graphs of Appendix B10, we find with $f = \infty$ that power $= .80$ if $\delta = 2.80$. The computed n is 122.5. With $f = 24$, power $= .80$ if $\delta = 2.90$ and the computed n is 131.4. Hence the minimum n is in the interval $122.5 \leq n \leq 131.4$, and we would probably take $n = 125$. (It can be shown that the actual minimum n is 125.)

9-4   We find $i = 8$, $j = 18$. Then $x_{(8)} = 45.85$, $x_{(18)} = 45.99$, and the interval is (45.85, 45.99).

9-5   We test $H_0 : Q_{.75} = 240$ against $H_1 : Q_{.75} > 240$. Letting p be the fraction of rats which mature in 240 days or less, this is equivalent to testing $H_0 : p = .75$ against $H_1 : p < .75$. Also, if we let q be the fraction of rats which do not mature in 240 days or less, then we test $H_0 : q = .25$ against $H_1 : q > .25$. If we let Y be the number out of a sample of size 50 that do not mature in 240 days or less, then the critical region is $y \geq 17$ with $\alpha = .09831$. Since the observed value of the statistic is $y = 14$, H_0 is accepted and the new variety is adopted.

9-6   The interval is $(x_{(32)}, x_{(43)})$.

9-13   We test $H_0 : Q_{.50} = 0$ against $H_1 : Q_{.50} > 0$ with $\alpha_0 = .01$. The critical region is $y = 0$ with $\alpha = .00098$. Since the observed value of the statistic is $y = 2$, H_0 is accepted.

9-14   Power $= \Pr(Y = 0) = \sum_{y=0}^{0} b(y; 10, .25) = .05631$.

9-17   Again we test $H_0 : \mu = 46$ against $H_1 : \mu \neq 46$. From Appendix B20 we find $t_1 = 21$ and $\Pr(T \leq 21) = .0247$. Then (9-22) gives

$t_2 = 84$ and (9-23) gives $\alpha = .0494$. Since the observed value of the statistic is $t = 16$, H_0 is rejected.

9-18 Test $H_0 : Q_{.50} = 50$ against $H_1 : Q_{.50} < 50$. The critical region is $t \leq 9$ with $\alpha = .0081$. Since the observed value of the statistic is $t = 23$, H_0 is accepted and apparently the chain is unwilling to buy.

9-22 Test $H_0 : Q_{.50} = \mu = 0$ against $H_1 : Q_{.50} = \mu > 0$ where μ is the mean of the differences. Since $\Pr(T \leq 43) = .0096$, (9-21) gives $\Pr(T \geq 167) = .0096$ and the critical region is $t \geq 167$. Since the observed value of the statistic is $t = 186$, H_0 is rejected and the psychologist has "proved" his theory.

9-23 Test $H_0 : \mu_X = \mu_Y$ against $H_1 : \mu_X \neq \mu_Y$ where variety A generates the X distribution. From Appendix B21 we find that the critical region is $t \leq 82$, $t \geq 128$ with $\alpha = 2(.0446) = .0892$. Since the observed value of the statistic is $t = 130$, H_0 is rejected. We would probably conclude that brand A has the higher mean.

9-25 Test $H_0 : \mu_X = \mu_Y$ against $H_1 : \mu_X > \mu_Y$ where the X distribution is generated by brand 2. From Appendix B21 we find that the critical region is $t \geq 172$ with $\alpha = .0096$. Since the observed value of the statistic is $t = 181$, H_0 is rejected and brand 1 is adopted.

9-27 Test $H_0 : \mu_X = \mu_Y$ against $H_1 : \mu_X > \mu_Y$ where the X distribution is associated with the first section. From Appendix B21 we find that the critical region is $t \geq 115$ with $\alpha = .0079$. Since the observed value of the statistic is $t = 125$, H_0 is rejected and we conclude that quizzes produce a higher mean.

9-29 Test H_0 : corn yields for the two varieties are independent of the localities against H_1 : large yields for each variety tend to occur together as do small yields. From Appendix B22 we find that the critical region is $r > .5515$. The observed samples yield $\sum_{i=1}^{10} d_i^2 = 48$, $r = .709$. Hence H_0 is rejected.

9-32 Test H_0 : scores are independent against H_1 : scores are not independent. From Appendix B22 we find that the critical region is $r < -.5179$, $r > .5179$. The observed samples yield $\sum_{i=1}^{15} d_i^2 = 664$, $r = .186$. Hence H_0 is accepted.

9-33 The critical region is $r > .4965$. The observed samples yield $\sum_{i=1}^{12} d_i^2 = 114$, $r = .605$. Hence H_0 is rejected.

INDEX